Pulsed Metal Vapour Lasers

NATO ASI Series

Advanced Science Institutes Series

A Series presenting the results of activities sponsored by the NATO Science Committee, which aims at the dissemination of advanced scientific and technological knowledge, with a view to strengthening links between scientific communities.

The Series is published by an international board of publishers in conjunction with the NATO Scientific Affairs Division

A **Life Sciences** B **Physics**	Plenum Publishing Corporation London and New York
C **Mathematical and Physical Sciences** D **Behavioural and Social Sciences** E **Applied Sciences**	Kluwer Academic Publishers Dordrecht, Boston and London
F **Computer and Systems Sciences** G **Ecological Sciences** H **Cell Biology** I **Global Environmental Change**	Springer-Verlag Berlin, Heidelberg, New York, London, Paris and Tokyo

PARTNERSHIP SUB-SERIES

1. **Disarmament Technologies**	Kluwer Academic Publishers
2. **Environment**	Springer-Verlag / Kluwer Academic Publishers
3. **High Technology**	Kluwer Academic Publishers
4. **Science and Technology Policy**	Kluwer Academic Publishers
5. **Computer Networking**	Kluwer Academic Publishers

The Partnership Sub-Series incorporates activities undertaken in collaboration with NATO's Cooperation Partners, the countries of the CIS and Central and Eastern Europe, in Priority Areas of concern to those countries.

NATO-PCO-DATA BASE

The electronic index to the NATO ASI Series provides full bibliographical references (with keywords and/or abstracts) to more than 50000 contributions from international scientists published in all sections of the NATO ASI Series.
Access to the NATO-PCO-DATA BASE is possible in two ways:

– via online FILE 128 (NATO-PCO-DATA BASE) hosted by ESRIN, Via Galileo Galilei, I-00044 Frascati, Italy.

– via CD-ROM "NATO-PCO-DATA BASE" with user-friendly retrieval software in English, French and German (© WTV GmbH and DATAWARE Technologies Inc. 1989).

The CD-ROM can be ordered through any member of the Board of Publishers or through NATO-PCO, Overijse, Belgium.

Series 1: Disarmament Technologies – Vol. 5

Pulsed Metal Vapour Lasers

edited by

Chris E. Little
School of Physics and Astronomy,
University of St Andrews,
St Andrews, Scotland

and

Nikola V. Sabotinov
Department of Metal Vapor Lasers,
Institute of Solid State Physics,
Bulgarian Academy of Sciences,
Sofia, Bulgaria

Kluwer Academic Publishers

Dordrecht / Boston / London

Published in cooperation with NATO Scientific Affairs Division

Proceedings of the NATO Advanced Research Workshop on
Pulsed Metal Vapour Lasers – Physics and Emerging Applications in Industry, Medicine and Science
St Andrews, U.K.
August 6–10, 1995

A C.I.P. Catalogue record for this book is available from the Library of Congress

ISBN 0-7923-4002-7

Published by Kluwer Academic Publishers,
P.O. Box 17, 3300 AA Dordrecht, The Netherlands.

Kluwer Academic Publishers incorporates the publishing programmes of
D. Reidel, Martinus Nijhoff, Dr W. Junk and MTP Press.

Sold and distributed in the U.S.A. and Canada
by Kluwer Academic Publishers,
101 Philip Drive, Norwell, MA 02061, U.S.A.

In all other countries, sold and distributed
by Kluwer Academic Publishers Group,
P.O. Box 322, 3300 AH Dordrecht, The Netherlands.

Printed on acid-free paper

Printed in the Netherlands

Table of Contents

Book cno: S214 480 **Count:** ____________

Rec'd date: 96/05/24 **Book collection type:** ____________

Book zero: HAK PAS S **Pub format:** P

****

done HAK

Cover title: Pulsed metal vapour lasers [SERIES: NATO Adv. Sci. Inst. Ser. Disarm. Tech.]

Namepub(m): Little, Chris E.
Sabotinov, Nikola V.

Kkey: St. Andrews,
Kyx: 1995

****

Publisher text:

Publisher: Kluwer Academic Publishers, Dordrecht
Extra publisher text:
Extra publisher:
Book pub year: 96
Book price: $225.00
0-7923-4002-7
Book source: Kluwer Acad. Publ.
Source code: 0-89838A

****

Collection title: ____________

Preface

This book is a collection of the papers given at the NATO Advanced Research Workshop (ARW) on Pulsed Metal Vapour Lasers - Physics and Emerging Applications in Industry, Medicine and Science, held at the University of St Andrews, 6–10 August, 1995. The ARW was a working meeting aimed at assessing present-day understanding of the physics and applications of pulsed metal vapour lasers (PMVLs), and identifying directions for future research and development. Specifically, it aims were:

1. To survey and critically assess the current state of knowledge of the physics of high-power PMVLs.

2. To discuss techniques relevant to making PMVLs more amenable to applications, namely wavelength agility (*e.g.* frequency doubling) and the characterization and improvement of beam quality.

3. To address future and emerging applications of these lasers for high-precision micro-machining, cutting, drilling, welding *etc.* of metals, ceramics and plastics; high-speed photography and holographic testing; selective photochemical processing of materials; production-line silicon chip manufacture; spectroscopy; adaptive-optics astronomy; medicine, *etc.*

The St Andrews ARW was the first gathering of the international community of specialists in PMVL physics and applications. Eighteen key speakers reported on the current status of PMVL research, and gave their views on future directions in those areas. An additional 31 participants contributed important and timely reports in the form of posters and short oral presentations. The four-day programme included three round-table discussions on pulsed metal vapour lasers, their physics and applications. Participation in the ARW was mainly by invitation. In all there were 69 registered participants.

The meeting brought together the pioneers who first developed PMVLs in the 1960's (Prof. W.T. Silfvast from the University of Florida who demonstrated the first high-gain PMVL, the lead vapour laser; and Dr W.T. Walter of Laser Consultants Inc. (USA) whose team first demonstrated the copper vapour laser, and who also started the first company to manufacture copper vapour lasers), and those in Russia who demonstrated the first practical high average power PMVLs (Dr A.A. Isaev and Prof. G.G. Petrash of the P N Lebedev Physical Institute, Russia).

The ARW was organized by two Directors, Dr Chris E Little (University of St Andrews, UK) and Prof. Nikola V Sabotinov (Institute of Solid State Physics, Bulgaria), with the assistance of an International Committee, that comprised Dr Jim J Chang (Lawrence Livermore National Laboratory, USA), Prof. Jim A Piper (Macquarie University, Australia), Prof. Gueorgii G Petrash (P.N. Lebedev Physical Institute, Russia) and Prof. Colin E Webb (University of Oxford, UK).

On behalf of the world community of scientists who work in the field of pulsed metal vapour lasers, the Directors and International Committee take great pleasure in thanking NATO for so generously sponsoring this meeting, and making it all possible.

Chris E Little
St Andrews, 27 October, 1995

NATO ARW
Pulsed
Metal Vapour
Lasers
1995
St Andrews

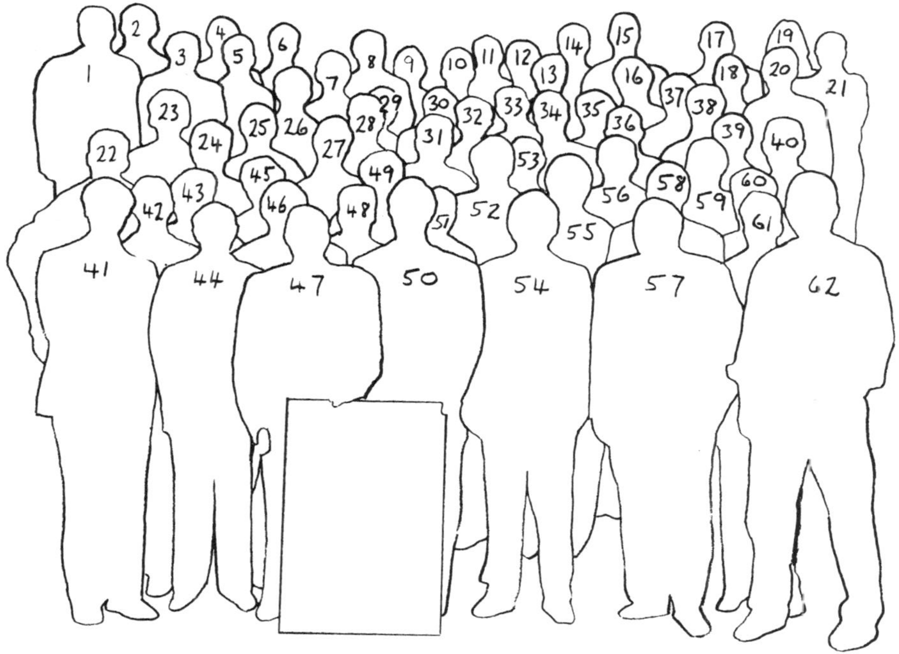

ARW Participants

1. A.N. Soldatov. 2. D.N. Astadjov. 3. J. Mizeraczyk. 4. K.I. Zemskov. 5. C. Konagai. 6. N.K. Vuchkov. 7. E.L. Latush. 8. I. Opachko. 9. R. Mayerhofer. 10. E.S. Livingstone. 11. I. Kostadinov. 12. P. Foster. 13. A.I. McIntosh. 14. D.W.Coutts. 15. B.E. Warner. 16. B.L. Ivanov. 17. A.A. Isaev. 18. W. Wadsworth. 19. C.G. Whyte. 20. J.B. Metzen. 21. R. Bhatnagar. 22. M. Withford. 23. T. Omatsu. 24. J.P. Pique. 25. A.J. Kearsley. 26. T. Karras. 27. G.P. Hogan. 28. R.Q. Fugate. 29. Y. Achiam. 30. G.J. Dickinson. 31. M.F. Sem. 32. G.S. Evtushenko. 33. R.C. Tobin. 34. P. Davis. 35. I. Smilanski. 36. B.L. Borovich. 37. J. Maury. 38. V. Chvykov. 39. W.T. Walter. 40. P. Blau. 41. R.J. Carman. 42. R. Mildren. 43. D.J.W. Brown. 44. G.G. Petrash. 45. S.A. Fairlie. 46. Yu.O. Shpenik. 47. N.V. Sabotinov. 48. D.R. Jones. 49. P.A. Bokhan. 50. C.E. Little. 51. M. Amit. 52. P.H. Butler. 53. M. Lesnoi. 54. J.J. Chang. 55. M.A. Kazaryan. 56. P. Lemaire. 57. C.E. Webb. 58. D. Doizi. 59. W.T. Silfvast. 60. J.M. Borgard. 61. L. Gigant. 62. J.A. Piper.

Introduction

High-power pulsed metal vapour lasers were first demonstrated in the USA in the 1960s, but it wasn't until the mid-seventies that high average output powers of tens of watts were achieved as a result of developments in Russia.

The technological breakthrough that led to high-power copper vapour lasers signalled the move for major deployments of manpower and funding to develop the technology further and improve the performance of copper vapour lasers, and also pulsed lasers utilizing other metals such as gold, manganese and barium, which produce high-power beams of light in the ultraviolet, blue, green, red and infrared regions of the spectrum.

The initiative for the move towards advanced technology came from the defence industry for isotope separation.

The technology of these lasers has continued to develop to the point today where multiple green/yellow laser beams of more than 1 kilowatt average power each are routinely generated at around 1% efficiency, 24 hours a day (Lawrence Livermore National Laboratory). These lasers are the highest power visible lasers ever developed.

Metal vapour lasers are now considered to be the enabling technology for a number of important processes. The copper vapour laser offers significant advantages in a number of areas, including very high-precision machining of metals such as copper and aluminium, which are traditionally very difficult materials to process. These lasers can also be used to manufacture high quality thin films of material produced by photoablation. For many applications, the present-day generation of CO_2, Nd:YAG and excimer industrial lasers would be better replaced by a new generation of high-power copper vapour lasers, provided long-lived, reliable and easy-to-use devices can be developed in the near future. With the application of the advanced technology, originally developed in isotope enrichment programmes, copper vapour lasers would certainly become key industrial lasers of the next century.

Other important and emerging applications of metal vapour lasers are to be found in medicine, spectroscopy and astronomy. These applications include diagnostic, therapeutic and forensic medicine, *e.g.* for cancer photodynamic therapies, removal of hæmangiomas (port wine stains) and other psychologically damaging skin abnormalities; the creation of laser guide stars for adaptive optics astronomy; and atomic and molecular spectroscopy for fundamental scientific studies.

With the ending of the Cold War, the most active leading and eminent researchers in the field of high-power metal vapour lasers, from the East and West, were able to meet at the St Andrews ARW for the first time. This Proceedings volume is an almost complete snapshot of the current understanding of PMVL physics, and world view of present and emerging PMVL applications.

1. Historical Viewpoints

PULSED METAL VAPOR LASERS: PAST AND FUTURE

WILLIAM T. SILFVAST
Department of Physics and
Center for Research and Education in Optics and Lasers (CREOL),
University of Central Florida
Orlando, FL 32826, USA

1. Historical Perspective

In the early days of lasers the use of metal vapors as a laser gain medium became a logical extension of the investigations of gas lasers since it gave a much wider variety of possible energy level arrangements and laser wavelengths. Gain was first reported in optically-pumped cesium vapor [1] however this system never achieved significant laser output and was therefore more of a research curiosity. In 1964 Earl Bell at Spectra Physics developed the first ion laser in mercury vapor [2]. Shortly thereafter, Bill Bridges, then at Hughes Research Labs, was operating a Hg ion laser with an argon buffer gas and inadvertently discovered the now famous argon ion laser [3]. The mercury laser, as far as metal vapor lasers were concerned, was relatively simple to construct because the Hg metal could be placed anywhere in the discharge tube and the necessary vapor pressure was automatically achieved at room temperature or slightly higher. A pulsed Hg ion laser operating in the red and green was sold by Spectra Physics for a while but then other lasers came along that replaced it.

In the Fall of 1964, I was looking for a PhD thesis project and approached Professor Grant Fowles about doing a project associated with lasers. He and Russell Jensen had just discovered the first ion laser in atomic iodine [4] and it looked like an exciting field for me to get involved in. We talked about what I might do for a thesis project and since Fowles was interested in studying the hyperfine structure of atoms we looked at the possibility of constructing a laser for that purpose. It soon became clear that we would have to use a metal vapor for our work because of the requirements associated with studying hyperfine structure.

The requirements for studying hyperfine structure involved the need for an atom with an odd number of protons in the nucleus to obtain a nuclear spin. The hyperfine splitting is the greatest for heavier elements. It is also desirable to have an atomic species that has only one naturally occurring odd isotope so that the spectra would not involve more than one isotope to complicate the data. Also an element with a vapor pressure high enough to operate in a quartz discharge tube would be desirable. Putting all of these requirements together led us to the element bismuth. We identified several potential laser transitions in the bismuth ion and commenced to think how we would implement the laser.

C. E. Little and N. V. Sabotinov (eds.), Pulsed Metal Vapour Lasers, 3–13.

Professor Fowles and Russ Jensen had already been using an all glass system including a glass diffusion pump for the iodine laser [4] and that system functioned very well for research investigations. It was also relatively inexpensive to put together if a good glass blower was available. I located a zone refining furnace from the Ceramic Engineering department and we were in business. Our idea was to sprinkle some bismuth throughout the bore region of the discharge, fill the tube with a buffer gas such as helium, and begin operating the discharge, either pulsed with a simple capacitive discharge triggered by an air spark gap or operating the discharge at 120 Hz with a commercial neon sign transformer. We would then begin heating the discharge to the temperature we estimated was necessary according to the vapor pressure charts, in order to reach a similar vapor pressure to that of the Hg laser. We ordered several sets of dielectric coated laser mirrors that could be readily interchanged on triangular rails at the ends of the discharge tube and thus we could readily cover a wide range of possible laser wavelengths. Because of the cost, we used microscope cover slips for our Brewster windows held on with high vacuum grease. We selected the flattest ones by using a sodium vapor lamp to examine the fringe patterns of the cover slips. We would also insert a He-Ne gas mixture in the tube and use our red reflecting mirrors to make sure the tube, including the windows, was functioning properly by operating the 632.8 nm He-Ne laser in the discharge.

By early January of 1965 we had our system operating and were ready to try to make a bismuth laser. We inserted the bismuth into the tube, pumped it down, heated it up to the desired temperature, observed the characteristic discharge color change to that of bismuth rather than helium, but never produced a laser. We tried many different buffer gases, various temperatures, various excitation conditions and never saw a laser. We had a technique of looking down the discharge bore through a transmission diffraction grating, a technique I would not advise to anyone for eye safety reasons, and it was possible to see when a laser output was beginning to develop, as we had learned when operating the 632.8 nm He-Ne laser in our discharge. We saw no indications of a laser even beginning to develop. After approximately two months of effort I began to get somewhat discouraged. We had tried everything and no laser. I could see my thesis project going down the drain.

I began thinking more about Bell's results in Hg vapor. Looking at the periodic table and recognizing the similarity of zinc and cadmium to that of mercury I began to think of them as possible laser candidates. The vapor pressure of both elements was relatively high and would therefore have to be heated only a few hundred degrees above room temperature. They were not particularly interesting candidates for hyperfine splitting measurements but at that point I was more concerned with just testing the concept of making a heated metal vapor laser rather than finding the ideal candidate for hyperfine structure measurements. I obtained some zinc and cadmium in granular form from the chemistry stockroom and placed zinc in the tube first, using exactly the same techniques we had developed for bismuth.

The first day I began to test zinc was on a saturday in March 1965 and I began heating up the discharge region with helium as a buffer gas. I believe the first time I heated the tube and applied the excitation pulse I saw this beautiful turquoise blue-green beam develop between the mirrors that became the first zinc laser [5]. I phoned Grant Fowles' office and was informed that he was in a committee meeting in the administration building. I ran to the administration building, interrupted his meeting and told him what I had seen. He immediately left the meeting and came back to the lab to see the zinc laser. We took spectrograms of the wavelengths and identified the transitions as belonging to the Zn ion. At that point we suspected that we also would most likely be able to make a cadmium laser on similar transitions.

The cadmium laser came next on the green transitions [5] that were similar to those of zinc (the blue transition did not come until almost a year later [6]). We then moved on to sulfur and phosphorous and obtained new visible transitions in the ions of both of those elements [7]. At that point we decided to submit our results to the upcoming Electron Device Research Conference in Urban Illinois that summer to announce the four new laser systems. Later when Fowles and I did attend that meeting, we found that we had scooped two groups at Bell Labs with our results, which was an amazing experience for both of us.

We made notes of many of the characteristics of these new lasers but decided we could return to them later to study them in detail. At this point we were having a good time generating new lasers. I examined the vapor pressure charts and decided that lead would be an element that we could heat to the necessary vapor pressure in our system. Examining the ion levels of Pb suggested that there might be a possible ion laser in the green portion of the spectrum as indicated in Figure 1. I made another trip to the chemical stockroom and found a jar of granular lead not realizing at that point the excitement that was in store for us. Lead required a temperature that was near the limits of our system since quartz discharge tubes begin to sag when the temperature is in the region slightly higher than 1100°C. With lead in the tube and the green laser mirrors in place we began to heat up the discharge region with the pulsed discharge operating. Observing the discharge through the transmission diffraction grating I began to notice an enhanced emission in the red portion of the spectrum. At first I didn't pay much attention to it because I knew we had the green high reflecting mirrors in the cavity and the reflection of those mirrors in the red was quite low. As the tube heated further and the lead vapor pressure increased the red emission became much more intense. It did not have the characteristics of the other lasers in that it wasn't as well defined spatially as the other lasers we had discovered. The emission appeared to come from a somewhat larger region around the discharge bore. I was perplexed and excited at the same time. At that point I realized that it definitely was a laser because it was very sensitive to mirror adjustment in the same way the other lasers were, however it was not quite as sensitive as the other lasers and of course I remembered that the reflectivity was very low for the red wavelengths. We definitely had something very interesting! Incidentally, we also saw the laser action on the green transition but at that point we were much more interested in the red output.

We quickly measured the wavelength. Our spectrographic film was not sensitive in the red but we were able to get an image because of the intensity of the laser output. We measured it to be 722.9 nm and Fowles concluded that it was a transition from the first excited state of the neutral atom to a level that was part of the ground configuration as shown in Figure 1. This was clearly something new! We readily deduced that it had to be a self-terminating transition since the lower laser level was part of the ground configuration and thus had essentially no radiative decay mode. The gain appeared to be very high since the laser would operate at the higher vapor pressures with just an optical window as the output mirror.

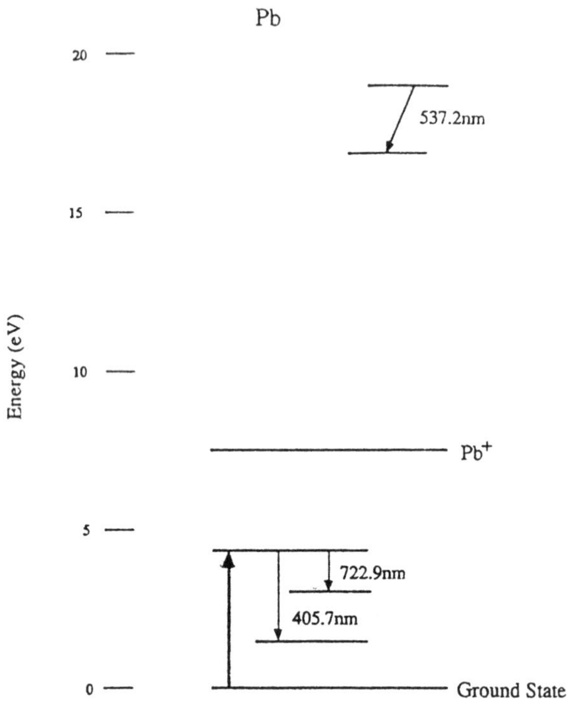

Figure 1. Energy level diagram of Pb laser transitions.

At about that time, Russ Jensen had made an inquiry to Professor Bill Bennett at Yale to see if there was the possibility of doing post-doctoral research at Yale. As a result, Bennett phoned and asked if he could stop by and visit us. Of course we were overjoyed and we now had all of these new metal vapor lasers to show him. That was like the mountain coming to Mohammed. Bennett visited us and seemed particularly interested in the new 722.9 nm Pb laser (we later found out why). We told him we had just submitted the result to Applied Physics Letters so he told us that he would call them and tell them to accept it immediately. It is probably one of the shortest times between submission and publication of any manuscript that has ever been submitted to APL [8].

We set out to look for other candidates for this type of laser that would be possible within the temperature constraints of our furnace. We selected manganese as a good candidate. When I put manganese into the system and heated it up it began to lase immediately in the green portion of the spectrum. We also observed transitions in the infrared. We took spectrograms of the laser output and began identifying the laser wavelengths. At that time Grant Fowles received a request to review an Applied Physics Letter that had been submitted by Marty Piltch, et al. from TRG and William R Bennett from Yale [9]. Bennett had taken our information and scooped us! I found out later that TRG had been working for some time trying to make a collisional laser in Mn according to an idea that Gordon Gould had developed. Hence they had a furnace and laser discharge system that was capable of heating manganese to the appropriate temperature. Marty Piltch later told me that Bennett flew back to New York from his visit with us and told them to start pulsing their manganese. I guess all is fair in love and war but I think it would have been more ethical if Bennett had told us that he was consulting with TRG on a metal vapor project before we showed him all of our techniques. We ended up publishing our measurements of hyperfine laser emission in the Mn laser [10]. We were also somewhat disappointed later when the group at TRG published their "cyclic laser" paper [11] that had very little mention of our work with lead. And when lead was mentioned, it was pointed out that lead was outside of their range of requirements for cyclic lasers. I was later told by one of the authors that this was done for patent reasons. The copper vapor laser (CVL) was reported in reference [11] to be one of the most efficient lasers and has remained so to date.

Figure 2 shows a diagram of several of the most efficient lasers including the CVL. They all have one thing in common. The ratio of the laser energy to the excitation energy is greater than 50%. We will come back to this point later.

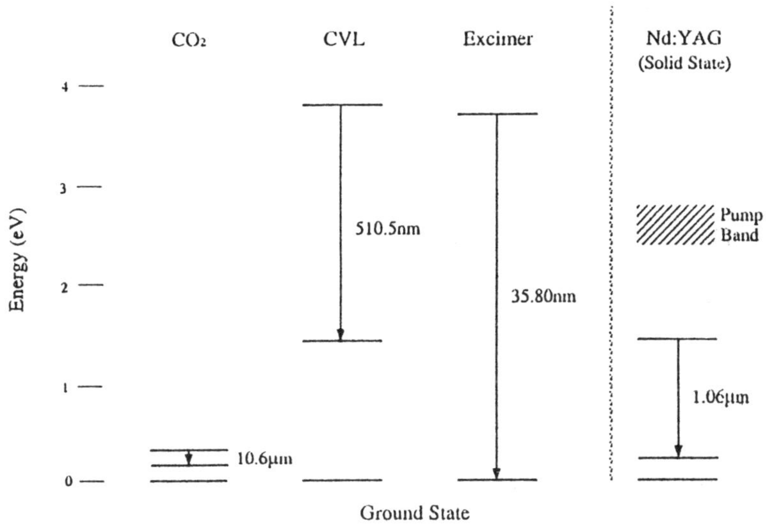

Figure 2. Energy diagrams of efficient lasers.

2. New Directions for Metal Vapor Sources

In this section I would like to compare the emission from a laser in the visible, such as a CVL, with the emission from an incoherent source in the extreme ultraviolet–soft-x-ray spectral region. I will show that an incoherent spontaneous emission source at short wavelengths can be as intense as a pulsed coherent laser source can be in the visible spectrum. This suggests that there may not be as much of a need for ultra-short-wavelength lasers since incoherent sources might be sufficient for many applications.

Consider a cylindrical gain medium that has a length L, a cross sectional area A, and a diameter d. Within that gain medium we assume that a population inversion exists. Let the population density in the upper level be N_u and that the radiative rate from level u to level l at frequency ν_{ul} is A_{ul}. Since this calculation is only an approximation, we will assume that the population inversion is large enough so that we can neglect the population density N_l in the lower level when considering the gain. In other words we are assuming that $\Delta N_{ul} \cong N_u$. This would essentially define an upper limit on the gain exponent, $\sigma_{ul}^H(\nu)\Delta N_{ul}L$.

We consider the beam starts via spontaneous emission in a small region of length d (equivalent to the diameter) at the end of the medium for simplicity. We will assume that the atoms in level u are radiating at a rate A_{ul} and with an energy per photon of $h\nu_{ul}$. Some of these photons are emitted in the elongated direction of the amplifier and therefore they would be enhanced by stimulated emission as they transit through the length L of the medium. Thus a beam would evolve as the radiation propagates down the length of the medium and the intensity grows exponentially. We will calculate the conditions that are required for the beam to reach the saturation intensity when it arrives at the opposite end of the amplifier. Of course, a similar situation would also happen at the other end of the medium for a beam traveling in the opposite direction. The calculation would be the same for both beams. We could also consider atoms in the next volume element next to the one we are considering, but the beam initiated by such a volume element would not quite reach the saturation intensity since it would not have traversed as much length as the beam originating from the end region. We will therefore consider this calculation as an approximation by neglecting effects from other volume elements to keep it simple. The result provides a reasonably good approximation compared to that of a more rigorous calculation.

We thus consider the energy radiated per unit time at frequency ν_{ul} into a 4π solid angle from within the volume $(A.d)$ as $N_u(A.d)A_{ul}h\nu_{ul}$. This is multiplied by the fractional portion of that energy that radiates within a solid angle $d\Omega$ that would reach the opposite end of the medium or $d\Omega/4\pi$. This fraction of the total solid angle can be expressed as $\left(A/L^2\right)(1/4\pi)$ or just $A/4\pi L^2$. We assume that the energy radiated from that volume element per unit time is amplified by an amount $e^{\sigma_{ul}^H(\nu)N_uL}$ by the time it reaches the other end of the medium. We divide that energy per unit time by the area A to obtain an intensity and then equate that intensity to the saturation intensity I_{sat}. This entire process is described below

$$(N_u.A.d)A_{ul}h\nu_{ul}.\frac{A}{4\pi L^2}\frac{e^{\sigma_{ul}^H(\nu)N_uL}}{A}=I_{sat}=\frac{h\nu_{ul}}{\sigma_{ul}^H(\nu)\tau_u}.$$

For the simplest situation in which the only decay process from level u is via radiative decay to level l we can express the decay time from level u as $\tau_u = 1/A_{ul}$ such that the above expression can be rewritten as

$$(N_u.A.d)A_{ul}h\nu_{ul}.\frac{A}{4\pi L^2}\frac{e^{\sigma_{ul}^H(\nu)N_uL}}{A}=\frac{h\nu_{ul}A_{ul}}{\sigma_{ul}^H(\nu)},$$

which reduces to

$$N_u.\left[\pi\left(\frac{d}{2}\right)^2\right].d.\frac{1}{4\pi L^2}.e^{\sigma_{ul}^H(\nu)N_uL}=\frac{1}{\sigma_{ul}^H(\nu)},$$

in which the area A has been rewritten as $A=\pi\left(\frac{d}{2}\right)^2$. Further simplification leads to

$$N_u\sigma_{ul}^H(\nu)Le^{\sigma_{ul}^H(\nu)N_uL}=16\left(\frac{L}{d}\right)^3.$$

This equation can be expressed in the following form as

$$xe^x=16\left(\frac{L}{d}\right)^3,$$

in which $\sigma_{ul}^H(\nu)N_uL \cong \sigma_{ul}^H(\nu)\Delta N_{ul}L$. The factor of 16 in the above equation is not particularly significant. It resulted from choosing a region of length d at the end of the gain medium that was equal to the diameter. An amount of either $d/2$ or $2d$ could also have been chosen without significantly altering the result because of the exponential nature of the equation. If too short a region were chosen, then it would not take into account a significant portion of the beam that contributes to the solution by being amplified through a length just slightly shorter than L. If a much longer initial region were chosen, a shorter exponential growth length would occur since the effective amplified length would be less over the remaining gain length. Both would have provided less realistic results.

From the above analysis we can see that it would be useful to have L/d be reasonably large so that the beam can develop a well defined direction before I_{sat} is reached. Typical lasers require an L/d ranging from approximately 10 to 1000 which from the above equation, suggests a desirable range of values of the gain of 10 – 20 in order to reach saturation or

$$\sigma_{ul}^{H}(\nu)\Delta N_{ul} L \cong 10-20.$$

Let us consider the amount of amplification of the spontaneous emission for gains of $e^{10} - e^{20}$:

$$e^{10} = 2.2\text{x}10^{4},$$

$$e^{15} = 3.3\text{x}10^{6},$$

$$e^{20} = 4.8\text{x}10^{8}.$$

That is the kind of amplification of spontaneous emission that we expect from a laser!

Now let us re-examine the issue of spontaneous emission with no amplification for an atomic system. We will compute Γ the power radiated per unit surface area A per unit frequency from a specific emission line of frequency ν_{ul}, wavelength λ_{ul}, and linewidth $\Delta\nu_{ul}$ in an atomic system of volume V.

$$\Gamma = \frac{N_u A_{ul} h \nu_{ul} V}{\Delta\nu_{ul} A}$$

Volume = Area A x depth b = Axb. However a plasma would be optically thin up to a depth of

$$b = \frac{1.46}{\sigma_{ul} N_l}.$$

Hence

$$\Gamma = \frac{1.46 N_u A_{ul} h \nu_{ul}}{\Delta\nu_{ul}\sigma_{ul} N_l}.$$

But

$$\sigma_{ul} = \left(\frac{g_u}{g_l}\right)\sqrt{\frac{\ln 2}{16\pi^3}}\frac{\lambda_{ul}^2 A_{ul}}{\Delta\nu_{ul}^D}$$

for Doppler broadening. Therefore Γ can be written as

$$\Gamma = \frac{1.46hc}{\left(\frac{g_u}{g_l}\right)\sqrt{\frac{\ln 2}{16\pi^3}}} \frac{1}{\lambda_{ul}^3} \left(\frac{N_u}{N_l}\right),$$

or

$$\Gamma = 7.7\times 10^{-24} \frac{1}{\lambda_{ul}^3} \left(\frac{N_u}{N_l}\right) \quad \text{(watt-sec/m}^2\text{)}.$$

This is a very simple expression that only depends upon λ_{ul} and the ratio $\left(\frac{N_u}{N_l}\right)$.

Let's remember that Γ is the power radiated per unit area per unit frequency. Thus to obtain the power over a given emission linewidth we must multiply by the frequency width. Hence, although the spontaneous emission isn't amplified in this case by having a gain region to pass through, as described above, if the wavelength is reduced by a factor of 50 in going from 500 nm to 10 nm, the spontaneous emission will increase by a factor 50^3 = 1.25x10^5 for a constant ratio of N_u / N_l. In addition the increased emission linewidth associated with the short wavelength will increase the spontaneous emission by at least another factor of 50 and generally much more than that since Doppler broadening varies in proportion to the frequency ν_{ul} and also the plasma temperature which must be significantly higher to access short wavelength transitions in ionized species. Hence the total increase in spontaneous emission can be a total of 50^4 = 6x10^6 and therefore there may be no need for a laser!

Let's take two examples: $\lambda_{ul} = 500$ nm, $\Delta\nu_{ul} = 2\times 10^9$ Hz

and $\lambda_{ul} = 10$ nm, $\Delta\nu_{ul} = 3\times 10^{12}$ Hz.

Also assume that $N_u/N_l \approx 1/e = 0.37$.

For λ_{ul} = 500 nm,

$$\Gamma\Delta\nu_{ul} = 4.6\times 10^4 \text{ watts/m}^2.$$

For λ_{ul} = 10 nm,

$$\Gamma\Delta\nu_{ul} = 8.5\times 10^{12} \text{ watts/m}^2.$$

For a 100 ns duration pulse and a 1 mm^2 area,

for λ_{ul} = 500 nm,

Energy/pulse = 5 x 10^{-9} joules.

For λ_{ul} = 10 nm,

Energy/pulse = 0.85 joules!

This result indicates a huge increase in spontaneous emission flux when using short wavelength transitions.

A much larger increase in flux than that described above can be achieved if a larger optical depth is considered. In that case the line will broaden but if multilayer-coated soft-x-ray mirrors, with a typical bandwidth of 0.3–0.4 nm, are used to collect and focus the emission then the flux will be increased in direct proportion to the optical depth. As a simple approximation, the flux will increase in proportion to the depth in units of the optical depth *b*. For a source at 10 nm and a density $N_l = 10^{18} / \text{cm}^3$ the optical depth is approximately 1 μm. Hence a source of depth of 0.2 mm will be 200 times more intense than the 0.85 joules mentioned above with a linewidth of 200 x 3 x 10^{12} Hz. This linewidth in wavelength units is 0.2 nm which is within the reflection bandwidth of a multilayer-coated mirror.

3. A New Source in Lithium Vapor

The H-like resonance transition in Li^{2+} at 13.5 nm is a possible intense and efficient soft-x-ray source [12]. Similar to that of the lasers in Figure 2 the 13.5 nm transition has a ratio of emission energy to excitation energy of 0.53 as shown in Figure 3. We have carried out calculations suggesting that the “wall plug” efficiency could be as much as 20% to 30%. Pulsing such a source at a repetition rate of 1 kHz could yield a 10 W average power source at 13.5 nm from a volume of the order of 250 μm in dimension. Such a source could yield a very intense source focused to a relatively small focal volume.

We have produced such a source in Li^{2+} ions in a capillary discharge plasma produced by ablation of LiH. We have observed 6 mJ emitted in a duration of 250 ns from such a source of up to approximately 0.5 mm diameter or a power of 25 kW which is comparable to the emission from a CVL. This result was obtained in a relatively inefficient discharge plasma with an electrical conversion efficiency of 0.1% which is at least an order of magnitude better than a laser-produced plasma source at this wavelength. Our source was contaminated with oxygen and thus was much less efficient than if the discharge were of pure lithium. We expect the flux and efficiency to improve significantly when we optimize our system.

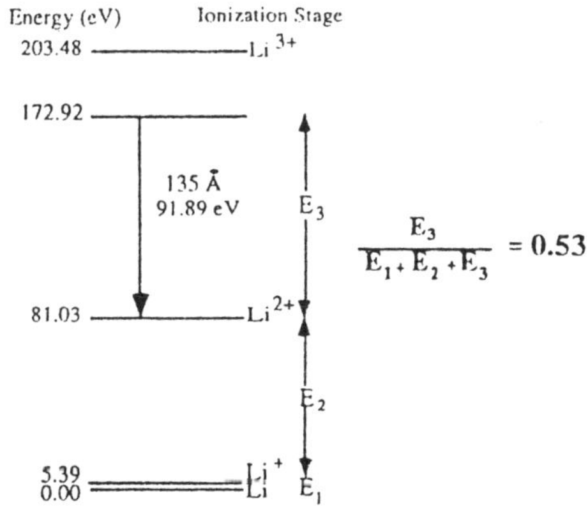

Figure 3. H-like Li^{2+} transition at 13.5 nm.

4. References

1. Rabinowitz, P. and Jacobs, S. (1964) The optically pumped cesium laser, in P. Grivet and N. Bloembergen (eds.), *Quantum Electron.* **Vol. III**, Columbia Press, NY, pp. 489-498.
2. Bell, W.E. (1964) Visible laser transitions in Hg^+, *Appl. Phys. Lett.* **4**, 34-35.
3. Bridges, W.B. (1964) Laser oscillation in singly ionized argon in the visible spectrum, *Appl. Phys. Lett.* **4**, 128-130.
4. Fowles, G.R. and Jensen, R.C. (1964) Visible laser transitions in the spectrum of singly ionized iodine, *Proc. IEEE* **52**, 851.
5. Fowles, G.R. and Silfvast, W.T. (1965) Laser action in the ionic spectra of Zn and Cd, *IEEE J. Quantum Electron.* **QE-1**, 131.
6. Silfvast, W.T., Fowles, G.R. and Hopkins, B.D. (1966) Laser action in singly ionized Ge, Sn, Pb, In, Cd, and Zn, *Appl. Phys. Lett.* **8**, 318-319.
7. Fowles, G.R. and Silfvast, W.T. (1965) Laser action in ionized sulfur and phosphorus, *IEEE J. Quantum Electron.* **QE-1**, 183.
8. Fowles, G.R. and Silfvast, W.T. (1965) High gain laser transition in lead vapor, *Appl. Phys. Lett.* **6**, 236-237.
9. Piltch, M, Walter, W., Solimene, N., Gould, G., and Bennett Jr., W.R. (1965) Pulsed laser transitions in manganese vapor, *Appl. Phys. Lett.* **7**, 309.
10. Silfvast, W.T. and Fowles, G.R. (1966) Laser action on several hyperfine transitions in Mn I, *J. Opt. Soc. America* **56**, 832.
11. Walter, W., Solimene, N., Piltch, M. and Gould, G. (1966) Efficient pulsed gas discharge lasers, *IEEE J. Quantum Electron.* **QE-2**, 474-479.
12. Silfvast, W.T., Bender, H., Eligon, A.M., O'Connell, D., Hanzo, A. and Richardson, M.C. (1993) Laser plasma source characterization for SXPL, in A.M. Hawryluk and R.H. Stulen (eds.), *OSA Proceedings on Soft X-Ray Projection Lithography*, Optical Society of America, pp. 117-126.

COPPER LASERS IN THE BEGINNING

W.T. WALTER
Laser Consultants, Inc.
344 West Hills Road
Huntington, New York 11743
USA

Abstract

In the 1960s vapors of metals were examined for low-lying energy-level structures to provide efficient laser action. Lead and manganese were earlier, but copper has a more ideal energy-level configuration that yields high-power green and yellow pulsed lasers. Copper lasers have demonstrated high gain, scalability to large diameters and a high-brightness output using unstable or high-loss optical resonators.

1.0 Efficient Gas Discharge Lasers

In 1965, five years after the first laser, the efficiencies of the visible gas lasers that had been developed were very disappointing. The CO_2 laser, operating in the far infrared, had demonstrated efficiencies > 0.1; however, the efficiencies with which rare gas and ion lasers converted electrical energy into visible laser output were less than 0.001. For applications requiring many watts of visible optical power, it is essential that lasers with higher efficiencies be developed. The development of more efficient gas lasers was a focus of a research team at TRG Inc. In 1967 the research moved to the Polytechnic Institute of Brooklyn which became Polytechnic Institute of New York and is now Polytechnic University.

The difference between the energy level structure of the visible rare gas lasers and the infrared CO_2 laser is striking because of the large gap between the ground and the excited levels. Gas laser media with appropriate low-lying energy levels are required for improved efficiency. For maximum efficiency, the active medium should possess appropriate laser levels which are both close to the ground level and among its first few excited levels. The low-lying energy levels of certain metals and transition elements make them particularly attractive as efficient laser media. Because these atoms have partially-filled electron shells and tend to bond strongly either to themselves or to other atoms, high temperatures are required to achieve a monatomic gas at a pressure of a few torr.

C. E. Little and N. V. Sabotinov (eds.), Pulsed Metal Vapour Lasers, 15–26.

Several high-temperature apparatuses were constructed at TRG to explore the possibility of cw collision laser action in the vapors of manganese and also in lanthanum. Excitation would be provided by electron-atom collisions, while inelastic atom-atom collisions would relax the lower level. If the requirement that both excitation and relaxation processes proceed simultaneously is dropped, then the efficient lasers would be cyclic [1] or pulsed. Excitation and relaxation processes can then be independently optimized in separate time periods. However, depleting processes which would terminate the population inversion must be slower than the risetime of the discharge excitation pulse. If the electron-atom collisions can be correctly characterized by the Born approximation, then the excitation cross sections are proportional to the squares of the electric dipole matrix elements; and energy levels strongly optically coupled to the ground level will be preferentially excited. To prevent rapid depletion by spontaneous emission, the density of the laser medium must be sufficiently high to produce radiation trapping lowering the Einstein A value from its normal value $\approx 10^8$ to a trapped value $< 10^6$ sec^{-1}. The A value of the proposed laser transition should also be less than the reciprocal of the risetime of the discharge pulse; and the laser medium density sufficient to make the branching ratio into the proposed laser transition ≈ 1.0. Thus, the energy level structure desired is a three-level structure with the upper level strongly coupled to the ground level and a metastable level lying between. The metastable level should have no electric dipole connection to the ground level and a reduced A-value to the upper level ($10^7 < A < 10^4$ sec^{-1}). Figure 1 illustrates a preferred energy-level structure for efficient pulsed gas lasers or cyclic lasers since excitation and relaxation are cyclic in operation.

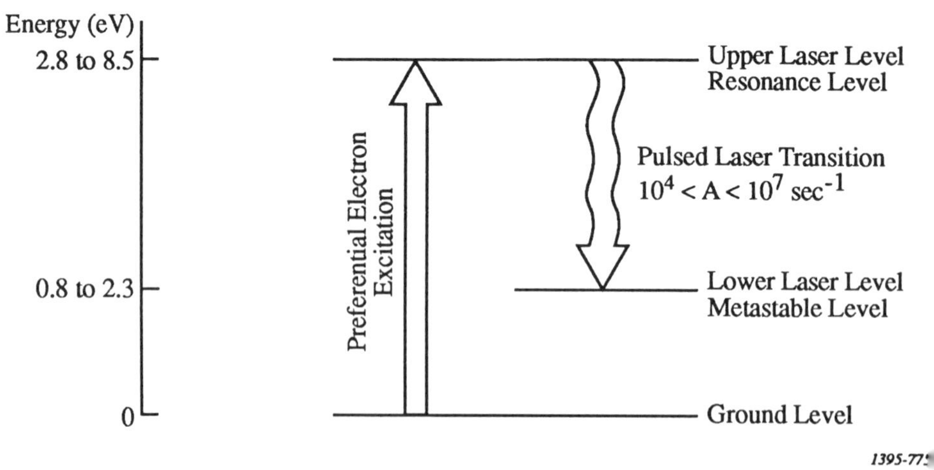

Figure 1. Efficient pulsed gas laser or cyclic laser.

2.0 Manganese and Lead Lasers

When the manganese discharge was pulsed, laser action was observed on five green lines as well as six lines in the infrared [2]. Examination of the manganese energy level structure, as shown in Figure 2, revealed that the electrons in the pulsed discharge primarily excited those levels that are most strongly optically coupled to the ground level. Of the first twelve energy levels above the ground level in manganese, the electrons predominately excited the fourth and twelfth levels, the two $^6P^o$ resonance levels which are strongly optically connected to the ^6S ground state of manganese and consistent with the Born approximation.. The high multiplicity of the resonant energy levels in manganese spreads the electron excitation energy among six excited sublevels and eleven laser lines. The energy level structures of other elements was therefore examined for an atomic level structure which could produce a similar, but more efficient pulsed visible laser.

Similar pulsed laser action at 722.9 nm in the vapor of atomic lead had been reported at the University of Utah. The $^3P^o{}_1$ upper laser level, however, is more strongly coupled to the $^3P^o{}_2$ and $^3P^o{}_1$ fine structure levels of the ground configuration. Indeed laser action on these two transitions, 405.8 and 364.0 nm were subsequently demonstrated [4] when discharge pulses with faster risetimes were utilized .

3.0 Copper Laser

3.1 PREDICTION

Of all the energy level structures examined, the level structure of atomic copper appeared to be closest to ideal. Copper appeared to have the potential of providing population inversions on only two lines in important parts of the visible spectrum while having the fewest potential parasitic loss mechanisms. As shown in Figure 3, the two sublevels of the $^2P^o$ resonance level, $^2P^o{}_{3/2}$ and $^2P^o{}_{1/2}$, could serve as upper laser levels leading to two laser lines 510.554 nm in the green and 578.213 in the yellow. Parasitic losses appear to be minimized. There is only one level, ^2D, between the ground ^2S level and the resonance $^2P^o$ level. The ^2D level does not have an electric-dipole coupling to the ground level and therefore should not be strongly excited. The next level is 1 eV above the $^2P^o$ resonance level. Indeed the level structure in copper appears to be remarkably free of other levels and their parasitic influence. Furthermore the configuration of the metastable ^2D level, $3d^9\ 4s^2$, makes the proposed laser transitions 2-electron transitions with A-values $\approx 10^6$ sufficiently slow that 50-nsec discharge pulses can build up large populations as long as the copper density is sufficiently high (10^{13} atoms/cm^3 in a 1-cm diameter tube) to radiation trap the copper resonance transitions.

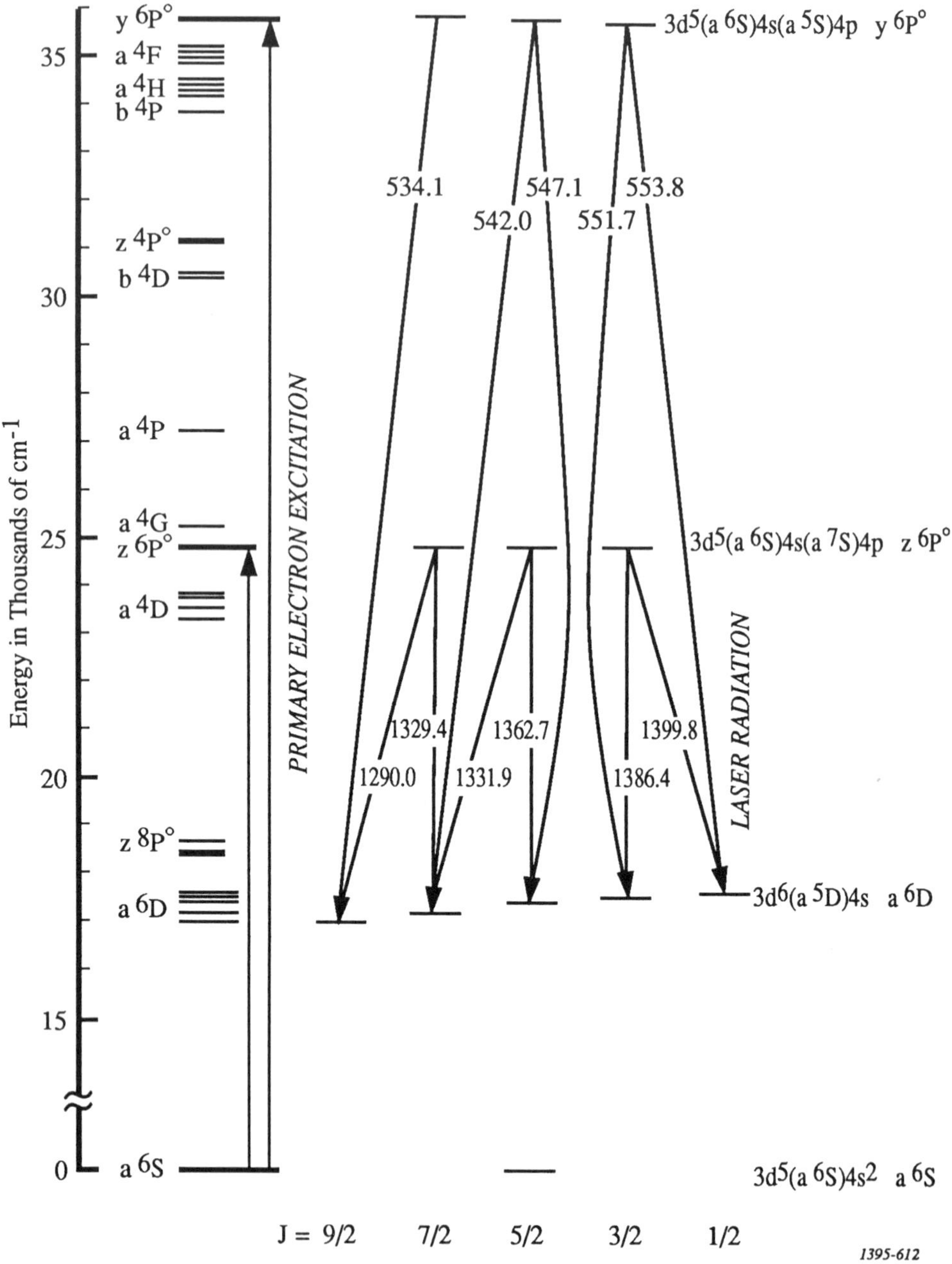

Figure 2. Energy level diagram of manganese on which the primary electron excitation and the laser transitions in nm are indicated.

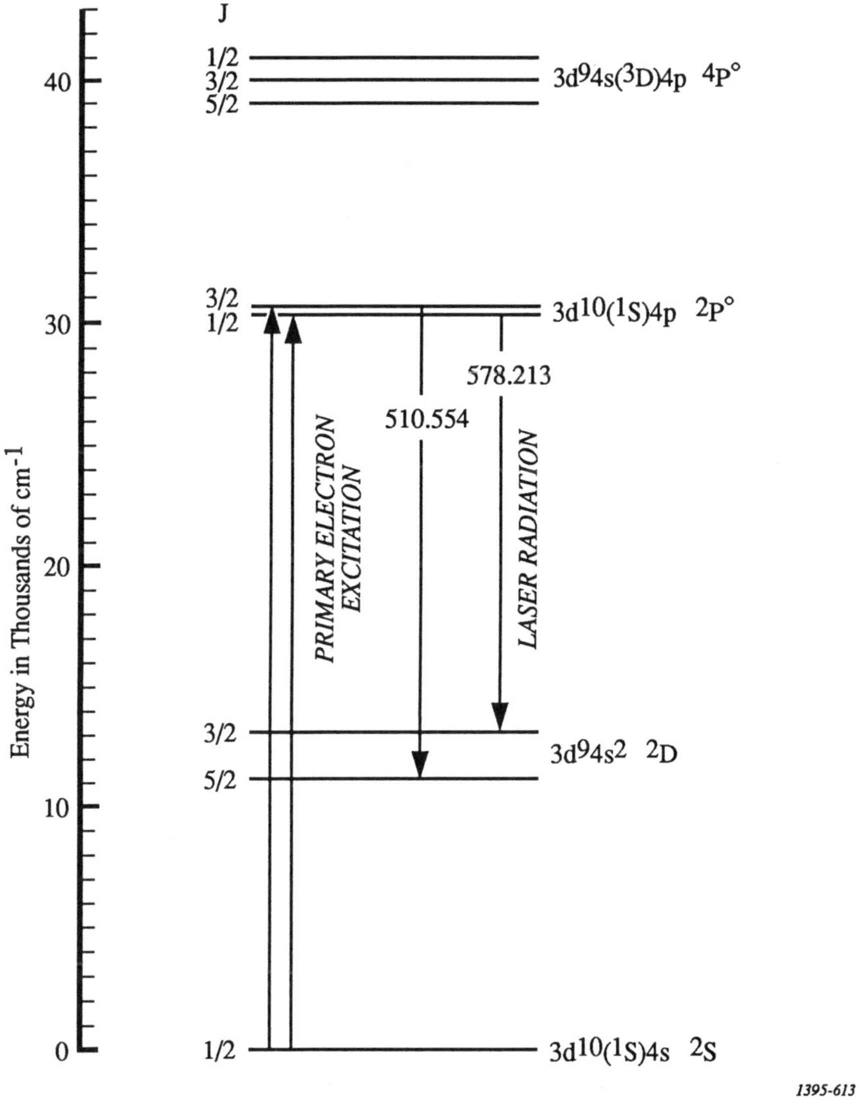

Figure 3. Energy level diagram of copper on which the primary electron excitation and the green and yellow laser transitions in nm are indicated.

3.2 FIRST LASER ACTION

Laser action in copper was discovered at the end of August 1965 [5] in the same apparatus [6] in which laser action in manganese had been discovered at the beginning of July. It is schematically illustrated in Figure 4. To attain the 1400C operating temperature two nested alumina tubes were used to provide the high-temperature confinement zone of the apparatus. The outer alumina tube, overwound with a 60% platinum-40% rhodium ribbon heater, provided vacuum confinement as well as serving as a support for the heating element. In the next apparatus constructed, a third nested tube was added to separate these functions and reduce the high-temperature stress on the vacuum tube. Copper was distributed inside a 1-cm diameter liner, the inner alumina tube. An approximately 80-cm length of the alumina tube could be maintained above 1400C. The windows were kept outside of the 1-m long oven. With this design optical windows could be cemented on, for the window temperature remained close to room temperature.

The tube was filled with 1-2 torr of helium to prevent the metallic vapor from depositing on the windows as well as to carry the discharge from the electrodes to the metal vapor region. The discharge was produced by a 7.5 kV transformer charging 10-m of RG8 coaxial cable at 60 Hz. An adjustable series spark gap was used to discharge the pulse-forming network through the alumina discharge tube. Later the transformer was replaced by a dc power supply, the RG8 coax by a 1.8 nF capacitor and the spark gap by a hydrogen thyratron.

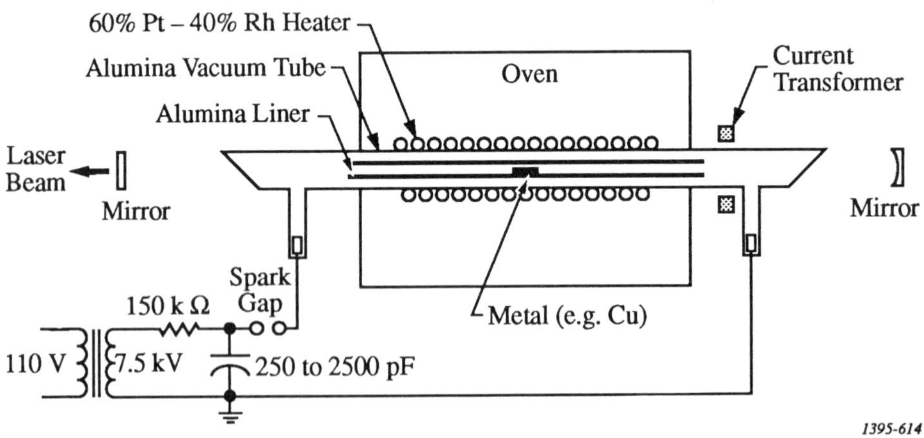

Figure 4. Schematic diagram of the original apparatus for producing laser action in the vapors of manganese and copper.

A 3-m spherical mirror and a flat mirror 158 cm apart formed the optical cavity about the alumina tube. Each mirror had an opaque gold coating. A barium fluoride pick-off plate within the cavity provided the output. When the temperature was raised above 1400C, green and yellow laser action was observed at 510.5 and 578.2 nm as predicted. Gains of the laser oscillations were measured by insertion of known losses. Because of superradiance, loss had to be inserted at each end of the laser cavity. Gain coefficients of 58 dB/m and 42 dB/m were measured for the green and yellow lines respectively. Subsequently an unsaturated single-pass gain of 3.91 dB/cm was measured [10] for the green line in a 1-cm diameter 10-cm long copper vapor amplifier tube.

When the Au flat mirror was replaced with a flat with a dielectric multilayer transmitting about 70% in the green, green laser pulses 20 nsec FWHM, 1.2 kW peak power were measured at a pulse repetition rate of 660 Hz. The average power was 15 mW in the green and 5 mW in the yellow. The efficiency computed from the energy stored in the capacitor was 0.001. Since the current pulse risetime and pulse width were 100-200 nsec, significant improvement could be anticipated. The limiting efficiency for the copper laser was projected to be about 0.23 [1]. Since most of the energy stored on the capacitor was dissipated in the discharge long after the laser pulse ended, an energy conversion efficiency of at least 0.01 seemed attainable.

3.3 DIAMETER SCALING

To achieve the high efficiency believed possible with the copper vapor laser, the apparatus must be self-heating. Higher power outputs and increased efficiency require increasing the number of copper atoms contributing to each laser pulse. Increases could be obtained in several ways. The copper density could be increased by raising the operating temperature and therefore the copper vapor pressure. Alternatively, the volume of the active laser medium could be increased. For order-of-magnitude increases the most practical approach is to enlarge the diameter of the laser tube. An increase in diameter is preferable to an increase in length because; 1. the number of vaporized copper atoms increases quadratically with diameter, 2. the electrical inductance of the circuit and, therefore, the current pulse risetime and width are kept to a minimum and 3. the surface-to-volume ratio of the heated region is minimized.

As a first step in investigation larger copper laser systems, the apparatus was scaled in diameter from 1 cm to 5 cm. Separate heater support (i.e. muffle), vacuum and liner alumina tubes with inner diameters of 11.4, 6.4 and 5.0 cm respectively were used. The additional muffle tube eliminated hot spots on the wall of the vacuum tube minimizing its porosity and leak up rate. It also made it possible to change the vacuum tube without changing the heater. Standard tapers were ground on each end of the vacuum tube. Mating tapers on the window-electrode assemblies provided easily demountable connections and convenient access to the liner with its copper charge.

The scaling in diameter from 1 to 5 cm enabled a 25-fold increase in the absolute number of copper atoms in the vapor phase. This could produce a comparable increase the laser output providing no new effects occur. One concern was that the discharge might not fill the entire volume. In addition, if the conductivity of the copper discharge was sufficiently high, the skin effect could confine the fields to an outer annular region of the vacuum tube. Then the penetration time for diffusion of the current into the tube center could appreciably lengthen the laser pulse and prevent attainment of high peak powers. However, the measured green laser pulse full width at half maximum was 16 nsec confirming that the diameter scaling did not appreciably lengthen the laser pulse.

A peak power of 57 kW was determined directly with a TRG 105B photodetector, whereas calculation from an average power measurement using a calibrated Eppley thermopile yielded 40 kW [7]. The efficiency, that is the energy of a laser pulse divided by the total energy stored on the capacitor was 0.012. This was the first visible gas laser to demonstrate an electrical conversion efficiency of 0.01.

The laser pulse was observed to begin about 100 nsec after the start of the current pulse. In the 1-cm apparatus the time delay was about 20 nsec. The longer delay in the larger tube is due to a lower current density. Lengthening the delay without sacrificing efficiency can only be accomplished in laser media where the upper laser level lifetime is even longer than this delay. This is possible in the copper laser where the upper laser level lifetime is ≈ 500 nsec [8] when the copper density exceeds a few millitorr to trap the resonance radiation.

The average power approached 0.5 W as the pulse repetition rate was increased toward 1 kHz. The operating temperature was about 1500C, corresponding to a partial copper pressure of 0.4 torr. The efficiency, that is, the energy of a laser pulse divided by the total energy stored on the capacitor was 0.012. This was the first visible gas laser to demonstrate an electrical conversion efficiency greater than one percent.

3.4 HIGH BRIGHTNESS OUTPUT

The output power of a laser alone often does not determine whether or not that laser can accomplish a particular application. In material processing applications the spot size is also critical. Single mode output is essential for holographic and long distance communication applications. Brightness [watts/cm^2-steradian] is a more appropriate figure-of-merit to determine the suitability of an individual laser for a given application.

Because of its high gain, the copper vapor laser can be operated in a superradiant (more correctly, an amplified spontaneous emission) mode by placing a single mirror at one end of the laser tube. Then the beam divergence is just the aspect ratio of the double-pass copper-vapor medium. When a 30% output reflector was placed at the other end of the laser tube completing the optical cavity, the beam divergence narrowed dramatically as shown in Figure 5. The curve at the top of Figure 5. plots the aspect ratio, the beam divergence with a single reflector. The addition of an output reflector narrowed the beam divergence of the copper laser by a factor of ten as shown in the middle curve. The lowest curve in Figure 5. indicates the beam divergence expected for diffraction-limited output. Aperture control is a standard method of discriminating against higher-order transverse modes. Figure 5. shows that closing an aperture inside the optical cavity reduced the measured beam divergence until diffraction-limited or single transverse mode output was obtained at a 2-mm diameter. The open circle at the right side of Figure 5. is the measured beam divergence when an unstable or high-loss resonator was used.

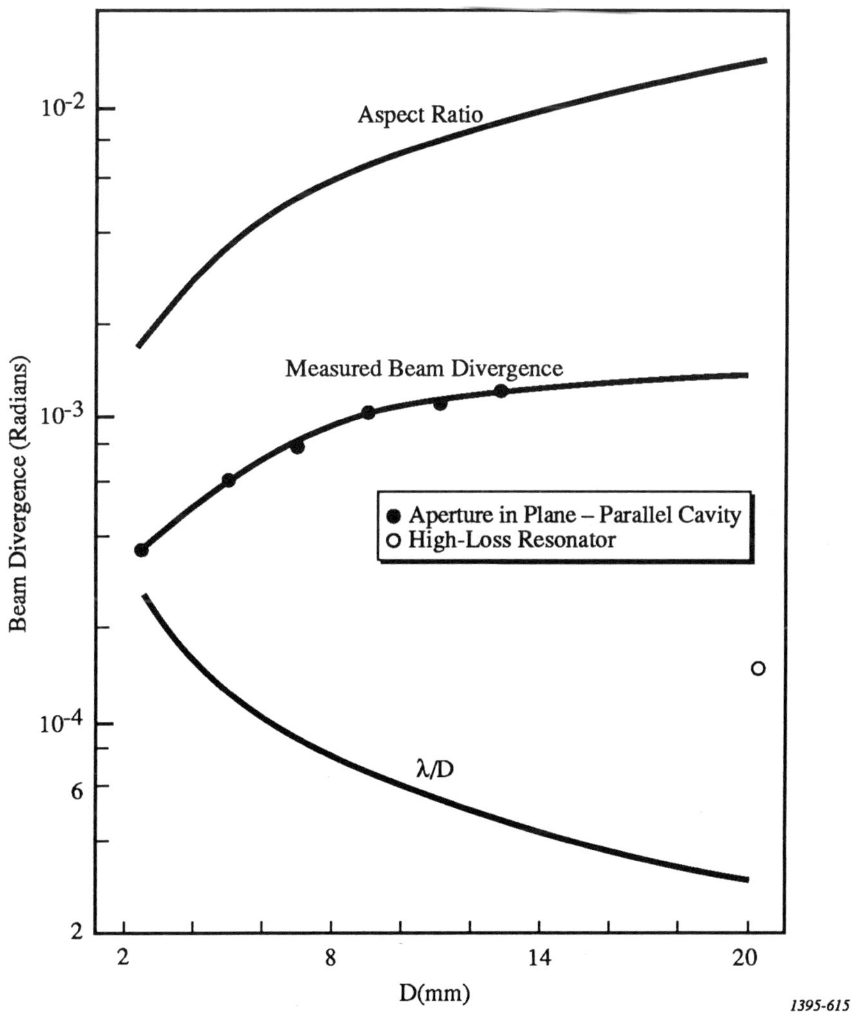

Figure 5. Coherence properties of the copper vapor laser. Output beam divergence as a function of output beam diameter D.

Figure 6. shows the effect on the power output and brightness of the laser output as the aperture was closed. Although the peak power output decreased from 1.4 kW to 100W as the aperture was reduced from 13 to 2 mm diameter, the brightness increased from 10^9 to 3 x 10^{10} W/cm^2-ster. The high small-signal gain of the copper laser enables an unstable resonator to be used to discriminate against higher-order transverse modes while utilizing the large-diameter gain volume available. The squares on the right side of Figure 6. indicate the peak output power and brightness obtained from a non-optimized unstable resonator, non-optimized because only 40% of the multimode output power was obtained. Brightness, however, was improved by a factor of 100 [9,10].

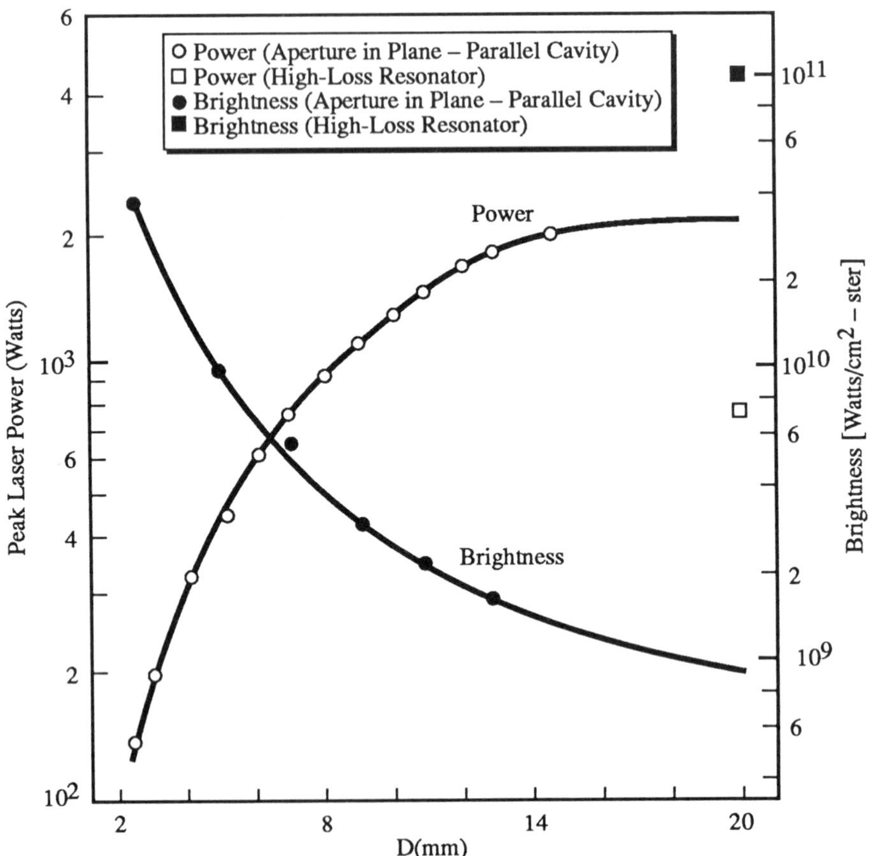

1395-616

Figure 6. Coherence properties of the copper vapor laser. Peak laser power and brightness as a function of output beam diameter D.

The unstable or high-loss resonator is shown in Figure 7. It consists of concave and convex reflectors with a common focus at the left of the figure producing a parallel beam output. The diameter of the convex reflector on the left was larger than optimum. Nevertheless, the peak output power was measured to be 770W and the brightness 1 x 10^{11} W/cm^2-ster. When a pick-off mirror was placed at a 45° angle in front of the convex reflector, a peak single-mode output power of 5 kW was measured, the same value as the multimode power output obtained when a plane-parallel optical cavity was used. The brightness obtained was 10^{12} W/cm^2-ster, which was the highest brightness obtained from any gas laser at that time.

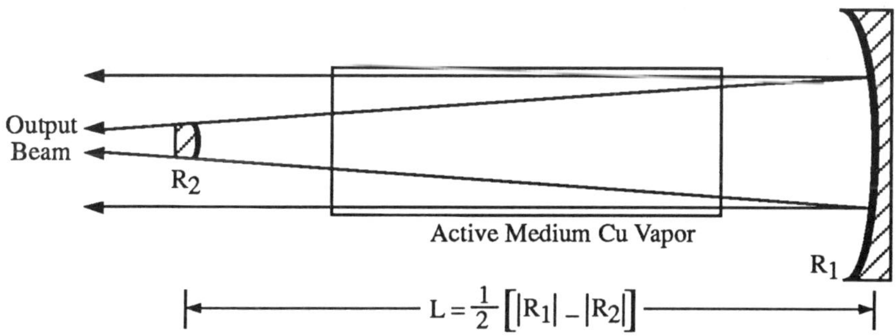

Figure 7. Unstable or high-loss resonator

3.5 LIMITING PROCESSES

At the time of its discovery, efficiencies of 10% were projected for the copper laser [1] based on a consideration of three loss processes; 1) the energy loss in relaxing the ^2D lower laser level, 2) the excitation energy required for upper laser level copper atoms remaining at the termination of the laser pulse and 3) excitation energy that does not go into excitation of the upper laser levels. An upper limit of 23% was estimated. Reported efficiencies for the copper laser; however, have not exceeded a few percent. A computer model was constructed [11,12] to examine limiting processes and aid in the development of pulsed gas lasers. Collisional mixing, substantial ionization and the influence of higher lying levels were examined. The model suggested that stepwise excitation to higher-lying copper energy levels acted as the rate-limiting process in the copper laser [12].

4.0 Other Metal Vapor Lasers

In addition to the manganese, lead and copper lasers, similar laser action was demonstrated at 627.818 nm in the vapor of neutral atomic gold [13] and at 854.209 and 866.214 nm in the ionized vapor of calcium [1]. Column 1b of the periodic table includes gold and silver as well as copper. The red line in gold is analogous to the yellow line in copper. The transition analogous to the green line is an ultraviolet line whose A value is too high to allow a large enough population inversion for a high power pulsed laser. In silver the metastable ^2D levels have been displaced upward and are too close to the resonance $^2P^o$ energy levels to provide laser transitions. With the removal of an electron, the energy level structure of ionized calcium is also analogous to that of copper. The corresponding transitions are high-gain transitions in the near infrared.

5.0 Acknowledgment

Members of the research team at TRG and the Polytechnic who have contributed to pulsed metal vapor research and development include Robert Chimenti, Gordon Gould, George Kull, James T. LaTourrette, Kyunglin Park, Martin Piltch, Walter Schmidt, Nicholas Solimene and William Woturski.

6.0 References

1. Walter, W.T., Solimene, N., Piltch, M., and Gould, G. (1966) Efficient pulsed gas lasers, IEEE J. Quantum Electronics **QE-2**, 474-479.
2. Piltch, M., Walter, W.T., Solimene, N., Gould, G., and Bennett, W.R. Jr. (1965) Pulsed laser transitions in manganese vapor, Applied Physics Letters **7**, 309-310.
3. Fowles, G.R. and Silfvast, W.T. (1965) High-gain transition in lead vapor, Applied Physics Letters 6, 236-237.
4. Isaev, A.A. and Petrash, G.G. New generation and superradiance lines of lead vapor (1969) JETP Letters **10**, 119-121.
5. Walter, W.T., Piltch, M., Solimene, N., and Gould, G. (1966) Pulsed-Laser Action in Atomic Copper Vapor, Bulletin American Physical Society **11**, 113.
6. Piltch, M., and Gould, G. (1966) High temperature alumina discharge tube for pulsed metal vapor lasers, Review Scientific Instruments **37**, 925-927.
7. Walter, W.T. (1967) 40-kW pulsed copper vapor laser, Bulletin American Physical Society **12**, 90.
8. Kock, M. and Richter, J, (1968) Experimentelle Übergangswahrscheinlichkeiten und die solare Häufigkeit des kupfers, Zeitschrift für Astrophysik **69**, 180-192.
9. Chimenti, R. and Walter, W.T. (1971) Coherence properties of the pulsed copper vapor laser, Bulletin American Physical Society **16**, 41-42.
10. Chimenti, R.J.L. [1972] The copper vapor laser, Ph.D. Thesis, Polytechnic Institute of Brooklyn.
11. Kull, G.M. [1980] Analysis of the copper vapor laser and development of a general laser plasma model, Ph.D. Thesis, Polytechnic Institute of New York.
12. Walter, W.T., Solimene, N. and Kull, G.M. [1981] Stepwise excitation - a limiting process in pulsed gas lasers, Proceedings International Conference on Lasers '81, 853-865.
13. Walter, W.T. [1968] Metal vapor lasers, IEEE J. Quantum Electronics **QE-4**, 355-356.

2. Physics of Pulsed Metal Vapour Lasers

COPPER LASER KINETICS - A COMPARATIVE STUDY

C.E. WEBB and G.P. HOGAN
The Clarendon Laboratory
Parks Road, Oxford, OX1 3PU, UK

1. Introduction

The work described in this paper was carried in the Clarendon Laboratory, Oxford over a period of years. It includes the work of several researchers, most notably the recent studies carried out by G.P.Hogan at Oxford and by C.P.Whyte and C.E.Little of the University of St Andrews.

Although several studies of the parameters of operating copper vapour lasers have been published in the literature hitherto, in nearly all cases the data refers to tubes of different diameters operated under widely different operating conditions of filling gas pressure and current density. The objective of the present series of experiments was to carry out a detailed study of one specific type of elemental copper vapour laser (CVL) operating under its optimum conditions and to compare that with a laser of the HyBrID (Hydrogen Bromide In Discharge) design originated by Maitland and co-workers [1] at the University of St Andrew's. The specific type of CVL which was the subject of the study was an Oxford Lasers Model Cu40 design operating at 6.5kHz pulse repetition frequency in a tube of diameter 42mm and 1.5m active length with a neon buffer gas pressure of 25mbar to provide an output of 40 watts on the 511 and 578 transitions combined. For all of the measurements detailed below the laser was operated at exactly this optimum set of conditions and its output power was monitored to ensure that the conditions were reproduced accurately from day to day.

The HyBrID laser was operated on a flowing gas mixture containing neon and HBr. It had a tube diameter of 25mm and an active length of 800mm. At its optimum repetition frequency of 17kHz it produced a total output power of 22W.

The study of the Cu40 device revealed several unsuspected features, particularly concerning the role of pre-ionisation resulting from the remenant electron density at the end of each discharge excitation cycle which is also responsible for the appreciable conductivity of the gas prior to the application of voltage to the discharge at the beginning of each excitation pulse when the laser is operated in a repetitive steady state manner. In contrast, the electron density and conductivity of the gas in the HyBrID device returns to near zero well before the beginning of the following pulse despite the fact that the interval between pulses in the HyBrID device is only 60μs compared to the 150μs of the CVL.

C. E. Little and N. V. Sabotinov (eds.), Pulsed Metal Vapour Lasers, 29–42.

This paper describes the experimental techniques used to measure the electrical parameters (discharge current and voltage) of the CVL including the observation of the pre-breakdown current - the so-called phantom current. The measurements of electron density in both CVL and HyBrID systems made using a two wavelength visible/infrared fast interferometer technique, are also described and their consequences for the understanding of the behaviour of the CVL plasma are discussed. Measurements of the populations of the ground state, lower and upper levels of the green and yellow laser transitions in both systems were carried out using a version of the classical Rozhdestvenskii hook technique [2] updated by the use of automated experimental control and data handling techniques [3].

2. Experimental

2.1 CVL: ELECTRICAL DISCHARGE PARAMETERS

The voltage across the discharge was monitored by a high voltage probe (rise-time < 5ns) connected directly to the HV flange of the plasma tube. Laser output and spontaneous emission were monitored by a vacuum photodiode with a 100ps rise-time, or a silicon PIN photodiode (rise-time < 1ns).

To measure the current flowing through the laser discharge, three different probes, of two basic types, were connected in different positions in the circuit. The first was a low inductance resistor (series resistance 20.1mΩ) comprised of a rectangular, titanium foil resistance element, clamped between two copper end terminals, with two thin insulating spacer layers of mylar sheet. The series resistor was inserted into the earth return path close to the HV electrode. Two different Rogowski coil type current probes (Pearson, models 110A and 2877) were also used. The traces from the three current monitors were essentially the same, however the electrical noise superimposed upon the signal from the two probes in the proximity of the HV electrode was considerably greater than that from the current monitor at the grounded end of the discharge tube. In addition, it was possible to obtain a quantitative result from this probe by assuming that the full current is divided equally among the 8 studs arranged symmetrically around the circumference of the tube. Such an assumption is not possible for the other two monitors. For these reasons all results presented are those from the monitor at the grounded end of the laser tube.

Care was taken to ensure that the coaxial leads used to connect each optical detector and discharge monitor to the oscilloscope were all of exactly the same length to allow reliable comparison of measured timings.

2.2 ELECTRON DENSITY MEASUREMENTS

The discharge tube of the laser was set up in one arm of a two wavelength Mach-Zehnder interferometer. The probe beam sources consisted of a 4W CW waveguide CO_2 laser to provide the 10.6μm radiation, and a 2mW 670nm laser diode for the visible. A change in n, the real part of refractive index of the plasma of length l, can be observed as a change of fringe intensity I on the detector monitoring the interferometer output since:

$$I = I_0 \{1 + \cos [2\pi l (n\text{-}1)/\lambda_0]\}/2$$

where I_0 is the intensity of the probe beam as it enters the interferometer. The refractive index is made up of contributions from the gas atoms and the free electrons. Because of their different dependencies on wavelength, these two contributions can be unambiguously separated by the two wavelength measurement technique. The contribution of the free electrons is related to the electron density n_e since

$$n^2 = 1\text{-} \omega_p^2/\omega^2 ,$$

in which $$\omega_p^2 = n_e e^2/\epsilon_0 m_e .$$

An important component in the system was the compensation pressure cell in the reference arm of the interferometer. Using this cell it was possible to adjust the phase shift in the reference arm and to lock the starting point of the fringe system any chosen point in the $\cos^2$ intensity pattern.

For the interferometer beam-splitters ZnSe could conveniently be used. Standard mirrors for use with 10.6μm laser radiation were also appropriate for use in the visible but to allow the transmission of both wavelengths through the CVL it was necessary to replace both the CVL windows and the cavity mirrors and uncoated BaF_2 disks were used for both windows and output coupler. The CVL head was mounted on a translation stage to allow the repetition of measurements at a number of radial positions without the necessity of realigning the interferometer.

Using the equipment as described it was possible only to measure a fringe shift, and therefore only a change in electron density, rather than absolute values. In order to convert the measurements obtained into final figures it was also necessary, at each radial position, to measure the pre-pulse electron density. This was accomplished by ceasing the trigger pulses to the CVL power supply unit for a time long enough to allow complete recombination of the free electrons down to the Saha value, but short enough to allow negligible change in plasma tube temperature. A momentary action switch (with bounce suppression) connected to an external trigger inhibit input to the pulse generator allowed such an interruption of about 150 – 200ms. In this way the fringe shift as the electron density fell to a negligible value could be observed and initial value thus calculated.

For each radial position four oscilloscope traces were recorded at each wavelength during the CVL pulse delay cycle, and four at each wavelength to measure the initial (pre-pulse) electron density. In order to provide good time resolution for the fast rising electron density during the current pulse, as well as enabling observation of its behaviour over the full inter-pulse period, two traces were acquired with different timebases on the oscilloscope. Immediately following this a second pair of traces were obtained under identical conditions except without the probe beam. In this way traces could be obtained containing all systematic, spurious optical and electrical signals, including residual background black body radiation, radiofrequency radiation from the discharge, and any offset in operating point of the detector.

2.3 MEASUREMENTS OF POPULATIONS OF ATOMIC ENERGY LEVELS

Figure 1 shows the equipment used [3] for measurement of populations of various excited levels of copper and neon as well as the copper ground state. The laser discharge tube is incorporated in one arm of a Mach-Zehnder interferometer and the fringes so formed are projected on to the entrance slit of a monochromator in the exit plane of which is a CCD camera connected to a frame-grabbing and storage system run by a desktop PC unit.

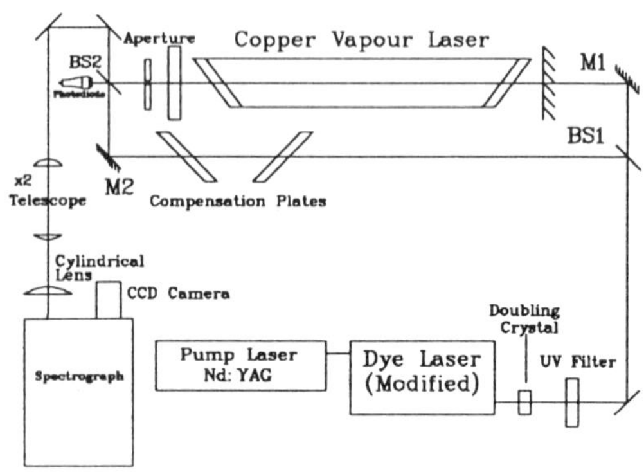

Figure 1. Schematic of optics for hook measurements

The light source which forms the fringes is a broad-band dye laser pumped by a Nd:YAG laser whose 1.06μm output is frequency doubled to 532nm. The broad-band output of the dye laser covering some 2-3nm is itself frequency doubled in a BBO crystal to access the ultraviolet resonance transitions of Cu at 324.8 and 327.4nm. To monitor transitions such as the 510.6 and 578.2nm laser transitions themselves, as well as transitions originating on the $4p^2P$ levels to higher lying levels such as $5s^2S$ and $4d^2D$ the visible output of the dye laser could be used directly.

The dye laser beam as it passed through the discharge tube was of less than 2mm diameter and thus by translating the entire CVL laser tube assembly on a translation stage, relative to the interferometer and other optical components which remained fixed relative to the optical table, it was possible to sample the atomic populations at various radial positions within the discharge tube.

The relative angular settings of the mirrors comprising the interferometer and the tilt of the compensator plates in the reference arm of the interferometer can be adjusted such that, with the discharge tube empty of copper vapour, a fringe system covers the entire region of the spectrum encompassed by the sensitive area of the CCD with the fringes sloping at about 45° across the field of view of the camera. When atomic copper is present in the tube and the monochromator/dye laser combination is tuned so that one of the absorption transitions lies within the spectral band pass, dispersion hooks of the type shown in Figure 2 are recorded. By measuring the wavelength range between the turning points of the dispersion hooks, and with knowledge of the calibration constants of the equipment, a value for N^*fl can be obtained. In this product N^* represents the population difference $(N_1-g_1N_2/g_2)$ where N_1 is the population of the lower level 1 (of degeneracy g_1) and N_2 is that of the upper level (of degeneracy g_2). The oscillator strength of the transition is represented by f and l is the length of the plasma. The relevant values of f are obtained from the data of [4] while the length of the plasma is estimated as 1.30m in the case of the CVL since copper vapour occupies only the central hot zone of the tube.

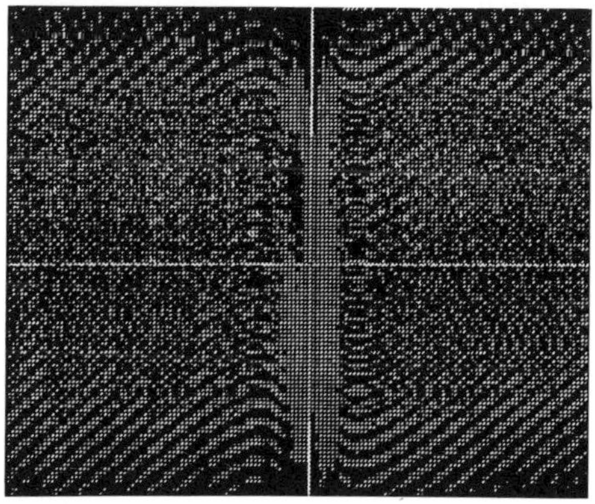

Figure 2. A typical hook pattern

The duration of the Nd:YAG pump laser pulse was 7ns, and that of the broad-band dye laser was only 5ns, so that by timing the probe laser pulse to occur at a predetermined instant within the CVL discharge cycle, time resolved population measurements could be made. An important feature of the experiments reported here is the accuracy and reproducibility of the timing event used as a fiducial marker of zero time.

Since all the measurements were performed on an operating CVL equipped with mirrors, reflective at the CVL laser wavelengths but not at the probe laser wavelengths, the leading edge of the green laser output at the axis of the CVL cavity provides a fiducial timing event which is accurately reproducible and was used as the reference point for all measurements carried out in the present experiment.

Data taking was fully automated The computer initiated a sequence of measurements which recorded a hook pattern for a given delay between CVL excitation pulse and probe laser pulse. The timing of the relative delay was then stepped by the computer through a sequence of predetermined delay settings with fringe patterns recorded at each delay value until the complete period between one excitation pulse and the next had been covered. The discharge tube position was then changed so that a new radial position was sampled and the whole process repeated for the transition in question. This process again repeated for each of the monitored transitions. During a day's operation, more than 1200 hook patterns could be stored on the computer's hard disk, which came close to filling its capacity. Overnight the data was reduced using a fast Fourier Transform technique [3] so that the *N*fl* values could archived to floppy disk and the hard disk cleared ready for the next day's operation.

3. Population Densities - Results

3.1 COPPER GROUND STATE DENSITY

Figure 3 shows a typical set of traces of ground state density on axis, discharge current, and laser emission with the CVL operating in steady-state under optimum conditions.

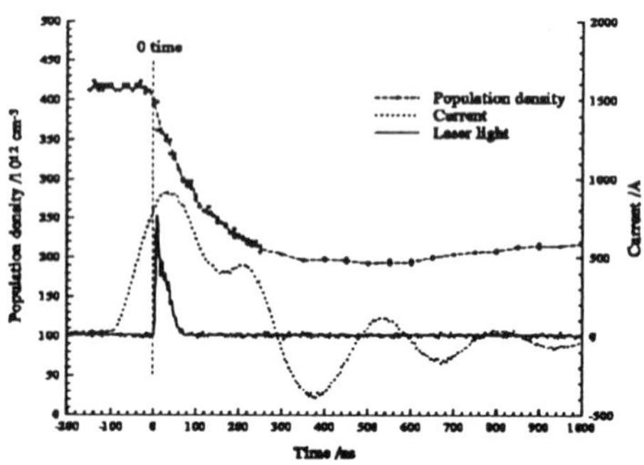

Figure 3. Copper ground state population on axis

It is evident that the ground state density of copper on the axis of the CVL starts to deplete some 10-20ns before the fiducial time origin defined by the onset of green lasing at the axis. The density of copper atoms falls from 4.2 to 2.0 x 10^{14}cm^{-3} over the first 500ns, about half of this depletion occurring before the end of the laser pulse at 60ns.

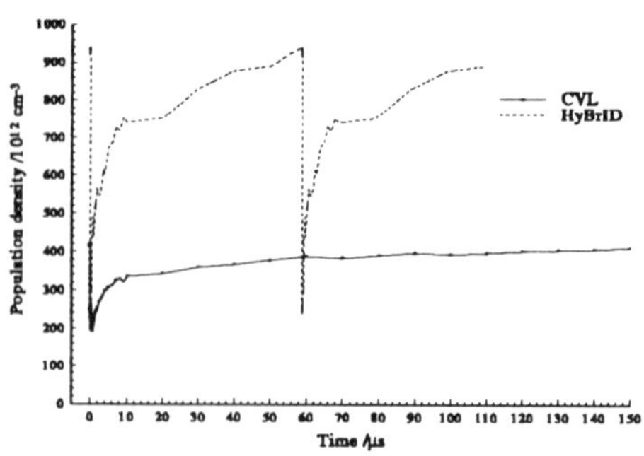

Figure 4. Copper ground state; CVL and HyBrID

Figure 4 compares the axial density of copper ground state atoms measured in the CVL and the HyBrID laser over the full interpulse period of each device.The density is initial higher in the HyBrID device 9.5 x $10^{14}cm^{-3}$, but although the depletion reduces it to a similar minimum value as in the CVL (about 2.0 x $10^{14}cm^{-3}$) it recovers to its pre-pulse value much more quickly than in the CVL.

3.2 UPPER LASER LEVEL POPULATIONS

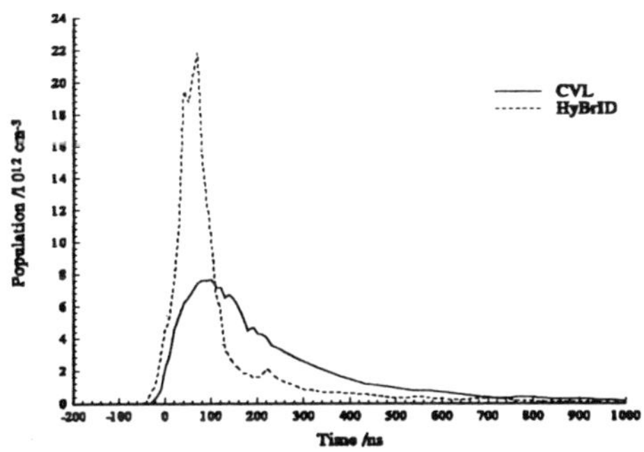

Figure 5. Copper 4p $^2P_{3/2}$ population on axis

Figure 5 shows the axial values of populations of the $4p^2P_{3/2}$ upper level of the green transition measured in both devices. (Values for the $4p^2P_{1/2}$ upper level of the yellow transition are about half as big in both cases, but otherwise show a similar behaviour.) The peak upper laser level population in the case of the 25mm diameter HyBrID device is 2.2 x 10^{13} cm^{-3}, some three times higher than in the 42mm diameter CVL. However the population of the upper laser level falls much more quickly in the case of the HyBrID laser so that for times later than 150ns after the time origin the population remaining in the upper laser level after the end of the current pulse is larger by about a factor of two in the CVL.

3.3 LOWER LASER LEVEL POPULATIONS

Figure 6 compares the populations in the $4s^2\ ^2D_{5/2}$ lower laser level of the green transition measured in the two devices. (Values measured for the $4s^2\ ^2D_{3/2}$ lower level of the yellow transition are about half those of the corresponding values of the $4s^2\ ^2D_{5/2}$ level in the case of the HyBrID laser, about one quarter in the case of the CVL.) The $4s^2\ ^2D_{5/2}$ population peaks at 1.1 x 10^{14} cm^{-3} in the HyBrID laser and although its decay is initially faster, its value at 10μs is higher than in the CVL case.

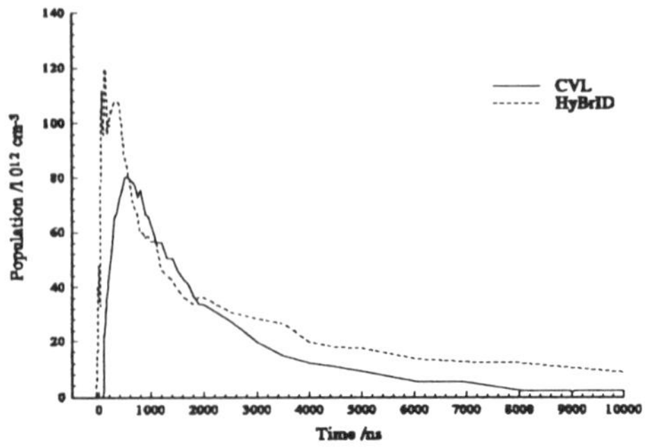

Figure 6. Copper $4s^2\ ^2D_{5/2}$ population on axis

4. Electron Density Measurements - Results

4.1 ELECTRON DENSITIES IN THE CVL

The electron density on axis in the CVL as a function of time over the early excitation phase of the laser cycle and at various radial positions over the full interpulse period are shown in Figures 7 and 8 respectively [5] The first point to note is the high value of the electron density (3.9 x $10^{13}cm^{-3}$) which exists at the end of each discharge excitation cycle immediately prior to the succeeding excitation pulse. This represents about 9% of the total density of atomic copper and implies that, at the onset of the discharge pulse, the plasma is already highly conductive. Comparing the pre-pulse electron density to the maximum value it attains at the end of the current pulse (1.3 x $10^{14}cm^{-3}$), it is evident that the electron density increases by only a factor of about 3 during the discharge. The electron density, as measured at the radial position 7mm off axis, shows that the electron density remains at its maximum value for the first 25μs before starting

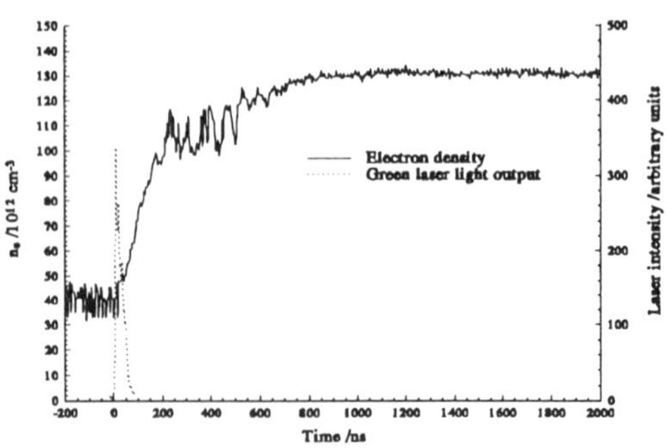

Figure 7. Electron density on axis

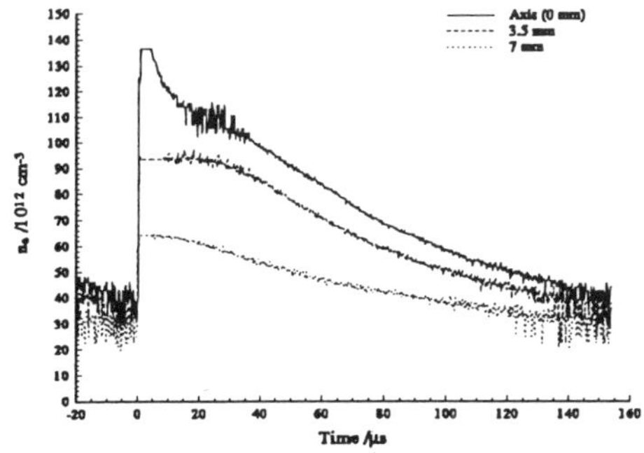

Figure 8. Electron density; 3 radial positions

its decline towards the pre-pulse value.

On axis there is a small peak which shows a much faster rate of decay in the first 20μs. The difference between these two types of behaviour we attribute to the fact that on axis the probe beam samples the electron density in the constricted discharge in the neon buffer gas in the cold end regions of the discharge tube comprising some 20cm at each end of the tube. Over the central hot zone, the dominant ion is Cu^+ whereas in the cold end regions there is a preponderance of Ne^+ ions.

At axial positions the electron density along the optical path sampled by the probe beam therefore includes a component due to the constricted discharge in the end region where recombination is relatively fast on account of the cold gas and hence low value of electron temperature once the tube voltage has collapsed. In contrast, at radial positions further than 7mm from the axis, the probe beam lies outside the current path in the cold end regions and samples only the electron density in the hot region of the tube where the recombination rate is correspondingly slower.

Another point to note from Figure 7 is that the maximum electron density is attained by 800 - 1000ns after the fiducial time origin, and the growth of electron density is essentially complete by the time that the current pulse has decayed to zero.

4.2 ELECTRON DENSITIES CVL/HyBrID COMPARISON

Perhaps the most dramatic difference between the CVL and the HyBrID laser devices is evidenced by the behaviour shown in Figure 9. While it might be expected that the peak value of n_e would be higher in the 25mm diameter HyBrID laser than in the 42mm diameter CVL, the difference in pre-pulse electron density is very marked and probably accounts for much of the difference in performance of the two types of device. As mentioned in section 4.1 above, the pre-pulse value of ne in the case of the CVL is only lower than the maximum value by a factor of about three. In the case of the HyBrID laser the election density is not only much higher at peak, but decays much more rapidly, reaching values too small for accurate measurement by the start of the next current pulse.

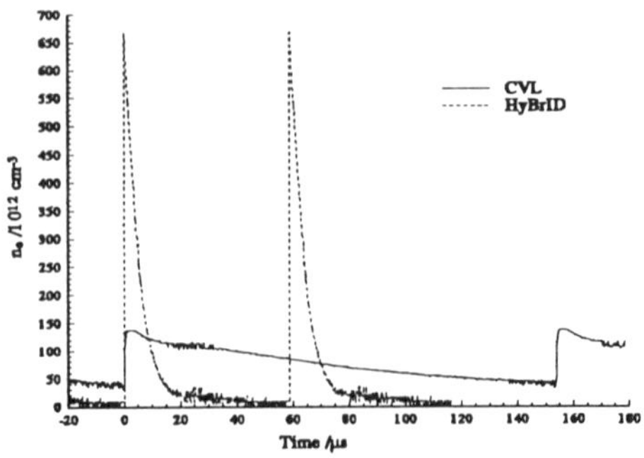

Figure 9. Electron density on axis; CVL and HyBrID

The fact that the CVL is effectively pre-ionized by the remenant electron density from the preceding pulse probably accounts for the fact that the CVL can be operated in wide-bore discharge tubes of 80mm diameter and larger with stable discharges uniformly filling the whole discharge tube cross-section. On the other hand the low value of pre-pulse electron density in the HyBrID laser almost certainly implies that the excursion of electron temperature during the discharge excitation pulse is larger, leading to a more favourable excitation of the upper laser level, and an ability to operate at higher pulse repetition frequency.

5. Pre-ionization and Breakdown - the Phantom Current

5.1 OBSERVATIONS

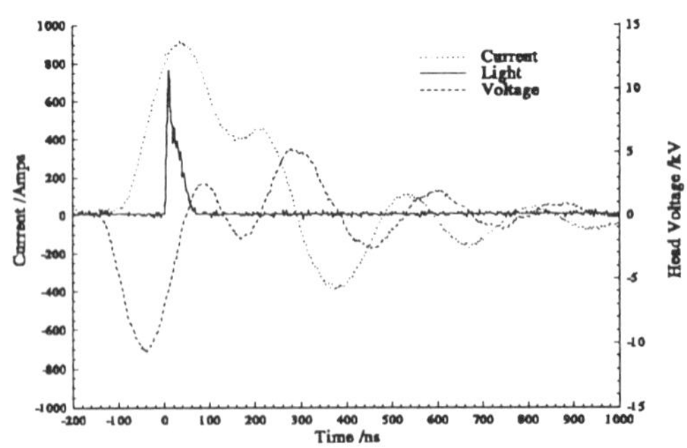

Figure 10. Voltage, current and laser light pulses

Figure 10 shows a typical set of traces of discharge current, voltage and laser emission. For a CVL operating in steady state under optimum conditions [6]. It can be seen that current begins to flow very early in the leading edge of the voltage pulse, and has reached over half its maximum value by the time that the voltage reaches its peak and starts to decrease – the event normally considered to signal "breakdown". Also it is evident that the start of the laser pulse occurs about 100 ns after the current starts flowing.

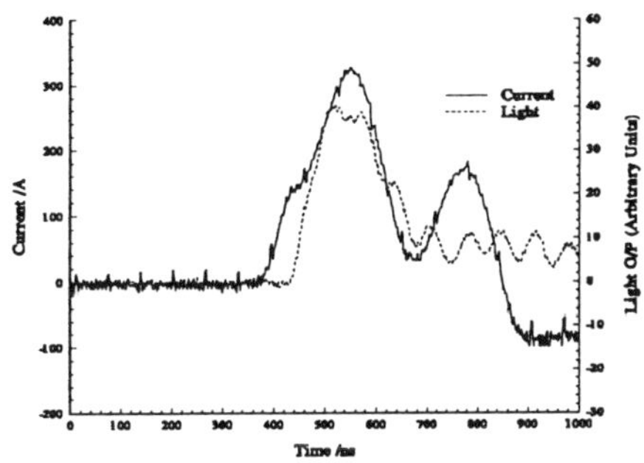

Figure 11. Spontaneous emission and current pulses

Figure 11 shows how the timing of the spontaneous emission pulse is correlated with that of the discharge current pulse. It can be clearly seen that there is no detectable emission during the early stages of the current pulse. In order to be able to observe the relatively weak spontaneous emission it was necessary to operate the discharge at a temperature below the optimum for lasing.

The leading edge of the current pulse shown in Figure 11 displays three distinct phases of current growth. At first the current shows a relatively rapid rise which is not accompanied by any light emission from the discharge. This is followed by a period of much reduced rise, giving the appearance of a "step", before the final period of current increase which has a growth rate similar to that in the initial period. The leading edge of the pulse of spontaneous emission appears to follow, at least approximately, the current waveform in the final phase of current rise. As there appeared to be an initial period of current flow during the discharge that did not give rise to any observable spontaneous emission, this component was dubbed the "phantom current".

It was observed that, as the laser tube warmed up towards its optimum operating temperature, the relative amplitude of the current within the step on the leading edge increased until, at full operating temperature, no separate "step" could be discerned. Under these conditions the "phantom current" component was an indistinguishable part of the current pulse as in Figures 3 and 10.

5.2 DISCUSSION

In order to test the hypothesis that the "phantom current" is a conduction current carried by the initially cold electrons remaining from the previous pulse (and which do not have sufficient energy to cause inelastic collisions at an appreciable rate) it was necessary to reduce the pre-pulse electron density without altering any other operating parameter of the laser.

Using the discharge trigger pulse inhibit facility described above it was possible to interrupt the excitation pulse train for long enough to ensure that the electron density had decayed to negligible values. The traces of voltage, current, and laser light emission were acquired for the first few discharge pulses following such periods of recombination.

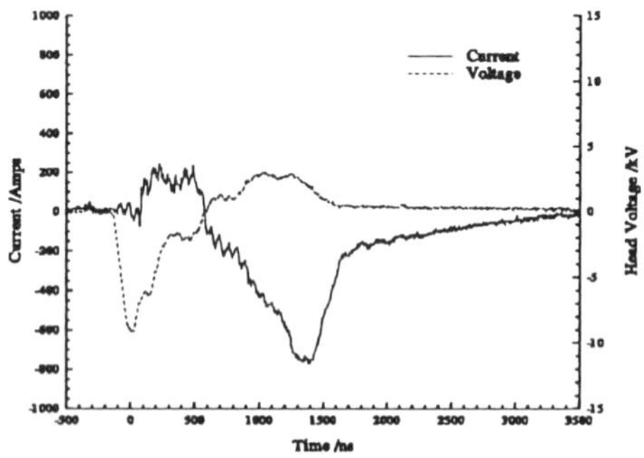

Figure 12. First discharge after blanking perod

Figure 12 shows the voltage and current traces characteristic of the first discharge pulse following the period of interruption to allow recombination. They are of strikingly different shape, amplitude and time-scale from those obtained under steady-state conditions, as would be expected. This graphically illustrates the significance of the pre-pulse electron density in pre-ionizing the gas and thus promoting a uniform and reproducible discharge. It can be seen immediately, however, that there is no trace of "phantom current", as no current starts to flow until the voltage has peaked, giving rise to a well defined "breakdown" of the gas.

The current and voltage traces of the second discharge pulse after blanking are very much more like those obtained under steady-state conditions. There are, however, a number of very significant differences. The first is that the "phantom current" begins to appear as a very small hump on the leading edge of the current pulse. Initially the current starts to rise at the same time as under steady-state operation, however it very soon reaches a limit before finally rising at a time more closely associated with "breakdown" as indicated by the voltage fall. Secondly, both current and, even more obviously, voltage, rise to higher peak values than in the steady-state case. This peak occurs in the voltage at the same time as usual, while that in the current is marginally later. Finally, the laser light output pulse commences significantly earlier than under steady state conditions. Combined with the fact that it finishes at the same time and is of only slightly lower amplitude, this means that the laser output pulse under these conditions contains significantly greater energy than a pulse under normal, steady state operation.In subsequent discharge pulses in a train, the size of the initial hump on the current trace is seen to increase, and the breakdown voltage decrease back towards steady-state operating conditions.

Further evidence that events in the discharge were correlated in time with the fall of voltage rather than the rise of current comes from our measurements of the populations of levels of atomic copper as functions of time and radial position discussed in section 3 above. As evident in Figure 3, depletion of the copper ground state population at the tube axis does not begin until around 70 – 80ns after current starts to flow. At radial positions close to the plasma tube wall where the discharge pulse commences earlier owing to the skin effect there is still a delay of around 60 – 70ns. Our population and emission measurements show that at most the skin effect can account for a relative delay between wall and axis of only about 8 ns in the CVL studied.

All of these observations fully support the hypothesis proposed that the "phantom current" consists of acceleration of the pre-pulse electron density prior to their attainment of sufficient energy to undergo inelastic collisions and thus give rise to spontaneous emission or stepwise excitation to ionization.The experimental measurements of free electron density further confirm that the pre-pulse conductivity of the plasma is high, and that, even following "breakdown", the increase is relatively small.In addition to the experimental evidence suggesting the significance of the pre-pulse electron density in giving rise to the observed "phantom current", initial results of recent computer modelling of the CVL plasma [7] confirms that delays of the order of 60 ns or more between the initial flow of ohmic conduction current in the plasma, and true "breakdown" may readily be reproduced.

The experiments described above, together with the computations performed by Greenland [7], lead us to conclude [6] that it is the pre-pulse electron density that is responsible for the observed "phantom current".

6. Conclusions

We have measured many of the internal plasma parameters of an operating CVL with accurate temporal and spatial resolution.The measurements apply to one and the same CVL operating under a well defined and reproducible set of conditions.

The atomic population densities in HyBrID and conventional CVLs have been measured using the same diagnostic equipment. The electron density in the HyBrID laser decays essentially to zero between each succeeding current pulse, unlike the case of the CVL in which the electron density decays to only about one third of its peak value before the onset of the next pulse.

The "Phantom Current" which flows at the outset of each current pulse in a CVL in normal operation prior to voltage breakdown of the discharge on each excitation pulse is carried by the large density of cold plasma electrons remaining from the previous excitation pulse.

7. Acknowlegements

It is a pleasure to acknowledge the financial support of British Nuclear Fuels plc and URENCO (Capenhurst) plc throughout the several years needed to complete the studies reported here.

8. References

1. Jones D.R, Akerboom F, Maitland A. and Little C.E.(1992) Copper HyBrID laser producuing 149 W at 2.4% efficiency and 112 W at 3.1% efficiency *CLEO '93* 460-462 Paper No. CThN8.

2. Rozhdestvenskii D.S (1912) Anomale dispersion im Natriundampf *Ann. Phys.,* **39** 307-345.

3. Hogan G.P, Svoboda M and Webb C.E (1993) The automation of the hook technique *Meas. Sci. Technol* **4** 263-269.

4. Bielski A. (1975) A critical survey of atomic transition probabilities for CuI *J. Quant. Spectrosc. Radiat. Transfer.* **15** 463-472.

5. Hogan G.P, and Webb C.E. (In preparation) Radially- and time-resolved measurements of electron density in an operating copper vapour laser.

6. Hogan G.P, and Webb C.E (1995) Pre-ionization and discharge breakdown in the copper vapour laser: the phantom current *Optics Commun.* **117** 570-579.

7. Greenland T, (1994) Private communication.

KINETICS OF METAL VAPOUR AND METAL HALIDE LASERS

G.G.PETRASH
P.N.Lebedev Physical Institute
53 Leninsky prospect, 117924 Moscow, Russia

1. Introduction

Pulsed metal vapour lasers (PMVLs), utilising transitions from resonance to metastable levels in atoms and atomic ions (r-m lasers), especially copper vapour laser (CVL), are now among the most efficient and powerful gas lasers in the visible and near IR regions of spectrum. They are widely used in many important applications. Consequently, they were subject of many investigations. Nevertheless, understanding of basic kinetic processes controlling operation of these lasers is still not satisfactory. As the result it is difficult to predict attainable output power, efficiency and other characteristics of the lasers. This is a result of considerable experimental difficulties associated with experimental investigations of kinetics in PMVL.

Besides pulsed metal vapour lasers, metal compound vapour lasers, particularly copper halide vapour lasers, attract now great attention. They have characteristics comparable and sometimes better than those of pure metal vapour lasers and often have some additional advantages, in particular they can operate at considerably lower temperature of discharge tubes. Kinetic processes in metal halide lasers are generally more complicated due to complex composition of their active media and presence of negative ions. Investigations of kinetic processes in these lasers are just at the beginning.

For the proper understanding of PMVL and metal halide laser operation it is the most important to measure, with good enough temporal and spacial resolution, population of upper (resonance) and lower (metastable) laser levels as well as density of the ground state of the atom or ion used in the laser action. It is also desirable to measure populations of some other levels of the atom or ion and some levels of the buffer gas used. For a metal halide laser it would be very important to measure the densities of major constituents of the active medium. And of cause it is desirable along with measurements of level populations to measure electron

C. E. Little and N. V. Sabotinov (eds.), Pulsed Metal Vapour Lasers, 43–54.

density and temperature because excitation and deexcitation of most levels under typical conditions of the laser operation is by interaction with electrons.

Since populations of all levels and their temporal behaviour, as well as parameters of a laser output, are strongly dependent on the laser discharge tube used and conditions of its excitation it is important to measure as many parameters as it is possible with one discharge tube and under the same, or at least comparable, conditions of operation. The conditions of any measurements as well as parameters of the laser output corresponding to these conditions of excitation should be fully described in publications. Without this a meaningful comparison of the results of different measurements is very difficult to do. Unfortunately, these requirements are not satisfied in many publications reporting kinetics measurements.

There are numerous publications reporting CVL kinetics measurements and only a few concerning other PMVLs. Unfortunately, many publications reporting kinetics measurements, even for CVL, do not satisfy the requirements mentioned above. Nevertheless, it seems still possible to formulate some general conclusions concerning kinetics of CVL. Therefore kinetics of CVL will be discussed here in greater detail. A comparison of CVL kinetics and kinetics of copper bromide vapour lasers and HyBrID laser will be also presented. Other PMVLs will be discussed briefly. Only lasers operating in regular pulses will be considered.

Two different well known methods of population measurement were used in studies of PMVLs kinetics: method of absorption spectroscopy and hook method. The first one usually allows to measure population of levels in wider range but needs some information concerning spectral shape of the lines used in the measurements, whereas the second one does not need such information but sometimes is not sensitive enough.

2. Copper vapour laser

2.1 GROUND STATE KINETICS

Measurements of atomic copper ground state kinetics were reported in [1-12]. Some examples of the results from these publications are shown in Fig.1. In all cases for any moment of time radial distribution of the ground state copper density $N_0(r)$ shows a decrease of density from the tube wall to the central zone and minimum at the tube axis. During an excitation pulse copper atoms are pumped out of the ground state by the discharge. This leads to a considerable depletion of the ground state ΔN_0. After the end of discharge a replenishment of the ground state takes place.

As it was shown experimentally in [9,12] and by computer modelling in [13] the decrease of N_0 at the tube axis is greater that can be attributed to gas heating.

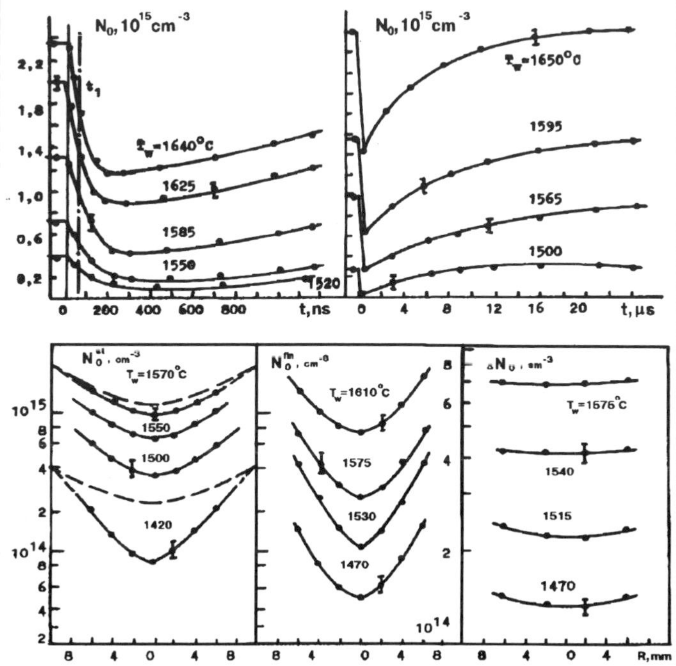

Lebedev Phys. Inst. **[12]**

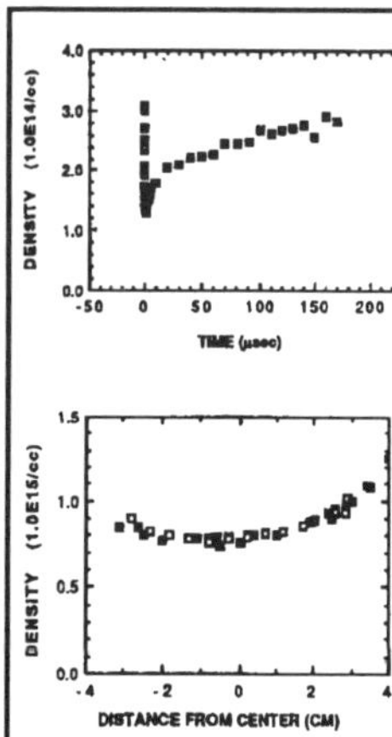

W. Molander, **[6]**

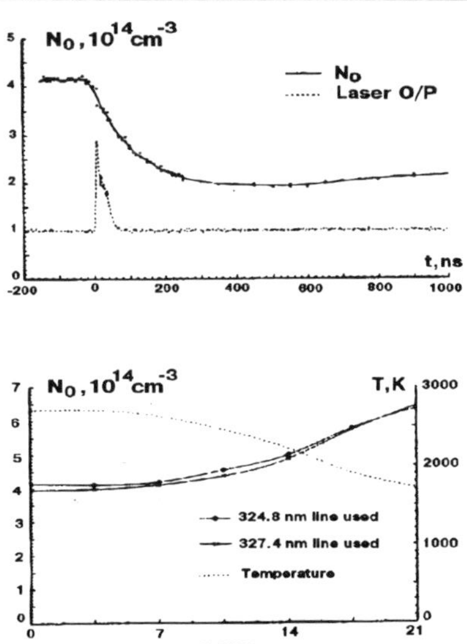

G. Hogan et al **[11]**

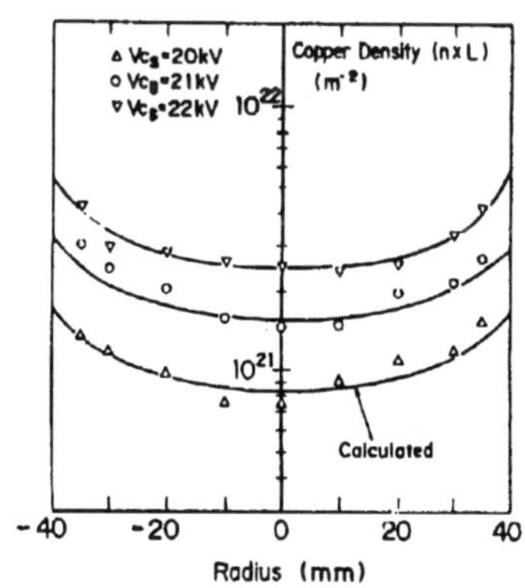

K.Hayashi et al, **[10]**

Fig 1. Temporal evolution and radial profiles of ground state density N_0 in CVL.

This is the result of the effect of radial cataphoresis of metal vapour in the afterglow when there is significant number of copper ions moving to the walls due to ambipolar diffusion. After recombination at tube walls back movement of neutral metal atoms into the central zone is slower since it is governed by normal diffusion. The effect of radial cataphoresis is more pronounced at smaller copper density and may be small under optimal conditions of laser operation.

In some cases the radial distribution of copper ground state density was used for calculation of the radial profile of gas temperature Tg. Due to the effect of radial cataforesis this may lead to significant overestimation of Tg, especially at low tube wall temperature T_w and high level of specific input power.

2.2. KINETICS OF THE RESONANCE LEVELS

Measurements of copper atom resonance level (upper laser level) populations N_r were reported in [2,4,8,12-22]. Typical dependencies from different publications are shown in Fig 2. As it is seen from the Fig 2, N_r is negligible prior an excitation pulse. It demonstrates a fast grow at the beginning of discharge, reaches maximum before the end of discharge current and then quickly decays. At the end of discharge pulse population of resonance levels is very low. If the discharge currents has some oscillations N_r also has similar oscillations closely repeating input power time dependence.

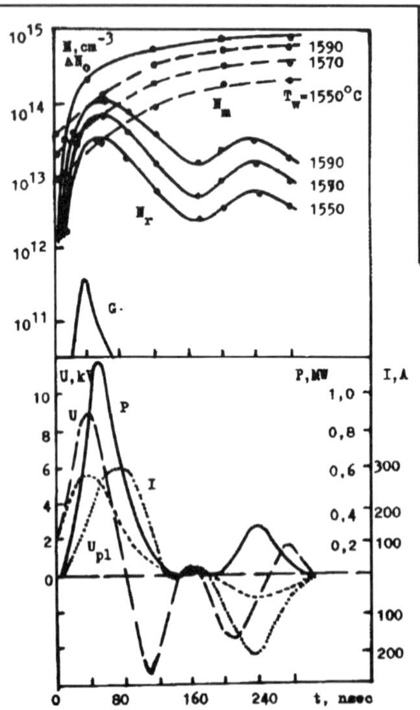

Lebedev Phys. Inst. [9]

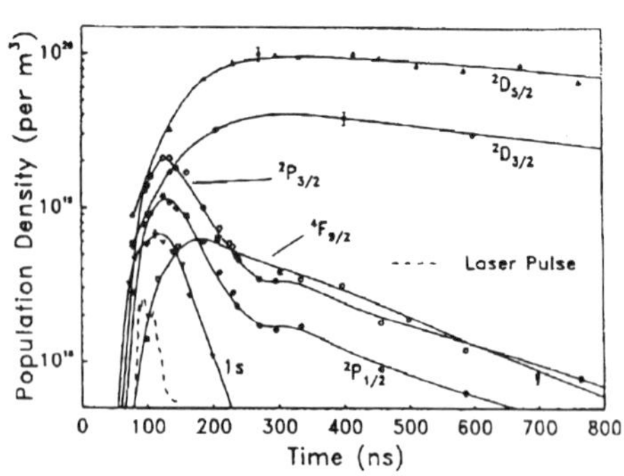

D. Brown et al [21]

Fig2. Typical temporal behaviour of N_r and N_m during exitation pulse. Parameters of exitation pulse is also shown (left)

Similar temporal behaviour was also observed for all higher lying excited levels of atomic copper. Maximum of their populations is usually reached slightly later than for the resonance levels. Time constants of the resonance levels decay, as well as for other higher lying levels, are typically smaller than their radiation life times and they decrease with increase of discharge current and electron density [12,15,21]

Measurements of the radial profiles of resonance level population $N_r(r)$ are described in [2,8,12,17,20,22]. It was shown that the radial profiles can have either maximum or minimum at tube axis, depending on condition of the tube operation, and they may change during excitation pulse. At the moment of maximum N_r the profiles had minimum at tube axis at lower tube wall temperature (copper density) and maximum at the temperature close to optimal [2,8,12,22]. Strong influence of skin effect on the N_r radial profiles was observed in large-bore laser discharge tubes [17].

2.3. KINETICS OF THE METASTABLE LEVELS

Measurements of copper atom metastable level (lower laser level) populations were reported in many publications [2,8,10,12,14,15,17,19-30]. Temporal behaviour of metastable level population N_m during excitation pulse is illustrated in Fig.2 where typical dependencies taken from different publications are shown.

Unlike the resonance levels atomic copper metastable levels have prepulse population which can not be neglected. During excitation pulse the metastable level population N_m grows steadily and saturates to the end of excitation. It does

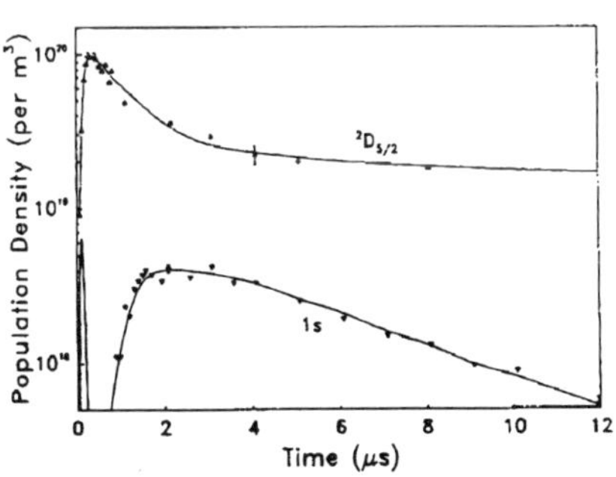

D.Brown et al, [21]

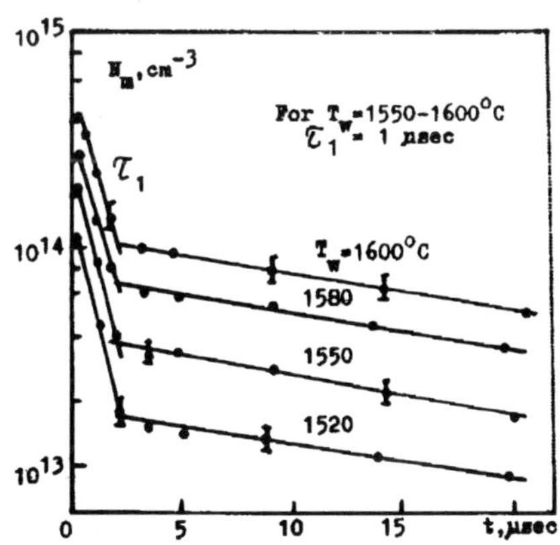

Lebedev Phys. Inst. [12]

Fig 3. Decay of copper meta stable levels in the afterglow of pulse discharge during interpulse periode

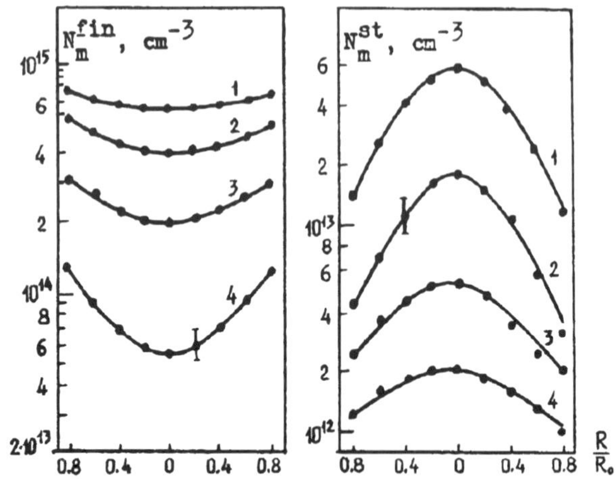

Lebedev Phys. Inst. [12]

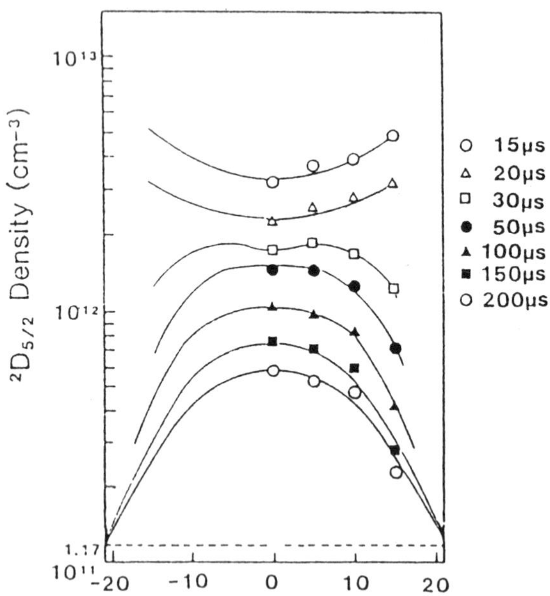

Y. Izawa et al, [29]

Fig. 4. Typical radial profiles of N_m

not follow oscillation of discharge current. At the end of excitation pulse population of metastable levels is much higher than that of resonance levels.

After the end of excitation decay of metastable levels takes place. The decay curves measured by different authors are illustrated in Fig. 3. Under typical conditions of laser operation the decay has two distinct stages: immediately after the end of discharge the decay is fast and then it becomes much slower. Usually the decay can be well approximated by two exponents. Time constants of the two stages of decay depend on discharge conditions. The time constant of the fast decay is usually from one to several microseconds and for slow decay 10-100 microseconds.

Radial profile of N_m has maximum at the tube axis during almost whole interpulse period and minimum at the tube axis during excitation pulse and in the early afterglow. Transformation of the N_m radial profiles from profiles with minimum at tube axis into ones with maximum takes place usually during fast decay stage. Typical radial profiles are shown in Fig. 4.

3. Kinetics of copper halide lasers

Recently rather good results have been obtained with copper bromide (with hydrogen additives) and HyBrID lasers [31-33]. Efficiency as high as 3.2% have been measured at 120 W output power with HyBrID laser [33]. It would be very important to understand what are the physical reasons responsible for the increase in efficiency with addition of hydrogen in CuBr laser and with HyBrID laser. Kinetics measurements and comparison of kinetics in CVL and in copper halide lasers might give the answer.

Unfortunately only a few measurements of level populations in copper halide lasers with good enough temporal resolution were reported to date. Measurements of kinetics in CuBr laser [12,34,35] showed that in general temporal behaviour of N_r, N_m in CuBr and CuBr with H_2 is similar to that of CVL. Addition of 0.3 Torr of H_2 leads to decrease of excitation rates both for N_r and N_m and shift of N_r maximum toward longer times. Decay of N_m in the afterglow in CuBr laser is also similar to that of CVL. Addition of H_2 does not change essentially the fast decay of copper metastables but substantially increase the time constant of slow decay.

Temporal behaviour of ground state density N_0 is similar for CVL and CuBr laser but addition of H_2 into active medium of CuBr laser considerably changes the ground state kinetics [12, 35]. Depletion of the ground state with H_2 is considerably smaller and recovery of N_0 much faster. This is illustrated in Fig. 5 where time dependence of depletion ΔN_0 for the three lasers is shown.

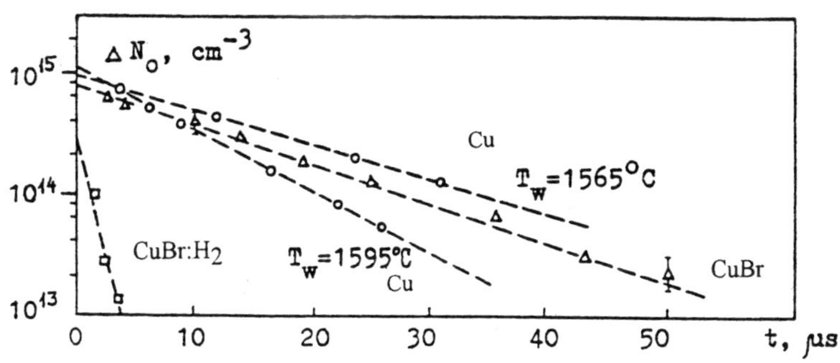

Fig. 5. Temporal behaviour of copper atom ground state depletion for CVL, CuBr and CuBr:H_2 lasers

4. Kinetics of other r-m lasers

Only decay of metastable levels has been studied for many other r-m lasers. The measurements of the decay were reported for BiI $^2D_{3/2}$ level [36], PbI $^1D_2$ level [37,38], for BaI $^1D_2$ level [39], and for $^1D_2$ level of SrII ion [40]. In most cases the decay of the metastable levels also had two stages and was similar to the decay of CuI metastables in copper and copper bromide lasers [22].

For barium vapour laser kinetics of upper and lower laser levels during excitation pulse and also ground state density were measured [41,42].

5. Physical processes determining the kinetics of r-m lasers

The results of CVL kinetics measurements discussed here are in a reasonable agreement. There exist some quantitative differences between some results but this is not surprising if one takes into account the difference in the conditions of laser operation. Temporal and spatial behaviour of N_0, N_r, N_m and population of some other copper atom levels is generally the same in all publications. Therefore, it seems useful to try and analyse, using the results of kinetics measurements, the main physical processes governing operation of CVL. Information regarding kinetics of other pure metal r-m lasers is not enough for detail discussion of physical processes in these lasers but experimental results available to date allow to assume that the main physical processes in these lasers are similar to those in CVL.

The results of kinetics measurements, as well as all other investigations of CVL, show that the most important is interaction of copper atoms with electrons. Therefore it is important to know electron density ne and electron temperature T_e and their temporal and spatial behaviour in CVL plasma. Estimates made by many authors show that under typical conditions of CVL operation T_e during a discharge pulse reaches several eV. It means that a strong ionisation nonequilibrium exists during excitation pulse; consequently during this period copper atoms are pumped from one big reservoir (copper ground state) to another big reservoir (ionised copper) through atomic excited state. After the end of discharge during interpulse period, in opposite, recombination nonequilibrium exists in CVL plasma , and there is a flux downward through atomic copper levels.

The behaviour of resonance and higher lying level populations can be explained by excitation and deexcitation by electrons. The fact that $N_r(t)$ follows oscillations of input power means that during almost whole excitation pulse N_r relaxation time is small compared to the duration of the excitation pulse itself. Consequently, resonance level populations, and probably populations of higher lying levels, may be considered, at least as the first approximation, as quasistationary. That means N_r is approximately proportional to the ground state density and the ratio of the resonance level rate of excitation from the ground state to the total rate of deexcitation of the level by electrons [12]. At the end of excitation pulse populations of resonance and higher lying levels are small.

Relaxation time of metasable levels, in opposite, is long compared to duration of excitation pulse. Therefore, N_m does not follow oscillations of discharge current and reaches its maximum at the end of discharge. Increase of N_m during excitation pulse is usually essential and must be taken into consideration. Ground state depletion also does not follow oscillations of discharge current and is saturated at the end of discharge.

After the end of excitation pulse a relaxation of N_m and ground state depletion and recombination of plasma begin. These processes are interconnected because repopulation of ground state is determined by plasma recombination and decay of metastable levels. The process of plasma recombination in CVL was analysed in [43]. The rate of recombination strongly depend on T_e. Just after the end of discharge decay of T_e is fast until the start of intense three-body recombination. The fast decay of T_e corresponds to the fast decay of N_m . When intense recombination begins the decay of T_e and N_m become slow due to recombination heating of electrons. This corresponds to the slow stage of N_m decay. During this stage relaxation time of N_m is short compared to duration of the stage and N_m is close to equilibrium with the ground state with the temperature T_e [12]. T_e in turn is slowly approaching gas temperature.

The main processes in other pure metal r-m lasers and probably in CuBr laser, as far as available data allow to conclude, are similar to those in CVL. But for CuBr laser with hydrogen and for HyBrID laser the processes of ionisation and electron recombination seem different [12]. In this case removal of electrons is much faster [44] probably due to strong attachment of electrons and formation of some negative ions. In this case electrons in discharge may be provided mainly not by ionisation of copper. Further investigations of these processes in CuBr:H_2 and HyBrID lasers are desirable.

6. Conclusion

Spatially and temporally resolved measurements of relevant level populations (kinetics measurements) in PMVLs are very important for understanding of physical processes in these lasers, as well as for improvement of the lasers performance. Up to date only kinetics of CVL was studied more or less thoroughly. Further investigations of PMVL kinetics are desirable, especially kinetics of CuBr:H_2 and HyBrID lasers. It may show the ways to improve efficiency and output power of the lasers. To have a complete picture of physical processes in PMVLs it is also important to curry out simultaneous measurements of T_e and n_e. In order to have useful comparison of different measurement results it is important in any publication to describe in detail the laser studied, condition of its operation and to measure under the same conditions as many parameters as it possible.

The author would like to express his gratitude to the colleagues who provided reprints of publications important for this review, especially to Dr. C.E.Little, Dr. D.J.W.Brown and Dr. I.Smilanski.

This work was supported in part by the International Science Foundation and by Russian Foundation for Fundamental Research.

7. References

1. Borovich B.L., Vasil'ev L.A., Gerz V.E., Negashev S.A., Regeda S.N., Tunitskii L.N. (1983), Determination of the copper vapor concentration and the degree of heating of the gas in a transverse-discharge laser, *Sov J.Quant. Electronics,* **11,** 1210-1212.
2. Isaev A.A., Mihkelsoo V.T., Petrash G.G., Peet V.E., Ponomarev I.V. and Treshchalov A.B, (1987), Spatial-temporal kinetics of excitation and relaxation of atomic levels in a plasma of pulsed copper vapour laser, *Preprint of P.N.Lebedev Phys. Institute,* No 171, 1-44.
3. Burlakov V.D., Gorbunova T.M., Loboda S.A., Michailichenko Yu.P., Osipova N.V., (1987), Investigation of atomic copper ground state population in a copper vapour laser with longitudinal discharge, *High Temperature*, **25**, 394-396. (In Russian)
4. Isaev A.A., Mikhelsoo V.T., Petrash G.G. *et al,* (1987), Experimental investigation of the kinetics of populations of atomic levels in a copper vapour laser, in R.A.Sorkina (ed), *Proc. Physics Institute of Estonian Academy of Sciences*, **v. 60,** *Laser Physics and Technique,* Tartu, 90-107.
5. Isaev A.A., Mikhelsoo V.T., Petrash G.G. *et al,* (1988), Kinetics of atomic copper ground state in a plasma of pulse-periodic discharge in copper vapour laser, *Preprint of P.N.Lebedev Phys. Institute,* No 98, 1-30.
6. Molander W.A., (1989), Measurement of ground state copper density using hook spectroscopy, in J.J.Kim, R.Kimball, P.J.Wisoff (eds), Proc. SPIE, **v. 1041**, *Metal Vapor, Deep Blue, and Ultraviolet Lasers,* 11-18.
7. Smilanski I., Gabay S., Karni Z., (1989), Ground state depletion in copper vapor lasers, ibid., 34-39.
8. Isaev A.A., Mikhelsoo V.T., Petrash G.G., *et al.*, (1989), Atomic levels population and depopulation kinetics in Cu-vapor laser, *ibid.,* 40-46.
9. Isaev A.A., Mihkelsoo V.T., Petrash G.G. *et al,* (1989), The Spatial-Temporal Kinetics of Population of the Copper Atom Ground State in a Copper-Vapor Laser, *Kvantovaya Elektronika (Quantum Electronics),* **16**, 1173-1183. (In Russian)
10. Hayashi K., Noda E., Iseki Ya., Watanabe I., Suzuki S., Morimiya O., (1992), Copper Density Measurement of a Large-bore CVL, in Proc. SPIE, **v. 1628,** *Intense Laser Beams,* 44-51.
11. Hogan J.P., Svoboda M. and Webb C.E., (1993), The automation of the hook technique, *Meas. Sci. Technol.,* **4,** 263-269.
12. Isaev A.A., Petrash G.G., (1993), Kinetics of excitation and physical processes in active

media of copper vapor and copper bromide vapor lasers, in G.G.Petrash (ed), Proc SPIE, **v. 2110,** *Metal Vapor Lasers and Their Applications,* 2-45.

13. Carman R.J., Brown D.J.W., Piper J.A., (1994), A Self-Consistent Model for the Discharge Kinetics in a High-Repetition-Rate Copper-Vapor Laser, *IEEE J. Quant. Electronics,* **30,** 1876-1895.
14. Smilanski I., (1979), Copper hook investigation in copper vapour laser, in V.J.Corcoran (ed), *Proc. Intern. Conf. Lasers'79,* Orlando, Fl, 327-334.
15. Smilanski I., Levin L.A., Erez G., (1980), Kinetics of population inversion in a copper-vapor laser investigated by a modified hook method, *Optics Letters*, **5,** 93-95.
16. Tenenbaum J., Smilanski I., Lavi S., Levin L.A., Erez G., (1981), Kinetic investigation of the upper laser levels of the copper vapor laser, *Optics Communications,* **36**, 391-394.
17. Kushner M.J., Warner B., (1983), Large bore copper-vapor lasers: kinetics and scaling issues, *J.Appl. Phys.,* **54**, 2970-2982.
18. Berik E.B., Isaev A.A., Mihkelsoo V.T. *et al*, (1986), Spectroscopy of copper vapor laser active medium, *Preprint of P.N.Lebedev Phys. Institute*, No 251, 1-43.
19. Isaev A.A, Mihkelsoo V.T., Petrash G.G. *et al*, (1988), Kinetics of excitation of copper vapor laser working levels in the twin-pulse mode, *Kvantovaya Elektronika (Quantum Electronics),* **15**, 2510-2513. (In Russian)
20. Brown D.J.W., Kunnemeyer R, McIntosh A.I., (1989), in J.J.Kim, R.Kimball, P.J.Wisoff (eds), Proc SPIE, **v. 1041**, *Metal Vapor, Deep Blue, and Ultraviolet Lasers*, 25-33.
21. Brown D.J.W., Kunnemeyer R., McIntosh A.I., (1990), Time-resolved measurements of excited state densities in a copper vapor laser, *IEEE J. Quant. Electronics*, **26**, 1609-1619.
22. Petrash G.G., Isaev A.A., (1991), Pulsed Gas Discharge Lasers, in G.G.Petrash (ed), *Proc. P.N.Lebedev Phys. Institute,* **v. 212,** *Optics and Lasers,* Nauka, Moscow, 93-108. (In Russian)
23. Miller J.L., Kan T., (1979), Metastable decay rate in a Cu-metal vapor laser, *J.Appl. Phys.*, **50**, 3849-3851.
24. Kelman V.A., Klimovskii I.I., Konoplev A.I. *et al*, (1984), Relaxation of copper atom metastable level in the afterglow of pulsed discharge in copper vapor laser, *Sov. J. Quant. Electronics*, **14**, 1460-1463.
25. Isaev A.A., Petrash G.G., Ponomarev I.V., (1985), Relaxation of metastable atoms in the afterglow of copper vapour laser, *Preprint of P.N.Lebedev Phys. Institute No 271*, 1-31.
26. Litvinenko A.Y., Kravchenko V.I., Egorov A.N., (1983), Measurement of the life times of the lower active levels of a copper vapor laser, *Sov. J. Quant. Electronics*, **13**, 778-781.
27. Isaev A.A., Petrash G.G., Ponomarev I.V., (1986), Relaxation of metastable atoms during the afterglow in a copper vapor laser, *Sov. J. Quant. Electronics*, **16**, 1512-1516.
28. Isaev A.A., Kazakov V.V., Lesnoi M.A., Markova S.V., Petrash G.G., (1986), Decay of metastable states and its influence on the output characteristics of a copper vapor laser, *Sov J. Quant. Electronics*, **16**, 1517-1524.
29. Izawa Y., Shimotsu T., Yamanaka Ch. *et al*, (1989), Density Measurements of the lower level of a copper vapor laser, in J.J.Kim, R.Kimball, P.J.Wisoff, (eds), Proc. SPIE **v. 1041**, *Metal Vapor, Deep Blue, and Ultraviolet Lasers,* 19-24.
30. Okamoto T., Tabata Y., Ueguri Sh., (1991), Spatially and temporally resolved measurements of a small-signal gain and the population of lower laser level in a large-bore copper vapor laser, *Intern. Conf. CLEO'91*, 518.

31. Astadjov D.N., Sabotinov N.V., Vuchkov N.K., (1985), Effect of hydrogen on CuBr laser power and efficiency, *Optics Communications*, **56**, 279-282.
32. Astadjov D.N., Vuchkov N.K., Sabotinov N.V., (1988), Parametric study of the CuBr laser with hydrogen additives, *IEEE J. Quant. Electronics*, **24**, 1927-1935.
33. Jones D.R., Maitland A., Little C.E., (1994), A high efficiency 200 W average power copper HyBrID laser, *IEEE J. Quant. Electronics*, **30,** 2385-2390.
34. Astadjov D.N., Vuchkov N.K., Isaev A.A. *et al*, (1987), Relaxation of metastable copper atoms in a copper bromide vapor laser emitting regular pulses, *Kvantovaya Elektronika (Quantum Electronics)*, **14**, 396-399. (In Russian)
35. Astadjov D.N., Isaev A.A., Petrash G.G. *et al*, (1992), Temporal and radial evolution of the populations of CuI levels in the CuBr vapor laser, *IEEE J. Quant. Electronics*, **28**, 1966-1969.
36. Kazakov V.V., Markova S.V., Petrash G.G., (1982), A decay of $6p^3$ $^2D_{3/2}$ Metastable Levels in Bismuth Atoms During Interpulse Interval in the Bismuth Vapor Laser, *Kvantovaya Elektronika (Quantum Electronics)*, **9**, 688-694. (In Russian)
37. Kazakov V.V., Markova S.V., Petrash G.G., (1983), Kinetics of decay of the $6p^2$ $^1D_2$ lower laser level in pulsed lead vapor laser, *Kvantovaya Elektronika (Quantum Electronics)*, **10**, 787-792. (In Russian)
38. Bokhan P.A., (1985), On the Mechanism of Limitation of the Optimal Pulse Repetition Rate in Lasers Due to Self-Terminating Transitions in Metal Vapor, *Kvantovaya Elektronika (Quantum Electronics),* **12**, 945-952. (In Russian)
39. Kazakov V.V., Markova S.V., Petrash G.G., (1984), A study of physical processes in a pulsed barium vapor laser, *Kvantovaya Elektronika (Quantum Electronics),* **11,** 949-956. (In Russian)
40. Prokop'ev V.E., Solomonov V.I., (1985), A study of a strontium vapor laser, *Kvantovaya Elektronika (Quantum Electronics),* **12**, 1261-1269. (In Russian)
41. Mildren R.P., Brown D.J., Carman R.J., Piper J.A., (1992), Investigation of the kinetics of a pulsed barium vapor laser using the hook method, *CLEO'92 Conference.*
42. Mildren R.P., Brown D.J., Carman R.J., Piper J.A., (1994), Spatio-temporal Population Density Measurements in a Barium Vapour Laser Operated in Burst Mode, *CLEO'94 Conference.*
43. D'yachkov L.G., Kobzev G.A., (1978), Electron energy balance in afterglow of metal vapour lasers, *J. Technical Physics,* **48**, 2343-2346. (In Russian)
44. Whyte C.J., Little C.E., Hogan J.P. Webb C.E., (1995), Spatially and temporally resolved measurements of electron density in a copper HyBrID laser, *CLEO'95 Conference.*

KINETICS OF METAL ION RECOMBINATION LASERS

M.F.SEM and E.L.LATUSH
Rostov State University, Physics Department,
Zorge st.5, Rostov-on-Don, 344104, Russia

Works are summarized on search of the recombination laser transitions mostly in ionic spectra of different metal atoms, and kinetics of pumping processes and population inversion are considered. Excitation mechanisms and time bechaviour of level populations and active media plasma papameters of the Sr^+ and Ca^+ recombination lasers are analysed in detail. The partial inputs of main populating and quenching processes are established for these lasers.

1. Introduction

Recombination pumping has a number of important advantages. First, in the recombination regime the flux of excited atoms is from high-lying levels downwards. This guarantees a high effective pump rate to the upper laser level, and eases the restrictions on energy-level structure required for an inversion compared with excitation in the ionization regime, where the flux of excitation is from lower levels upwards. As a result, in the recombination regime, stray pumping of the lower laser level is found to be smaller. Second, the potential energy of ionized atoms is effectively utilized - this energy can be stored to a significant degree during the course of a current pulse. Third, in the recombination regime, collisional transfer between levels are established due to impacts with slow electrons - these transfer considerably increase the range of possible laser transitions, and widen the regime that permits significant population inversion to be attained. Because absorption of radiation does not affect the lasing process, it is possible to volume-scale the active medium to increase the available output power.

The first mention of the possibility of using recombination to pump laser levels and sustain inversion was made in [1]. Theoretical studies of plasma recombination lasers are summarized in [2].

Up to date, laser oscillations in recombining plasma has been obtained in afterglow of pulsed gas discharge, in laser-induced plasma, in segment plasma excited system, in plasma-dynamic devices, in plasma produced by e-beem or nuclear pumping.

C. E. Little and N. V. Sabotinov (eds.), Pulsed Metal Vapour Lasers, 55–66.

In this article, we discuss only metal ion lasers pumped by recombination in pulsed discharge afterglows, mostly those which we have been investigating in our laboratory.

2. Pumping and Inversion in Recombining Plasmas.

For the pumping of ion laser transitions it is most appropriate to use the high rates of collisional or collisional-radiative recombination found in dense plasmas.

In three-body recombination with two participating electrons,

$$A^{z+} + e + e \rightarrow A^{(z-1)+} + e, \tag{1}$$

the rate of decay of the concentration of multiply charged ions after terminating the current pulse (neglecting diffusion losses) is given by the equation

$$dn^{z+}/dt = -\alpha n^{z+} n_e^2, \tag{2}$$

where α is the recombination coefficient, leading to

$$n^{z+} = n_0^{z+} \exp(-t/\tau_r) \; ; \quad \tau_r = 1/\alpha n_e^2 \; . \tag{3}$$

The coefficient of three-body recombination depends on the charge of the ion, z, and the electron temperature, T_e:

$$\alpha = 9.21 \cdot 10^{-27} z^3 L T_e^{-9/2}, \tag{4}$$

where $L = \ln(z^2+1)^{1/2}$ is the Coulomb logarithm, T_e is in eV, α is in cm^6 s^{-1}, n^{z+} and n_e are in cm^{-3}. It is clear from (4) that the rate of recombination, and therefore the pump rate of levels, greatly increases with reducing T_e and with increasing n_e and n^{z+}, and also with the charge of the recombining ions. Thus, the dobly charged ion recombination rate is roughly 18 times of that for singly charged ions.

Even in the discharges of pulsed gas lasers it is difficult to obtain significant concentrations of ions with $z > 2$, and the rate of decay of singly charged ions due to recombination is small, so that the creation of population inversion in the recombination regime pumping favours transitions in excited singly charged ions; in this case the initially high concentration of doubly charged ions required can be realized with currents typical of pulsed discharges, and fast collicional recombination can readily be established.

The period of recombination non-equilibrium is characterized by large values of n_e and low T_e. Under these conditions de-excitation processes via impacts with slow electrons strongly influence the excited level populations [3,4].

With this in mind, the general requirements for establishment of inversion in the collisional-recombination regime can be formulated as follows [3,4]:

1. the upper laser level should be one of the lowest in an upper group of closely spaced levels;

2. the lower laser level should be one of the highest in a lower group of closely spaced energy levels;

3. the electron concentration should be sufficiently high that the probability of collisionally induced transitions within groups exceeds the probabilities for radiative transitions; and

4. the electron temperature should be as low as possible.

With these criteria, electron deexcitation causes pumping to be channelled to the upper laser level, and effectively clear the lower laser level.

3. Metal Ion Recombination Lasers.

The wavelengths of ion transition excited by recombination obtained and investigated so far in our laboratory are listed in Table 1.

TABLE 1. Wavelengths of ion laser transitions excited by recombination in the positive columns of pulsed gas discharges [3,4].

Ion	λ,nm	Ion	λ,nm	Ion	λ,nm	Ion	λ,nm
BeII	527.2	AlII	1063.3	SrII	1087.0	BaII	1158.1
MgII	924.4		1063.6		1123.0		1193.5
	921.8		1063.9	CdII	533.7		1247.8
AlII	622.8	CaII	370.6		537.8	HgII*	567.7
	623.3		373.7	SnII	558.9	TlII	473.7
	624.3		994.0		579.9		498.1
	683.0	ZnII	491.2		1062.0		507.9
	747.1		492.4		1074.0		515.2
	835.4	SrII	416.2	BaII	677.1	PbII	537.2
	836.1		430.5		687.6		1162.8
	836.5		870.0		1109.2		1322.1

* HgII laser lines with upper level potential above that of He^+ (λ =734.66, 862.2, 855, 739.7, 1058.3, 1117.9 nm) are presumably due to recombination as well.

The analysis of the laser level arrangement in spectra of the studied elements and experimental data are in good agreement with the above outlined general principles and confirm the importance of electron de-excitation in the ion level population kinetics in the discharge afterglow. The best results were obtained for 430.5 and 416.2 nm SrII; and 373.7 nm CaII lines [4]. The level arrangement of these laser transitions is near optimum from the standpoint of population inversion obtaining in the recombining plasma (Fig.1): the upper laser level in strontium ion spectra 6S - is the lowest among 4F, 5D, 6S group of levels, and the lower laser level - 5P is the highest in 5P, 4D, 5S group of levels. The electron de-excitation effectively populates upper and depopulates lower laser levels.

Nowdays, the Sr^+ and Ca^+ recombination lasers, first discovered in [5,6], are probably the only ones among pulsed gas lasers to deliver average powers of several watts in blue-violet and UV at p.r.f. of an order ten kHz, thus luckily complementing lasing in green and red by copper and gold metal vapour lasers. That explains a growing interest to their research in later years [7-10]. These lasers are far more studied then other recombination lasers and therefore we shall consider kinetics of metal ion recombination lasers on the example of these lasers, and in particular, strontium ion laser, because the distribution of levels, excitation mechanisms and output parameter are, in most features, similar for strontium and calcium laser.

4.He-Sr Recombination Laser Kinetics

4.1.EXCITATION MECHANISMS

Proposed in [3,5,6] pumping and population inversion mechanisms in these lasers can be briefly described as follows (see Fig.1).During the discharge pulse in a mixture of helium with metal vapors, doubly charged strontium ions are produced due to direct and stepwise electron-impact ionization. As the energy needed for single and double ionization of alkaline-earth metals is small,the stepwise ionization of theirs proceeds ruther easily even with great amounts of buffer gas present. On termination of current pulse, rapid cooling of plasma electrons occurs mainly due to elastic collisions of electrons with helium atoms and ions, which pressure is typically amounts hundreds of Torrs,and as soon as electron temperature T_e goes below ~0.5 eV a collisional-radiative recombination of electrons and ions starts-up

$$\mathrm{Sr}^{++} + 2e \rightarrow \mathrm{Sr}^{+*} + e, \tag{5}$$

which populates the strontium ion levels, the upper laser level among them.

Therefore, fast recombination of Sr^{++} ions stored through the current causes a short (< 1 μs) intense recombination flash in early afterglow (Fig.2), during which lasing on the $6^2S_{1/2} - 5^2P^o_{1/2,3/2}$ (430.5, 416.2 nm) SrII transitions takes place.

In addition to electron-impact ionization, the Sr^{++} ion production can occur in the discharge afterglow due to reaction of a charge transfer with ionization (CTI) on atomic helium ions [4,5,6]

$$\mathrm{Sr} + \mathrm{He}^{+} \rightarrow \mathrm{Sr}^{++} + \mathrm{He} + e, \tag{6}$$

and on molecular helium ions [4,9]

$$\mathrm{Sr} + \mathrm{He}_2^{+} \rightarrow \mathrm{Sr}^{++} + 2\mathrm{He} + e. \tag{7}$$

The reactions (6) and (7) are not resonant, since an excess potential energy is taken away by released electron,consequently,their cross-sections yield significant values of 10^{-15}-10^{-14} cm^2 [11].

Some extra Sr^{++} ion production can be provided by a multistage Penning reaction (MP) with helium metastabile atoms [12]

$$\mathrm{Sr}^{+} + \mathrm{He(m)} \rightarrow \mathrm{Sr}^{++} + \mathrm{He} + e, \tag{8}$$

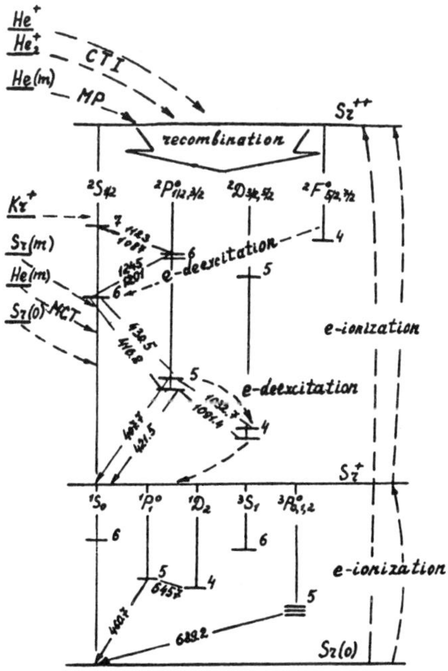

Fig.1. Relevant diagram of strontium with the main pumping and population inversion mechanisms

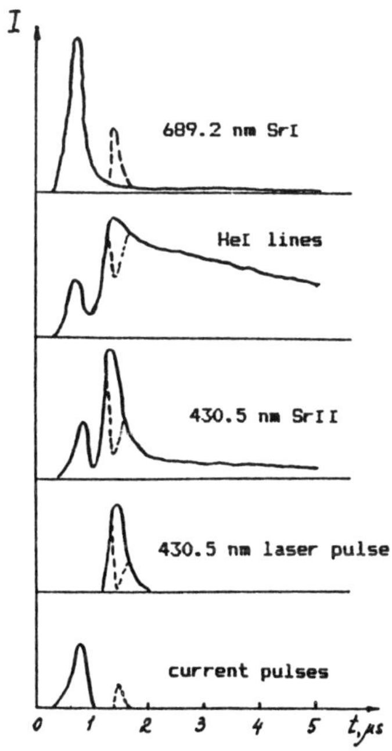

Fig.2. Typical oscilloscope traces of the line intensities. The dashed curve represents the intensity observed on application of an auxialiary current pulse.

since, due to low strontium ionization energy, almost complete ionization of strontium occurs [7,8,12,13].

Population invertion is mostly ensured by superelastic collisions of excited Sr^{+*} ions with slow electrons (electron deexcitation) resulting in preferential filling of the upper laser level and intense emptying the lower ones through the 5P - 4D - 5S channel.

Our conclusion on preferential role of the recombination agrees quite well with experimental and theoretical results of many indepedant studies of these lasers [7,8,10,14] that allowed to consider recombination pumping mechanism for the Sr-He and Ca-He lasers as profoundly established.

However in [15 and 16] an alternative excitation mechanism was proposed for the Sr-He laser,which is due to multistage charge transfer (MCT) reactions of the following type:

$$Sr^{++} + Sr(m) \rightarrow Sr^{+*} + Sr^{+}, \quad (9)$$

$$\mathrm{Sr}^{++} + \mathrm{He}(\mathrm{m}) \rightarrow \mathrm{Sr}^{+*} + \mathrm{He}^{+}. \quad (10)$$

These reactions are essentially resonant,for there no released electrons appear at the right side. Note, that for efficient pumping by the MCT, high densities of both doubly charged strontium ions and stpontium and helium metastables are needed simultaneously, that is more difficult to provide compared to only high needed ion densities in (5).

In order to discriminante between recombination and MCT mechanisms, it should be established, first, a temporal behaviour of plasma parameters, such as n_e and T_e, in early afterglow, and second, time dependences of ion, metastable atom and ground state atom concentrations. Indeed, the recombination rate is strongly dependant on T_e (see Eq.4), whereas cross-section of the MCT is practically not. On the other hand, for efficient recombination pumping,only big amounts of Sr^{++} ions are needed,whilst for MCT, in addition, high metastable atom concentrations are critically important. Therefore, investigations of an Sr-He laser active medium were undertaken in two ways: the diagnostics of plasma parameters (n_e, T_e), and kinetic studies of the long-lived particle densities (ions, metastables,and ground state atoms) in Sr-He discharge afterglow.

4.2. ELECTRON TEMPERATURE AND DENSITY IN THE PLASMA AFTERGLOW OF STRONTIUM ION RECOMBINATION LASER

A low T_e is the most important factor for the existence of inversion on the recombination laser transitions of alkaline-earth metals, because only this condition ensures both strong recombination pumping and electron deexcitation. According to the analysis in [3,4] of the behaviour of T_e during the discharge afterglow in the mixture of a metal vapour and a noble gas, after the current pulse T_e falls rapidly initially, and then reaches a quasi-steady state value which is several times higher than the gas temperature. The experimental results obtained by us and other autors confirm this point. In our experiments on Sr-He discharge diagnostics electrical probe and spectroscopic diagnostic techniques were employed.

Probe measurements in the Sr-He mixtures were carried out in a BeO-ceramic tube with discharge length of 25 cm and 1.1 cm in diameter. Double probes were inserted sideways. Strontium was vaporized from pieces placed inside the tube, by external heating. Two different pairs of probes had been used: 5 and 60 μm in diameter and with equal lengths of 1 mm. Underheated, optimum for lasing, and overheated regimes were studied. Typical results of measurements are submitted in Fig.3, which evidence that T_e drops sharply after current pulse over time interval $\Delta t < 0.75$ μs down to 0.2-0.3 eV at laser pulse period and afterwords remains at quasi-static level which, however, considerably exceeds gas temperature. Noteworthy, that the quasi-static level of T_e builds up when Sr pressure increases beyond optimum for lasing. Determined by probe method n_e = (3-5)·10^{14} cm^{-3} in laser pulse time interval.

In spectroscopic studies, T_e was found from measurements of relative population distribution of high-lying helium energy levels assuming that model of a partial local thermodynamic equilibrum was valid. In the Sr-He discharge conditions, that was proved to be the case for excited helium levels with principal numbers in the range 5 < n < 9. Accordingly, appropriate singlet and triplet helium lines were chosen, which emission was registered out of axis zone of the discharge tube. n_e was spectroscopically measured using Stark broadening of hydrogen lines. The results obtained by spectroscopic technique are presented in Fig.4. They show that T_e = 0.25± 0.1 eV and n_e = (2-5)·10^{14} cm^{-3} during the 430.5 nm laser pulse. The values and kinetics of T_e and n_e measured by both methods (Figs.3 and 4) are in good agreement that presumably improves the validity of

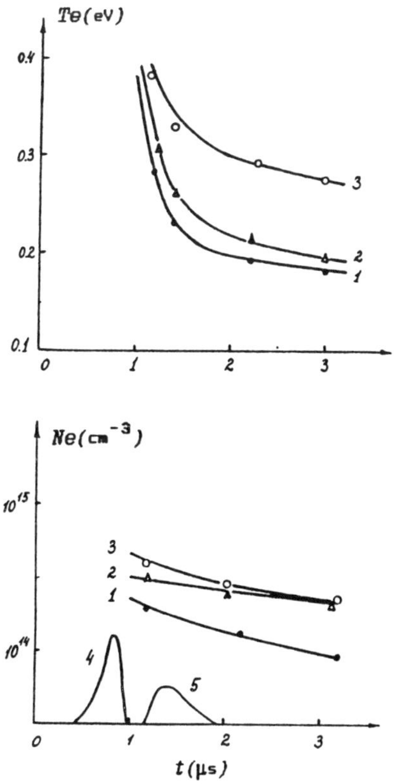

Fig.3. Results of probe diagnostics of the laser discharge in (1) underheated-p(Sr)≤ 0.001 Torr, (2) optimum-p(Sr)≈0.1 Torr, and (3) overheated-p(Sr)≈1 Torr regimes. Helium pressure is 250 Torr. (4) Current pulse, (5) the 430.5 laser pulse

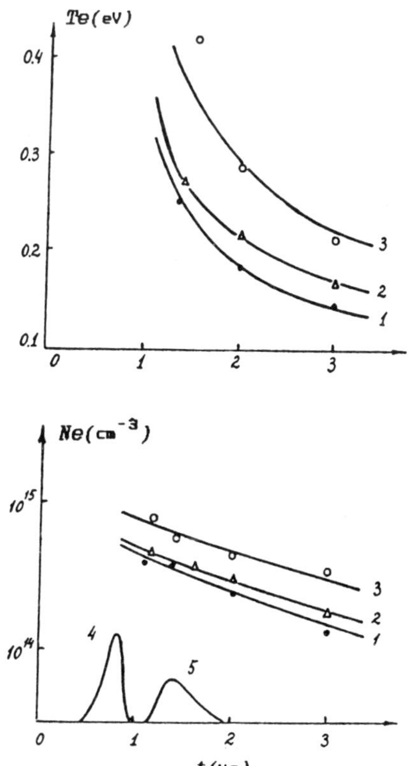

Fig.4. Results of the spectroscopic measurements of Sr-He laser plasma parameters in regimes similar to those in Fig.3. Helium pressure is 250 Torr. (4) Current pulse, (5) the 430.5 nm laser emission pulse.

obtained data.

Some results on Sr-He laser plasma parameters obtained by other authors can be presented for comparison. In [17] electron density $n_e = 3 \cdot 10^{14}$ cm^{-3} was measured in optimum for lasing discharge conditions at p(He)= 500 Torr; in [10] laser interferometric studies had yielded $n_e = (6.8 \pm 0.4) \cdot 10^{14}$ cm^{-3} at p(He) = 250 Torr and $n_e = (7.5 \pm 0.4)\ 10^{14}$ cm^{-3} at p(He) = 380 Torr. T_e deduced from these data turned out to be 0.25-0.35 eV at p(He) = 250 Torr, and 0.39 eV at p(He) = 380 Torr.

4.3. KINETICS OF THE LONG-LIVED PARTICLE DENSITIES

The measurements of level populations were performed by means of reabsorption method of two lengths [12] and by the method of resonant absorption of a tunable due laser emission [13]. The employed in these studies BeO discharge tube consisted of two sections 13 cm long each which could be fired separately or simultaneously. Besides, line emission could be registered sideways. In two-lengths method, the measured ratio of intesities of self-absorbed lines at two different discharge lengths allowed to determine the absorption coefficient and to deduce out of it the lower level population density of the investigated line, provided it far exceeded the population of the upper level.

An excimer pumped dye laser employed in resonant absorption method generated laser pulses of up to 10 mJ energy and 10 ns pulselength at p.r.f. 100 Hz. By changing dyes and by non-liniar optic methods the spectral range of 250-700 nm could be overlaped by the dye laser. The intensity of dye laser probe radiation in resonant absorption experiments was always established low enough in order not to saturate the investigated transition. The results of reabsorption and resonant absorption studies are depicted in Figs.5 and 6.

The obtained data evidence of strong depletion (~ 90 % and even more) of ground state strontium atom population in the course of current pulse. The triplet Sr metastable density reaches maximum value of 8 10^{12} cm^{-3} in the middle of current pulse and then falls abruptly to a current pulse end. In afterglow,it rises again up to $(7\text{-}8) \cdot 10^{11}$ cm^{-3} , but in the period of 430.5 nm laser emission Sr(m) concentration is only $(2\text{-}5) \cdot 10^{11}$ cm^{-3}. The Sr^+ ground state $5^2S_{1/2}$ density shows rather complicated time behaviour, that is due to fast Sr^+ second-step ionization into Sr^{++}. Therefore, the Sr^+ density reaches maximum of $5.5 \cdot 10^{13}$cm^{-3} not in the course of current, but in the afterglow after most of the Sr^{++} ions rapidly recombine, and following slow decay of Sr^+ concentration occurs now due to slower recombination of singly charged Sr^+ ions.

As for total helium metastable density during the 430.5 nm laser pulse, it was measured to amount (3-5) 10^{12}cm^{-3}.

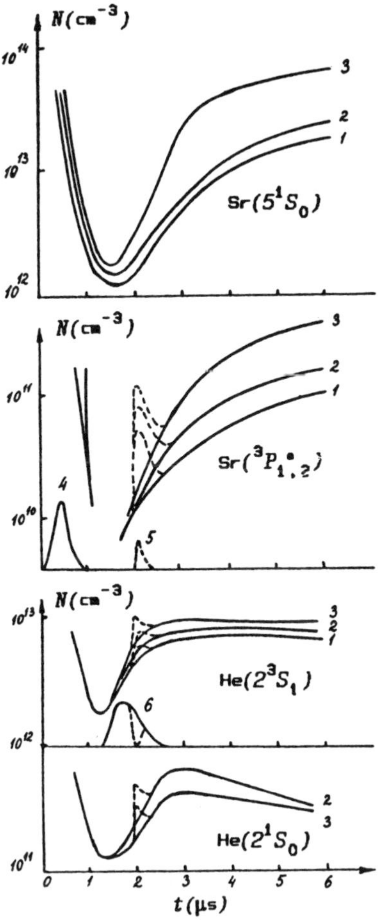

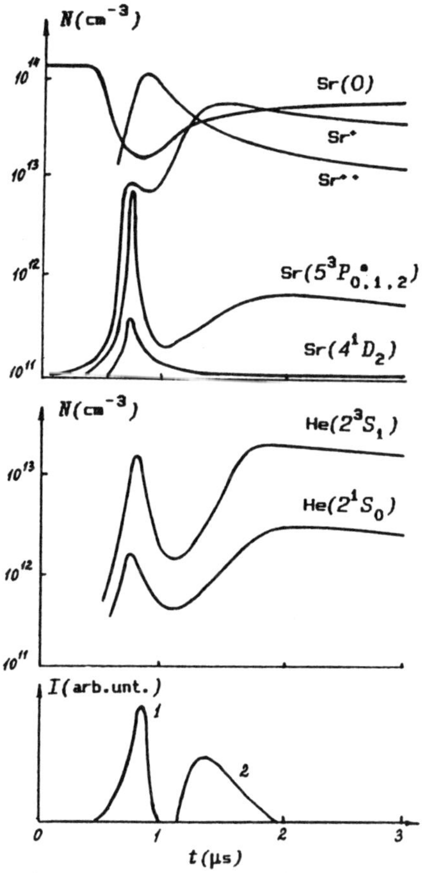

Fig.5. Measured by a reabsorption the populations of levels at p(He)= 30 Torr (1), 100 Torr (2), and 250 Torr (3); (4) and (5) are main and auxiliary current pulses; (6) laser pulse. The dashed curves show the influence of the second pulse.

Fig.6. Time dependences of Sr and He level populations obtened by resonant absorption method. 1)Current pulse; 2) laser pulse at 430.5 nm. Helium pressure is 250 Torr.

Noteworthy,that the results of reabsorption and resonant absorption measurements agree with each other rather good, although absolute values obtained with the latter technique should certainly be considered more reliable. Good qualitative agreement also exists of our data with those obtained in [7] where hook method of measurements was used. All these studies demonstrate that strontium and helium metastable atom densities during laser pulse time interval are one-two orders of magnitude smaller than those used in [15,16] for calculations in proof of the MCT mechanism, and therefore its role should be considered as highly overestimated.the main one. The second pulse somewhat heated up the electron gas and led to consequent growth of metastable densities,which was registered both by reabsorption (Fig.5) and by 689.2 nm SrI line spontaneous radiation increase (Fig.2). The latter line originates from the $5^3P_1^0$ strontium metastable level. Then, if the MCT were the principial excitation mechanism, both spontaneous and laser radiation at 430.5 nm would be enhanced during the second current pulse. But quite opposite effect was actually observed (see Fig.2).

More selectively, metastable density can be influenced by resonant optical pumping with dye laser radiation, since in that case T_e remains almost unperturbed. Dye laser radiation of 10 mJ energy per pulse precisely tuned into 689.2 nm SrI line was enough to saturate the given transition, that increased the population of strontium metastable states by one-two orders of magnitude. However, no noticable change in the 430.5 nm spontaneous or laser radiation intensities was registered [12,13]. Thus, the experiments with an additional current pulse and optical pumping also demonstrate a negligible role of the MCT in the Sr-He laser exitation mechanisms.

5.Discussion of the Excitation Mechanisms

Relevant concentrations and plasma parameters of Sr-He laser are listed in Table 2 for two time delays in afterglow, i.e., at $t = 1.35$ μs that corresponds to laser pulse maximum (see Fig.6), and at $t = 3$ μs when lasing is ended. These data can be used to estimate numerically relative importance of the discussed excitation mechanisms. The cross-section $Q = 2.7\cdot10^{-15}$ cm^2 was used for CTI and MP reactions [11], $Q = 3.4\cdot10^{-14}$ cm^2 for MCT reaction (9) and $Q = 10^{-14}$ cm^2 for MCT reaction (10). The MCT cross-sections were deliberately taken very large in order to obtain the upper limit for possible inputs of MCT. Corresponing rate constants will be $K_1 = 2.1\cdot10^{-9}\,T^{1/2} = 5.86\cdot10^{-10}$ cm^3 s^{-1} for CTI and MP; $K = 8\cdot10^{-9}\,T^{1/2} = 2.23\ 10^{-9}$ cm^{3}s^{-1} for MCT, where T is gas temperature in eV. Recombination rate coefficient of Sr^{++} was calculated with the formula (4). Pumping rates by different processes calculated with formulas:

$$W(\text{rec}) = \alpha n_e^2 N\left(\text{Sr}^{++}\right), \qquad W(\text{MCT}) = K_2\left[N(\text{He(m)}) + N(\text{Sr(m)})\right]N\left(\text{Sr}^{++}\right)$$
$$W(\text{CTI}) = K_1 N\left(\text{He}^+\right)N(\text{Sr}), \qquad W(\text{MP}) = K_1 N(\text{He(m)})N\left(\text{Sr}^+\right), \tag{11}$$

are also listed in Table 2. At considered p(He) = 250 Torr, noticable convertion of He^+ into He_2^+ for a period of 1-3 μs does not occur. Therefore, the reaction (7) is neglected in the estimates. Calculations show that both in the laser pulse and in later afterglow the recombination pumping prevails. The CTI reaction gives one third of the Sr^{++} ion yield in the afterglow, and the MP ~3%. It should be kept in mind that CTI and MP do not pump directly the excited strontium-ion levels but create Sr^{++} ions which then recombine and add to the *W*(rec). Thus, as the estimates demonstrate, the fraction of pumping rate via MCT does not exceed several per cent even with such great cross-sections for the MCT used in calculations. Computer modelling of the Sr-He laser [12] has demonstrated even smaller influence of the MCT.

TABLE 2. Plasma parameters and pumping rates by different mechanisms in the Sr-He laser afterglow for conditions of Fig.6: helium pressure is 250 Torr, total density of strontium species is $1.4\ 10^{14}$ cm , gas temperature is 0.078 eV (900 K).

Parameter	Moment on time axis in Fig.6	
	1.35 μs	3 μs
Electron temperature T_e	0.3 eV	0.25 eV
Densities of species (cm^{-3})		
electrons	$3\ 10^{14}$	$2\ 10^{14}$
He	$1.76\ 10^{14}$	$1.37\ 10^{14}$
He(m)	$3.9\ 10^{12}$	$2\ 10^{13}$
Sr(o)	$4.1\ 10^{13}$	$6.3\ 10^{13}$
Sr(m)	$3.9\ 10^{11}$	$5.6\ 10^{11}$
Sr^+	$5.6\ 10^{13}$	$3.8\ 10^{13}$
Sr^{++}	$3.4\ 10^{13}$	$1.25\ 10^{13}$
Pumping rates ($cm^{-3}\ s^{-1}$)		
W (rec)	$4.09\ 10^{19}$ (100%)	$1.52\ 10^{19}$ (100%)
W(CTI)	$4.23\ 10^{18}$ (10.3%)	$5.06\ 10^{18}$ (33%)
W(MP)	$1.28\ 10^{17}$ (0.3%)	$4.45\ 10^{17}$ (3%)
W (MCT)	$3.27\ 10^{17}$ (0.8%)	$5.73\ 10^{17}$ (3.8%)

In summary, the discussed above kinetics studies speak definitely for the recombination pumping mechanism prevailing in the strontium and calcium violet and UV ion lasers.

6.Acknowledgements

The autors are grateful to S.V.Gumberg and L.T.Latush for their valuable help in preparing this article.

The research discribed in this publication was made possible in part by Grant No JG1100 from the International Science Foundation and Russian Goverment.

7. References

1. Gudzenko, L.I., Shelepin, L.A. (1964) *Sov.Phys.JETP* **18**, 998-1000.
2. Gudzenko, L.I., Yakovlenko, S.I. (1978) *Plasma lasers*, Atomizdat, Moscow (in Russian).
3. Zhukov, V.V., Latush, E.L., Kucherov, V.S., Mikhalevskii, V.S., Sem, M.F. (1977) Recombination lasers utilizing vapours of chemical elements. I,II, *Sov.J. Quant.Electron.* 7, 704-714.
4. Ivanov, I.G., Latush, E.L., Sem, M.F. (1990) *Metal-vapour Ion Lasers*, Energoatomizdat, Moscow (In Russian).
5. Latush, E.L. and Sem, M.F. (1973) Laser recombination transitions in CaII and SrII, *Sov.Phys.JETP* **37**, 1017-1018.
6. Latush, E.L. and Sem, M.F. (1973) Stimulated emission due to transitions in alkaline-earth metal ions, *Sov.J.Quant.Electron.* **3**, 216-219.
7. Kunnemeyer,R., McLucas,C.W., Brown,D.J.W., and McIntosh A.J. (1987) Time-resolved measurements of population densities in a Sr^+ recombination Laser, *IEEE J.Quant.Electron.* **23**, 2028-2033.
8. Carman, R.J. (1990) A self-consistent model for a longitudinal lischarge ecited He-Sr recombination laser, *IEEE J.Quant.Electron.* **26**, 1588-1608.
9. Little, C.E. and Piper, J.A. (1990) Average-power scaling of self-heated Sr^+ afterglow recombination Laser, *IEEE J.Quant.Electron.* **26**, 903- 910.
10. Loveland, D.G., Webb, C.W. (1992) Measurement of the electron density in a strontium vapour laser, *J.Phys.D.* **25**,597-604.
11. Zhukov, V.V., Latush, E.L., and Sem, M.F. (1977) *Izv. Vyssh. Uchebn. Zaved.Fiz.*, **7**, 137 (in Russian).
12. Latush, E.L., Koptev, Yu.V., Sem, M.F., Chebotarev, G.D., and Korogodin, D.A. (1991) Role of multistage collisions of the second kind in the mechanism of pumping of a helium-strontium recombination laser, *Sov.J.Quant.Electron.* **21**, 1314-1320.
13. Koptev, Yu.V., Latush, E.L., Sem, M.F. (1992) Population kinetics of the levels in the active medium of a He-Sr laser, studied by resonance absorption, *Optics and Spectroscopy*, **72**, 681-684.
14. McLucas, C.W., McIntosh, A.I. (1987) Investigation of laser emission in Sr^+ and Ca^+, *J.Phys.D.* **20**, 591-596.
15. Grinchenko, B.I. (1982) *Sov.Phys.Tech.Phys.* **27**, 1163.
16. Grinchenko, B.I. (1985) *Preprint No.5-155*, Institute of High Temperatures, Academy of Sciences of the USSR, Moscow (in Russian).
17. Klimkin, V.M., Prokop'ev, V.E., Fadin, L.V. (1979) *Sov.J.Quant.Electron.*, **9**, 357-359.

EXPERIMENTAL STUDIES OF CVL KINETICS

G.P. HOGAN, C.E. WEBB, C.G. WHYTE* & C.E. LITTLE*
Laser Group, Clarendon Laboratory,
University of Oxford, Parks Road, Oxford OX1 3PU, UK

**University of St Andrews, North Haugh, St Andrews, Fife KY16 9SS, UK*

1. Introduction

An extensive programme of experimental diagnostics has been undertaken to further our understanding of the underlying kinetics of the copper vapour laser (CVL). This has involved the direct measurement of population densities in various atomic levels as a function of time through the discharge cycle and radial position across the plasma tube, electron density, similarly resolved, and measurement of various "external" parameters such as voltage on the laser head, discharge current through the laser, output laser pulse etc. Comparison of the results of such measurement made on the conventional CVL, the same laser with hydrogen added to the buffer gas and the copper HyBrID laser have allowed us to draw a number of conclusions concerning the differences between these systems. One of the greatest values of these results, however, is in conjunction with computer modelling of the CVL plasma, and to this end a number of collaborations with modellers around the world are proving highly fruitful.

2. Experimental

A computer controlled automation [1] of the "fast" hook technique, using a Nd:YAG pumped dye laser giving 5 ns pulses was used to measure the atomic population densities, and a two colour interferometric technique, measuring change in refractive index at both 10.6 μm and 670 nm using CW sources, was used to measure change in electron density [2,3]. A full description and overview of the first stage of the project, together with all the results obtained, is given in reference [4].

The standard CVL used for the baseline series of measurements and the hydrogen experiments was an old Oxford Lasers CU-25 device upgraded to CU-40 specifications, with a 42 mm diameter, 1500 mm long plasma tube, giving 40 W average power at 6.5 kHz pulse repetition frequency (prf) (36 W with 3% added hydrogen). The discharge circuitry was a standard capacitor transfer configuration. The copper HyBrID laser was designed and built at St Andrews University, had a 25 mm diameter, 800 mm

C. E. Little and N. V. Sabotinov (eds.), Pulsed Metal Vapour Lasers, 67–72.

long plasma tube, and gave an output power of 22 W at 17 kHz prf.

A gas mixing manifold with independent metering needle valves and flow tubes for gas lines to cylinders of pure neon, and a mixture of 10% hydrogen in neon (or 1% hydrogen in neon for calibration) were installed for the hydrogen addition experiments. The proportion of each gas could be controlled and monitored separately to give an arbitrary percentage of hydrogen, up to 10%, whilst keeping the overall flow rate constant.

With a pure neon gas supply a low residual concentration of hydrogen was observed in the gas outflow of the laser that was traced to the porosity of the (standard) nylon gas tubes and is assumed to be from water vapour. This was measured to be around 0.5%.

3. Results

Although an extensive range of measurements was made on each device, for the purposes of the current paper it will be sufficient to study closely just four parameters: the specific output power, the copper ground state density, the copper 4p $^2P_{3/2}$ upper level of the green laser transition and the electron density. In addition, the different forms of the discharge current pulse will be noted.

Looking first at the output power available from the devices as a guide to "performance": although the 40 W from the standard CVL is the highest total power, this only represents 19 $mWcm^{-3}$, in contrast to the 22 W total from the HyBrID, which corresponds to 59 $mWcm^{-3}$. The 36 W from the CVL with 3% added hydrogen corresponds to only 17 $mWcm^{-3}$. In terms of specific pulse energy, however, the difference between the CVL and the HyBrID is considerably less marked, only 10%, at 3.0 μJcm^{-3} and 3.3 μJcm^{-3} respectively. In these two cases it can be seen that the increased specific power output of the HyBrID can be attributed almost entirely to its ability to run at higher prf.

In contrast to the results of other studies [5], it was found that the output power of the CVL fell as successively higher concentrations of hydrogen were added, both at the "standard" prf of 6.5 kHz, and at 9.15 kHz (at constant input power). A prf of 6.5 kHz with an admixture of 3% hydrogen was therefore chosen for subsequent experiments.

In Figure 1 the pre-pulse copper ground state population density is plotted as a function of relative radial position for the three situations. Despite the fact that it was not possible to make hook method measurements in close proximity to the plasma tube wall in the case of the HyBrID as a result of the high gradients present and smaller tube bore, a number of observations may be made. The copper density in the case of the HyBrID is considerably greater than in that of the CVL with added hydrogen, which is in turn greater than in that without. In these latter two cases it is realistic to consider the thermal information this yields as at this time the populations should be approximately at thermal equilibrium.

From the copper number density at the plasma tube wall, and copper vapour pressure data, it is possible to calculate the plasma tube wall temperature. In the case of the CVL without hydrogen this has been confirmed using an optical pyrometer, with

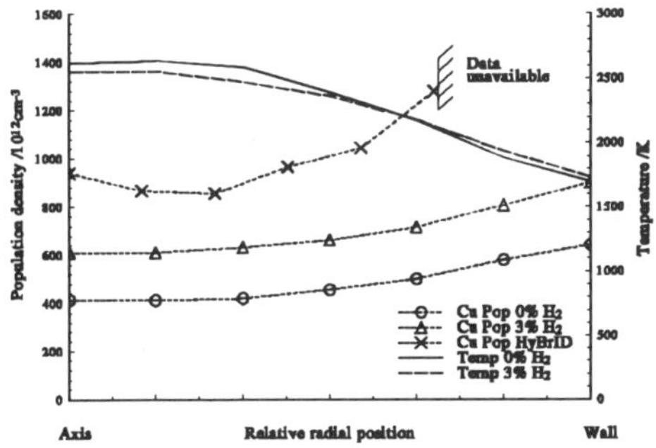

Figure 1. Copper ground state density and gas temperature

excellent agreement. From these values, and assuming a constant vapour pressure of copper across the tube radius, an approximate gas temperature profile can be calculated. This assumes a negligible contribution from ambipolar diffusion, which is not necessarily justified [6], however any such correction would not change the inference discussed below.

Although the wall temperature in the case of the added hydrogen is slightly higher than without (about 30 K), giving rise to the higher vapour pressure of copper observed, the thermal profile is flatter, giving rise to a lower gas temperature at the tube axis.

Figures 2, 3 & 4 show surface plots of the population density of the 4p $^2P_{3/2}$ level as a function of time and radial position (data mirrored at the axis to give a full diameter). The scales are identical in each case. Once again the peak values are significantly higher in the case of the HyBrID than the CVL with hydrogen, which was greater in turn than the CVL without. In addition it can be seen that excitation to this level in the HyBrID is very much more highly constrained to the tube axis than in the conventional CVL.

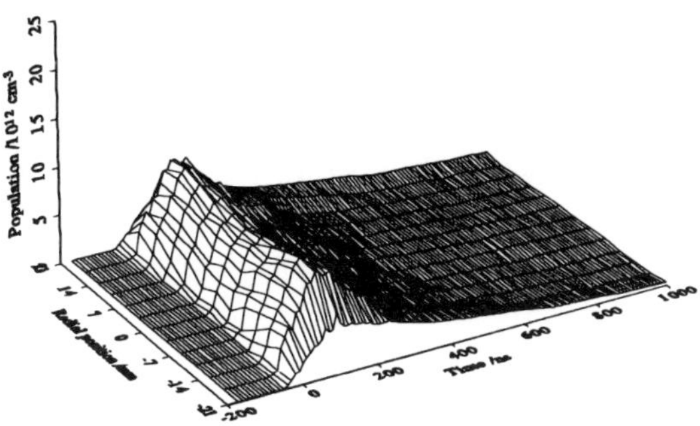

Figure 2. Copper 4p $^2P_{3/2}$ level; plus 0% H_2

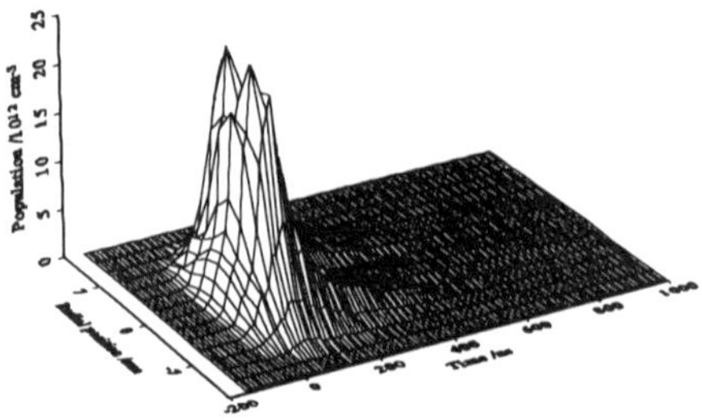

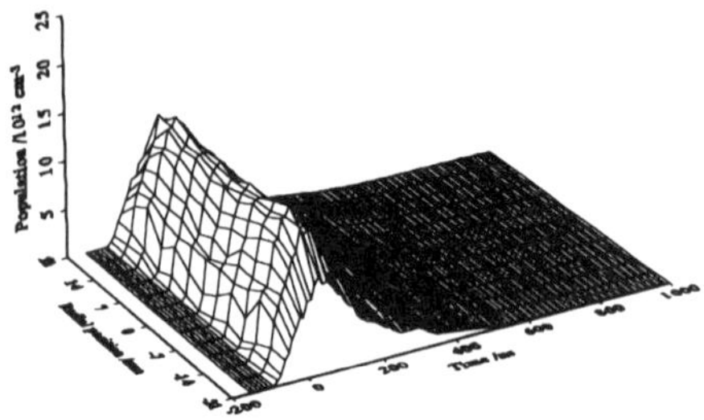

Figure 3. Copper 4p $^2P_{3/2}$ level; HyBrID *Figure 4.* Copper 4p $^2P_{3/2}$ level; plus 3% H_2

Finally, Figure 5 shows the measured electron density on the tube axis in the three devices. The HyBrID shows by far the highest peak value, with the CVL with added hydrogen significantly below the one without. What is even more striking, however, is that, whereas the pre-pulse electron densities in the cases of the CVL with and without hydrogen are not significantly different, though slightly lower with hydrogen, and in each case represent a significant fraction of the peak value the pre-pulse value in the HyBrID is negligible. The very high peak electron density can be seen to decay very fast in comparison to the conventional CVLs.

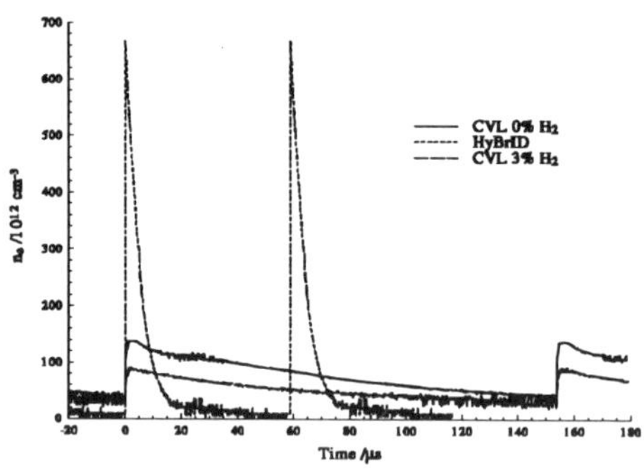

Figure 5. Electron density on axis; CVL (0% & 3% added H_2) and HyBrID

4. Discussion

In the case of the HyBrID many of its observed characteristics may be attributed to the electron density results. The very fast decay of electron density in the early afterglow has been attributed [7] to electron attachment by the highly electronegative bromine (and possibly hydrogen) atoms and two body recombination. The very low pre-pulse electron density, and hence pre-pulse conductivity, gives rise to improved voltage hold off, improved breakdown characteristics, and a reduction of the phenomenon known as the "Phantom Current" [7,8]. This is an ohmic conduction current that is associated with no spontaneous emission occuring during the rising edge of the voltage pulse in the conventional CVL, and is reduced to a vestigial hump in the HyBrID. In the HyBrID, therefore, the electron temperature may be expected to rise faster, to a higher peak value, and fall faster (with the increased population densities in excited (and ground) states). In the conventional CVL the increasing output power with increasing copper density (wall temperature) ceases at high copper density as the peak electron temperature is reduced. It would appear that in the case of the HyBrID the improved breakdown characteristics allow this to be maintained in the presence of considerably higher copper density, giving ground state depletion or 73% against 55% for the conventional CVL, or 51% with the hydrogen addition.

In the conventional CVL the high pre-pulse electron density gives rise to a highly uniform discharge. In the case of the HyBrID this can be seen to have suffered, giving rise to the more axial nature of the excitation observed in Figure 3. This gives rise to the more closely Gaussian nature of the beam from the HyBrID laser.

The slightly reduced pre-pulse conductivity in the CVL with added hydrogen in comparison with that without, is likely to have improved the matching between the discharge circuitry and the laser head, resulting in more power deposited into the laser plasma, giving rise to the slightly higher plasma tube wall temperature. The fact that there is still 51% depletion of the (higher) ground state density during the discharge suggests that an adequate electron temperature can still be maintained in its presence. This is further supported by the greater peak population density in the 4p $^2P_{3/2}$ upper laser level.

The other effect of the hydrogen, it would appear, is thermal. Its high mobility increases the thermal conductivity of the plasma, giving rise to the flattened temperature profile observed. This reduced gas temperature on axis will have a number of beneficial effects: it will give rise to a more uniform radial distribution of copper; for a given ground state density it will reduce the pre-pulse thermal population in the lower laser levels; and it will increase the rate of electron stabilised three body recombination.

Two questions remain: if adding hydrogen gives rise to a higher peak population density in the upper laser levels, and a lower gas temperature on axis, why do we observe a slight decrease in output power in our laser? and: why do some groups observe an increase in output power, but we observe this decrease?

The answer to the first question is that despite the lower gas temperature on axis, the higher ground state density still leads to an overall increase in pre-pulse thermal populations, thus decreasing the overall population inversion.

The answer to the second question seems likely to lie in the level, and design, of thermal insulation used, and hence the specific input power. Lightly insulated devices with a very high specific input power, such as that of [5], generate a comparatively steeper thermal profile, with a consequently lower ground state density on axis, and a higher thermal pre-pulse metastable population. This gives rise to the annular beam profile sometimes seen, even on relatively small bore lasers. In this case the improved thermal conductivity provided by an admixture of hydrogen could be highly beneficial, in addition to any advantages in reducing electron temperature.

The optimisation of insulation is a complex issue and depends upon the overriding criteria concerning the laser, whether it is to optimise beam quality, output power, or overall efficiency.

5. Conclusions

The greater output power and efficiency of the copper HyBrID laser, in comparison with the conventional CVL would appear to be primarily concerned with its very low pre-pulse electron density as a result of electron attachment and two body recombination during the afterglow. This gives improved discharge characteristics and excitation efficiency, and, together with a lower average gas temperature giving lower pre-pulse metastable populations, allows operation at higher prf.

The benefits of low levels of hydrogen addition to the buffer gas of conventional CVLs would appear to be associated with its high thermal conductivity and its ability to reduce the electron temperature. Whether the net result of adding hydrogen to a given device is positive or negative, and, if positive, where the optimum lies, will depend upon the thermal insulation and specific input power of that device. It would appear to be possible, however, to design a laser that was optimised for use with added hydrogen. The use of a sophisticated computer model should theoretically allow the optimum hydrogen concentration, and insulation level, to be calculated.

6. References

1. Hogan GP, Svoboda M & Webb CE (1993) The Automation of the hook technique, *Meas. Sci. Technol.* **4**, 263-269
2. Hogan GP & Webb CE (1993) Radially and time resolved measurements of electron density in a copper vapour laser, *LEOS '93* 6th Annual Meeting: IEEE Lasers and Electro-optics Society
3. Hogan GP & Webb CE (in preparation) Radially- and time-resolved measurements of electron density in an operating copper vapour laser, *Meas. Sci. Technol.*
4. Hogan GP (1993) A study of the kinetics of copper vapour lasers, *DPhil Thesis* Oxford University
5. Withford MJ, Brown DJW & Piper JA (1994) Investigation of the effects of hydrogen and deuterium on copper vapour laser performance, *Optics Commun.* **110** 699-707
6. Carman RJ, Brown DJW & Piper JA (1994) A self-consistent model for the discharge kinetics in a high-repetition-rate copper-vapor laser, *IEEE J. Quantum Electron.* **30** (8) 1876-1895
7. Jones DR, Halliwell SN & Little CE (1994) Influence of remanent electron density on the performance of copper HyBrID lasers, *Optics Commun.* **111** 394-402
8. Hogan GP & Webb CE (1995) Pre-ionization and Discharge Breakdown in the Copper Vapour Laser: The Phantom Current, *Optics Comm.* **117** (5-6) 570-579

PHYSICS OF TRANSVERSE-DISCHARGE COPPER VAPOUR LASERS

B.L. BOROVICH
N.I. YURCHENKO
Special Design Bureau of High Energy Lasers «Granat»
95 Volokolamskoye shosse, Moscow 123424, Russia

Brief results of creation of the transverse-discharge copper vapour laser (CVL) with average output power exceeded 100 W at $f = 2$ kHz and divergence $2x10^{-4}$ rad are presented. Numerous calculations based on the physical-mathematical model of laser kinetics with automatic optimization of all laser parameters are developed. These results demonstrate potential possibilities of a CVL and satisfactory correspond to all big scale experiments fulfilled up to date.

Powerful transverse-discharge CVL's (~100 W) were first made in 1976 at SDB «Granat» of SPA «Astrophysica» [1-3].

The laser chamber consists of a water-cooled metallic body, thermoinsulation, ohmic heaters and a discharge unit mounted on a chamber cover (Fig.1 and Fig.2). The discharge unit includes discharge gap formed by two electrodes; high temperature high voltage ceramics (pyrolytic boron nitride) for the discharge space restriction and prevention of high voltage breakdown onto the chamber construction metal elements; return current guides providing discharge minimal inductance; high voltage joints connecting cable lines of the chamber feeding circuit from a pulsed power supply (PPS). To reduce the circuit inductance the discharge unit is divided along the chamber into sections each including a joint and a pair of electrodes. The PPS was connected with the discharge chamber by cable reducing transformer. Thyratron was used as a commutator. The thyratron current pulse duration was 1-2 mcs, then the pulse being repeatedly passed in nonlinear C-L-C circuits was compressed down in PPS to a required duration of about 100 ns. Thus the PPS's of output energy per pulse W_0 up to 10-20 J at pulse repetition rate f up to 5 kHz with one thyratron TGI1-2500/50 only were created. The reached CVL characteristics are presented in table 1.

Multiunits version of CVL was tested in 1981. Four 10-section discharge gap 17x120x1080 mm were placed in one power amplifier body M-40. Laser pumping was carried out by two PPS with 10 J per pulse each. Photos of power amplifier body and his discharge unit are presented in Fig.1 and Fig.2.Output power was 160 W at $f = 4{,}5$ kHz in case when all four «slits» were used in common

C. E. Little and N. V. Sabotinov (eds.), Pulsed Metal Vapour Lasers, 73–78.

TABLE 1

Chamber index	3B2	MP1-40
Year	1976	1979
Discharge space dimentions (mm)	60x60x1000	90x100x1600
Section quantity	10	8
Average output power (W)	85	104
Pulse repetition rate (kHz)	3	2
Enrgy of pulse (mJ)	30	50
Peak power (MW)	1	2
Specific output energy (mcJ/cm3)	8	4
Efficiency (related to PPS power to matched loading) %	1.4	1.25
Connecting cable wave resistance (ohm)	2	6
PPS pulse amplitude (kV)	up to 3	up to 8
Neon pressure (optimum) (atm)	0.8	1
Resonator	plane mirrors	unstable with spherical mirrors, M=50
Divergence (mrad)	5	0.24

oscillation regime. Output power of one «slit» with plane resonator was 65 W at f=3 kHz, and 120 W in amplification regime. The test of power amplifier M-40 was not finished completely because of financial problems.

Physical-mathematical model of laser was worked out for analysis of CVL generation mechanism [4-7]. One can distinguish two characteristic periods of plasma kinetics: excitation pulse and following relaxation of plasma. In accordance with this calculation was divided into two periodically repeated stages (regimes of «pulse» and «relaxation») which are partly distinguished by choice of kinetic processes taken into consideration . In the kinetic model it is reasonable to consider five basic states in a copper atom: *1* - ground, *2* - united metastables, *3* - united resonance levels, *4* - high lying levels block and *5* - ionized state. Three energetic states (*1'* - ground, *2'* - block excited and *3'* - ionized) are distinguished in the buffer gas (neon or another inert gas). The following transitions due to electron - atom collisions are taken into consideration in both periods: *1-2, 1-3, 2-3, 3-4, 4-5, 2'-3', 2'-1'*, corresponding reverse processes, and in pulse regime in addition *1-4, 1-5, 2-4, 3-5, 1'-2', 1'-3'*. This set of processes allows for a correct modeling of kinetics of working states excitation and kinetics of direct and many-stage ionization of both components of active mixture.

Excitation rates of high lying levels block are sum up most essential transitions between concrete levels. The diffuse approximation was used for calculation of many-stage ionization. The Maxwell electron energy distribution function was used for an averaging of collision cross-sections, exept of the neon excitation and ionization rates, calculated with taking into account of two thresholds of screening due to copper and neon direct excitation.

Fig.1. Power amplifier M-40.

Fig.2. Discharge unit of power amplifier M-40.

Other rate constants were approximated by analytical functions of electron temperature (T_e) which, in turn, is determined from nonstationary thermal balance equation including heat by external field, nonelastic, elastic and diffusion cooling of electrons. The electric field, in its turn, is determined from external circuit equation taking into consideration variable discharge resistance and circuit parameters: capacity C, inductance L and charge voltage U_0. Output power is calculated in rate equation approximation for green line only.

In pulse regime spontaneous transitions between working levels (3-2) are taken into consideration only. At the relaxation regime radiating transitions acquire more significance because they can have effective influence on recombination flow distribution, recombination heating of electrons and on a gas heating. Some of principal such processes for Cu and Ne are included into the model too with account for self-absorption of radiation.

Recombination processes through formation of molecular and negative copper ions, triple recombination and metastable quenching due to collisions with neon also can be essential at a pressure above atmosphere or at high copper concentrations. Diffusion processes: diffusion of metastables, charged particles and diffusion cooling of electrons are essential at a low pressure. These processes are included in the model too. The energy transferred to heavy particles by elastic collisions was being integrated in the course of calculations. It permits to calculate gas average temperature T_a at pulse-periodic regime at known geometry.

Selection of cross-sections of processes for calculations see in detail in [5].

The elaborated model and calculation program of CVL parameters describes all complex plasma and laser characteristics of CVL transverse - and longitudinal - discharge satisfactorily (the comparison was fulfilled with the most of known up to date experimental results). Analysis results in detail including «thing effects» displaying at use different buffer gas He, Ar and Ne see in [5]. On the basis of this model direct multiparametric CVL optimization at automatical

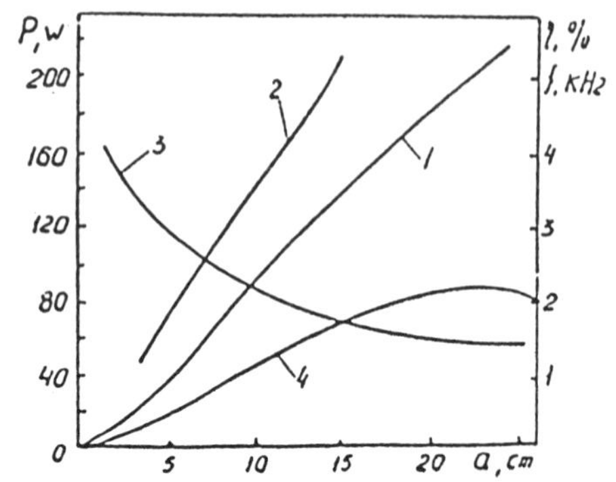

Fig.3. Main results of optimization calculations (transverse-discharge CVL).
1,3,4 - PPS-B, *2* - PPS-A, *1,2* - output power, *3* - optimal *f*, *4* - optimal η.

regime was fulfilled using computer БЭСМ-6. «Run down» and «gradient down» methods were used and following main initial parameters were varied to optimize the average output power *P* from whole volume: copper and neon concentration N_{Cu} and N_{Ne}, pulse repetition rate *f*, initial PPS pulse amplitude U_0 and specific energy w_0 (it is supposed this pulse has triangle form with long on zero level τ), active length *l* and cross-size of active volume *a* (square or circle), PPS exit resistance *R*. In Fig. 3 dependencies of the transverse-discharge CVL output power on cross-size *a* which can be reached at a green component of radiation at optimum laser parameters are presented. Here τ = 200 ns, *l* = 1 m, resonator length l_r = 2 m, inductance *L* = 10 nH are fixed. PPS-A and PPS- B mean two different possible schemes of PPS: PPS-B - energy reflected from discharge during the first current half-wave is deposited into discharge and heats gas; PPS-A - this energy is absorbed in PPS elements.

At the transversal pumping scheme optimum parameters are: P_{Ne} = 1 atm and T_a = 2800±200 K practically everywhere. A high pressure is necessary here for enough high efficiency of matching of PPS with the discharge η_d and it is restricted by injurious effects of recombination processes by triple collisions. For the partial compensation of these processes an optimum gas temperature T_a is increased to a top high level permissible by a limit of not too much thermal metastable population.

The numerical optimization shows that the growth of efficiency of matching PPS with discharge η_d and a physical efficiency η is possible simultaneously by increasing of an active volume and some reducing of the frequency. The low limit of the reasonable specific stored pumping energy is 0.1 J/*l* because of approaching to the generation threshold. In this limit η_d = 70 %; η =2 %. In case of inductance increasing by 1,5-2 times, optimum output power is reduced by 15-20 %. Fast decrease of τ down to 40-70 ns occurs during optimization, it results to a doubling output power as compared with τ = 200 ns at the same others conditions. However the growth of generation power is accompanied by essential decrease of radiation pulse duration and increase of energy losses on the chamber walls at the expense of nonaxial beams [8]. Under such conditions it is necessary to transfer a laser in the MOPA system. The main results of the optimization calculations of CVL with longitudinal discharge are presented in Fig.4 and Fig.5. A principal difference of this CVL is reduced to the change of parameter $X' = lS/d$ [6] characterizing the ratio of inductance to ohmic discharge resistances

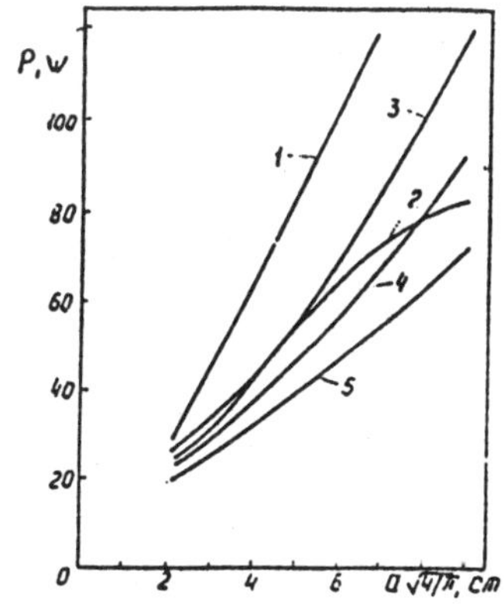

Fig.4. Main results of optimisation calculations (longitudinal-discharge CVL, pumping through cable line).
1,5 - PPS-B, *2,3,4* - PPS-A, *1* - without limitations, *2* - W_0 = 1 J, *3,4,5* - U_0 = 10 kV, *3* - L = 0,5 mcH, *1,2,4,5* - L = 1 mcH.

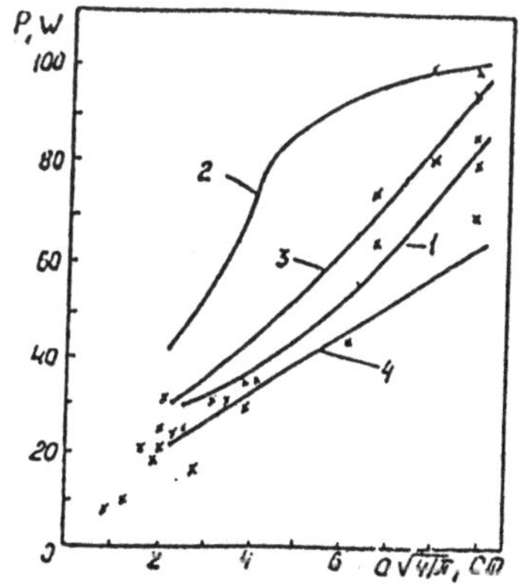

Fig.5 Main results of optimization calculations (longitudinal discharge CVL, pumping by discharge of oscillation circuit directly).
1 - PPS-B, *2,3,4* - PPS-A, , *2,4* - L = 1 mcH, *1,3* - L = 0,5 mcH, *2* - W_0 = 1 J, *1,3,4* - U_0 = 10 kV.

(S - cross-section, d - length of discharge gap). At a transverse discharge geometry X' doesn't practically depend on cross-section of an active volume just as at longitudinal discharge X' increases proportionally to cross-section value : X' - a^2. At $a = d$ = 10 cm values X' are equal at both schemes and optimization results coincide naturally.

A transverse discharge has the definite advantage for X' if $a > 10$ cm but this advantage is small because of the increased a reduces optimal specific energy deposition per pulse w_0^{opt} and influence of inductance in general. On the contrary longitudinal discharge has the advantage if $a < 10$ cm moreover the difference is significant because of the great w_0^{opt} and if a =3,5 cm it reaches 30-40%. In general optimization results are similar for both geometry of pumping.

However the optimum neon pressure remains about atmosphere at longitudinal discharge too. It requires the significant greater PPS pulse amplitude that is realized with difficulty in practice. Therefore optimization results with restriction upon U_0. and W_0 are interest. Optimum values R, N_{Ne} and N_{Cu} with *fixed* U_0 = 10 kV or W_0 = 1 J decreases considerably with comparison to a case without restrictions. Some results of a laser parameters optimization are presented in Table 2. Here PPS-B were used, l = 1 m and fixed parameters is concluded in brackets.

Results of optimization were compared with experimental data (crosses in Fig.5) from the most of published works ([9-13] and others,full list is on cited in [5]) According to calculations and experiments the output power of a green component is about 40-70 % of the total output power at optimum frequency So calculated output in Fig.4 and Fig.5 is doubled for convenience of the comparison with experiments. It is seen from Fig.5 that results which were calculated and reached are matching with each other fine in spite of the difference of experimental constructions, optimum regimes and etc. The main qualitative

TABLE 2

Type of discharge	a, cm	L, mcH	C, nPh	τ, ns	NNe, 10^{18} cm^{-3}	NCu, 10^{14} cm^{-3}	$U0$, kV	$W0$, J	f, kHz	P, W	ηd, %	η, %	Imax, kA	w, mcJ x cm^{-3}	Ta, meV
Transv	(6)	(0,01)	-	(200)	12	5,0	3,6	2,3	3,3	56	52	1,4	11	4,7	247
Transv	(10)	(0,01)	-	(200)	4,9	4,8	3,6	3,6	2,2	90	51	2,3	14	4,2	245
Transv	(20)	(0,01)	-	(200)	4,5	3,9	5,0	6,0	1,4	181	62	3,5	16	3,3	241
Transv*	(6)	001	l=366 cm	(200)	5,0	3,5	2,7	(5)	5,2	128	50	1,0	35	1,9	236
Transv**	15	(0,01)	l=205 cm	(200)	3,1	2,5	3,8	(5)	3,4	232	67	2,1	23	1,5	238
Longitud	(2)	(0,5)	7,2	-	0,01	8,2	(15)	0,8	8,1	19	88	0,3	1,5	6,0	168
Longitud	(4)	(0,5)	9,7	-	0,26	4,0	(15)	1,1	3,7	21	90	0,57	1,6	3,5	181
Longitud	(6)	(1)	1,5	-	3,6	5,7	58	2,5	4,3	71	82	0,8	1,5	4,6	259
Longitud	(8)	(0,5)	17	-	0,83	2,4	(15)	1,9	3,1	44	93	0,8	1,9	2,2	211
Longitud	(16)	(0,5)	31	-	0,69	1,5	(15)	3,6	1,3	56	93	1,3	2,7	1,7	197

* - l is free parameter; ** - a and l are free parameters.

conclusion is:for increasing of the average specific (per length) output power it is necessary the approximately proportional increasing of diameter of active volume with the simultaneous optimization of other laser parameters.

References.

1. Artem' ev, A.Yu., Babeiko, Yu.A., Bakhtin, O.M., Borovich, B.L. et al. (1980) Energetics of a transverse discharge copper vapor laser, *Kvantovaya electronika* **7**, 1948 - 1954.

2. Artem' ev, A.Yu., Borovich, B.L., Vasil' ev L.A. et al. (1982) A multi-sectional copper vapor laser, *Kvantovaya electronika* **9**, 738 - 743.

3. Borovich, B.L. (1993) Transverse-discharge copper-vapor laser, Metal Vapor Lasers and Their Applications: CIS Selected Papers, G.Petrash, Editor, *Proc.SPIE* **2110**, 46 - 63.

4. Arlantsev, S.V., Buchanov, V.V., Vasil' ev L.A., Molodykh, E.I., Tykotskii, V.V., Yurchenko, N.I. (1980) A Theoretical Study of a Repetitively Pulsed Copper Vapor Laser, *Kvantov. electronika* **7**, 2319 - 2325.

5. Borovich, B.L., Yurchenko, N.I. (1984) An Analisis of the Excitation and Relaxation Kinetics in a Longitudinal-Discharge Copper Vapor Laser, *Kvantovaya electronika* **11**, 2081 - 2095.

6. Borovich, B.L., Nalegach, E.P., Rybin, V.M., Yurchenko, N.I. (1984) Recombination in the Cu-Laser Plasma under the High Buffer Gas Pressure, *Kvantovaya electronika* **11**, 2471 - 2479.

7. Yurchenko, N.I. (1993) Theoretical Study of Copper Vapor Lasers, Metal Vapor Lasers and Their Applications: CIS Selected Papers, G.Petrash, Editor, *Proc.SPIE* **2110**, 64 - 77.

8. Borovich, B.L., Yurchenko, N.I. (1985) Dumping Efficiency in Lasers with a Short Emission Pulse, *Kvantovaya electronika* **12**, 1377 - 1386.

9.Grove, R.E. (1982) Copper-Vapor Lasers Come of Age, *Laser Focus* **18**, 45 - 49.

10. Kushner, M.J.,Warner,B.E. (1983) Large-bore Copper-Vapor Lasers: Kinetics and Scaling Issuer, *J. Apple. Phys.* **54**, 2970 - 2980.

11.Gabay,S., Blau,P., Lando,M., Drikkman,I., Horvar,Z., Yprah,I., Heu,I.,Muron,E., Smilanski,I. (1991) Stabilization of high power copper vapor laser, Opt. and Quant.electronics **23**, N 4, S485-S492.

12. Tanaba,Y., Hara,K., Veguri,S. (1992) Lasing characteristics of a large-bore copper vapor laser. Proc. SPIE **1628**, 32-43.

13. Chang,J.J., Boley, C.D., Martinez,M.W., Molconder,I., Warner,B.E. (1994) Beam characteristics of a large-bore copper laser with a radiatively cooled plasma, Proc. SPIE **2118**, 2-8.

INVESTIGATION OF BISMUTH VAPOUR LASERS

V.A. KELMAN*, I.I. KLIMOVSKI†, Yu.O. SHPENIK*
** Institute of Electron Physics, Ukrainian National Academy of Sciences, Universitetskaja 21, Uzhgorod, 294016, Ukraine*
† Institute of High Temperatures, Russian Academy of Sciences, Izhgorskaja 13/19, Moscow, 127412, Russia

1. Introduction

The bismuth-vapour laser (BVL), that oscillates on the 472.2 nm of atomic bismuth ($6p^27s^4P_{1/2} \rightarrow 6p^3\ ^2D^o_{3/2}$ transition) belongs to the class of self-terminating lasers (STL). The population inversion is created during a short (50 - 200 ns fwhm) excitation pulse due to faster population of the upper laser level (resonant) than the lower one (metastable). These lasers operate in cyclic mode with pulse recurrence frequencies (PRF) from single pulses to hundreds of kHz.

The BVL system is of interest because it radiates blue light and can supplement the green/yellow copper-vapour and red gold-vapour lasers. Moreover, this laser have a number of interesting features (lower limit of PRF, narrow temperature interval of lasing, low average power of lasing, ring character of the radiation spot at buffer gas pressure over 6 Torr, etc).

Since 1977, when the first report about the observing of stimulated emission was published [1] by G.G.Petrash *et al,* a number of investigations was made. Laser oscillation was observed in the form of a thin ring, adjacent to the inner wall of the discharge tube (DT) for PRF < 6.25 kHz, neon buffer-gas pressures < 32 Torr and DT wall temperatures 750 ÷ 900 ^oC. Average power of up to 17 mW was achived. This group investigated the excitation conditions of lasing, measured the decay of the lower laser level during the interpulse period and examined the reasons that influence so dramatically the lasing characteristics [2,3]. K.A.Peard and R.C.Tobin [4] reported about the unsuccessfull attempt to realize lasing in BVL. They investigated the kinetics of the BiI ground state and the lower laser level population and examined the processes, that may prevent lasing. Stimulated emission on BiI self-terminating transition was demonstrated by D.R.Jones and C.E.Little in the Bi-

C. E. Little and N. V. Sabotinov (eds.), Pulsed Metal Vapour Lasers, 79–84.

Ne-HBr system [5]. Metal atoms in this mixture are provided by electron-impact dissociation of bismuth-bromide molecules in a high-PRF discharge. The laser output was similar to that of the conventional BVL, but the laser beam profile changes with the increasing of HBr concentration from 3 - 4 mm wide annulus to an almost flat across the beam profile.

This paper summerizes our investigations of the BVL that were provided during 10 years. Based on these experimental and calculational studies we grounded the mechanism of the BVL lasing and explained the extraordinary features of it.

2. Discharge-excited BVL

The first step of our investigations was the calculational examination of the possibilities of BiI levels and transitions parameters in obtaining laser oscillation. The model of saturated power was used for the calculation of the excitation kinetics of pulsed laser oscillation in bismuth vapour [6]. We calculated the specific laser pulse energy, efficiency, terminating values of lower laser level population and gas temperature dependencies on initial excitation conditions (i.e. Bi-vapour pressure, electron temperature, electrons concentration). The computed kinetics of the levels population (n_{gr}, n_m, n_r), electrons concentration (n_e), laser power density (P) and efficiency (η) during the excitation pulse with constant electron temperature T_e = 3 eV are presented as an example at Fig.1.

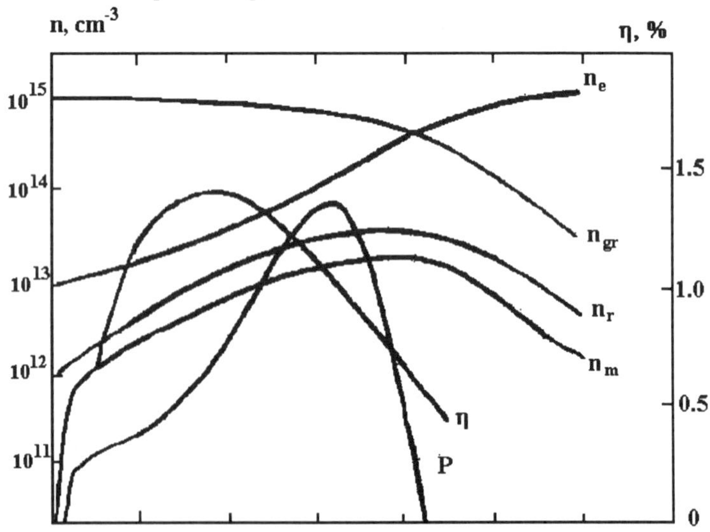

Fig.1. Calculated kinetics of levels population, electrons concentration, laser power density and efficiency.

Computer modelling of excitation, ionization and deexcitation processes during the excitation pulse using six-level scheme shown, that ensuring proper conditions of plasma excitation laser oscillation with good output parameters can be expected.

This results assured us to begin experimental and calculational investigations to find the reasons. that prevents effective lasing in BVL plasma. At the early beginning of experimental investigations we obtained pulsed laser oscillation in the Bi - Ne mixture on the BiI 472.2 nm transition. For the first time we observed lasing from the whole volume of the DT for Ne pressure < 6 Torr [7].

We investigated the dependences of laser power on PRF, Ne pressure, wall temperature, rectifier voltage and optimized this excitation parameters for maximum laser power. This results together with the laser average power distributions across the DT radius are presented at Fig.2. It should be mentioned that traditional methods of excitation conditions optimization doesn't lead to

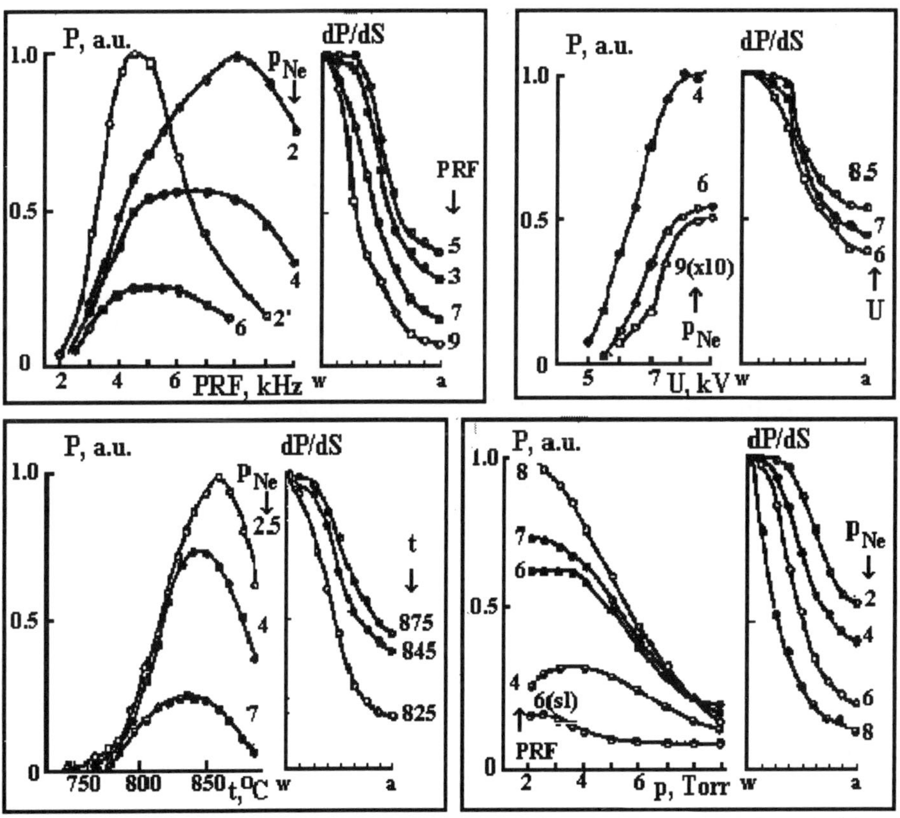

Fig.2. Bismuth-vapour laser average power (P) and power density (dP/dS) dependencies on excitation conditions (PRF, rectifier voltage (U), DT wall temperature (t) and Ne pressure (p)).

essential improvement of BVL parameters. The largest value of average power was ~ 20 mW. We found another unique for atomis STL feature - lower limit of PRF (2.4 - 1.0 kHz depending on Ne pressure). This feature is typical for STL, where population inversion on selfterminating transition is produced due to electron-impact dissociation of metal-halogen molecules.

We investigated the relaxation of lower laser level in the interpulse period [8] near the DT wall and on it axis. The measured values of metastable atoms concentration before the next excitation pulse and their decay time in this two cases show, that the processes of metastable level relaxation doesn't prevent effective lasing from the whole volume of the DT and can't give the explanation to the BVL poor parameters.

Analyzing all the results obtained by other groups and by us, we proposed that the most probable reason of it is the presence of Bi_2 dimers in the laser plasma and their main influence on the inversion formation.

3. The influence of Bi_2 dimers on population inversion formation

The presence of Bi_2 dimers in the heated vapour is confirmed by a number of experimental investigations and calculational evaluations. At BVL operating temperature the concentration of Bi_2 dimers in equilibrium vapour is ~ 50 % of the whole amount of bismuth particles. In our study of bismuth vapour radiation and absorption in the 200 - 600 nm region [9] we don't observe the radiation of molecular bands in the emission spectrum of the BVL discharge. But it doesn't mean that Bi_2 dimers are absent in the laser plasma. Measurements of the Bi_2 dimers concentration (n_d) show that dimers are presented in the BVL plasma and their concentration correlates with the laser power (P). We found that;

a) n_d fells with the increase of P; it's minimal value is ~ $5{\cdot}10^{-14}cm^{-3}$ at maximal P;

b) n_d increases with the reduction of rectifier voltage and for U < 4 kV (lasing is absent) it's value is ~ 1/3 of dimers concentration in the heated tube without discharge;

c) n_d increases with the PRF reduction to the same value at 2 kHz (lasing is absent);

d) there is no considerable absorption round the laser 472.2 nm and resonant 306.8 nm line (at optimal for lasing conditions);

e) the averaged over the bore of DT value of n_d in the interpulse period has no considerable changes (i.e. exists stationary value of n_d that is settled after a number of excitation pulses and depends on PRF and U).

The next step was the study of the spatial profiles of distribution (SPD) of bismuth atoms and molecules in DT. Such spatial profiles of metal atoms and metal-halide molecules are formed in DT of metal-halide lasers with predominant location of molecules near the DT wall. We calculated the SPD in BVL DT solving the diffusion equations for Bi and Bi_2 relaxation in the interpulse period [10]. Fig.3 shows the calculated SPD of dimers M (1-4) and atoms N (5-8) for different values of frequency factor $\Omega = fR^2/D_m$ (f - PRF, R - DT radius, D_m- dimers diffusion coefficient). Analizing the calculated SPD and comparing them with the experimental dP/dS (see Fig.2) we made a conclusion that lasing occurs from those parts of the DT where dimers are concentrated. This means that the processes with bismuth molecules are responsible for lasing excitation. Such fast processes in low pressure plasma that results lasing pulse during a short excitation pulse, may be only the process of dissociative excitation of dimers with creation of Bi atoms in upper laser and ground states. Analizing the energy diagram of Bi_2 molecule electronic states (Fig.4) we see, that probably the insufficiently known F- and N- states, that dissociates in this way, may be of our interest. The pumping of the BVL upper level predominantly takes place as:

$$e + Bi_2(X') \rightarrow e + Bi_2^* \rightarrow e + Bi(^4P_{1/2}) + Bi(^4S_{3/2})$$

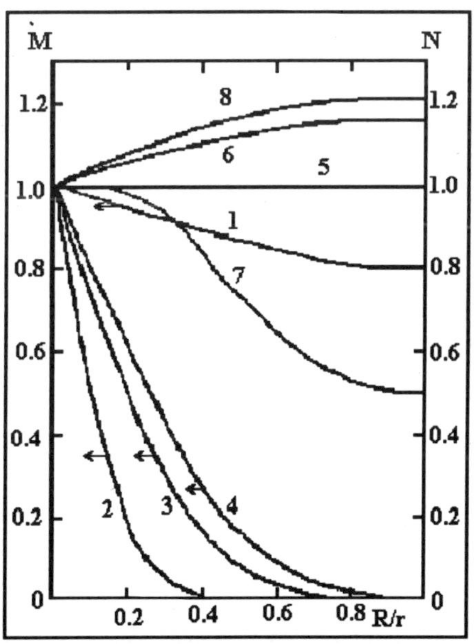

Fig.3. SPD of Bi and Bi_2 relative densities (Ω = 4 (1,5); 33 (2,6); 109 (3,7)).

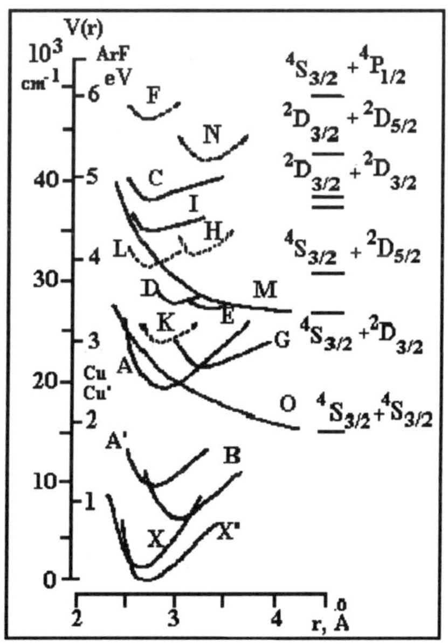

Fig.4. Diagram of Bi_2 energy states.

The energy threshold of this reaction is ~ 6.5 eV. This is considerable higher than the excitation of $Bi(^4P_{1/2})$ from the ground state (4.04 eV). Therefore it is necessary to keep up considerably higher T_e in BVL plasma than optimal for other STL.

The direct evidence of dimers distribution in the DT was experimentally obtained by us at longitudial optical pumping of the BVL plasma by ArF-laser (193 nm). The idea was, that the optical pumping of dimers causes the same dissociation way as at electron-impact dissociative excitation with the same energy. When bismuth vapour was pumped only by ArF laser radiation, homogeneous lasing across the DT bore 472.2 nm was observed. Applying discharge excitation simultaniously with optical pumping causes spatial distribution of laser power density like in the traditional BVL.

Thus, we based the mechanism of inversion formation in the discharge-excited BVL that can explaine it's unique features. For the first time lasing on BiI selfterminating 472.2 nm transition was obtained due to photodissociative pumping of bismuth dimers.

4. References

1. Markova, S.V., Petrash, G.G. and Cherezov, V.M. (1977) Pulsud lasing on the 472.2 nm line of the bismuth atom, *Soviet J. Quantum Electronics* **4**, 1154-1155.
2. Markova, S.V., Petrash, G.G. and Cherezov, V.M. (1979) Investigation of the lasing mechanism in pulsed bismuth-vapour laser, *Soviet J. Quantum Electronics* **6**, 1200 -1207.
3. Kazakov, V.V., Markova, S.V. and Petrash, G.G. (1982) Decay of metastable $6p^3\ ^2D^o{}_{3/2}$ levels of bismuth atoms in the period between pulses in a bismuth-vapour laser, *Soviet J. Quantum Electronics* **9**, 688-694.
4. Peard, K.A. and Tobin, R.C. (1984) Difficulties associated with stimulated emission in bismuth-neon discharge, *IEEE J. Quantum Electronics* **20**, 765-771.
5. Jones, D.R. and Little, C.E. (1992) A 472.2 nm bismuth halide laser, *Optics Communications* **91**, 223-228.
6. Shpenik, Yu.O. and Kelman, V.A. (1989) Calculation of the active medium plasma parameters and bound characteristics of the bismuth vapour laser, *XIX International Conf. on Phenomena in Ionized Gases* Belgrade, 502 - 503.
7. Vochmin, P. A., Zapesochny, I. P., Kelman, V. A. and Shpenik, Yu. O., (1987), Investigation of pulse lasing on the 472.2 nm self-bounded transition of the bismuth atom at low buffer-gas pressures, *Soviet J. Quantum Electronics* **24**, 1655-1657.
8. Shpenik, Yu.O. and Kelman, V.A. (1991) Relaxation of the $6p^3\ ^2D^o{}_{3/2}$ metastable Bi atoms in the bismuth vapour laser, *XIX Intrenational Conf. on Phenomena in Ionized Gases* Pisa, 1221-1222.
9. Zapesochny, I.P., Kelman, V.A. and Shpenik, Yu.O. (1989) About the mechanism of Bi_2 dimers influence on the bismuth vapour laser output parameters, *Soviet J. Optics & Spectroscopy* **66**, 711-714.
10. Kelman, V. A., Klimovski, I. I. and Shpenik, Yu. O. (1994) Generation-excitation mechanism in bismuth-vapour laser, *Journal of Soviet Laser Research* **15**, 69-73.

3. Laser Characteristics and Operation

PLASMA PARAMETERS OF METAL VAPOUR LASERS

I. SMILANSKI
Nuclear Research Centre - Negev
P.O. Box 9001, Beer Sheva , Israel.

The CO_2 laser dominated the thought behind the original patent [1] and the first descriptions of metal vapor lasers (MVL) [2,3,4]. Efficient, powerful and continuous, it was the star of those days. It became very desirable to have something like it in shorter, visible wavelengths.

In the CO_2 laser, operation has two distinct agents: The upper level is excited by collisions with electrons, while the lower laser level (LLL) is de-populated by collisions with helium atoms. This lower level is populated mainly by radiative decay of the upper laser level (LLL). It was supposed that metal vapour lasers would work in a similar way. In the words of Walter [2]:

"Excitation would be provided by electron-atom collisions, while inelastic atom-atom collisions would relax the lower levels. If the requirement is dropped that the excitation and relaxation proceed simultaneously, then a class of efficient pulsed lasers becomes possible which is cyclic in operations. Collisional excitation and relaxation occur sequentially rather than simultaneously. Therefore each process may be considered separately and optimized virtually independently of others".

Another "plasma parameter" that was considered by the pioneers was the gas temperature. This was regarded as constant across the laser tube - a reasonable assumption in externally heated, low repetition rate lasers such as theirs. Again the assumption was that only the lower laser level is affected by the gas temperature.

Looking back a generation later, it is amazing how influential these assumptions were, and for how long they guided the CVL research. They set the "narrative" of the MVL stories, and even an ever growing number of other "plasma parameters" only slowly caused them to fade away. This paper, however, is not intended to give a comprehensive account of the evolution of concepts behind the operation of MVLs, but rather simply to describe NRCN's involvement in this evolution.

Some early works took this simple model literally. One immediate conclusion that can be deduced from this model is that the metastable decay would govern the maximum laser power and its highest repetition rate [5,6].

It had seemed plausible to scale up the laser energy by increasing the metal density in the plasma tube [7]. Estimates had been made [8] that up to 85 W/cm could be extracted from copper at 511 nm, provided that helium would be the buffer gas. Neon was predicted to give less than 1/4 of that power, the limiting mechanism being the thermal population of the LLL.

C. E. Little and N. V. Sabotinov (eds.), Pulsed Metal Vapour Lasers, 87–99.

A major milestone in MVL progress was the development of a self heated, high power, high repetition rate device as demonstrated by Isaev, Kazaryan and Petrash in 1973 [9].

In the same year, a new class of MVL's emerged: the metal halide lasers [10]. Here two or more discharge pulses were used to excite lasing in He-MX admixtures - the first to dissociate the halide molecules into metal (M) and halide (X) atoms, and the rest to excite the laser and keep the halide dissociated.

An optimum time interval between the pulses was found to exist. The explanation? - "this time interval is thought to be the time required for the depletion of the LLL following dissociation" [10].

For myself, the first sign that things may not be that simple was seen in the Shukhtin *et al.* paper [11] of 1975.

Using the hook method, they had examined Cu ground state kinetics in the pulsed dissociation of CuBr. They suggested (although by extrapolation) that Cu $^2S_{1/2}$ was built up slowly in the after-glow of the dissociation pulse and reached peak population about $200\mu s$ after trigger. It was also suggested that very high $(7\times10^{16}\,\mathrm{cm}^{-3})$ density of Cu $^2S_{1/2}$ could be generated in that way, almost metastable free — and in a low $(700\,^0C)$ gas temperature. It all meant that there might be a different reason for the minimum and optimum delays - other than the LLL population.

Another mystery that the paper emphasized was the large quantities of Cu $^2S_{1/2}$ at $700\,^oC$ - because at that time it was clear that CuBr would not lase at that temperature [10, 12, 13, 14].*

The possibility of less successful dissociation at higher temperatures was now ruled out.

The attractiveness of copper halide lasers was caused not only by the possibility of a larger variety of laser building materials due to lower temperatures, but also by the promise of lower gas temperature - high density of ground state Cu with negligible metastable population.

We therefore tried to study carefully the kinetics of the $^2S_{1/2}$ and $^2D_{3/2}, ^2D_{5/2}$ in CuX lasers. Previous measurements had suffered from various problems - insufficient temporal resolution [11], total absorption, and fluorescence in short delay times [15]. We overcame total absorption by measuring the transmittance transversely to the tube axis [16,17]. Typical results are shown in Fig. 1.

Several conclusions were drawn from these measurements:

1) Ground state and metastable copper are created and their populations are built up together;

2) The decay rate of the metastable states is much faster then the decay rate of the ground state - meaning that diffusion to the walls is not the only quenching mechanism..

3) There must be other important "plasma parameters" that influence the laser operation other than the LLL.

* *The only exception is the Winiaczyk and Krause experiments* [48].

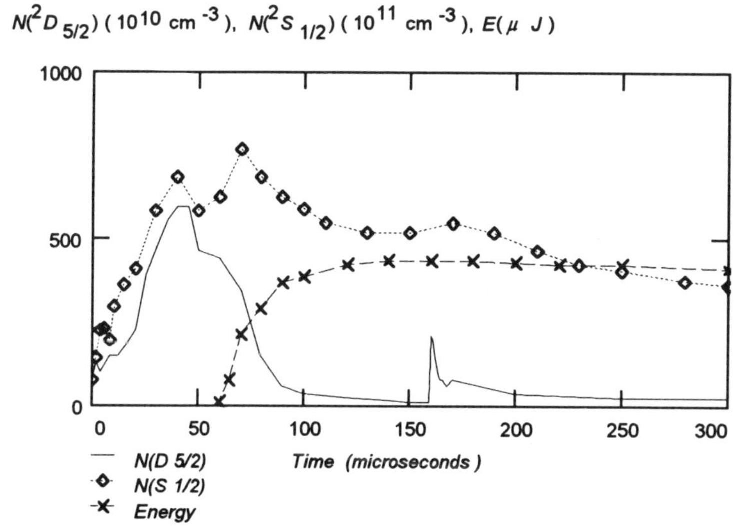

Fig 1. $^2S_{1/2}$, $^2D_{5/2}$ densities and laser energy as function of the delay from the dissociation pulse. The $^2D_{5/2}$ curve also shows the population rise due to an excitation pulse at the optimum delay - 160 μs.

Soon afterwards Vetter and Nerheim [18] showed that the addition of a small amount of HCl to a CuCl laser has significant effect on both minimum delay and maximum power. That was followed by Tobin's observation that the decay rates of metastables in CuBr lasers "are increased significantly by decomposition products which arise from the presence of cupric oxybromide as an impurity" caused by a reaction which contributes HBr to the plasma [19].

These two papers gave the first clues to the successful HyBrID laser which was yet to come, as well as demonstrated the importance of the gas composition, while still concentrating on the traditional "pre-pulse metastable" concept.

However, these measurements did make it clear that metastable quenching was a volumetric process rather than the result of diffusion to the walls. The small quantities of impurities could not have so dramatically shortened the diffusion rate; furthermore, each ^2D level had its own decay rate, contrary to what the diffusion theory would have implied. The conclusion was that quenching had to be made by an internal process.

An important paper published at that time helped us to depart further from the "two parameters" approach. Eletski *et al.* [20] pointed out the dynamics of the population of the ULL, which are not only populated by electron collisions but also de-populated by them. This sets an upper limit on the available energy density of an MVL.

By requiring that:

1) The electron collision de-excitation rate of the ULL would not exceed the de-excitation by spontaneous emission, the authors obtained the optimum electron density n_e.

$$\overline{n_e} \cdot k_1 \approx \frac{1}{\tau_{12}} \Rightarrow \overline{n_e} = \frac{1}{\tau_{12} k_1}, \tag{1}$$

where τ_{12} is the spontaneous decay time of the laser transition, and $k_1 = k_1' + k_1''$ is the electron de-excitation rate to the LLL (k_1') and to upper levels (k_1').

2) n_e would stay close to its optimum value during the laser pulse, they set a limit to the ionization rate of copper during that time.

$$\frac{d\bar{n}_e}{dt} = \bar{n}_e n k_{ion}, \tag{2}$$

where n is the ground state density and k_{ion} is the ionization rate. For a given k_{ion} there should be an *optimum ground state density*

$$\bar{n} \approx \frac{1}{\tau_{12} k_{ion}} \tag{3}$$

and therefore a limiting energy density available from the laser

$$\overline{E} \approx \bar{n}_e \bar{n} k_{01} \hbar\omega\tau_{12} \approx \frac{\hbar}{\tau_{12}} \frac{\omega}{k_{ion}} \frac{k_{01}}{k_1}. \tag{4}$$

Here k_{01} is the excitation rate of the ULL, and $\hbar\omega$ denotes the energy of a laser photon.

For a CVL they had obtained

$$10\mu J/cm^3 \leq \overline{E} \leq 100\mu J/cm^3, \tag{5}$$

which is a practical, even an optimistic estimate. This was the first time that important new plasma parameters, n_e and n, were emphasized. Fig. 2 shows the competing processes.

During the mid-seventies, a series of inspiring papers was published by a group at the Institute for Atmospheric Optics in Tomsk, headed by P.A. Bokhan. It seemed that, under favorable conditions, electrons can rapidly excite substantial population into metastable levels. Klimkin discovered lasing in ytterbium vapour [21], in a transition whose ULL were not optically connected to the ground state, but the LLL were connected (see Fig. 3).

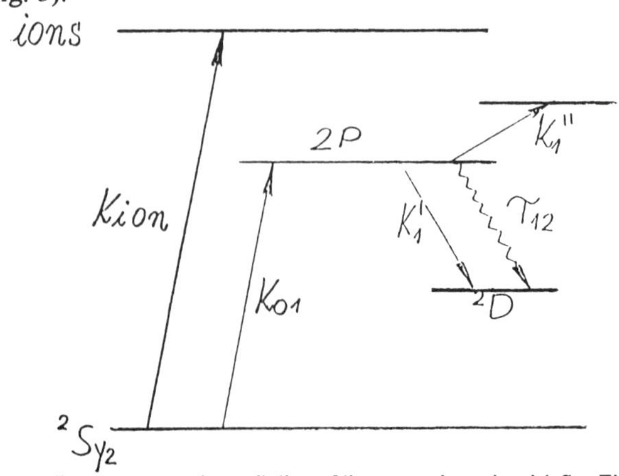

Fig 2. Competing processes in excitation of the upper laser level (after Eletskii [20]).

The recognition that electrons can excite and de-excite both laser levels, and that electrons are as important in the afterglow as during the current pulse, was slowly being developed.

In 1979 Solomonov stated [22]:

"The suggestion is put forward that it is mainly the recombination rate of the electrons and not the population of the lower metastable active level which exerts an influence on the maximum pulse repetition frequency".

But he referred also to another aspect. The green and IR transitions in the Mn vapour laser have separate ULLs, but share the same LLLs. And yet, in dual-pulse experiments the IR lines have shorter delays; which suggests that what is recovering gradually as the delay increases is the electron temperature.

The author also added a remark that would be manifested clearly many years later: "It was observed that as the second excitation pulse approaches the first, its current and voltage phases become closer and practically coincide for a delay time of less than $30\mu s$. This is evidence for a high electrical conductivity of the decaying plasma, and consequently that a high free-electron density remains after the end of the first current pulse, since the electrical conductivity is proportional to the electron density".

The lower plasma impedance would overload the exciter, which would result in a lower electrical field, cooler excitation electrons and loss of green power relative to the IR power. The parameter of plasma conductivity had arrived to stay.

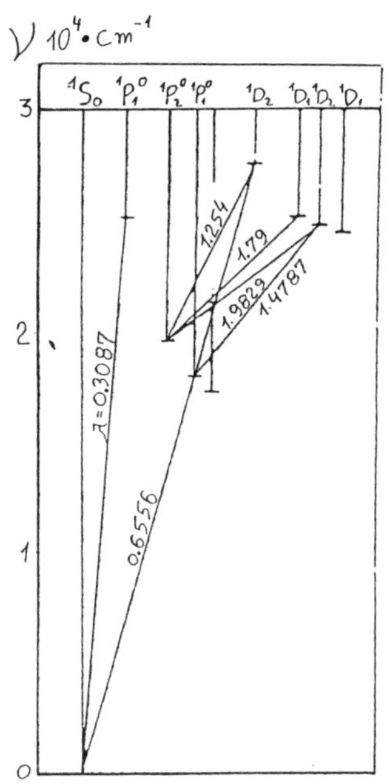

Fig 3. Ytterbium energy levels (after Klimkin [21]).

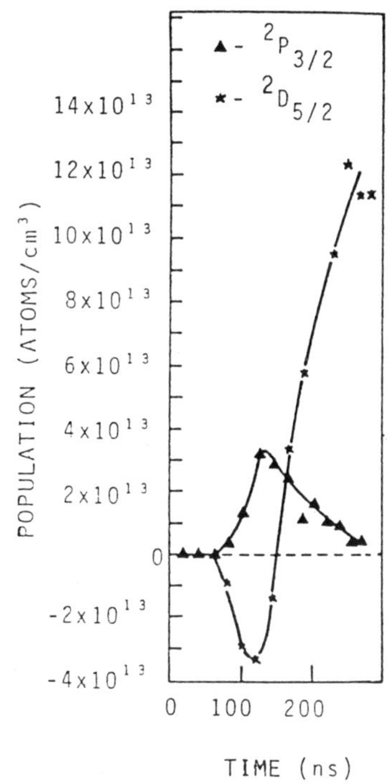

Fig 4. Kinetics of a CVL laser levels.

The seventies saw great technological developments in self-heated CVLs [9,23,24]. High neon pressure and large bore lasers appeared [25,26]. On one hand, measuring the laser output and speculating about its internal kinetics had not seemed adequate any longer. On the other hand, the absorption method [6] or the fluorescence method [27] had not yet been able to give satisfactory results - among other things by inability to include plasma parameters in the broadening mechanisms. Their limitations had tempted people to extrapolate, resulting in rather expensive mistakes [7,28].

An improved hook spectroscopy system was demonstrated around that time by Miyazaki and Fukuda [29]. The hook method is insensitive to the fine details of both line broadening and total absorption. The light source in the new system was a wide-band N_2 pumped dye laser that couples high brightness with nanosecond resolution (a wide band dye laser is a dye laser without the telescope and the etalon - simple indeed). At that time we were already using a video camera to monitor, record, and analyze spectra of dye lasers [30]. So it actually required only a few days to get our first “hook” spectrograms on a laser under typical operating conditions.

With that early system we were able to monitor the kinetics of the ULL, LLL, some “displaced” copper levels, neon excited states - and sodium impurities... at least on the tube axis. Typical results are shown in Fig 4. It is seen that the early decay of the ^2P states, for a density ratio of over a decade, is very closely exponential. Hence in this interval the decay is dominated by a single process. Since during the decay the electrical field has already vanished, the free electrons cool down rapidly by elastic collision (populating the “displaced” levels and the ions) and the most plausible process is a common electronic and photonic decay:

$$\frac{dN_p}{dt} = -N_p(Sn_e + A), \tag{6}$$

where S is the electronic transition rate. Since the measurement gives us dN_p/dt and N_p, we can calculate $Sn_e + A = 10^7$ for both the $^2P_{3/2}$ and $^2P_{1/2}$ levels.

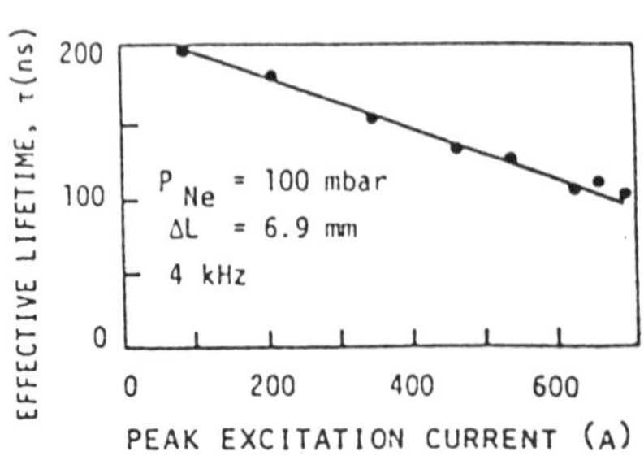

Fig 5. Effective lifetime of the $^2P_{1/2}$ level populati-on as a function of the peak excitation current [31].

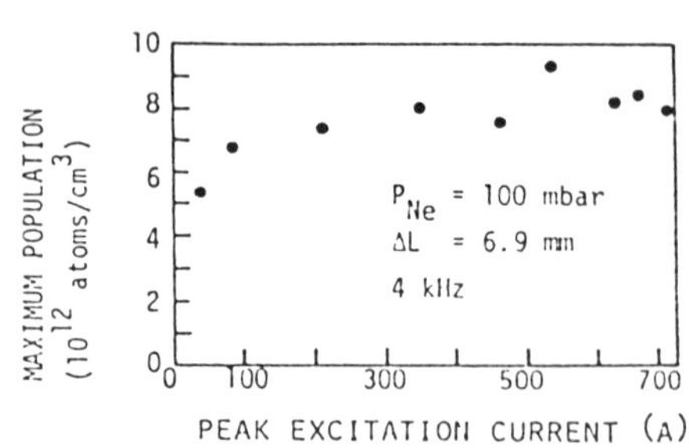

Fig 6. Maximum $^2P_{1/2}$ level population as a function of the peak excitation current [31].

Because of radiation trapping, $A \approx 2 \times 10^6$ for both transitions, and we then have

$$S \cdot n_e = 8 \times 10^6 . \tag{7}$$

Therefore radiative transitions are negligible, in comparison with collisional transitions.

Other measurements that we performed on a similar laser are shown in Fig. 5 and Fig. 6 [31]. They indicate that the population of the ULL saturates at a rather low excitation current, and that the effective lifetime of the ULL is inversely proportional to the current. We shall return later to these results.

An inspiring paper by Batenin *et al.* that had reached us in those days was a paper on the electron density in a CVL [32]. It showed that an electron density as high as $10^{13} cm^{-3}$ remains in the gas prior to the next pulse, which implies that the laser tube under typical conditions contains a conducting plasma, and that there is no room to consider a "gas breakdown" as, for example, in pulsed TEA CO_2 lasers. It makes the design of an appropriate exciter for MVL more difficult than for lasers with gas breakdown, which have higher initial impedance and can rely on the discharge tube itself as the fast switch. It also meant that measurements by the "double pulse" method are misleading.

We too had demonstrated the last point with a pre-ionized oven-heated laser [33]. That experiment also helped to explain the formation of homogeneous longitudinal discharge in a large bore MVL.

It might look odd today, but as late as 1980 we had not found in the literature any reference to ground state depletion. The first computer model of a CVL, by Leonard [34], was a transposition of the N_2 laser model into CVL terms [35]. That model "assumed constant ground state density, justified by the results".

However, measuring the ground state kinetics in a CVL required a UV tunable laser which was unavailable to us at that time. Instead, we fed our tube with barium and ran it at a lower repetition rate in order to reduce its temperature. Barium has a lower ionization potential then copper, and all the energy level intervals are "squeezed". As such all the important transitions are in the visible (with large f values - which enhances the accuracy of the hook method) so that it leant itself to be the subject of a ground state kinetics study.

The results [36] are presented in Fig 7. They indicate a large ground state depletion, between 72% to 85%. At least 5×10^{14} cm^{-3} atoms are removed from the ground state by each electrical pulse. Only $1.5 \times 10^{13} cm^{-3}$ ($\approx$ 3%) of them produce laser photons. When modeling the laser, such a process cannot be ignored; it might be as important as the LLL pre-pulse population.

This conclusion was further substantiated in 1982 (and published years later) by measurements of CVL ground state depletion [37]. The most interesting finding here was that a fraction of the copper ground state population has not yet recovered back to its initial level when the next pulse arrives, i.e., the actual ground state population is less than what is determined by the thermal conditions. The "missing" atoms are clearly ionized. (Ironically, the only species which are "metastable" in the laser - meaning that their population is diffusion controlled - are the ions).

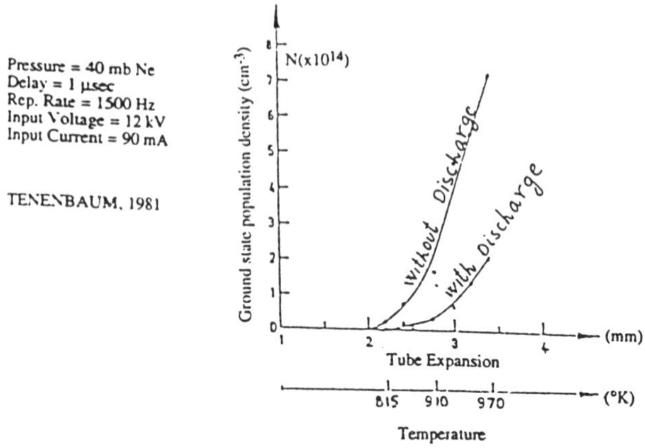

Fig 7. Barium ground state concentration as a function of temperature with and without a discharge current in the laser tube. The 1 μs delay refers to the measurement with a discharge [36].

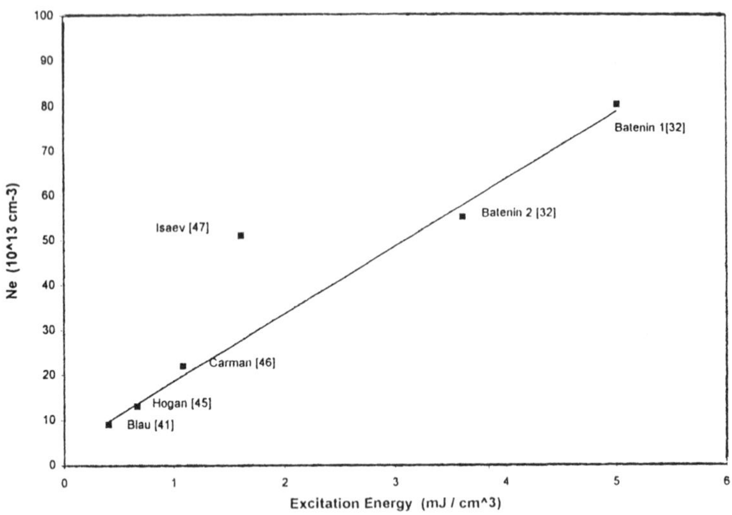

Fig 8. Electron density as a function of excitation energy.

Our "window of opportunity" for that research was closed in 1980, before we were able to use the tools we had developed to perform a comprehensive study of the CVL.

Luckily, in 1989 Pinhas Blau joined the CVL group and undertook the task of measuring the electron density and gas temperature - this time on a 100 W , 8 cm laser tube. Other parameters of a similar laser had been published elsewhere [38,39], and we hoped to add complementary information.

Blau used the emission of trace hydrogen present in the laser's plasma and measured the broadening of its Balmer lines. By using careful de-convolution and correlating different lines, the Stark and Doppler broadening could be separated and the electron density and gas temperature obtained [40,41].

I still remember my surprise on the first day Pinhas ran his experiment. The line broadening was so small that our spectrograph could not resolve it. Only later, by using a high quality scanning etalon, could its width be measured. The reason for this surprise is explained by Fig 8. It shows the peak electron density as a function of the excitation energy, as measured by different authors over the years. Basically, it shows that the average energy required to ionize a copper atom in a CVL is 30eV.

By the time Pinhas began his work the only published results were those of Batenin et al. [32], and I failed to notice the large difference between the excitation energy density in the two cases.

It is plausible that at least part of the large spread between the results of different measurements of the gas temperature in CVLs is due to the same reason: the spread in excitation energy density. Blau's gas temperature results match very closely Molander's results on a similar laser taken by a different method [39], the two independent sets of measurements thus supporting each other.

Blau then conducted the following experiment which illuminated the interaction between the laser and its excitation circuit. The laser was operated at normal excitation energy, but at repetition rates lower than, and up to, the nominal 5.5 kHz. The wall temperature and the gas temperature rose gradually with the repetition rate. At each repetition rate the maximum and minimum on-axis electron density was measured. The results are shown in Fig. 9.

As the interval between successive pulses gets shorter, there is less time for ion-electron recombination, so that the minimum density increases. This, in turn, has a significant effect on the tube resistance, which is estimated to drop from 1.7 Ω at 1kHz to 0.9 Ω at 5.5 kHz [41]. Since the laser head inductance is ≈ 240 nH (≈ 5 Ω for our 150 ns current pulse), this means almost a factor of three reduction in the electrical field during excitation! This results, among other things, in a lower ionization increment. However, under the experimental conditions of higher repetition rates, copper dominates as the ionization agent, and, having a considerably lower ionization

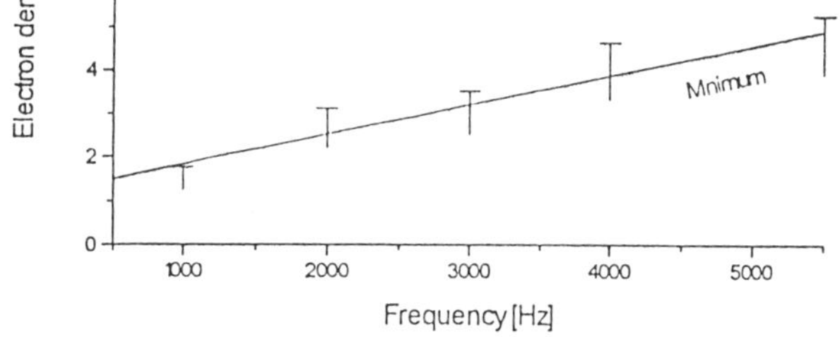

Fig 9. Minimum and maximum electron density, as a function of the repetition rate (the temperature of the tube rises as the repetition rate rises) (Blau [41]).

potential than neon, peak electron density rises again.

The resistive part of the tube impedance is given by:

$$\rho = \frac{1}{\sigma} = \frac{m_e \nu}{n_e e^2}, \tag{8}$$

where m_e is the electrons' mass, ν is the electrons' elastic collision frequency (mainly with neon atoms), and e is the electron's charge. ν is given by $\nu = N_g \sigma_m \langle V \rangle$, where N_g is the neon density, σ_m is the momentum transfer cross-section between the electrons and neon atoms, and $\langle V \rangle$ is the average electron velocity. We can express N_g as

$$N_g \cong 3.5 \times 10^{16} \frac{273}{T_g} P_{Ne} \qquad (cm^{-3}), \tag{9}$$

where P_{Ne} is the neon pressure. Combining all this together we get:

$$\sigma(r) = \frac{n_e}{m_e} \frac{e^2}{3.5 \times 10^{16}} \frac{T_g(r)}{273} \frac{1}{P_{Ne}} \frac{1}{\sigma_m \langle V \rangle}, \tag{10}$$

so that

$$\Sigma = \int_0^R \sigma(r) \cdot 2\pi r dr, \tag{11}$$

where Σ is the tube conductance. Hence the tube resistance is

$$R = kLP_{Ne} \frac{1}{\int_0^R n_e(r) T_g(r) r dr}, \tag{12}$$

where k is a constant (which depends on T_e). The measured $N_e(r)$ and $T_g(r)$ are shown in Fig 10.

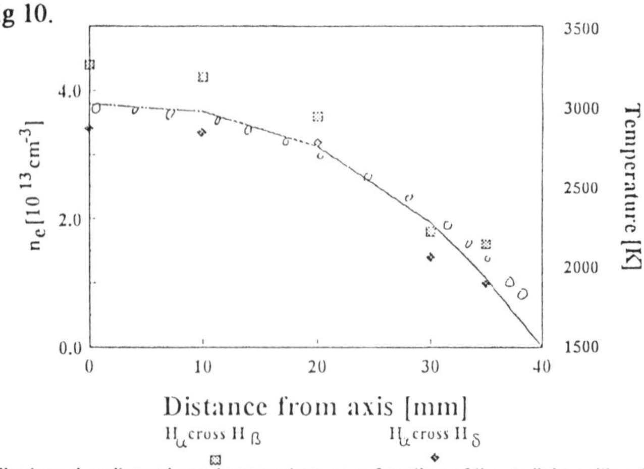

Fig 10. Electron density and gas temperature as a function of the radial position (after Blau [41]).

If we go one step further and assume that n_e and T_g radial distribution are independent of their peak value, e.g.,

$$n_e = n_e(0) f_1(r) \ , \qquad T_g(r) = T_g(0) f_2(r) \ , \tag{13}$$

we shall get:

$$R = KL \frac{P_{Ne}}{T_g(0) n_e(0)} . \tag{14}$$

We can conclude that the resistance of the tube will increase linearly with its length and the neon pressure, and inversely with the gas temperature and the electron density.

For large bore tubes, where their resistive component can be smaller than the reactive ones and their characteristic impedance, lengthening the tube and increasing the neon pressure would result in better matching and higher efficiency.

Increasing the excitation energy, or the repetition rate with a fixed excitation energy, even under constant wall temperature, would result in higher n_e and T_g , lower resistance , and eventually lower electron temperature through excitation.

Now let us consider two lasers: one CVL and the other HyBrID, otherwise identical. Since the presence of HBr or its products in the discharge of the HyBrID laser reduces n_e significantly [42,43], and T_g is also lower compared to those at the CVL, the HyBrID would match better at high repetition rate and would produce more efficiently more power.

References

1. Gould, G., Bennet, W.R., and Walter, W.T. (1965) Low-level laser with cyclic excitation and relaxation, US Patent 3,576,500.
2. Walter, W.T. (1966) High power copper laser final report, TRG-082-FR, TRG Inc, A subsid. of Control Data Corp., Mellville, N. Y.
3. Gould, G. (1965) Collision lasers, *Appl. Opt. Suppl.* **2**, 59-67.
4. Petrash, G.G. (1972) Pulsed gas-discharge lasers, *Soviet Physics Uspekhi* **14**, 747-765.
5. Bokhan, P.A. and Solomonov, V.I. (1974) Mechanism of laser action in copper vapor, *Sov. J. Quant. Electr.* **3**, 481.
6. Miller, J.L. and Kan, T. (1979) Metastable decay rates in a Cu-Metal-Vapor laser, *J. Appl. Phys.* **50**, 3849-3851.
7. Chimenti, R.J. (1974) Heat pipe copper vapor laser, Report AD/A 005, Exxon Research and Engineering Co., Linden, NY.
8. Isaev, A.A., Kazaryan, M.A., and Petrash, G.G. (1974) Possibility of generation of high average power in the visible parts of the spectrum, *Sov. J. Quant. Electr.* **3**, 521-523.
9. Isaev, A.A., Kazaryan, M.A., and Petrash, G.G. (1973) "Copper vapor pulsed laser with a repetition frequency of 10 kHz," *Opt. Spectrosc.* **35**, 307-308.
10. Che Jen Chen, Nerheim, N.M., and Russell, G.R. (1979) "Double discharge copper vapor laser with copper chloride gas a lasant" *Appl. Phys. Lett.* **23**, 514-515.
11. Shukhtin, A.M., Mishakov, V.G., Fedotov G.A., and Ganeev, A.A. (1975) "Interference method for observing the dissociation of copper halide molecules in a pulsed discharge" *Opt. Spectrosc.* **39**, 444.
12. Gabay, S. , Smilanski, I., Levin, L.A., and Erez, G. (1977) "Comparison of CuCl, CuBr and CuI as lasants for copper vapor lasers", *IEEE J. of Quant. Electr.*, 364-368.
13. Shukhtin, A.M., Fedotov, G.A., and Mishahov, V.G. (1975) "Lasing with CuI lines using copper bromide vapor" *Opt. Spec.* **39**, 681.
14. Piper, J.A. and Webb, C.E. (1975) "A TE Copper Iodide Laser," *IEEE J. Quant. Electr.* **11**, 95D .
15. Liu, C.S., Feldman, P.W., Pack, J.L., and Weaver, L.A. (1977) "Kinetic Processes in Continuously Pulsed Copper Halide Lasers" *IEEE J. Quant. Electr.* **13**, 744-751.
16. Tenenbaum, J., Smilanski, I., Gabay, S., Erez, G., and Levin, L.A. (1978) "Time dependence of copper-atom concentration in Ground and metastable states in pulsed CuCl laser", *J. Appl. Phys.* **49**, 2662-2665.

17. Tenenbaum, J., Smilanski, I., Gabay, S., Erez, G., Levin, L.A., Katriel, K., and Speizer I. (1978) "Buffer gas effect on Ground and Metastable Population in a Pulsed CuBr Laser" *IEEE J. Quant. Electr.* **14**, 680-685.
18. Vetter, A.A. and Nerheim, N.M. (1977) "Addition of HCl to the double-pulse copper chloride laser", *App. Phys. Lett.* **30** , 465-407 .
19. Tobin, R.C. (1980) "Rapid differential decay of metastable populations in a copper halide laser" *Optics Comm.* **32**, 325-330.
20. Eletskii, A.V., Zemtsov, Yu.K., Rodin, A.V. and Starostin, A.N. (1975) "Optimum characteristics of a laser with high-Pressure metal vapors", *Sov. Phys. Dokl.* **20**, 42-44.
21. Klimkin, V.N. (1975) "Investigation of an ytterbium laser," *Sov. J. Quant. Electr.* **6** 326-329.
22. Solomonov, V.I. (1979) "Some characteristics of self-terminating manganese and lead vapor lasers" *Sov. J. of Quant. Electr.* **9**, 736-739.
23. Anderson, R.S., Springer, L., Bricks, B.G. and Karras, T.W. (1975) "A discharge heated copper-vapor laser," *IEEE J. Quant. Electr.* **11**, 172-174.
24. Burmakin, V.A., Evtyunin, A.N., Lesnoi, M.A. and Bylkin, V.I. (1978) "Long-life sealed copper-vapor laser" *Sov. J. Quant. Electr.* **8**, 574-576.
25. Smilanski, I., Erez, G., Kerman, A., Levin, L.A. and Tenenbaum, J. (1978) "Longitudinal high power high-Ne-pressure copper vapor laser," *J. Opt. Soc. Am* **68** , 713-714.
26. Burmakin, V.A., Evtyunin A.N., and Lesnoi, M.A. (1979) "Sealed Copper-Vapor laser operating at atmosphere pressure of Ne buffer gas," *Sov. J. Quant. Electr.* **9** , 939-940.
27. Chimenti, R.J. (1972) *The Copper-Vapor Laser*, Ph.D. Dissertation, Polytechnic Institute of Brooklyn, NY.
28. Kan, T., Bull, D., Schmit, E. and Hill, J. (1979) "Annular discharge copper vapor laser" *Appl. Phys. Lett.* **35**, 676-677.
29. Miyazaki, K. and Fukuda, K. (1977) "Time resolve population measurement with N_2-laser-pumped dye lasers" *J. Phys. D. Appl. Phys.* **10**, 1905-1910.
30. Lavi, S., Miron, E., and Smilanski, I. (1978) "Spectral distribution measurement of single laser pulses" *Opt. Commun.* **27** , 117-120.
31. Tenenbaum, J., Smilanski, I., Lavi, S., Levin, L.A., and Erez, G. (1981) "Kinetic investigation of the upper laser levels of copper vapor laser" *Opt. Commun.* **36** , 391-394.
32. Batenin, V.M., Burmakin, V.A., Vokhmin, P.A., Evtyunin, A.I., Klimorski, I.I., Lesnoi, M.A. and Selezenevii, L.A. (1977) "Time dependence of the electron density in a copper vapor laser" *Sov. J. Quant. Electr.* 7, 891-893.
33. Gabay, S. and Smilanski, I. (1980) "Effect of pre-ionization on a copper vapor laser" *IEEE J. Quant. Electr.* **16** , 598-601.
34. Leonard, D.A.(1967) "A theoretical description of the 5106A pulsed copper vapor laser," *IEEE J. Quant. Electr.* **3**, 380-381.
35. Gerry, E.T. (1965) "Pulsed-molecular-nitrogen laser theory, " *Appl. Phys. Lett.* **7** , pp. 6-8.
36. Tenenbaum, J., Smilanski, I. and Levin, L.A. (1981) "Investigation of the kinetics of a metal vapor laser by an interferometric method," *Optics Commun.* **36**, 395-398.
37. Smilanski, I., Gabay, S. and Karni, Z. (1989) "Ground state depletion in copper vapor lasers," *Proc. SPIE* **1041** , 34-39.
38. Kushner, M..J. and Warner, B.E. (1983) "Large-bore copper vapor lasers: Kinetics and scaling issues," *J. Appl. Phys.* **54**, 2970-2982.
39. Molander, W.A. (1989) "Measurement of ground state copper density using hook spectroscopy," *Proc. SPIE* **1041**, 11-18.
40. Blau, P., Smilanski, I., and Rosenwaks, S. (1992) "Simultaneous time-averaged measurements of gas temperature and electron density in a copper-vapor laser using hydrogen emission spectroscopy" *J. Appl. Phys.* **72**, 849-854.
41. Blau, P. (1994) "Investigation of Copper Vapor Laser Plasma Parameters and Electrical Discharge", Ph.D. Thesis, Ben-Gurion University, Beer-Sheva.
42. Jones, D.R., Halliwell, S.N., and Little, C.E. (1994) Influence of remanent electron density on the performance of copper HyBrID lasers, *Optics Comm.* **111**, 394-402.
43. Whyte, C.G., Little, C.E., Hogan, G.P., and Webb, C.E. (1995) Temporal and spatially resolved measurements of electron density in a copper HyBrID laser, CTHG4, CLEO, Baltimore, MA.
44. Gabay, S., Blau, P., Lando, M., Druckmann, I., Horwitz, Z., Yfrah, Y., Hen, I., and Smilanski, I. (1991) Stabilization of high-power copper vapour laser, *Opt. and Quant. Electr.* **23**, 5485-5492.
45. Hogan, G. (1993) A Study of the Kinetics of Copper Vapour Lasers, Ph.D. Thesis, Oxford.
46. Carman, R.J., Brown, D.J., and Piper, J.A. (1994) A self-consistent model for the discharge kinetics in a high-repetition-rate copper vapour laser, *IEEE J. Quant. Electr.* **30**, 1876-1895.

47. Isaev, A.A. and Petrash, G.G. (1993) Kinetics of excitation and physical processes in active media of copper vapour and copper bromide vapour lasers, in *Metal Vapour Lasers and Their Applications: CIS Selected Papers*, G.G. Petrash (Ed.), *Proc. SPIE* **2110**, 2-45.
48. Winiaczyk, W. and Krause, L. (1982) Temperature effects in the double-pulsed CuCl laser, *Appl. Opt.* **15**, 2659-2660.

HIGH-POWER COPPER VAPOUR LASERS AND APPLICATIONS

J.J. CHANG, B.E.WARNER, C.D. BOLEY AND E.P. DRAGON
Lawrence Livermore National Laboratory
P.O. Box 5508, M/S L-463, Livermore, CA, 94551, USA

Abstract

The expanded applications of copper vapor lasers have prompted increased demand for higher power and better beam quality. This paper reports recent progress in laser power scaling, MOPA operation, and beam quality improvement, along with applications in precision laser machining and second harmonic generation. Issues such as gas heating, radial delay, discharge instability, and window heating are also discussed.

1. Introduction

The applications of high-radiance, short-pulse copper vapor lasers (CVLs) in second harmonic generation [1,2], precision laser machining [3.4], and laser beacons [5] have prompted increased demand for diffraction-limited high-power CVLs. The copper vapor lasers developed for atomic vapor laser isotope separation (AVLIS) are not quite suitable for these applications mainly because of their large beam divergence. Furthermore, the need of a simple multi-hundred-watts master-oscillator-power-amplifier (MOPA) system for most industrial applications makes the power scaling of a CVL amplifier more important. In the past few years, LLNL has developed several new technologies that have successfully expanded high power CVLs to non-AVLIS applications. This paper reports our works on CVL power scaling and beam quality improvement. High-power CVL applications in precision laser machining and second harmonic generation will also be discussed.

2. Near-Diffraction-Limited Oscillator

A near-diffraction-limited CVL oscillator is essential to achieve a low-divergence high-power output through amplification. A conventional CVL unstable resonator is not suitable because only the latter part of the output pulse reaches near-diffraction-limited beam quality [6,7]. With a self-filtering unstable resonator [8], the resonator beam quality generally improves, but the front of the output pulse still contains a large amount of ASE that shortens the pulse and lowers its usable low-divergence laser light for efficient amplifier extraction. To circumvent these difficulties, injection locking techniques [9,10,11] have been applied to obtain low divergence output for the entire duration of the laser pulse.

C. E. Little and N. V. Sabotinov (eds.), Pulsed Metal Vapour Lasers, 101–112.

Figure 1 illustrates an injection controlled oscillator used to achieved near-diffraction-limited output for high-power amplifier extraction. A 5-10 w low-divergence injection beam is generated with a 2-cm bore master oscillator (MO) that has a confocal negative-branch unstable resonator with magnification 100. Its beam quality is 2 times diffraction limited (XDL) based on a power-in-the-bucket measurement [7], which is similar to a M^2 knife-edge method. Note that a self-filtering unstable resonator is not appropriate for MO because of the possibility of injection feedback that may generate parasitic oscillation and damage the intracavity pinhole. The 4-cm bore injection controlled oscillator (ICO) is a self-imaging unstable resonator with magnification 15. The cavity relays the feedback aperture to the laser output on the scraper mirror that effectively reduces the diffraction effect of the hard aperture and substantially improves ICO output beam quality. A time delay is applied between the discharges of MO and the ICO such that the ICO gain is seeded only by the latter part of the MO output that has a better beam quality; the ICO is essentially a free-running oscillator after the injection beam terminates itself.

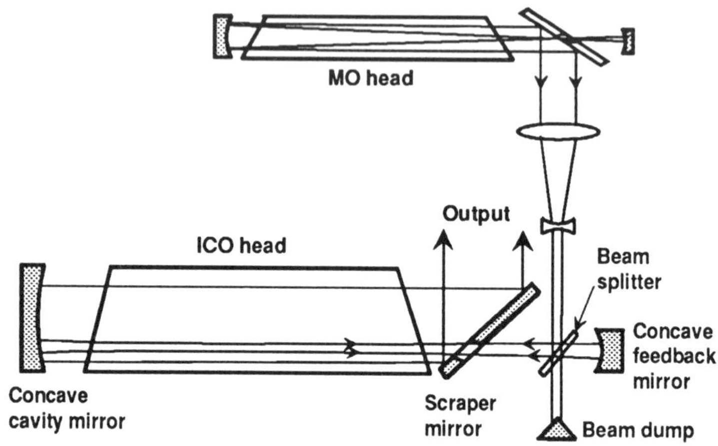

Fig. 1. A schematic layout of the injection controlled oscillator.

The demagnified MO beam is injected into the ICO cavity via a beam splitter inside the ICO cavity, as shown in Fig. 1. An adjoint-coupled injection scheme is applied to match the injection signal with the ICO adjoint mode (i.e., convergent mode). In this scheme the collimated injection beam, upon reflecting from the concave cavity mirror, first travels inward but then gradually spreads out because of Fresnel diffraction. This process speeds up cavity mode formation, and the conversion of the injection signal to an ICO cavity mode becomes possible before the ICO laser gain starts. To avoid strong pre-pulse plasma absorption and to take the advantage of CVL radial delay, an off-axis unstable resonator is applied to accommodate an off-axis injection as illustrated in Fig. 1. In this scheme, the seeding beam is injected into the cavity along the wall of the plasma tube where the laser gain starts earlier and the plasma absorption is weaker. To obtain a better seeding signal, the 2-XDL injection beam is improved through a spatial filtering technique that accommodates the injection and ICO cavity configurations. In this method, the large concave cavity mirror and the small feedback aperture on the scraper serve as a spatial filtering system. The size of the injection beam is controlled such that its diffraction-limited spot size, focused by the large concave mirror, is the same as the hole size on the scraper, which is a spatial filter and a cavity feedback aperture. This technique effectively improves the seeding signal that leads to an ICO beam quality better than that of MO. Figure 2 illustrates pictures of two reverse-shearing fringes of the ICO output with and without injection control. The improvement of beam spatial coherence with injection is evident, and the uniformity of fringe visibility at various shear is clearly shown in Fig. 2b. A power-in-the-bucket measurement indicated that ~84% of ICO output is within 1.3 XDL beam divergence.

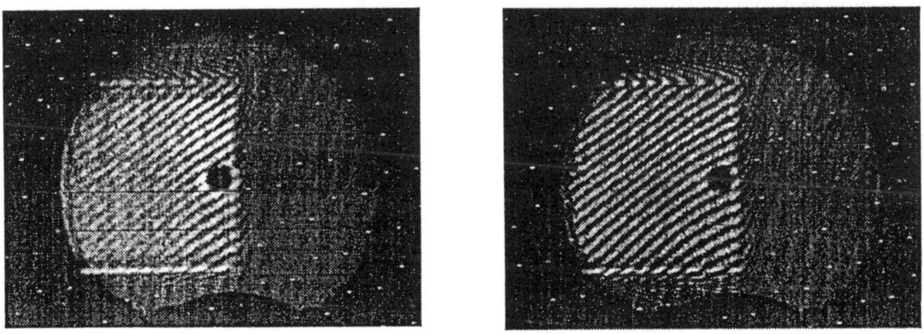

(a) (b)

Fig. 2. Shear fringes of the 4-cm ICO output beam measured with lateral reverse shearing techniques, (a) without injection, (b) with injection. Due to limited size of shearing optics, only half of the beam was sheared.

3. High-Power Amplifiers

The expanded applications of high-power CVLs have prompted considerable effort directed towards improving their performance. Volumetric scaling for higher laser power can be attempted either by lengthening the discharge tube or by increasing the tube diameter. Length scaling, however, is limited by high voltage constraints and pulsed power technology. Attempts to scale in diameter are limited by excessive central gas heating, which degrades the laser efficiency by thermally populating the metastable lower laser levels, $^2D_{3/2}$ and $^2D_{5/2}$ states [12]. In addition, the decreased central gas density lowers the plasma impedance, thus lengthening the time for the discharge field to diffuse to the center. This leads to insufficient axial pumping and a nonuniform beam profile. The increased input power requirement also puts significant thermal load on discharge electrodes that worsens discharge instability, resulting in large discharge jitters, severe window contamination, and short electrode lifetime. In addition, the large window heating resulting from either window contamination or tube IR radiation introduces severe amplifier wavefront distortion that limits applications. Recently, considerable progress has been made in overcoming these issues. .

3.1. GAS TEMPERATURE AND RADIATIVE COOLING

A new method has been developed recently to reduce gas heating of large-bore CVLs and to improve their performance through radiative cooling [13]. In this approach, several segmented plates (septa) are placed along the length of the tube, as illustrated in Fig. 4. The efficient radiative heat exchange between the side wall and the septa effectively lowers the septa temperature close to the tube temperature. The gas in the axial region is then in turn cooled through heat conduction to the septum plates.

In a CVL, the gas temperature changes little during a pulse because of the large thermal inertia of the gas. The gas temperature distribution *T(r)* can thus be described by the steady state heat conduction equation,

$$\nabla \cdot (k \nabla T) = -P, \tag{1}$$

with k the conductivity of buffer gas, and P the time-averaged power per unit volume deposited in the gas, given by a sum over momentum transfer collisions between electrons and heavy species. Under the reasonable assumptions of uniform power deposition and uniform wall temperature, one can calculate the temperature profile. The difference between a cylindrical (C) laser and a septum (S) laser is illustrated in Figure 5, which shows calculated radial temperature distributions at 75 torr buffer gas pressure (Ne with 0.5% H_2). The presence of septa lowers the central temperature by about 1370 K, and it lowers the temperature at a/2 (above the radial center of a septum) by 440 K.

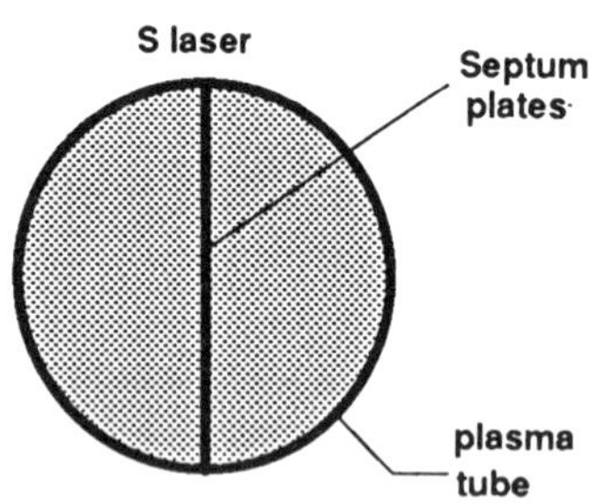

Fig. 4. Schematics of the cross sections of a septum laser (S laser).

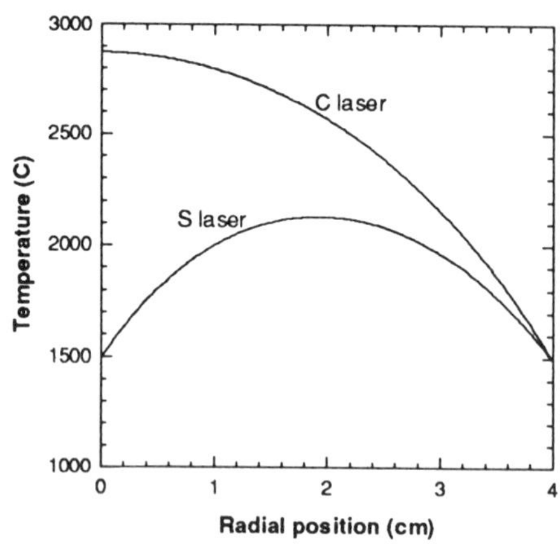

Fig. 5. Computed gas temperature profile of a C laser and a S laser at a buffer gas pressure of 75 torr

In the case without septa, the profile calculations agree with the measured copper ground state density profile [14]. The decreased gas temperature associated with the septa plays two beneficial roles. First, the prepulse electron temperature is allowed to equilibrate to a lower value, thus lowering the prepulse metastable densities. Second, the increased gas density gives rise to an increased discharge impedance. This is manifested, for example, in a shorter radial delay in field penetration and a better coupling between the laser head and the pulse power modulator.

3.2 FIELD PENETRATION

Because of the high conductivity of the CVL plasma, the plasma skin effect predominates during the initial stage of its gas discharge. The effect of gas temperature and gas pressure on field penetration follows from the behavior of the resistivity, as manifested in the field diffusion equations. In the absence of axial gradients, the field components are $B_\theta(r,t)$ and $E_z(r,t)$. The magnetic field diffuses according to

$$\frac{\partial B_\theta}{\partial t} = \frac{\partial E_z}{\partial r}, \qquad E_z = \frac{D}{r}\frac{\partial (rB_\theta)}{\partial r}, \tag{2}$$

where the diffusion coefficient is $D=\rho/\mu_0$. The penetration time varies inversely with the resistivity, $\rho = m_e\nu/e^2n_e$, which in turn is proportional to the electron momentum transfer frequency ν and inversely proportional to the electron density n_e. During the pulse, ν is dominated by electron-neon collisions and hence increases as the gas density is increased at lower temperature or higher pressure. In addition, our simulations show that, during the time of initial field penetration, the electron density is closer to its quiescent value than to its peak value, and that the latter is not too sensitive to the gas temperature. Since a lower gas temperature gives rise to a lower quiescent electron density, this is another mechanism for raising the resistivity. According to the model, the two effects are of comparable importance. The effect of a cooler plasma and higher gas pressure on field diffusion is illustrated in Fig. 6, which shows the measured radial delay for both a C laser and an S laser. The reduction in radial delay at higher gas pressures and lower gas temperatures (i.e., S laser) is evident because of higher gas density [15].

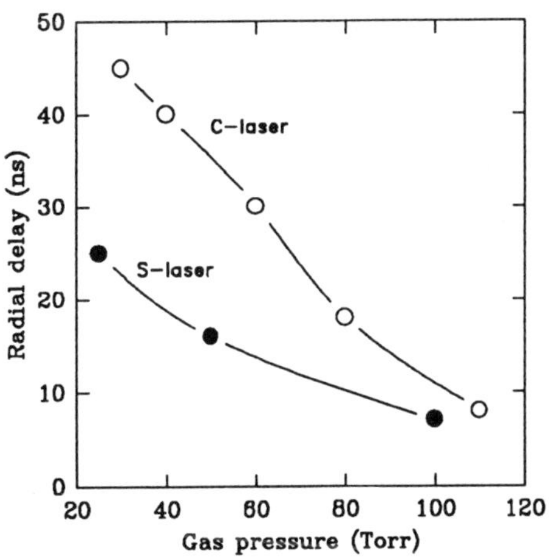

Fig. 6. Comparison of measured radial delay between an S laser and a C laser at various buffer gas pressures.

3.3 SATURATION FLUENCE AND LASER POWER

This amplifier was driven by a 4.4 kHz pulse modulator with 3-stage magnetic compression that delivered a peak voltage of approximately 40 kV with a rise time of less than 50 ns. To determine the S-laser saturation flurence, its energy has been extracted with various signal intensities as illustrated in Fig. 7. This shows the extracted energy levels at injection of approximately 2 mJ for the yellow output (578 nm) and 3 mJ for the green output (511 nm). Thus injection fluences of ~40 μJ/cm^2 for the yellow and 60 μJ/cm^2 for the green are required for efficient energy extraction of the 8-cm amplifier.

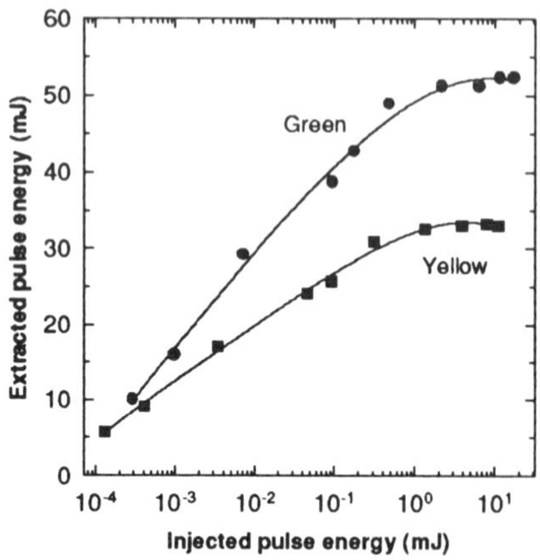

Fig. 7. Measured energy extraction at various injection energy for both green (511 nm) and yellow (578 nm) output. The solid lines are least-square curve fits to the data points.

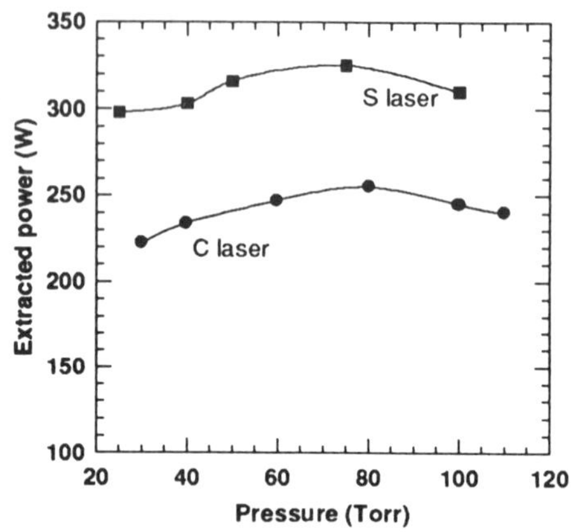

Fig. 8. Extracted output power of 8-cm copper laser amplifiers with septa (S-laser) and without septa (C-laser) at different buffer gas pressures.

Figure 8 illustrates the measured amplifier power of both S and C lasers. In this experiment, the optimized laser input power needs to be lowered at higher pressures because of increased head impedance. The optimized gas pressure in each case is about 70-80 torr. At this optimized lasing condition, the extracted laser power increased from 255 W with a C-laser to 325 W with an S-laser, a 27% improvement. A small reduction (~5%) in required input power was also observed when septa were used. This is expected because of lower gas temperature associated with the S laser that improved coupling between the laser head and the pulse modulator. An improvement of laser efficiency, based on the energy stored in the front capacitor, from 1.1% to 1.38% was achieved with the addition of septa,. As a final remark, Iseki et. al. recently report a nearly 57% increase of laser power (i.e., running with a plane-plane resonator) with an 8-cm CVL by the addition of two cooling plates [16]. Their measurement of Cu ground state density confirms the reduction of axial gas temperature with the addition of septa.

3.4. ELECTRODES AND WINDOW CONTAMINATION

The peak discharge current of a CVL is typically very high for efficient pumping. For example, it is about 2.5 kA for the 8-cm 300-W amplifier. The large discharge current induces an abnormal cathode fall as high as 1-2 kV within a cathode sheath that is smaller than 100 μm [17]. The inevitable high discharge impedance using conventional electrodes leads to a large amount of thermal energy deposited into the cathode-fall region within a time duration of 200-500 ns. Additionally, the advancement of pulse-power electronics enables CVL amplifiers to be operated at higher Ne pressures that put more stress on electrodes because the volumetric thermal loading at the cathode increases quadratically with gas pressure. This high rate of energy deposition can lead to a thermal instability [18] that usually results in a constricted discharge at the cathode accompanied by severe electrode sputtering and erosion. In a CVL this phenomenon degrades the laser beam quality due to induced window contamination and instability of the far-field spot. This instability also leads to large discharge jitters (10-20 ns) which lower amplifier extraction and cause fluctuation of MOPA output energy.

One approach to resolve these problems is the use of a hollow cathode, because of its more efficient generation of free electrons for a high-current discharge. However, conventional hollow-cathode electrodes are mostly applied to low-power, low-pressure devices because of the requirement of small hollow cathode space that limits the electrode size. A new type of hollow cathode was thus developed [19] for high-power CVLs that typically operate at 40-100 torr of buffer gas pressures. The electrode employs a uniform-field profile that significantly enlarges the effective discharge area. Although it is generally believed that a profiled electrode is not necessary for longitudinal discharge devices, the short-pulse high-peak-current discharge of a CVL differs from conventional longitudinal discharge that have either CW or long-pulse operation. We found that the electrode profile is essential to improve the CVL discharge stability and electrode lifetime. Multiple hollow-cathode slots attached to the profiled electrode surface as illustrated in Fig. 9, were found to enhance electron emission substantially. Copper was used for the electrode construction because of its lower fabrication cost and less oxidization induced window contamination than refractive materials.

Fig. 9. A profiled multi-slot hollow cathode electrode for high-power CVL amplifiers.

This new electrode design not only satisfies the requirement of small hollow cathode space for efficient hollow-cathode discharge at elevated gas pressure, but also fulfills the high-power requirement by applying multiple slots. The gap size of the slots is designed to have an effective hollow-cathode effect under CVL operational gas pressure but with minimized discharge erosion. A gap size of a fraction of a mm was found to be adequate for effective electron emission and electrode lifetime. Our experiments indicated that discharge constriction started at about 30 torr for a profiled electrode without hollow-cathode slots. The addition of multiple slots effectively extended the stable glow discharge to ~100 torr. Window contamination and discharge jitters were greatly reduced with this new electrode design. The lifetime of this multi-slot hollow cathode electrode has been found to be around 15,000 hours for 8-cm amplifiers.

3.5. WINDOW HEATING AND AMPLIFIER BEAM QUALITY

The installation of hollow-cathode copper electrodes has eliminated most of the window heating caused by window contamination. However, the unavoidable IR radiation from the high-temperature plasma tube introduces noticeable window heating that results in beam focus and wavefront aberration, if care is not taken in window design. Fused silica windows, due to their lower cost and negligible laser light absorption, have been widely used in CVLs. Nonetheless, the relatively high-IR absorption of fused silica makes it unsuitable for applications that require near-diffraction-limited output. To diminish IR heating, CaF_2 and MgF_2 windows are better suited than fused silica windows. The IR transmission ranges (i.e., greater than 50% transmission) for a 1 inch thick window are 8.8 μm and 6.8 μm for CaF_2 and MgF_2, respectively. These are considerable improvements over the case of fused silica windows, which also has a ~0.5 μm OH absorption band centered at 2.7 μm.

As listed in Table 1, both CaF_2 and MgF_2 have low absorption in CVL wavelengths. They are almost insoluble in water and will retain an optical polish with no deterioration under normal atmospheric conditions. With a typical tube temperature of 1700 K, it is estimated that 40% of the tube radiation that reaches the window is absorbed when a fused silica window is used; the OH absorption band absorbs 10% of the radiation. This absorption translates to a heat deposition of 4W to the laser window, assuming an 8-cm plasma tube and a distance of 60 cm between the window and the tube opening. This heat deposition reduces to 0.6 W and 0.3 W for MgF_2 and CaF_2, respectively. Although the window heating of MgF_2 is slightly higher than CaF_2, its high thermal conductivity and low dn/dT represent significant advantages in lowering window heating induced beam focus and wavefront distortion. MgF_2 is also more resistant to color centers that may appear in CaF_2 due to the large amount of UV radiation from the laser plasma. Most MgF_2 windows are cut perpendicular to its optical axis due to its birefringence characteristic. Sapphire windows have the highest thermal conductivity, but they are not considered because of their high dn/dT and the cost of material and optical polish.

Table 1. Comparison of window materials for CVL [20]

Material	Thermal conductivity (W/m/K)	Refractive index at 500 nm	dn/dT (10^{-6}/K)	Cut-off wavelength at 50% T* (μm)	Estimated window absorption** (W)	Absorption coefficient at 500-600 nm (10^{-5}/cm)	Radiation resistance (color center)
Fused silica	1.38	1.462	12.8	3.3	4 (40%)	17	Very good
CaF_2	9.71	1.437	-10.6	8.8	0.3 (3%)	7.2	Fair
MgF_2	21	1.383 (O) 1.395 (E)	2.1 (O) 1.7 (E)	6.8	0.6 (6%)	11	Good
Sapphire	27.21	1.77	13	5.5	1 (10%)	-	Excellent

* Assuming a window thickness of 1 inch.

** Assuming a distance of 60 cm between the window and the 8 cm tube opening

We have tested and compared fused silica, MgF_2 and CaF_2 windows in terms of window focusing and wavefront distortion by using wavefront shearing technique. The near-diffraction-limited oscillator described in Section 2 was used for the amplifier extraction. A window thermal isolator between the aluminum housing and the window was added to lower the window edge cooling and flatten the radial temperature profile. Table 2 lists the wavefront curvatures based on measured shear fringes of x-axis and y-axis of the output beam for fused silica, CaF_2, and MgF_2 windows.

Table 2. Measured window focusing of an 8-cm amplifier with different window materials

Material	Fused silica	CaF_2	MgF_2
X-axis wavefront radius (m)	250	850	∞
Y-axis wavefront radius (m)	360	1200	∞
P-V astigmatism*	1.6 waves	0.5 waves	0

*Estimated based on the difference in X & Y wavefront curvatures.

The astigmatism caused by window heating shown in Table 2 is believed to be mainly a result of free convection cooling of the IR heated window (i.e., 10-15 K higher than the ambient air for fused silica windows) that in effect shifts and tilts the curved wavefront. Table 2 shows that the improvement of output wavefront with MgF_2 is significant because of its smaller window heating, larger thermal conductivity, and lower dn/dT. The higher dn/dT of CaF_2 windows made it less effective than MgF_2 windows for wavefront improvement. Based on shear fringe patterns, some spherical aberration induced by window heating still exists even with MgF_2 windows. A C-laser amplifier with MgF_2 windows was extracted with a magnified ICO beam. Its far-field spot illustrated in Fig. 10 demonstrates a good beam focusability because of the use of MgF_2 windows that considerably reduced window heating induced wavefront distortion. The far-field energy spread measured with pinholes indicates larger than diffraction limited energy spread, as plotted in Fig. 11 at different laser power levels. Note that some wavefront distortion was introduced by the oscillator beam expander. At nearly full power level, 50% of the output has a diffraction-limited beam divergence, and about 84% the output is within 2.2 XDL beam divergence. This is a significant improvement over an amplifier with fused silica windows, which had a beam quality of 5-6 XDL.

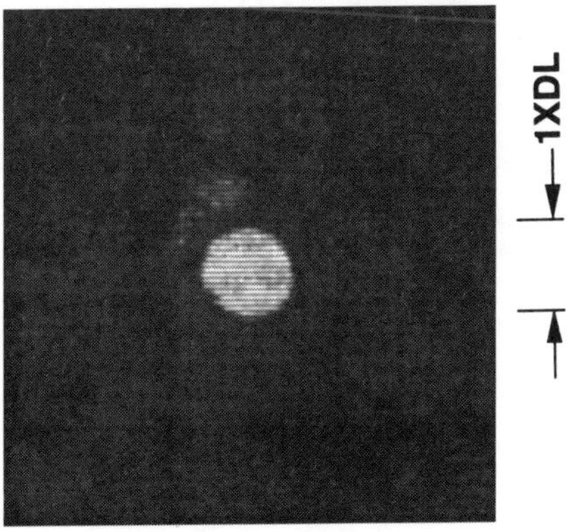

Fig. 10. An amplifier far-field spot.

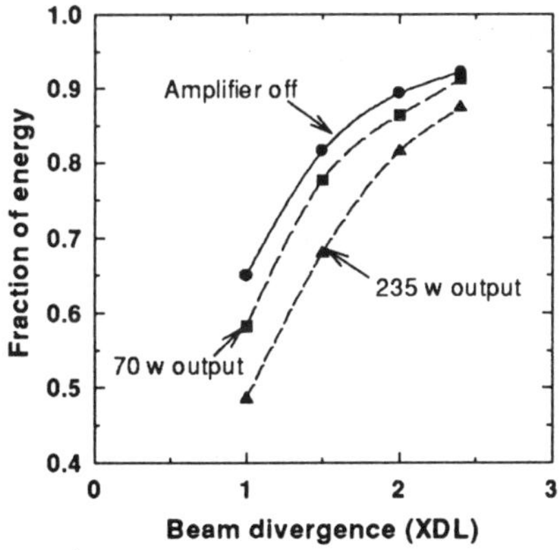

Fig. 11. Power-in-the-bucket measurements of the 8-m amplifier far-field spot at various power levels

4. Applications

4.1 PRECISION LASER MACHINING

Most industrial laser machining systems are based on CW or long pulse CO_2 and YAG laser, and material removal with these lasers is mostly through melt expulsion. This mechanism normally leads to poor dimensional control and sizable recast layer that are not acceptable in precision machining applications. A CVL provides a significant opportunity for precision laser machining [3,4] because of its near-diffraction limited beam quality, short pulse output, and high repetition frequency. A peak power density of 10^8 W/cm^2 to 10^{10} W/cm^2 can normally be obtained on the material surface. This power density effectively removes material in a ablative fashion such that little recast layer is formed on the material surface. The short pulse also ensures micron scale heating depth and minimal heat affected zone. The CVL visible output greatly improves the laser-material coupling over IR lasers because of higher surface absorption and lower plasma shielding.

Figure 12 illustrates 185-μm holes drilled on 1 mm thick stainless steel by a CVL, based on a precision trepanning technology [3]. The holes are straight and their repeatability is excellent, with hole size variation in the order of a few microns. Sections of CVL drilled holes show little recast layer or heat affected zone. Fig. 13 illustrates grooves and holes machined on a piece of silicon carbide (1.25 mm thick) with a high-power CVL amplifier. The grooves are 120-μm wide and 120-μm deep in the part. Detailed examination of the grooves showed straight side walls and fairly flat bottoms. The holes in the part have diameters of 1.25 and 2.5 mm. The hole roundness is nearly perfect, and no material cracks have been observed. Micromachining on alumina and silicon nitride has also been demonstrated with promising results.

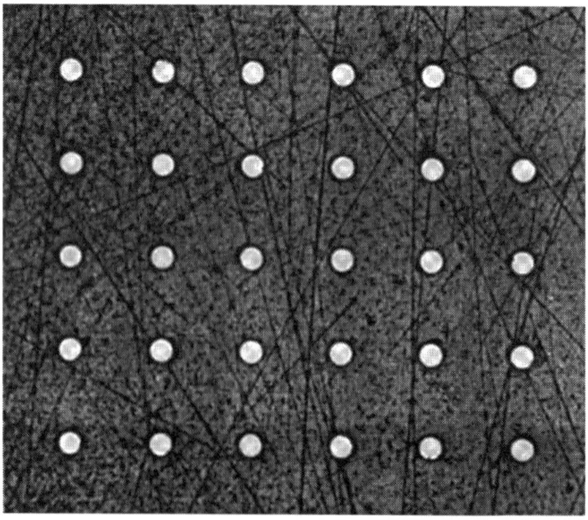

Fig. 12. Precision trepanned holes with 185 μm diameter on 1-mm thick stainless steel using injection controlled CVL oscillator.

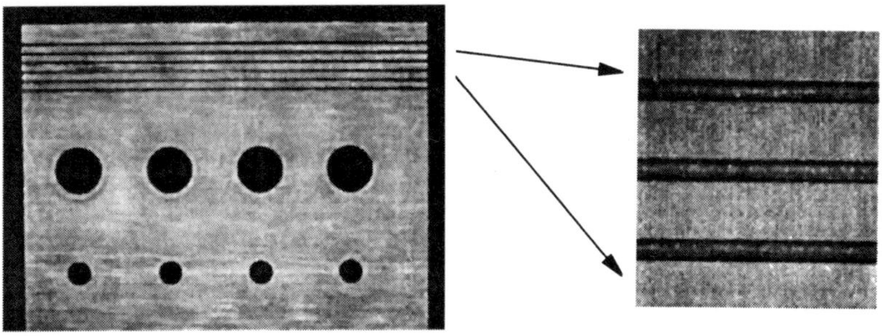

Fig. 13. Precision machined grooves and holes on a piece of SiC using a high-power CVL amplifier. The hole diameters are 1.25 and 2.5 mm; the grooves are 120 μm wide by 120 μm deep.

4.2 SECOND HARMONIC GENERATION

The ability to generate high average power UV by frequency doubling the CVL output would greatly extend its applications in material processing [21] and microlithography. The frequency doubled CVL could replace the KrF laser, which has a wavelength (248 nm) similar to that of doubled CVL 511 nm. The doubled CVL output has higher repetition frequency and near-diffraction-limited beam quality at a competitive cost. Second harmonic generation of an injection controlled CVL oscillator and a high-power amplifier has been demonstrated. A conversion efficiency of 18% has been achieved for the 511 nm output of ICO with a single BBO crystal [2]. This conversion efficiency was improved to nearly 30% with dual crystals oriented in an alternating-z configuration [22]. The conversion efficiency of the 8-cm amplifier (with fused silica windows) output drops to less than 10% because of beam quality degradation and severe thermal dephasing in the crystal. However, with an amplifier producing 113 W at 511 nm, a record of 9.0 W at 255 nm has been demonstrated [22].

5. Conclusion

Significant progress has been achieved in the past years in CVL beam quality improvement, power scaling, and system reliability. These advancements have put high-power CVLs in a very competitive position with respect to non-AVLIS applications such as laser material processing and UV generation. However, some further system improvements are still needed to make the CVL a widely acceptable laser tool. Issues such as maintenance cost (i.e., copper reload, tube lifetime), time requirement for laser on and off, and reliability of CVL pulse-power electronics are some of the main concerns normally raised by potential users. Progress in these areas will greatly enhance CVL applications. This work was performed under the auspices of U.S. Department of Energy by Lawrence Livermore National Laboratory under Contract W-7405-Eng-48.

6. References

1. Coutts, D., Ainsworth, M., and Piper, J. (1990) Enhanced efficiency of UV second harmonic and sum frequency generation from copper vapor lasers, *IEEE J. Quantum Electron.* **QE-26**, 1555-1558.
2. Molander, W., Norton, M., and Chang, J. (1991) Improved efficiency of second-harmonic generation in BBO using the 511 nm line of an injection-seeded copper laser, *Digest of Conference on Lasers and Electro-Optics*, 176.
3. Chang, J., Martinez, M, Warner, B., Dragon, E., Huete, G., and Solarski, M. (1995) Precision micro drilling with copper vapor lasers, *Proceeding ICALEO'94*, 323-332.
4. Knowles, M., Kearsley, A., Foster-Turner, R., Abbott, J., Boaler, J., and Errey, K. (1994) Visualization of small hole drilling using a copper laser, *Proceeding ICALEO'94*, 352-361.
5. Fugate, R., Fried, D., Ameer, G., Boeke, B. (1991) Measurement of atmospheric wavefront distortion using scattered laser light from a laser guide-star, *Nature* **353**, 144-146.
6. Omatsu, T. and Kuroda, K (1992) Time resolved measurement of spatial coherence of a copper vapor laser using a reversal shear interferometer, *Opt. Commu.* **87**, 278-286.
7. Chang, J. (1994) Time-resolved beam quality characterization of copper-vapor lasers with unstable resonators, *Applied Optics* **33**, 2255-2265.
8. Gobbi, P. and Reali, G. (1984) A novel unstable resonator configuration with a self filtering aperture, *Opt. commu.* **52**, 195-198.
9. Warner, B. (1986) Injection locked copper laser oscillator, *Digest of Conference on Lasers and Electro-Optics*, 160.
10. Nayor, G., Lewis, R., and Kearsley, A. (1988) *SPIE* **894**, 110-115
11. Chang, J. (1995) Copper laser oscillator with adjoint-coupled self-filtering injection, *Optics Letter* **20**, 575-577.
12. Kushner, M. and Warner, B. (1983) Large-bore copper-vapor lasers: kinetics and scaling issues, J. Appl. Phys. **54**, 2970-2982.
13. Chang, J., Boley, C., Martinez, M., Molander, W., and Warner, B. (1994) Beam characteristics of a large-bore copper laser with a radiatively cooled plasma, *SPIE* **2118**, 2-8.
14. Molander, W. (1989) Measurement of ground state copper density using hook spectroscopy, *SPIE* **1041**, 11-18.
15. Chang, J. (1993) Pressure dependence of copper laser output characteristics, *Applied Optics* **32**, 5230-5235.
16. Iseki, Y, Watanabe, I., and Noda, E. (1995) Output characteristics and improvement in beam-profile of large-bore copper vapor lasers, *SPIE* **2380**, 172-182
17. von Engle, A. (1965) *Ionized gases,* Clarendon Press, Oxford.
18. Jacob, J. and Mani, S., (1975) thermal instability in high-power laser discharge, Appl. Phys. Lett. **26**, 53-55.
19. Chang, J., Alger, T, Anderson, A., and Arnold, A. (1994) New hollow-cathode electrode for high-power copper lasers, *Digest of Conference on Lasers and Electro-Optics*, 402.
20. The majority of data are from Optovac's Crystals Handbook. Absorption data are from Hass, M. and Davisson, J. (1977) Visible and UV absorption in laser windows, Proceeding of The High Power Laser Optical Components and Component Material Meeting, Arlington, Visginia, 152.
21. Glover, A., Coutts, D, Ramsay, D., and Piper, J. (1995) Progress in high-speed UV micro-machining and high-repetition rate frequency-doubled copper vapor lasers, *Proceeding ICALEO'94*, 343-351.
22. Molander, W. (1994) Improved efficiency of second-harmonic generation in BBO using the 511 nm line of an injection-seeded copper laser, *Digest of Conference on Lasers and Electro-Optics*, 363.

COPPER BROMIDE LASERS

N.V. SABOTINOV
Department of Metal Vapor Lasers, Institute of Solid State Physics,
Bulgarian Academy of Sciences
Tzarigradsko Chaussee 72, Sofia 1784, Bulgaria

1. Introduction

Today, a series of pulsed lasers is known, which operate at transitions from a resonance to metastable level in metal atoms and ions. The most efficient and producing the highest output power in the visible region among all, is the copper vapor laser (CVL). Practically an efficiency of up to 1.5% and an average power of the order of 500 W are already achieved. One of the main difficulties for CVLs is imposed by the necessity to provide a sufficient concentration of copper atoms in the active volume. The temperature guaranteeing such copper atom concentration is ~1500C and that requires high-temperature materials for constructing laser tubes. The high temperature causes not only technical difficulties but also complications of a more general nature, since the lower laser levels that are metastable, can be significantly populated at high temperatures. The difficulties associated with introducing the Cu atoms into the active volume by heating have compelled researchers to seek other methods for creating a high density of copper atoms in the active volume.
Dissociation of copper compounds in a pulsed discharge is a way to get enough copper atoms into the active volume. This idea was suggested for the first by W.T.Walter *et al.* [1] in 1966. Using this method it is possible to achieve relatively good laser action in continuously pulsed regime. Self-terminating lasing in copper atom via dissociation of copper halide compounds - copper iodide was obtained for the first time by C.S.Liu *et al.* [2] in 1973. Lasing was achieved in the copper chloride vapors as well by C.J. Chen *et al.* [3] in 1973.

Self-terminating lasing on transitions in copper was attained in 1974 by N.V.Sabotinov *et al.* using third copper halide compound - copper bromide. The first papers reporting about the laser action in copper bromide issued almost simultaneously in 1975 [4], [5], [6]. Unlike the used technique of double pulses in the initial studies of CuJ and CuCl, in [4] and [6] regular current pulses are introduced into the active volume of the laser tube at a high PRF (pulse repetition frequency). Double-pulse technique is employed in [5].

The comparative studies conducted on copper bromide vapor lasers (CBrVL) and CVL [7] showed that at high PRF (10 kHZ - 20 kHz) the basic output lasing

C. E. Little and N. V. Sabotinov (eds.), Pulsed Metal Vapour Lasers, 113–124.

parameters are practically relatively high and identical. Compared with the other copper halide lasers (CuCl and CuI) the CBrVLs give higher output power: P_{CuBr} : P_{CuCl} : P_{CuI} = 6 : 3 : 2 [8]. That's why attention of our research laboratory as well as of other ones is directed to the copper bromide laser.

CBrVLs present one different and quite prospective alternate of CVL. The CBrVL operates at relatively low temperature ~ 500^0C that immediately removes many difficulties when constructing the laser tubes and allows quartz tubes of a simple construction to be used.

However to make the CBrVL competitive to the CVL several important problems connected with the development of the CBrVL have to be solved. Such ones are stable operating regime, long lifetime of a sealed-off laser tube, optimal conditions for maximum output power and efficiency.

2. Physical View of Processes in a Copper Bromide Vapor Laser

The processes leading to inverse population are more complex in a laser with copper halide lasant compared with processes taking place in the metallic copper vapor laser. These processes include not only copper atom excitation but production of these atoms from the copper bromide molecules as well.

The working medium of the CuBrVL at the operation start contains molecules of copper bromide and buffer gas atoms. The copper bromide turns into gaseous phase as a result of thermal heating of copper bromide which is a solid at room temperature.

According to [9] copper halides in gaseous phase occur largely in the form of trimer molecules Cu_3Br_3. At higher temperatures of the gas into the laser active zone the trimer molecules decay to CuBr as a result of thermal dissociation. Bearing in view these phenomena we consider that in the active medium of the CuBrVL exist only CuBr mono molecules.

All processes taking place in the plasma including copper atom production, can be conventionally divided into two groups. The first group of processes take place during the current pulse. They continue several microseconds and result in creation and support of nonequilibrium plasma. This process group includes: dissociation of copper bromide molecules, copper atom excitation, production of ions and electrons. The second group of processes occur between current pulses and last significantly longer - from tens to hundreds of microsec. Relaxation processes take place there, and they are directed to establishment of equilibrium state. This process group includes: the decrease of ion and electron densities , the decrease of density of copper and bromine atoms respectively, and the increase of concentration of copper bromide molecules.

Physical processes resulting in laser generation in a CuBrVL are shown in Fig.1 ly schematically via simplified energy level diagram for copper atom and copper bromide molecule. In the active medium of CuBrVL five basic processes occur: dissociation of copper bromide molecules, copper atom excitation and upper laser level population, laser generation, decay of copper metastable levels and recombination of copper and bromine atoms to copper bromide molecules. Dissociation of copper bromide molecules is accomplished by electron impacts. According to [10] the following dissociation processes are possible:

$$e + CuBr \rightarrow Cu + Br^- \quad (1)$$
$$e + CuBr \rightarrow Cu + Br^* + e \quad (2)$$
$$e + CuBr \rightarrow Cu + Br + e \quad (3)$$

The cross-section of process (1) is, e.g., about 1.10^{-18} cm^2. The energy for dissociation of CuBr-molecule is about 3.4 eV. Dissociation process can occur either during a previous dissociating pulse under double-pulse excitation, or during any pulse under regularly recurrenced pulses of more than 5 kHz PRF. The last value is not exactly defined and may depend on the conditions under which the active medium is.

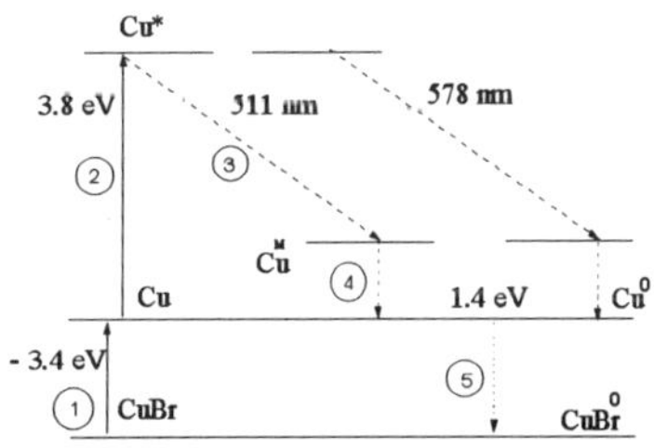

Figure 1.

(1) $CuBr + e \rightarrow Cu + Br + e$ Dissociation
(2) $Cu + e \rightarrow Cu^* + e$ Excitation
(3) $Cu^* \rightarrow Cu^M + h\nu(511nm)$ Lasing
(4) $Cu^M + e \rightarrow Cu + e$ Deexcitation
(5) $Cu + Br \rightarrow CuBr$ Recombination

The copper atoms obtained in the ground state $4s^2S_{1/2}$ as a result of electron collisions are excited to resonance levels $4p^2P_{3/2}$ and $4p^2P_{1/2}$. Levels $4p^2P_{3/2}$ and $4p^2P_{1/2}$ are in fact the upper laser levels for the two self-terminating transitions:

$$Cu\,(4s^2S_{1/2}) + e \rightarrow Cu(4p^2P_{3/2}) + e \quad (4)$$
$$Cu\,(4s^2S_{1/2}) + e \rightarrow Cu(4p^2P_{1/2}) + e \quad (5)$$

Conditions under which these two processes occur are more complicated compared with these ones in the copper vapor laser (CVL). In the active medium composition the molecules of copper bromide, bromine atoms, ions and molecules take an active part. That affects processes (4) and (5) mainly in terms of electron concentration and average electron energy. Because of presence of electronegative element - bromine, the electron concentration in the laser plasma with copper bromide vapors is lower.

Laser emission processes (6) and (7) are consequence of excited copper atoms-transitions from the two resonance levels $^2P_{3/2}$ and $^2P_{1/2}$ to the metastable copper levels $^2D_{5/2}$ and $^2D_{3/2}$ as a result of induced transitions:

$$Cu(^2P_{3/2}) \rightarrow Cu(^2D_{5/2}) + h\nu(511nm) \quad (6)$$
$$Cu(^2P_{1/2}) \rightarrow Cu(^2D_{3/2}) + h\nu(578nm) \quad (7)$$

Laser generation in the active medium with copper bromide vapors depends strongly on relaxation of copper atom resonance levels, their deexcitation by electron collisions of the IInd kind. Emission capture is closely related to dissociation processes (1-3). As it can be seen a simple realization of laser emission mechanism cannot be expected there compared with its occurring in a copper vapor laser.

The IVth process is relaxation of the metastable copper atoms:

$$Cu(^2D_{5/2}) + e \rightarrow Cu(^2S_{1/2}) + e \quad (8)$$
$$Cu(^2D_{3/2}) + e \rightarrow Cu(^2S_{1/2}) + e \quad (9)$$

We can already assume with confidence that the main mechanism of metastable copper atom deexcitation consists of IInd kind electron collisions. Relatively small is the part of metastable copper state relaxation as a result of collisions with atoms of the buffer gas neon or with the walls of the gas-discharge tube. In respect to the significance of (8) and (9) the influence of bromine and copper bromide is also involved there because of their effect on the electron concentration and temperature.

The last group of processes are recombination processes of copper atoms with bromine atoms. These processes occur after the current discharge pulse.

$$Cu + Br + Ne \rightarrow CuBr + Ne \qquad (10)$$

$$Cu + Br^- \rightarrow CuBr + e \qquad (11)$$

Recombination and dissociation are inherent processes to the lasers with lasant of copper halide vapors. They determine the copper atom concentration. These processes go in a way that determine laser energy characteristics.

A mix of copper bromide molecules and noble-gas neon exists in the gas-discharge tube before the current pulse running. The concentration ratio of the two components is of about two orders in favor of neon. Dissociation of the copper bromide occurs during running of the discharge current pulse for a time about 100 ns. The plasma contains copper atoms in ground and metastable states, bromine atoms, initial molecules, electrons and ions after the first current pulse. Three basic processes described above are taking place during the excitation current pulse: dissociation, population of the copper resonance levels, laser generation. During the interpulse time the plasma relaxes. The rest two basic processes take place here: deexcitation of the metastable states and recombination.

3. Basic Problems in Development of the Copper Bromide Vapor Laser

CuBrVL possesses numerous valuable features allowing it to gain an advantage over the copper vapor laser (CVL) and over lasers with other copper halides as it was found later. The positive feature comes first from working temperature of the active medium, that is 1000-1100C lower than that of the copper vapor laser (CVL). From here numerous advantages are coming. The laser tube can be made of quartz glass with no use any special ceramics of vacuum properties. CuBrVL can get into operation regime in 10-15 minutes if its pre-heating is turned on together with service electronics to start lasing simultaneously with the discharge switch-on. That time is significantly shorter than the time for getting into laser regime of the copper vapor laser (CVL) which is usually from 30 minutes to 1 hour.

Some shortcomings about the CuBrVL were observed during the first experiments carried out in our laboratory. Hence some problems have to be solved for getting a laser applicable in the practice in spite of attractiveness of the idea for a low-temperature CuBrVL.

The first problem is connected with creation of a stable lasing laser with low fluctuations in output power. The complex composition of the laser plasma containing as neon, copper and a halogen element (bromine), as well as molecules of copper bromide, requires appropriate discharge conditions which can be accomplished in a special design of the laser tube.

laser applicable in the practice in spite of attractiveness of the idea for a low-temperature CuBrVL.

The first problem is connected with creation of a stable lasing laser with low fluctuations in output power. The complex composition of the laser plasma containing as neon, copper and a halogen element (bromine), as well as molecules of copper bromide, requires appropriate discharge conditions which can be accomplished in a special design of the laser tube.

The second also important problem connected with development of the CuBrVL is the laser tube lifetime. This question is in connection with the possibility for practical application of the laser. As a matter of fact the lifetime is a common question for the copper lasers. The CuBrVL operates under alleviated temperature conditions. However numerous questions arise from here. They are connected with the influence of bromine on the discharge tube elements, electrode and glass strength. Moreover copper bromide losses and its recovering are also a serious problem. The dissociated bromine influences the plasma parameters and respectively the lasing mechanism. All these questions outlined the second basic problem connected with the lifetime of the CuBrVL.

The third essential problem is about achievement of high power from the CuBrVL. Comparative studies carried out in [10] in the range of output power up to 15 W showed that in the active medium of copper bromide the same output power can in principle be achieved as in the CVL. This conclusion proved the significance of CuBrVL development and the studies of our laboratory connected with enhancement of average output power and efficiency of the laser.

4. Lasing Stability Problem

To attain stable and efficient laser operation, it is necessary, first, to design a laser tube in which it is possible to control the plasma parameters. In several different laser tubes the optimal lasing conditions are determined under continuously pulsed excitation and by comparing the results, the best laser tube design for further experiments is chosen.

4.1. GAS DISCHARGE TUBES WITHOUT LIMITING THE DISCHARGE ZONE

Our investigations of CBrVL began in 1974 when we obtained laser action at continuously pulsed mode of operation in a gas discharge tube of the simplest construction with the lasant in the discharge zone (Fig.2a). Discharge self-heating was used to evaporate the metal compound. Lasing occurred in a finite temperature range: it began at an external tube wall temperature of 350 C and ceased at 500 C. The discharge plasma was very unstable and it easily went into a pinch.

Our investigations as well as the results of other authors show that the instability of the discharge plasma and its pinching are induced by uncontrollable variations of the copper bromide pressure in the active volume. To control the lasant vapor pressure in the discharge zone independently of the discharge parameters, the lasant is placed in separate reservoirs, heated in special ovens (Fig.2b).The wall of the laser tube is discharge heated and is kept at temperature higher than the temperature of the reservoirs by external thermal insulation. One of substantial drawbacks of these discharge tubes is the uncontrollable pinching of the discharge. Usually the diameter of the output beam at optimal conditions is about 0.6-0.7 of the tube diameter. This indicates that the pinching of the discharge leads to a narrowing of the zone in which the population inversion is created and thereby does not permit the entire volume of the discharge tube to be used.

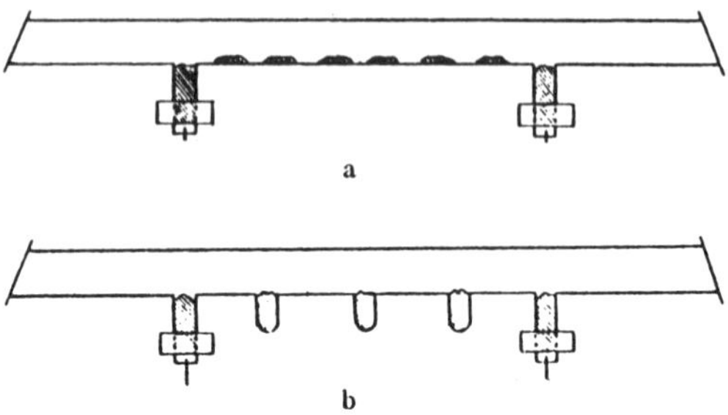

Figure 2.

4.2. GAS DISCHARGE TUBES WITH A LIMITED DISCHARGE ZONE

In 1975 we reported experiments [11] with a laser tube in which the discharge volume was limited by ring apertures, (Fig.3a). Later [12] it was shown that pinching could be completely eliminated by isolating the discharge zone from the tube walls. In Fig.3(b,c) tubes with different designs using discharge zone confinement, are shown. In these tubes the required vapor pressure of the lasant is created either due to self-heating by the discharge (Fig.3b) or the lasant is in special reservoirs heated separately (Fig.3c).

We combined the positive properties of the tube with limited discharge zone with the possibility to control the CuBr vapor pressure independently by evaporating the lasant in separate reservoirs (Fig.3c). Thus a stable and efficient lasing was obtained at tube wall temperature somewhat higher than the temperature of the reservoirs with the lasant.With this construction of the laser tube a maximum stability of 1% for the output power was obtained during 8 hours continuos operation.

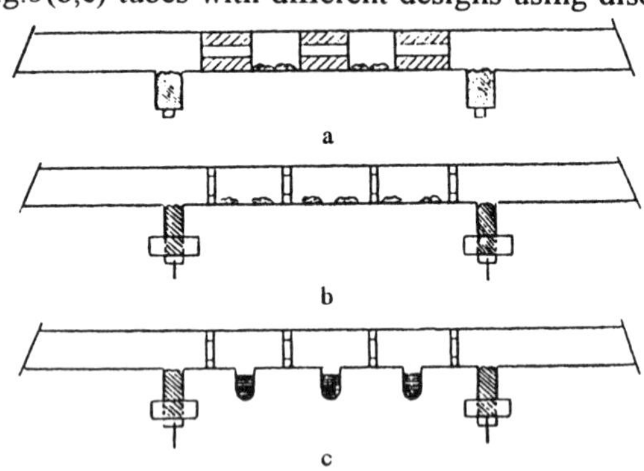

Figure 3.

The discharge tube electrodes are very important for obtaining stable operation for a long enough time. During the investigations different electrodes were experimented: copper and tantalum coaxial electrodes, water cooled electrodes and electrodes from commercial pulsed flash lamps filled with copper turnings. The best results were obtained with electrodes filled with copper turnings and that is why such porous electrodes were used when studying the lifetime of CBrVL.

5. Lifetime Problem

Discussing the laser tube lifetime, two main regimes of operation should be considered: a)with periodical change of the buffer gas and b)a completely sealed off laser tube.

In [13] CBrVL operation of 734 hours at 3 W average power on flowing neon is reported. Changing periodically the neon in the laser tube, we have achieved a laser tube lifetime of over 2000 hours. At such a regime of operation the laser tube lifetime depends mainly on the mechanical strength of the quartz tube electrodes and the quantity of CuBr in the tube.

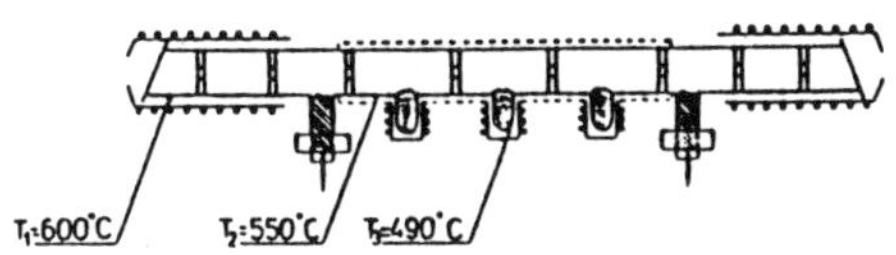

Figure 4.

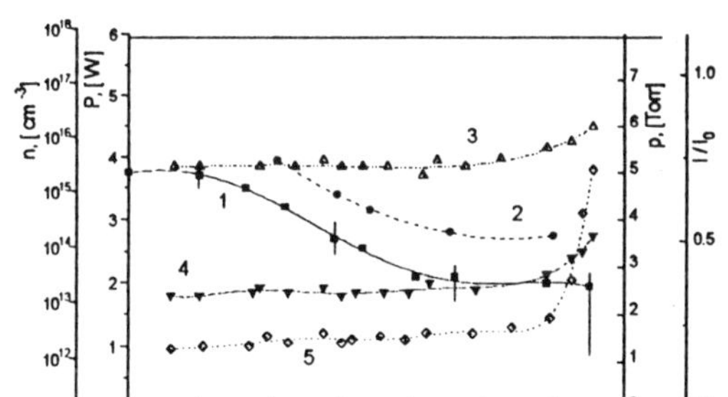

Figure 5.

In our opinion, such a regime of operation with a gas flow or a periodical change of the gas is not very convenient for practicable lasers. We had aimed at achieving a continued lifetime of a sealed-off laser tube and all our experiments were oriented in this direction. We have studied the factors determining the CuBrVL lifetime in a whole heated sealed-off quartz laser tube, shown in Fig.4.

The lowest temperature is in the CuBr reservoirs. The total average laser power as a function of time during 335 h of laser operation is shown in Fig.5 (curve 1). Curve 2 shows the time dependence of the laser windows transmittance, measured using a He-Ne laser. The bromide concentration is measured by an absorption method during laser operation. Curves 3,4 and 5 in Fig.5 show the time dependence of atomic and molecular concentration of bromine and total bromine pressure, respectively. These examples allow us to determine the following main processes, taking place during laser operation and limiting its lifetime. In the first place the lifetime of sealed-off laser tube depends strongly on free bromine accumulation in it. At bromine partial pressure over 2 Torr the tube destroys. In second place is the problem with the laser windows pollution as in the whole-heated laser tube this pollution consists of metallic copper layer and cannot be removed by external heating. In third place is the consumption of substance from the reservoirs as it is minimum at a whole-heated laser tube: 25 mg/hour.

The best results on continued lifetime we achieved in the laser tube, shown in Fig.6 In this tube the zones between electrodes are cold and in them the consumed copper bromide condenses. Recombination process between copper and bromine atoms is actively running in the zones outside the discharge, i.e., in the area between the discharge volume and the laser tube wall. Bromine that could not recombine with the copper atoms is captured first by the porous copper electrodes which have a special for this purpose construction, and second by the condensed copper bromide layer. In this type tube bromine accumulation was not observed during an operation at average of 1000 h.

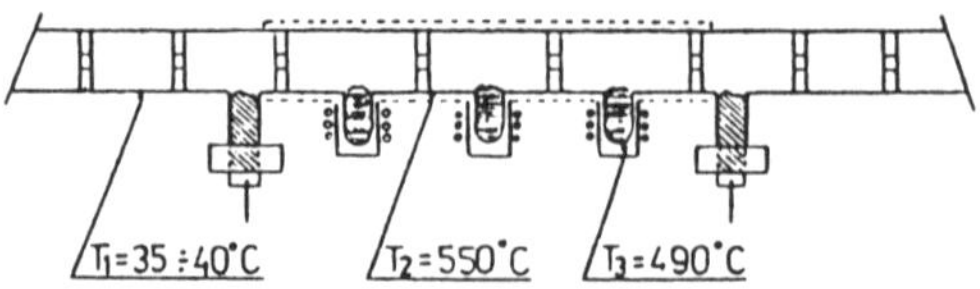

Figure 6.

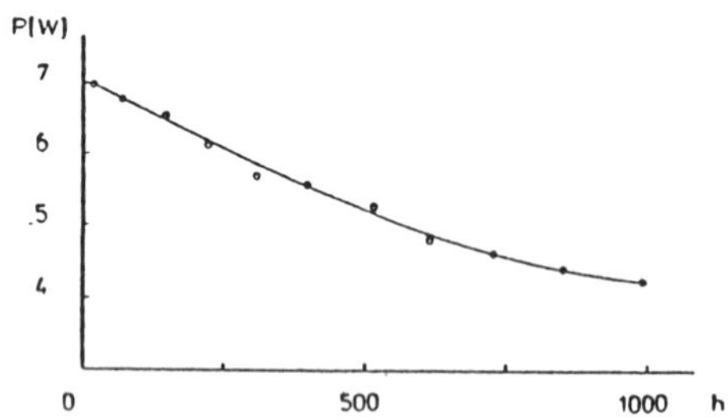

Figure 7.

Fig.7 shows variation of the average laser power during 1000 h - test. The output power drops from 7 W to 4.5 W during the test owing to the laser windows pollution by a CuBr layer. This CuBr layer can be easily removed during the laser operation [14]. The maximum copper bromide consumption in this type tube is 0.1g/h.

6. High Average Output Power and Efficiency of CuBrV Laser

Development of the CuBrVL is strongly related to achievement of high average output and efficiency no less than these ones of CVL. We have made two essential steps in this direction, namely: finding out the effect of hydrogen in the CuBrVL and using original IC-circuit for excitation of CuBrVL.

6.1. EFFECT OF HYDROGEN

In connection with creating a long-lifetime CuBrVL we have used two precise technologies to clean the laser tube, its electrodes, and CuBr employed. However good cleaning led to a surprising result: the laser output power significantly dropped and radial distribution of the laser beam intensity took a circle shape with "black hole" at the center. It has been supposed that unidentified impurity species exist in the gas discharge and they have a positive effect on the laser output parameters. Checking the effect of different possible impurities in the laser gaseous medium we found that adding small amounts of hydrogen to the neon-buffer gas caused a doubling of the average output power and efficiency from the laser [15]. Thus we found for the first time a strong effect of H_2 in the CuBrVL. In [16] a similar but weaker effect of hydrogen in CVL was before us reported. However nobody after that was successful in confirming this result, and the authors themselves did not use this effect in their subsequent works that gives rise to some doubts.

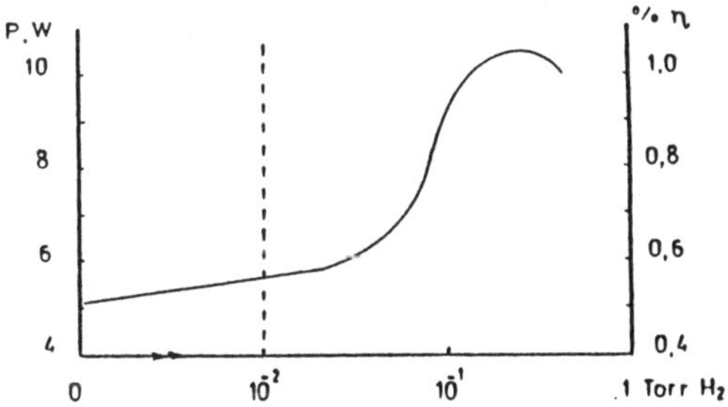

Figure 8.

Our experiments in which we observed for the first time this effect of hydrogen were carried out in a laser tube as it is shown in Fig.6 with an active length of 500 mm and four diaphragms with inner diameter of 20 mm. The input power was 1.2 kW at a pulse recurrence frequency (PRF) 15.5 kHz and 16 Torr neon pressure. The effect of H_2 -partial pressure on the output power is given in Fig.8 Our experimental findings indicate that small amounts of H_2 added to the neon buffer gas cause a doubled output power and a doubled efficiency of the continuously - pulsed CuBrVL. The output laser power has a maximum at 0.3 - 0.4 Torr H_2 - partial pressure. Fig.9 shows variation of the radial distribution of the laser beam intensity with increasing the pressure of H_2-additives. It can be seen that H_2-adding modifies strongly intensity distribution that is approximately "Gaussian" at 0.3 Torr optimum pressure.

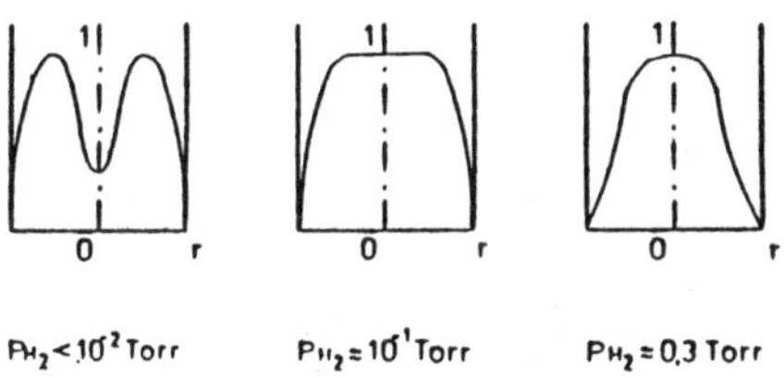

Figure 9.

Effect of hydrogen is very well expressed at discharge tubes of different kinds and it is easily repeatable under clean experimental conditions. Our findings led to studies in this direction and in other laboratories. First an effect of hydrogen was obtained in copper vapor laser [17] and somewhat later - in gold vapor laser [18]. This effect is widely employed currently in the practice of different laboratories. The strong effect of H_2 in CuBrVL lies at the base of creating one of the most high-power and high-efficiency variants of the copper laser: CuHyBrID laser [19].

Using the effect of hydrogen we attained an average output power of 20 W of CuBrVL with a sealed-off laser tube having active medium length of 500 mm and diameter of 20 mm at a PRF of 16 kHz and efficiency of 1.5% [20]. The highest average output power (112 W) achieved so far in CuBrVL is also obtained at H_2 adding to the neon buffer gas [21].

6.2. A NEW CIRCUITRY FOR CUBRVL EXCITATION

Essential advance in increase the average output power of CuBrVL was the creation of a novel electrical circuitry (IC) for discharge excitation in the laser tube [22]. The excitation circuitry is presented in Fig.10. A thyratron T is used as a commutator. Electrical circuitry is made up of two interacting paths (IC), and each of them consists of a conducting metallic plate connected to the tube electrodes via two capacitors and the tube itself. Four capacitors $C_{1\text{-}4}$ make up a capacitor set. Metallic plates are placed near the tube along its length. The laser power and efficiency depend primary on capacitances of the capacitors making up the capacitor set, secondly on the shape and size of the metallic plates, and third on the space between them and the tube. The best results are obtained with the same metallic plates of half-cylindrical shape.

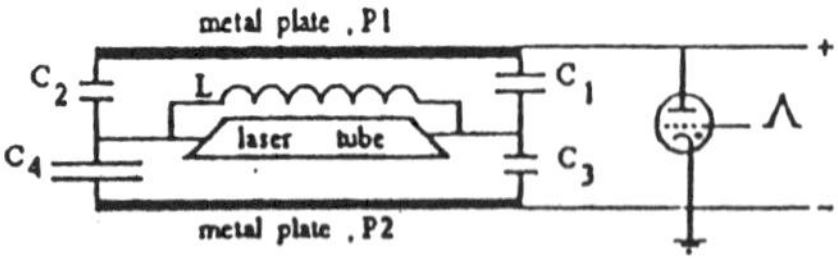

Figure 10.

One of the main IC-circuit advantages is possibility to make a very low-inductance circuit. On the other hand, however, maximum laser power and efficiency are obtained at not too low inductance of the excitation circuitry as reported in [23]. Because of that an inductor L is placed between the thyratron anode and capacitor set.

An enhancement of 45-50% for both the average output power and efficiency of CuBrVL is attained in comparison with using the conventional wide-spread circuitry - viz. the combination of a charging capacitor and a peaking capacitor. Better laser performance was obtained owing to a very good matching to the laser tube provided by the IC. As a result of that it was observed: 1) an increase of both the current pulse amplitude through the laser tube and of the pulse power introduced into it.; 2) an increase of the pulse energy introduced into the discharge and in particular of the pulse energy introduced into the laser tube by the end of the laser pulse. Electric IC circuitry analysis showed that the improvement of the electric parameters had resulted from specific interactions between the different parts of the electric circuitry.

6.3. HIGH SPECIFIC AVERAGE OUTPUT POWER OF CUBRVL

One of the interesting questions in laser physics, especially from the point of view of high-power laser development, is how much laser power can be extracted per unit volume of the active medium of a laser. This question concerns all kinds of lasers, including the continuously pulsed CuBrVL. To demonstrate the possibilities of CuBrVL to produce high specific average power (SAP) we have undertaken an experimental investigation of a small-bore CuBrVL.The construction of the laser tube that was studied is shown in Fig.11.

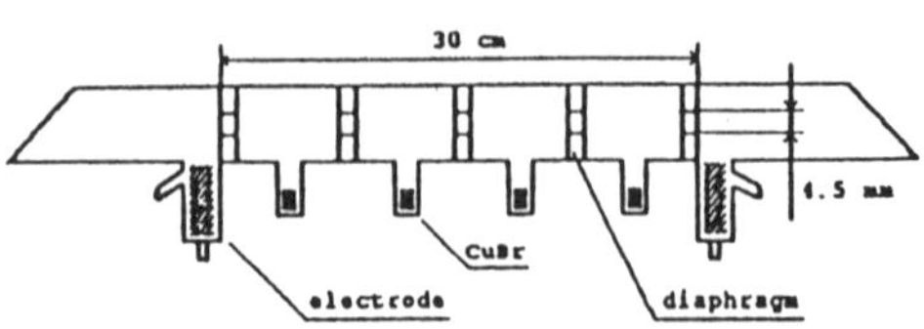

Figure 11.

laser powers. The optimization of the small-bore CuBrVL required a multiparametric investigation. The highest laser power of 6.7 W was obtained at 20 Torr neon-pressure, with a capacitor set C_1= 330 pF, C_2=C_3= 100 pF, and C_4= 1000 pF and by-pass inductor L=0.3 mH. The pulse recurrence frequency was 52 kHz, and the input electric power was 1.kW. Thus the active volume as defined by the tube dimensions was 4.77 cm^3, and it yielded 1.4 W/cm^3, that is higher SAP compared with CVL. A maximum average output power 3.1 W [24] was obtained of the same active volume of 4.77 cm^3 in CVL. Hence better conditions to obtain high SAP can be realized in CuBrVL.

The results of these experiments are important fór the future development of small practical high-power lasers for the visible. The results are also relevant to addressing the question of how such high specific output powers can be obtained from large active volumes.

7. Conclusions

A new version of the CVL is developed - a CuBrVL working at a lower temperature of the active medium: about 500^{0}C, with a simple and compact construction of the sealed-off laser tube and quick start-up time (5-10 min.) to specified power. A high stability of the output laser power is attained. The above results demonstrate that a sealed-off CuBrVL of 1000 h operation is already realizable. High output power (~20 W) and efficiency (~1.6%) are reached owing to the effect of hydrogen addition. A high specific average output power 1.4 W/cm ^{3} is obtained of CuBr active medium. The laser tube design is simple and cheap. Owing to its advantages the CuBrVL can very successfully replace the CVL in applications for which total average output power up to 20 W is enough.

Currently two firms have already been manufacturing CuBrVL. The first firm is “Norseld Pty Ltd” from Australia, and the second one is “Pulssvet” from Bulgaria.

Future studies on CuBrVL will be directed to development of sealed-off laser of 100-200 W output power, and to opportunities for application in the most advanced technologies for materials processing.

8. References

1. Walter, W.T., Soliemene, N., and Piltch, M. (1966) Efficient pulsed gas discharge lasers, *Journal of Quantum Electronics* **QE-2**, 474-479.

2. Liu, C.S., Sucov, E.W., and Weaver, L.A. (1973) Copper superradiant emission from, pulsed discharges in copper iodide vapor, *Appl. Phys. Lett.* **23**, 92-93.

3. Chen, C.J., Nercheim, N.M., and Russell, G.R. (1973) Double-Discharge Copper Vapor Laser with Copper Chloride as a Lasant, *Appl. Phys. Lett.* **23**, 514 - 515.

4. Sabotinov, N.V., Telbizov, P.K., and Kalchev, S.D. (1975), Bulg. Patent N28674.

4. Sabotinov, N.V., Telbizov, P.K., and Kalchev, S.D. (1975), Bulg. Patent N28674.

5. Shuhtin, A.M., Mishakov, V.G., Fedotov, G.A., *et al.* (1975) Observation of Dissociation of Copper Halide Molecules in a Pulsed Discharge by the Interference Method, *Optica and Spectroskopy* **39**, 785 - 786.

6. Akirtava, O.S., Dzhikiya, V.L., and Olainik, Yn.M. (1975) Laser Based on CuI Vapors in Copper Halide Vapors, *Kvant. Electron.* **2**, 1831-1832.

7. Kazaryan, M.A., Petrash, G.G., and Trofimov, A.N. (1980) Comparative Characteristics of Copper, Copper Hloride, and Copper Bromide Vapor Lasers, *Kvant. Electron.* 7, 583 - 592.

8. Gabay, S., Smilanski, I., :Levin, L.A., and Erez, G. (1977) Comparison of CuCl, CuBr, and CuJ as Lasants for Copper Vapor Lasers, *IEEE J. Quantum Electron.* **QE-13**, 364 - 366.

9. Guido, M., Baldueci, G., and Giglli, G. (1971) Mass spectrometric study of vaporization of cuprous chloride and the dissociation on energy of Cu_3Cl_3, Cu_4Cl_4, and Cu_5Cl_5, *J. Chem. Phys.* **55**, 4556- 4572.

10. Kazaryan, M.A., Petrash, G.G., and Trofimov, A.N. (1987) Pulsed Copper Halide Vapor Lasers, *Trudi FIAN* **181**, 78-170.

11. Sabotinov, N.V., Kalchev, S.D., and Telbizov, P.K. (1975) Copper Vapor Laser with a high-pulse repetition frequency, *Kvant. Electron.* **2**, 1822-1834.

12. Kazaryan, M.A., and Trofimov, A.N. (1978) Gas-Discharge Tubes for Metal Halide Vapor Lasers, *Kvant. Electron.* **5**, 2471 - 2472.

13. Wang, Y., Sun, W., Yao, Z., and Shi, B. (1985) A Practical CuBr Laser with Flowing Buffer Gas, *Optics Commun.* **55**, 345- 346.

14. Davis, P.E., Sabotinov, N.V., Vuchkov, N.K., and Astadjov, D.N. (1989), Commonwealth of Australia, Patents Act 1952.

15. Astadjov, D.N., Sabotinov, N.V., and Vuchkov, N.K. (1985) Effect of Hydrogen on CuBr Laser Power and Efficiency, *Optics Communication* **56**, 279- 282.

16. Bokhan, P.A., Silantiev, V.I., and Solomonov, V.I. (1980) On a Mechanism of Limitation of the Lasing Pulse Repetition Rate in a Copper-Vapor Laser, *Kvant. Electron.* 7, 1264-1269.

17. Huang, Z.G., Namba, K., and Shimizu, F. (1986) Influence of Molecular Gases on the Output Characteristics of a Copper Vapor Laser, *Japan J. Appl. Phys.* **25**, 1677-1679.

18. Huang, Z.G., Shan, H.Y., and Wang, H.H. (1987) A Gold-Vapor Laser Using Ne-H_2 as a Buffer Gas, *Appl. Phys.* **B 44**, 57-59.

19. Livingstone, E.S., Jones, D.R., Maitland, A., and Little, C.E. (1992) Characteristics of a copper bromide laser with following Ne-HBr buffer gas, *Optical and Quantum Electronics* **24**, 73-82.

20. Sabotinov, N.V., Vuchkov, N.K., and Astadjov, D.N. (1990) Copper Bromide Lasers - Discharge Tubes and Lifetime Problems, *High-Power Gas Laser*, Proceedings SPIE 1225, Los Angeles.

21. Elaev, V.F., Lyakhana, G.D., and Pelenkov, V.P. (1989) CuBr Laser with Average Lasing Power Exceeding 100 W, *Atm. Opt.* **2**, 1045-1047.

22. Vuchkov, N.K., Astadjov, D.N., and Sabotinov, N.V. (1991) A new circuit for CuBr laser excitation, *Optical and Quantum Electronics* **23**, S549-S553.

23. Vuchkov, N.K., Astadjov, D.N., and Sabotinov, N.V. (1994) Influence of the Excitation Circuits on the CuBr Laser Performance, *IEEE J. of Quantum Electronics* **30**,.750-758.

24. Vorobev, V.B., Kalinin, S.V., Klimovski, I.I., Kostadinov, I., Krestov, V.A., Kabasov, N.V., and Marasov, . (1991) A copper vapor laser with average specific emission power above 1 W/cm^3, *Kvant. Electronika* **18**, 1178-1180.

METAL HYBRID LASERS

C.E. LITTLE, D.R. JONES, S.A. FAIRLIE AND C.G. WHYTE
School of Physics and Astronomy, University of St Andrews
North Haugh, St Andrews, Fife KY16 9SS, Scotland

Abstract

A brief history of the development of HyBrID lasers is given, and the mechanisms that govern the metal seeding process in metal HyBrID lasers are discussed, with specific reference to the operating regime of the copper HyBrID laser.

1. Introduction

Copper HyBrID (Hydrogen Bromide In Discharge) lasers have been developed at the University of St Andrews as a variant of conventional elemental copper vapour and copper halide lasers. The efficiencies and output powers of HyBrID lasers can be double those of conventional elemental copper vapour and halide lasers of similar dimensions. Average powers of 120 W can be generated from a single laser tube with the record conversion efficiency, for any high power visible laser, of 3.2% (based on stored energy) [1]. The same device gave a maximum average output power of 201 W with an efficiency of 1.9% [1]. The highest specific average output power for copper lasers of any type, 2.0 $W.cm^{-3}$ (9.5 W average output power), was also achieved with a copper HyBrID laser [2]. Recently, 130 W average power of near-diffraction limited output was achieved using an injection-seeded HyBrID oscillator [3].

In metal HyBrID lasers, the metal halide lasant donor species are formed on demand in the active zone by reaction of metal pieces with the discharge products of a halogen-bearing vapour, usually HBr, that is added to the flowing buffer gas. These lasers usually operate hundreds of degrees below the temperatures required for lasers utilizing the same metal and which depend on vaporization of that metal to provide the requisite metal atom density in the active region. As a result, the designs of HyBrID lasers are simpler than those of their elemental metal cousins. The low operating temperatures of HyBrID lasers also allow the warm-up time to be as short as a few minutes, compared with an hour or more for elemental lasers.

Clearly, a knowledge of the details of the seeding mechanism is very important for the future development and understanding of metal HyBrID lasers. We shall discuss this after the brief description of the HyBrID laser's development which now follows.

C. E. Little and N. V. Sabotinov (eds.), Pulsed Metal Vapour Lasers, 125–136.

2. HyBrID Laser Development

In 1982, Vuchkov *et al.* [4] reported a sealed-off copper laser where copper pieces were placed along the length of the discharge region, and reacted with the vapours of $BiCl_3$, $FeCl_2$ or NaCl, each supplied from a number of reservoirs attached to the discharge region. This device was essentially a HyBrID laser, since a vapour containing chlorine-bearing molecules was made to react with copper pieces in the active region to generate *in situ* the volatile CuCl molecules which introduced copper to the pulsed discharge. Similar experiments were conducted by Saito and Taniguchi [5-7] on copper lasers where volatile compounds of copper, of the type $Cu_lAl_mCl_n$, were produced through reactions of metallic copper powder and $BiCl_3$, $AlCl_3$ or $AlBr_3$ at 90°C, 70°C and 25°C, respectively, in a reservoir. The volatile copper compound so-produced entered the discharge, carried by a flow of He buffer gas.

It was established by Bokhan *et al.* [8, 9] in 1980 that the addition of small quantities of H_2 to the gas mixture of conventional copper vapour lasers can enhance output power and efficiency. The enhancement for copper bromide and copper chloride lasers was shown by Astadjov *et al.* [10] to be even greater, where output powers and efficiencies can be doubled by the addition of H_2.

Output power and efficiency can also be enhanced by the addition of hydrogen-bearing molecules [11, 12] to copper vapour and copper halide lasers. One of the earliest hydrogen-bearing molecular additives which were found to enhance the output characteristics of copper lasers was HCl, demonstrated by Vetter and Nerheim in 1976 with a double-pulse copper chloride laser [12]. Vetter and Nerheim showed that addition of HCl increased the output pulse energy of their laser, whereas H_2 caused it to fall and led to considerable plating of metallic copper on the discharge tube wall. In 1980, Tobin [13] considered that $CuBr_2 3Cu(OH)_2$, commonly present as an impurity in the CuBr compound used in CuBr lasers, decomposes to products which can aid the deactivation of the metastable lower laser levels. The dissociation products of $CuBr_2 3Cu(OH)_2$ are CuO and the vapour-phase, hydrogen-bearing species H_2O and HBr [13].

The idea to react copper metal with a halogen-bearing gas, which has been added to a flowing inert buffer gas, to produce a volatile copper donor compound *in situ* in the active zone was patented in 1988 by EEV Ltd [14]. In the first of these lasers, Livingstone and Maitland used Cl_2 and Br_2 [15], and then HBr [16] entrained with the flowing Ne buffer gas to produce volatile copper halides. Copper cylinders were used to provide the wall of the active region. Although these devices were powerful and simple to operate, they were not suited to long-term operation - laser oscillation would often become unstable after only a few minutes running - and the behaviour of the long copper cylinders in the tube introduced further engineering problems.

In 1992, to obtain stable operation of these metal-halogen reaction lasers, two of the authors (Little and Jones) began experiments with a smaller, simpler laser tube of quartz, used with or without an inner alumina liner. Small metal pieces were loaded into the tube at intervals along the active region, as is done in conventional elemental metal vapour lasers. The laser tube itself was essentially a scaled-down version of the device used at Macquarie University, in the late 1980's for Sr ion recombination laser development [17].

We also wanted to see how self-terminating metal transitions other than those in copper would behave in metal-halogen reaction lasers. In the beginning, we demonstrated lasing in lead [18] and then bismuth [19, 20] using the technique with HBr added to the flowing Ne buffer gas in the 12.5-13 mm bore devices. In the case of bismuth, the idea was to cause bismuth dimers to be displaced by bismuth bromide molecules, and so remove the dimers that were believed at that time to be detrimental to laser oscillation. Unfortunately, the bismuth laser did not behave any better in this form than it does as an elemental metal laser. However, we did see how easy it was to obtain laser oscillation using the hydrogen halide reaction scheme, in what was generally thought to be a difficult laser system. We pressed on to characterize the same small-bore laser system as a copper laser [21], and followed it up with iron [22] and manganese [23] laser characterization studies.

Copper was of course the most powerful and efficient lasant, although manganese also operated very well - we attained specific output powers of up to 125 mW.cm^{-3} in the latter case. We also tried rhenium [22] - which would need a temperature of 3000°C to operate in an elemental system - in the same small laser by producing ReBr in a discharge at around 800°C wall temperature, but we couldn't get it to lase on the predicted yellow transition [24] in our 30 cm long tube.

With the support of BNFL and then Urenco, we scaled our copper HyBrID lasers to larger devices, and continued to improve their design [25, 26]. In particular, we used a cylindrical quartz insert called a 'spot-lock' to line the discharge-facing surface of each cylindrical copper electrode. The spot-lock has a small hole (0.5 to 4 cm^2, depending on the size of current required) drilled in one side, which has two major advantages. First it restricts the attachment of the discharge at each electrode to the small area defined by the hole, reduces unwanted reaction of the HBr with each copper electrode by shielding that part of it not supplying current to the discharge. Second, the spot-lock stabilizes the discharge at the electrodes to give flicker-free output laser beams, and virtually eliminates the production of CuBr 'dust'. The scaling work culminated in the demonstration of 200 W average output power at just under 2% efficiency from a tube of dimensions 2 m active length and 60 mm bore [1].

Most recently, we have pursued the following lines of research in the laboratories at St Andrews. First, we have been investigating the possibility of sealed-off operation of the HyBrID laser, which is important to be able to dispense with the cylinder of toxic gases and thereby make the laser attractive to use in future industrial applications. Second, the use of saturable magnetics in excitation circuits for copper HyBrID lasers has been examined. Third, we have investigated the use of unstable resonators for the HyBrID laser to produce high-quality radiation for nonlinear generation in BBO in a collaboration with Prof. Petrash's group at the Lebedev Physical Institute, Moscow. Fourth, we have produced high-power (130 W), high beam quality (near-diffraction limited) output by using injection-seeding and MOPA arrangements in a collaboration with Prof. Piper's group at Macquarie University, Sydney. Finally, we have just completed measurements of ground and excited copper populations and electron density in a copper HyBrID laser using hook spectroscopy and refractive index measurements in a collaboration with Prof. Webb's team at the University of Oxford [27]. We shall report on these developments in the near future.

3. Metal Seeding

The production of volatile metal halides by the reaction of halogens or hydrogen halides with a heated metal surface is the first step of the well-known 'iodide' process or 'hot filament' technique of metal purification [28]. In this process, elemental iodine and crude metal are reacted to form a volatile metal iodide, which can be thermally decomposed on a hot filament back into the solid metal and gaseous iodine. The freed iodine is then available to react with more of the crude metal in a cyclic manner.

The thermodynamics of the iodide process is applicable to the operation of metal HyBrID lasers, since the metal halide is generated in the same way in each case. When the solid or liquid metal is present in a halogen atmosphere, the rate of production of metal in the gas phase, as either atoms or part of a chemical compound, depends on four main factors: the metal vapour pressure, the metal halide vapour pressure, the thermal stability of the halide and the partial pressure of the halogen [29].

The purification of metals by the iodide process can be carried out using either the flow method or the sealed-bulb method [30]. In the first technique, the metal halide vapour produced is entrained in a gas and transported to the site of deposition (which is usually a heated filament). The second method requires the continuous formation of the halide by reaction, its diffusion to the deposition site and subsequent decomposition of the halide within a sealed environment. The technology of the flow method of the iodide process is of course applicable to flowing-gas metal HyBrID lasers, whereas the sealed-bulb method may form the basis of a sealed-off metal HyBrID laser.

With the flowing gas method, the vapour pressure of the halide that is sustained depends on the flow rate of the carrier gas used, and the rate of reaction at the surface of the metal, which in turn depends on the metal surface temperature and the partial pressure of the halogen or hydrogen halide.

Deposition of the metal on the hot filament may be accelerated by a reducing reaction. For example, lower temperatures are required to dissociate metal chlorides in the presence of hydrogen than in its absence [30].

If the halogen concentration is allowed to increase, the rate of metal deposition will diminish [29]. An increase in halogen partial pressure can be prevented by placing a small quantity of the metal near the deposition region, which is at a lower temperature than the filament. Excess halogen can then be removed by the formation of the solid metal halide [28]. This procedure is of relevance to controlling the halogen partial pressure in sealed-off metal halide and HyBrID lasers.

4. Processes in Flowing Gas Metal Vapour Lasers

In the HyBrID laser, metal pieces are usually placed at regular intervals along the laser tube active length, and the volatile metal halide is formed by reacting the clean metal surfaces with the vapours of HBr, Br_2, HCl, Cl_2 or SF_6 or their dissociation products produced in the continuously pulsed gas discharge. The halogen donors are premixed with the inert buffer gas that flows through the laser.

In HyBrID lasers that make use of the hydrogen halides, there is a one-to-one ratio of added hydrogen atoms to added halogen atoms. It is therefore expected that hydrogen reduction of the metal halides will be an important process controlling the equilibrium vapour pressure of the metal halide in the laser tube.

In choosing reactant additive for a flowing-gas laser, the conditions to be satisfied are:

1) The metal lasant atoms should form a volatile compound at a readily accessible temperature.

2) The compound so-formed should have a ratio of metal-to-halogen atoms close to unity, or better still greater than unity (*e.g.* ideally monohalides or sub-halides), so that the halogen atom concentration does not exceed levels above which discharge stability is compromised, whilst maintaining the metal halide concentration at that level required for strong laser oscillation.

3) The metal halide should have a decomposition/reduction temperature well above the wall temperature needed to maintain the halide at the vapour pressures that are necessary for laser operation, *i.e.* a few tenths of a torr.

4) The metal halide should be readily dissociated in a pulsed gaseous discharge, suitable for the production of a population inversion in the metal atoms.

If there is metal or metal halide deposition near the outlet of the laser tube, then the thermal characteristics of the laser should be such that deposition takes place at controlled locations where it will not interfere with laser operation or affect the life-time of the laser tube.

The reaction processes at the surfaces of the metal pieces in metal HyBrID lasers can be classified according to surface temperature. In the case of copper:

1) *Metal surface temperature < 450°C.* The species HBr, Br_2, Br *etc.* present above the copper surface will react slowly with it to produce solid CuBr (and H/H_2). If the rates of arrival of HBr, Br and Br_2 are faster than the evaporation rate of CuBr at the surface temperature, then the copper piece will become coated with CuBr. The rate of removal of copper atoms from the surface will be the evaporation rate of CuBr. The CuBr coating may react further to produce $CuBr_2$, for which the evaporation rate of $CuBr_2$ will determine the rate of removal of copper atoms from the surface.

2) *Metal surface temperature > 450°C, <800-900°C.* At a sufficiently high surface temperature, the evaporation rate of CuBr may be higher than the rate of

arrival of Br-bearing species, in which case no surface layer of CuBr will form, and any surface layer formed at lower temperatures will be removed. The rate of production of CuBr vapour will then depend on the rate of arrival of HBr, Br and Br_2 at the surface. This is the working regime of a flowing-gas HyBrID laser, where the metal surface is maintained in a 'clean' state by the removal process. The reaction rates are governed primarily by processes at the gas-solid interface, and will not depend on diffusional transport through a layer of condensed product, as is the case in many oxidation reactions.

3) *Metal surface temperature > 900°C.* As the temperature of the metal surface is raised higher still, a point is reached where CuBr ceases to be stable at the metal surface. As temperature is raised further, the rate of removal of copper atoms from the surface will fall until the rising evaporation rate of copper takes over, and the copper vapour density above the metal surface begins to rise again. The removal rate of copper will have a minimum at an elevated temperature (~900°C when the halogen is iodine [31]). The addition of hydrogen to the halogen will lower the temperature at which the rate of removal of copper will begin to fall by introducing hydrogen reduction of the halide as a metal deposition mechanism.

In the steady state, the rate of loss of metal atoms to the tube wall by condensation, the rate of loss of metal atoms and copper bromide molecules out the end of the laser tube with the gas flow, and the rate of loss of metal halide molecules by condensation and by thermal dissociation and hydrogen reduction, must sum to be equal to the rate of removal of metal atoms from the metal surface and the alumina tube wall by reaction with halogen atoms/molecules. Since the inside wall of the alumina tube remains clean of metal deposits during ordinary operation of a metal HyBrID laser, and the metal pieces slowly lose their mass and appear to have clean surfaces at all times, we can assume that the rate of supply of metal atoms to the discharge is limited by the rate of reaction of the halogen-bearing species with the metal surfaces to yield volatile metal halides.

Frommer and Polanyi [32] examined the reaction of chlorine with copper surfaces in temperature regime 2. At chlorine pressures of a few millitorr and surface temperatures between 400°C and 700°C, the reaction is rapid (collision yield ~0.3), with an activation energy of less than 12.6 $kJ.mol^{-1}$ (0.13 eV per molecule). This rapid temperature independent rate was said to be characteristic of the reactions of many metals and alloys with chlorine and bromine. However, clearly, the reaction rate cannot be independent of temperature, otherwise the iodide process would not work. Thus, the results of their experiments can be taken to be a maximum collision yield for the reaction of solid copper and gaseous chlorine.

Brewer and Lofgren [33] studied the reaction of solid copper with HCl vapour, and found that Cu_3Cl_3 trimers were present above the metal surface in a greater number than CuCl molecules. This means that a full description of the kinetics above and on the copper metal surface would require the trimers and higher-order polymers to be included in the analysis. In the case of copper iodide, the number of CuI monomers is thought to dominate over the number of polymers, although they are still present [28]. The situation for the bromide is similar to that for the chloride [28].

The reactions that govern the formation of copper bromide monomers at the tube wall will include:

$$Cu(s) + Br_2(g) \rightarrow CuBr(g) + Br(g)$$
$$Cu(s) + Br(g) \rightarrow CuBr(g)$$
$$Cu(s) + HBr(g) \rightarrow CuBr + H$$
$$Cu(s) + H(g) + Br(g) \rightarrow CuBr(g) + H(g)\,.$$

The removal of copper molecules from the gaseous phase on a hot surface will include:

$$\text{thermal dissociation:}\ CuBr(g) \rightarrow Cu(s) + Br(g)\,,$$
$$\text{hydrogen reduction:}\ CuBr(g) + H(\text{or } H_2)(g) \rightarrow Cu(s) + HBr(g)\,.$$

5. Chemical Equilibrium in the Flowing Gas HyBrID Laser

We will attempt to estimate the copper production and loss rates in the 200 W copper HyBrID laser described in [1], after making a number of observations.

During operation of the laser, the entire inside wall of the alumina laser tube is available as a reaction surface. That this surface is free from copper deposition under normal running conditions implies that any solid copper formed on the alumina surface is removed faster than it can be deposited. Hence, the effective source of copper under steady state running will remain to be the copper pieces originally placed in the tube.

By visual inspection, the sidelight emitted by the laser discharge is uniform along the full length from anode to cathode. This implies that equilibrium is established near the point of entry of the gases to the discharge zone near the anode, and that this equilibrium is maintained as the seeded gas flows along the tube. We have found that around 5 cm^2 area of copper surface facing the discharge, and located in the tube near the anode, is sufficient to ensure a CuBr vapour density close to optimum for laser oscillation along the length of the discharge. Since tan equilibrium mixture of Cu, CuBr, HBr vapours *etc.* is established by reaction at the very first copper pieces, reactions at subsequent copper pieces will merely act to restore the copper density as the gas passes along the tube. Hence, upstream pieces of copper are used up more quickly than downstream pieces.

Copper is found to be removed only from the sides of the pieces that face the discharge volume, and not from the underneath of the pieces, even though there may be a gap of several millimetres between the copper and the alumina. Therefore, the reaction to remove copper from the pieces proceeds by the action and/or mediation of an active species from the discharge, which is not likely to be found on the other side of the copper pieces. Such active species could include electrons and ions which would recombine at the surface, atomic bromine or hydrogen, and/or UV light from the discharge. Stable species such as HBr and Br_2 will be less likely to be involved. For the purposes of the following calculations we shall assume that CuBr is formed at the copper surface by the reaction of Br atoms, and that the collision yield is 0.3, as Frommer and Polyani found for the Cu-Cl_2 system [32].

The presence of hydrogen in copper halide lasers is known to increase the rate of loss of copper from the discharge. For example, when H_2 is added to conventional copper chloride lasers, the deposition of metal on the tube wall is observed to greatly accelerate [12]. This could be due to any of the following mechanisms: conventional hydrogen reduction of the halide at the copper surface; hydrogen reduction above the copper surface, followed by condensation of the copper vapour; or thermal dissociation of CuH formed near the wall. Condensation of CuBr on the wall as a loss process in HyBrID lasers is ruled out because the tube wall temperature (≥600°C) is sufficient to allow a vapour pressure of >1 torr of CuBr to be maintained [34].

We will consider hydrogen reduction on the copper surface to be the dominant loss mechanism in our case. Since the ceramic wall remains free of copper under normal operating conditions, the reaction there to produce CuBr must be faster than the deposition process. It is likely that the copper surface catalyses the reduction process, either by adsorption of hydrogen there, or by possessing a temperature which is higher than the ceramic wall temperature. This last fact is indicated by the growth of dendrites preferentially at sites on copper surfaces when the tube temperature or hydrogen concentration is too high.

In the case of hydrogen reduction on the copper surface, the rate of loss of copper from the discharge will depend on the collision rate of CuBr on the copper surface. The CuBr concentration in conventional copper bromide lasers is ~0.3 torr [35]. We shall assume a similar concentration of copper bromide to be present near the wall in the HyBrID laser, although the amount of copper in the HyBrID discharge may be greater [36]. Taking the reduction of CuBr to occur at gas kinetic rates, the loss of CuBr at the copper surface is 2.8×10^{19} $cm^{-2}.s^{-1}$. If, as for the reaction of chlorine with copper surfaces [32], every third collision of Br leads to the formation of a copper bromide molecule, then the density of atomic bromine near the surface can be estimated to be 6.6×10^{15} cm^{-3}, or 0.7 torr at 700 K. There is 1.2-1.7 torr added HBr in a gas mixture of 24 torr total pressure in the 200 W laser [1], so that 40-60% of the added bromine is present in atomic form near the tube wall. The balance of the bromine is presumably locked up as CuBr (0.3 torr), HBr and, to a lesser extent, as Br_2.

The flow rate of gas through the laser tube is 3-4 litre.atm.hr^{-1}, which translates to 8-11 x 10^{22} molecules/atoms per hour, or 2.2-2.9 x 10^{19} particles/s. From the partial pressure of CuBr (around 0.3 torr in 24 torr Ne-HBr), we see that the loss of copper from the end of the tube is 4-5 x 10^{16} atoms s^{-1} (~0.01 g.hr^{-1}). The rates of forward and reverse reactions that determine the equilibrium CuBr density of 0.3 torr in the tube, taking a 5 cm^2 copper reaction area, is 1.4×10^{20} s^{-1}. This rate is over two orders of magnitude greater than the rate of loss of copper atoms from the end of the tube. Hence, reactions at the wall completely dominate over gas entrainment in establishing the equilibrium CuBr density in the discharge zone.

In general, it is found that there is an optimum gas flow rate associated with maximum laser output power in the laser. It is also clear, however, that equilibrium is established as soon as the gas has traversed the first few copper pieces at the beginning of the tube. This is consistent with the typical <1 $cm.s^{-1}$ flow rates used in the 200 W laser, which we would expect to give ample time for chemical equilibrium to be established. If the gas flow was raised to such an extent as to prevent equilibrium from being attained near the beginning of the laser tube, it would simply be established further downstream after encountering more copper pieces, *i.e.* a larger copper area would allow the equilibrium to be established, although the output power would be lower. Our experiments show that if the flow rate is increased by a factor of 3 then the output power falls by ~25%. However, equilibrium is established for the same 5 cm^2 area of exposed copper surface as for the optimum flow rate. Hence, although the laser power falls at higher flow rates, it is not because of dilution of the CuBr molecules in the buffer gas as a result of higher gas flow, since the effect could be reversed if a larger area of copper was provided near the inlet of the tube. One possible mechanism which could act to lower laser power if the flow rate was too high is a reduction in the local gas temperature at the inlet end of the laser tube.

6. Growth of Metal Dendrites in HyBrID Lasers

In small bore (13 mm) copper and iron HyBrID lasers [21, 22], we found that metallic dendrites can grow on the metal pieces in the laser active zone over 1 to 2 hours, which can reduce the transverse cross section of the laser gain medium. On the other hand, with a manganese HyBrID laser, which was operated for periods of up to 7 hours at a time, we found no sign of the growth of dendrites on the pieces of manganese in the laser tube, and laser power remained at its initial level over the course of the day's experiments.

Contrary to what might be expected, copper is not deposited on the coolest parts of the laser tube, but on the hottest sections. The dendrites are believed to be formed by the thermal decomposition and hydrogen reduction of the metal halide molecules on hot surfaces in the laser tube, although there is evidence that charged particles also play some role in dendrite growth. In our copper and iron HyBrID lasers, the lasant metal pieces did not have good thermal contact with the tube wall, and lay in the discharge channel. As the dendrites grew into the discharge, further heating of the tips closest to the discharge occurred, and the deposition of metal by decomposition and reduction accelerated.

However, unlike copper and iron iodides, pure manganese has not been obtained from the thermal decomposition of manganese iodide using the iodide process because of its thermal stability and the high vapour pressure of manganese metal [28]. This explanation probably also holds for the bromides, thus explaining why manganese dendrites were not observed in our manganese HyBrID laser.

In order to prevent the growth of dendrites in HyBrID lasers, and hence to prevent aperturing of the beam, the uniformity of tube temperature and the hydrogen concentration must be controlled carefully. Local hot areas in the laser tube, due to uneven insulation, non-uniform discharge channel (particularly near the electrodes) *etc*. will promote the growth of dendrites and should be designed out of the laser. By ensuring that the hydrogen concentration is not too high, the rates of reduction reactions will be controlled.

In HyBrID lasers where the metal lasant is in liquid form, dendrites will not form, since there is then no mechanical strength of support. This is the case in lead and bismuth HyBrID lasers [18-20].

The copper HyBrID laser with very high specific input powers (370 $W.cm^{-3}$) and average output powers (2 $W.cm^{-3}$) described in [2] was designed with internal silica apertures (4.5 mm) such that the inside tube wall diameter (12 mm) was larger than the discharge width (4.5 mm). Dendrites did indeed grow up quickly towards the discharge channel. However, the input power deposition in the discharge was such that the dendrites melted when they came close to it, so that those growths did not cause aperturing of the laser beam, and were benign in that case.

7. Conclusions

The metal HyBrID laser is a logical development of earlier metal halide laser work, in particular that of Astadjov *et al.* [4], Saito and Taniguchi [5-7] and Vetter and Nerheim [12].

The HyBrID laser's characteristics depend on a balance between (1) reactions of halogen-bearing species with copper metal to generate volatile metal halides, and (2) the reverse reaction of hydrogen reduction, which causes metal to be deposited on the wall of the discharge tube. The operating temperature regime is rather broad (500 - 800°C) because the rates of the forward and reverse reactions can be controlled to a large extent by the amount of HBr added, so as to control the density of metal in the discharge. 'Hot' spots or excess hydrogen in the laser tube can cause accelerated loss of metal from the active volume in the form of dendrites. These metal seeding processes should be considered in any future kinetic model of metal HyBrID lasers.

8. Acknowledgements

The authors thank Urenco (Capenhurst) Ltd for supporting their work on copper HyBrID lasers.

9. References

1. Jones, D.R., Maitland, A., and Little, C.E. (1994) A high efficiency 200 W average power copper HyBrID laser, *IEEE J. Quantum Electron.* **QE-30**, 2385-2390.
2. N V Sabotinov, D R Jones, A Maitland and C E Little (1995) A copper HyBrID laser with 2.0 W/cm^3 specific average output power, *IEEE J. Quantum Electron.* **QE-31**, 747-753.
3. Results of collaborative experiments between the University of St Andrews (Scotland) and Macquarie University (Australia), announced by D.J.W. Brown during his talk at this NATO ARW (paper in preparation).

4. Vuchkov, N.K., Astadjov, D.N., and Sabotinov, N.V. (1982) A copper vapor laser utilizing pure copper with an admixture of other metal chlorides, Opt. Comm. 42 199-200.
5. Saito, H., and Taniguchi, H. (1985) Low-temperature operation of copper-vapor lasers by using vapor-complex reaction of metallic copper and metal halide, *IEEE J. Quantum Electron.* **QE-21**, 1308-1309.
6. Saito, H., and Taniguchi, H. (1985) Generation of the metal vapor-complex and low temperature operation of the metal-vapor-laser, *Rev. Laser Engineering* **14**, 677-685.
7. Kano, T., Taniguchi, H. and Saito, H. (1987) A copper vapor laser by using a copper-vapor-complex reaction at a low temperature, *Trans. IEICE* **E70** 312-314.
8. Bokhan, P.A., Silant'ev, V.I., and Solomonov, V.I. (1980) Mechanism for limiting the repetition frequency of pulses from a copper vapor laser, *Sov. J. Quantum Electron.* **10** 724-727.
9. Bokhan, P.A., Mal'tsev, A.N., and Silantiev, V.I. (1980) A new mechanism for limiting generation mean power of metal-vapour pulse lasers, *IX Conf. on Quantum Electronics and Nonlinear Optics, Apr. 23-26, 1980, Poznan, Poland, summaries* 333-334.
10. Astadjov, D.N., Vuchkov, N.K., and Sabotinov, N.V. (1985) Effect of hydrogen on CuBr laser power and efficiency, *Opt. Comm.* **56** 279-82.
11. Huang, Z., Namba, K., and Shimizu, F. (1986) Influence of molecular gases on the output characteristics of a copper vapor laser, *Jap. J. Appl. Phys.* **25** 1677-1679.
12. Vetter, A.A., and Nerheim, N.M. (1977) Addition of HCl to the double-pulse copper chloride laser, *Appl. Phys. Lett.* **30** 405-407.
13. Tobin, R.C. (1980) Rapid differential decay of metastable populations in a copper halide laser, *Opt. Comm.* **32** 325-330.
14. EEV Ltd (1988), UK patent, Laser apparatus, Application No 8812276, 24 May 1988, Patent Application No 2219128 A, inventors: Maitland, A. and Livingstone, E.S.
15. Livingstone, E.S. and Maitland, A. (1989) A low temperature, segmented metal, copper vapour laser, *J. Phys. E: Sci. Instrum.* **22** 63.
16. Livingstone, E.S. and Maitland, A. (1991) A high power, segmented metal, copper bromide laser, *Meas. Sci. Technol.* **2** 1119-1120.
17. Little, C.E. and Piper, J.A. (1990) Average-power scaling of self-heated Sr^+ afterglow recombination lasers, *IEEE J. Quantum Electron.*, **QE-26** 903-910.
18. Jones, D.R. and Little, C.E. (1992) A lead bromide laser operating at 722.9 nm and 406.2 nm, *IEEE J. Quantum Electron.* **QE-28** 590-593.
19. Jones, D.R. and Little, C.E. (1991) Laser action at 472.2 nm in a bismuth halide laser, *III European and X Nat. UK Quantum Electronics Conf., Heriot-Watt University, Edinburgh, 27-30 Aug. 1991*, Technical Digest, 57.
20. Jones, D.R. and Little, C.E. (1992) A 472.2 nm bismuth halide laser, *Optics Comm.* **91** 223-228.
21. Livingstone, E.S., Jones, D.R., Maitland, A., and Little, C.E. (1992) Characteristics of a copper bromide laser with flowing Ne-HBr buffer gas, *Opt. Quantum Electron.* **24** 73-82.
22. Jones, D.R. and Little, C.E. (1992) A self-heated iron bromide (λ452.9 nm) laser with Ne-HBr buffer gas", *Opt. Quantum Electron.* **24** 67-72.
23. Jones, D.R. and Little, C.E. (1992) A compact, high-power, fast start-up manganese bromide laser (λ534-554 nm, 1.29-1.40 μm)" *Opt. Comm.* **89** 80-87.
24. Walter, W.T., Solimene, N., Piltch, M., and Gould, G. (1966) Efficient pulsed gas discharge lasers, *IEEE J. Quantum Electron.* **QE-2** 474-479.
25. Jones, D.R., Sabotinov, N.V., Maitland, A., and Little, C.E. (1992) A high-power high-efficiency Cu-Ne-HBr (λ 510.6, 578.2 nm) laser *Opt. Comm.* **94** 289-299.
26. Jones, D.R., Maitland, A., and Little, C.E. (1993) A copper HyBrID laser of 120 W average output power and 2.2% efficiency *Opt. and Quantum Electron.* **25** 261-269.
27. Whyte, C.G., Little, C.E., Hogan, G.P., and C. E. Webb (1995) Spatially and temporally resolved measurements of electron density in a copper HyBrID laser, *Conf. on Lasers and Electro-Optics '95, Baltimore, Maryland, USA, 21-26 May 1995*, OSA Technical Digest.

28. Rolsten, R.F. (1961) *Iodide Metals and Metal Iodides* (New York: John Wiley & Sons).
29. Waymouth, J.F. (1971) *Electric Discharge Lamps* (Cambridge, MA: The MIT Press).
30. Shelton, R.A.J. (1957) The deposition of metals other than those of the titanium group by the hot filament technique, *Metallurgia* **55** 283-289.
31. Van Arkel, A.E. (1934) Über die Herstellung von hoch-schmelzenden Metallen durch thermische Dissoziation ihrer Verbindungen, *Metallwirtschaft* **13** 405-408.
32. Frommer, L. and Polyani, M. (1928) Über heterogene Elementar-reaktionen I. Einwirkung von Cl_2 auf Cu *Z. physik. Chem.* **137A** 201-208.
33. Brewer, L. and Lofgren, N.L. (1950) The thermodynamics of gaseous cuprous chloride, monomer and trimer, *J. Am. Chem. Soc.* **72** 3038-3045.
34. Srigouri, K., Ramaprabhu, S., and Prasada Rao, T.A. (1986) Vapour pressure equation for metal halides used in double pulse lasers, *Current Science* **55** 974-975.
35. Astadjov, D.N., Vuchkov, N.K., and Sabotinov, N.V. (1988) Parametric study of the CuBr laser with hydrogen additives, *IEEE J. Quantum Electron.* **QE-24,** 1927-1935.
36. Jones, D.N., Halliwell, S.N., and Little, C.E. (1994) Influence of remanent electron density on the performance of copper HyBrID lasers, *Optics Comm..* **111** 394-402.

OUTPUT LIMITING MECHANISMS AND THE PROSPECTS FOR METAL VAPOUR LASERS WITH AVERAGE OUTPUT POWER ABOVE 1 KW/M.

P.A. BOKHAN
Institute of Semiconductor Physics, 630090, 13 Lavrentiev AV, Novosibirsk, Russia
E.I. MOLODYKH
Special Design Bureau of High-Energy Lasers "Granat" 95 Volokolamskoye shosse, Moscow, Russia

1. Introduction

Pulse metal vapour lasers, specifically copper vapour lasers, are the most powerful and efficient among the visible lasers. Meanwhile their characteristics are far from those which were predicted in the first stage of investigation [1,2]. This paper has pursued the goal to make a critical review of investigations which were fulfilled to elucidate the rate-power restriction mechanisms and provide an answer to methods of producing the output characteristics which were presented in the papers [1,2] on the foundation of the metals energy levels properties, which are the lasers operating media. These are: efficiency of *eff* ~ 23 % [1] and mean linear output power of more than 1 kW/m [2].

2. Specific output energy and efficiency.

Output power P_l of a pulse-periodic laser is defined by the well-known equation:

$$P_l = wVF \tag{1}$$

where *w*-specific output energy; F-pulse repetition rate; V-active volume.

Laser efficiency *eff* is defined by the well known equation as well [1]

$$eff \leq \eta \frac{g_{ms}}{g_r + g_{ms}} \frac{E_r}{E_{ms}} \tag{2}$$

where η is a share of total energy, injected into active volume, which is expended to pump an upper laser level; g_r, g_{ms}; E_r, E_{ms}; - accordingly the statistical weights and excitation energy of the resonance RS and metastable MS states.

C. E. Little and N. V. Sabotinov (eds.), Pulsed Metal Vapour Lasers, 137–148.

Among the quantities from equations (1-2) w, F and *eff* are the complex functions of parameters and properties of an active medium and a pumping pulse as well as a geometrical factor V. It has been just this circumstance which makes difficult a correct analysis of the reasons for the rate-power characteristics RPC restrictions of a such seemingly simple system as a laser on the self-terminated transition.

A character of lasing with the lasers being considered is such that under ordinary conditions the lasing period is $\tau_l \leq 100$ ns. Another important peculiarity is that the active particle lase not more than one time over a pulse. These properties force performing fast pumping of relatively dense active media. Typical concentration of active particles at gas discharge excitation is $N_a \sim 10^{15} \div 10^{16}$ cm^{-3}. A short lasing period and relatively high concentration of easily ionized particles of great excitation, direct and stepwise ionization cross section forces of use a powerful low inductance power supply, combined with gas discharge volume in a unified low inductance set. Under these conditions, along with useful RS excitation, powerful direct and stepwise ionization is realized, which consumes the greatest part of power supply energy. Strong excitation of MS is realized as well, therefore, $\eta << 1$.

This sets a limit to the laser efficiency. Under the best experimental conditions *eff*~3% [3,4] and only slightly depends on a geometrical factor V. Similar results are produced by a computer simulation [5], analytical theory [6] and analysis by the similarity theory [7]. Using of an electron beam to pump the lasers reduces a share of energy spent to excite the MS and for direct and especially stepwise ionization. As a result physical efficiency increases as much as approximately one order of magnitude and approaches the quantity predicted in paper [1] both under experimental data [8,9] and computer simulation investigations [10], Fig 1.

So upon determination of one from the most important characteristics, that is efficiency, accordance between theory and experiment is observed. It is clear that there are some ways to attain its limiting quantities.

Similar things have been on the way of improving a second characteristic - a specific output energy. Under the proper established synthesis laser tube + power supply, which is operating at the greatest or close to greatest efficiency, specific energy quantity is defined by the quantity of lasing particles and depends on N_a, electron temperature T_e and electron concentration N_e. T_e in turn depends on the operating strength of an electric field, the upper limit of which is determined from the matching conditions of a laser cavity with power supply.

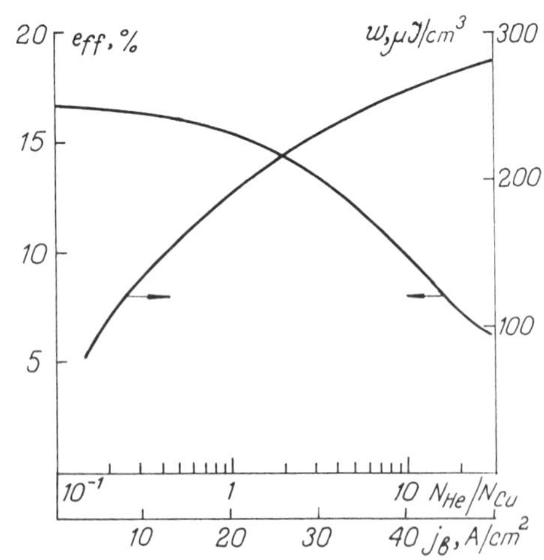

Fig.1. Lasing efficiency versus of active mixture composition (1) and specific output enegy w versus of a beam current density of a pulse duration 70 ns. $N_{Cu}=10^{16}cm^{-3}$, 1-j_b=10 A cm^{-2};2-N_{He}/N_{cu}=10:1

The totality of parameters listed here defines a strong upper limitation on specific pumping energy and accordingly on specific lasing energy. A typical specific lasing energy of well established lasers *w* is usually ranged from 1 to 10 μ J/cm^3 respectively from the longest wave lasers (Ca, Sr, Eu) to a Cu-laser. A specific energy, which is realized in a large body of research, is in a good agreement with the data of computer simulation, analytical theory and a simulation by the similarity methods. It only slightly depends from geometrical factor V and, consequently, is considered to be a more or less determined quantity as well. It should be only pointed out that because of larger efficiency under an electron beam pumping method an optimal specific energy approximately one order of magnitude greater as well and can be as much as from 10 to 100 μ J/cm^3, Fig. 1.

3. Lasing pulse repetition rate.

Essential discussions have been raising by the issue about an optimal F_{opt} and limiting F_{lim} repetition rate of lasing pulses and about the mechanisms which restrict them. It has been securely established that both F_{opt} and F_{lim}, contrary to the *w* and *eff*, have turned out to be complex and critically depend upon geometrical factor V and therefore could not be discussed in isolation from it.

In the case of a cylindrical geometry factor V is defined from equation:

$$V = \pi r^2 l \tag{3}$$

where *r* and *l* accordingly radius and length of an active cavity. The parameters *w* and *eff* both depend from *r* and *l*, as F changes. It is obvious that we are interested in the dependence on *r*, and we have been concentrating emphasis on that.

Early in the development of lasers on self-terminated transitions restrictions of F accepted as obvious should be caused by the rate of deexcitation of metastable states. MS are accepted as obvious as well should be quenched in collisions with a wall. Therefore a distinctive pulse repetition rate must depended on *r* as $F \sim r^{-2}$, which is to say at r~1cm F~1 *kpps*.

But direct measurements which were carried out in the study [11] with tube diameter d=15 *mm*, have demonstrated that essential copper vapour laser characteristics are valid at least up to F=15 *kpps*. As the authors of the papers [12,13] have concluded only deexcitation mechanisms involving electrons are capable of achieving such a high pulse repetition rate. Rate coefficients of the MS have been determined in paper [14]. Other fast deexcitation processes have been discovered as well, but an electron deexcitation has happened to be the most effective at self-terminated lasers conditions, except for case of state Eu (b^8D^0), Table 1.

From papers [13,14] and from numerous measurements by different authors the electron concentration at the near afterglow it is uniquely determined that typical rate of deexcitation exceeds $\sum k_i N_i \rangle 10^6 s^{-1}$, table 1. Direct measurement of MS concentration which have been fulfilled in the paper [15] and just earlier of this in paper [16] allowed the authors of the paper [15] to conclude that already after

$\tau = 10 \div 15\mu s$, MS concentration in a Cu-laser becomes so negligible that it is not liable to exert perceptible influence on the lasing parameters.

TABLE 1. Rate coefficients and deexitation rates of MS on typical self-terminated lasers conditions.

N^0	MS	k_e 10^{-8} cm^3s^{-1}	k_eN_e 10^6 s^{-1}	k_b 10^{-12} $cm^3\ s^{-1}$	k_bN_b 10^{-6} s^{-1}	k_a 10^{-12} $cm^3\ s^{-1}$	k_aN_a 10^6 s^{-1}	$\sum k_iN_i$ $10^6\ s^{-1}$
1	$Cu(^2D_{5/2})$	1.4	2.8	0	0	0	0	2.8
2	$Ba(^1D_2)$	7.5	15.0	0	0	17	0.08	15.1
3	$Ba^+(^2D)$	>3	>6	0	0	500	0.25	>6.2
4	$Ca^+(^2D)$	>3	>6	6	1.2	80	0.04	>7.2
5	$Eu(a^8D^0)$	>5	>10	2.5	0.5	900	4.5	>15.0
6	$Eu(b^8D^0)$	>5	>10	10	2	4000	20	>32
7	$Eu^+(a^9D^0)$	>5	>10	30	6	1400	0.7	>16.7
8	$Mn(a^6D)$	1.8	3.6	0	0	0	0	3.6
9	$Pb(^1D_2)$	21	42	0	0	45	0.22	42.2

Note: k_b,N_b;k_a,N_a -rate coefficients and concentration accordingly with the participation of buffer gases and operating atoms; $N_e=2\times10^{14}cm^{-3}$, $N_b=2\times10^{17}cm^3$, $N_a=5\times10^{15}cm^{-3}$ for atom lasers and $N_a=5\times10^{14}cm^{-3}$ for ion lasers.

The criterion of low-influence is fulfillment of the condition:

$$\frac{g_r}{g_{ms}} N_{ms}^{pp} << N_{ps} < N_r^{max} \tag{4}$$

where N_{ms}^{pp} - prepulse MS concentration; N_{ps}- specific photons quantity, emitted on the operating transition under a condition when N_{ps} does not depend on frequency; N_r^{max} - RS maximal concentration within a pumping pulse. Equitation (4) on physical nature is equivalent other inequality:

$$k_{pp} << k_p^{max} \tag{5}$$

where k_{pp} is a prepulse absorption coefficient at a lasing transition, which is determined by absorption of residual MS, and k_p^{max} is the greatest amplification coefficient, which is realized within an excitation pulse under conditions when k_p^{max} does not depend on F, i.e. at the single pulse mode operation.

Quantities from equation (5) are easily measured. On the other hand, it is well known from lasing theory that if an active medium has the prepulse absorption coefficient k_{pp}, conditioned by absorption from non relaxing states, and the greatest amplification coefficient is within excitation pulse k_p^{max}, then the lasing abatement ξ owing to the residual MS will be defined by the following equation:

$$\frac{\xi}{w}=\frac{k_{pp}}{k_p^{\max}}=\frac{N_{ms}^{pp}}{N_{ps}}\frac{g_r}{g_{ms}} \tag{6}$$

From the above reasoning, if the restriction of the RPC has exclusively been conditioned by the residual MS, then the mean lasing output, in accordance with equation (1,6), would not be less than:

$$P_l > \begin{cases} w\pi r^2 lF \\ 10^{-5}\times 3.14\times 1.82\times 10^2\times 10^5 = 572\ W/m \end{cases} \tag{7}$$

in accordance with the data from paper [15] for tube diameter d=2.7 cm, and from [17]

$$P_l > 4\times 10^{-6}\times 3.14\times 16\times 10^2\times 10^5 = 2010\ W/m \tag{8}$$

for tube diameter d=8 cm, if P_l would be conditioned by the specific energy from position r=3.5 cm from tube axis, and

$$P_l > 1430\ \ W/m \tag{9}$$

if P_l would be conditioned by the specific energy from position *r=0* (tube axis). But reality, as is well known, is much more insignificant:

$$P_l \le 50\text{W/m}\ \ (\text{d}=2.7\text{ cm})\text{ and }P_l\sim 100\text{W/m}\ \ (\text{d}=8\text{cm}) \tag{10}$$

From above reasoning we can uniquely determine that even for tube diameter d=8 cm, prepulse population of MS is of minor importance on restrictions of RPC.

Late in the 1977 we propounded a new mechanism for restriction of RPC. This mechanism is connected with an insufficient plasma recombination rate over the interpulse period [13,15]. It had been originally supposed that a high prepulse electron concentration causes decrease in an electron temperature T_e within an excitation pulse, and that this causes in turn the redistribution of pumping rates of the lower and upper operating levels in favor of the lower ones. This is just the mechanism of a restriction of the RPC.

Just later the hypothesis of influence of insufficient plasma recombination on the copper atoms concentration have been propounded [18,19]. This mechanism in combination with the decrease in atoms concentration, by way of gas heating in a central region of an active volume results lasing decay. An increase of these mechanisms for lasing decay is obvious upon decrease of an interpulse period. And at last third mechanism which is sufficiently often considered and connected with N_{eo} is a skin-effect [20].

In this paper for reasons of space we can not analyze these three last mechanisms. We only touch on this problem and conclude that by to-day's experimental data [17,21,22] it is easily to verify that their concrete total influence does not exceed 30 % even for the tube diameter 8 cm. This, of course, can not be compared with total lasing decrease, which follows from relations (7-10) and exceed at least one order of magnitude.

In the mid-80s we undertook direct experiments on measuring the $N^{pp}{}_{ms}$ concentration of lead vapour [8] and Ba vapour [23] lasers. But contrary to the paper [15] by way of additional gas discharge excitation we have independently changed $N^{pp}{}_{ms}$. By doing so we have looked more closely at their influence on the lasing characteristics. It was evident from these experiments that an influence of $N^{pp}{}_{ms}$ on the lasing characteristics is out of the question.

In the early 80s we have initiated two methods of overcoming the restriction, conditioned N_{eo}:

- an intensification of a plasma recombination rate over an interpulse period (a)
- going from pulse discharge to the pulse electron beams pumping method (b)

With aim of intensification of a recombination three methods have been suggested :

- adding hydrogen in mixture Cu - Ne [15] (c)
- adding a low-ionized molecular admixture in Cu - Ne medium [24] (d)
- improvement of the excitation circuits [15,25,26] (e)

Adding H_2 accelerates an electron cooling and consequently the rate of their

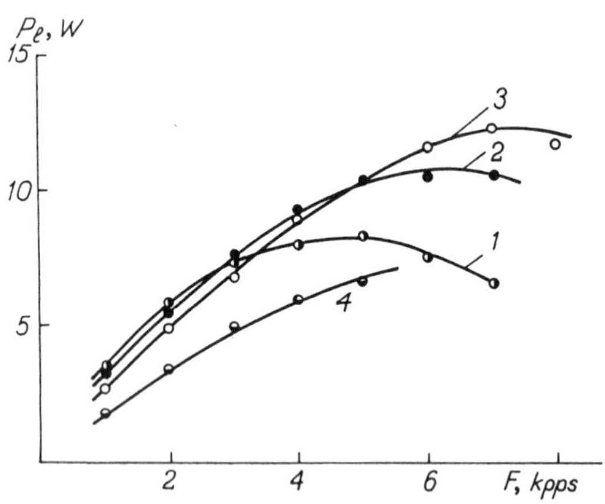

Fig.2. Influence of H_2 admixture on the rate - power characteristics of Cu laser. 1-without H_2; 2-4 accordingly H_2 admixture 13 Pa, 18 Pa, 32 Pa, l=68 cm, d=4 cm, U_r=4.5 kV.

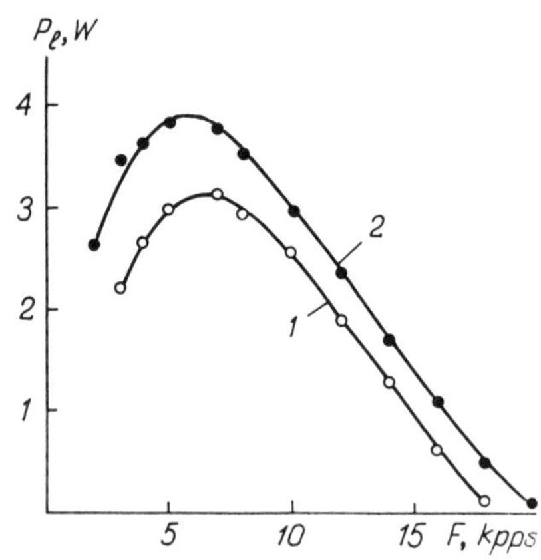

Fig.3. Influence of BaO admixture on the rate-power characteristics of Cu-laser. 1-without BaO; 2-1 Pa BaO admixture. l=25 cm, d=2.7 cm.

recombination over an interpulse period. In Fig.2 we demonstrate our results on investigation of lasing characteristics of a copper vapour laser with tube diameter 40 mm, partly published in the paper [27]. Optimization of a mixture composition made possible to obtain lasing output 60 W for tube diameter 40 mm as compared with 36 W without H_2 [29] for the same tube [28,8]. Currently all of the most powerful Cu-lasers exploit active media with H_2 admixture.

Adding of low-ionized molecular admixture makes it possible to replace slow three-body-particles recombination for fast dissociating recombination through a charge exchange mechanism, for example:

$$Cu^+ + BaO \rightarrow BaO^+ + Cu + \Delta E \tag{11}$$

$$BaO^+ + e \rightarrow Ba + O + \Delta E \tag{12}$$

We also have carried out the experiments with BaO adding in mixture Cu - Ne and established that improves the output characteristics, especially at the reduced frequency (Fig.3). The later derives from the fact that the process (12), contrary to the three-particles recombination, depends only to the first power on electron concentration. Therefore it is more effective for a far afterglow.

And, at least, a third method is as following: reduction of the tube inductance and inductance closing the lasing tube [3,15]. The latter adjustment reduces the current via tube while capacity is charged. Other methods of reduction of this current use the application of interacting exciting circuits [25], a transformer pumping circuit [26] and power supply with magnetic compression of the pumping pulse. Also, the quantity of energy, entered into a tube over an interpulse period is negligible in comparison with the latter within pumping pulses, this energy having a pronounced effect on the recombination process, particularly at the high repetition rate.

But also adding H_2 and improving of exciting circuits significantly improves Cu-vapour rate-power characteristics, especially in a CuBr laser [30], as well as in a gold and barium lasers, nonetheless their characteristics are very far yet from those predicted based on the MS deexcitation rates and even based on the thermal conductivity of an active medium. True, this should not be applied to a CuBr laser, which has yet unknown limiting characteristics.

By way of illustration we demonstrate our latest data for determining the limiting pulse repetition rate of a CuBr+ H_2 laser. These results are interesting because they have been obtained using a new experimental procedure. The principle of the procedure is as following.

Let us have an active volume with identical areas 1 and 2 (Fig. 4). Area 3 is dedicated to reducing the influence of a discharge in one tube part on the other part. The tube itself can be placed in a resonator with 100% - mirror 4 and glass plate 5.

If we delay pumping of the first part of laser relative to the second, lasing development of delayed pulse comes under the influence of N^{pp}_{ms} initial conditions both from the first and second areas. If τ_d is a delay period and T is a regular interpulse period and $\tau_d << T$, then $N^T_{ms} << N^\tau_{ms}$, where $N^{T,\tau}_{ms}$ accordingly in the first and second areas. That is to say $N^\tau_{ms} + N^T_{ms} \sim N^\tau_{ms}$, and lasing development of a delayed pulse comes under initial conditions from second area. It is easily shown [31], that an influence of N^τ_{ms} from the second area is precisely the same as if N^τ_{ms} had occurred in the first area. To be specific energy losses are the same as was written in equation (6), but $k_{pp} = k^\tau_{pp} + k^T_{pp} \sim k^\tau_{pp}$, where k^τ_{pp}, k^T_{pp} accordingly absorption coefficient in the second and first area.

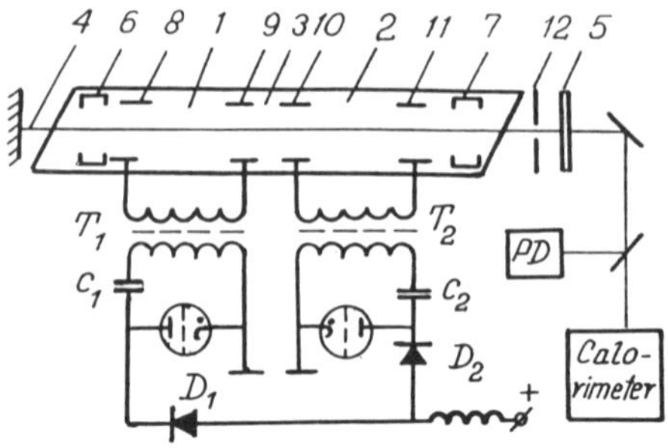

Fig.4. Operating set diagram for the measuring of PRR by a method of the space - separated absorption and amplification areas. 1,2 active areas; 3-isolating area; 4,5 - mirrors; 6,7,12 - diaphragms; 8-11 electrodes

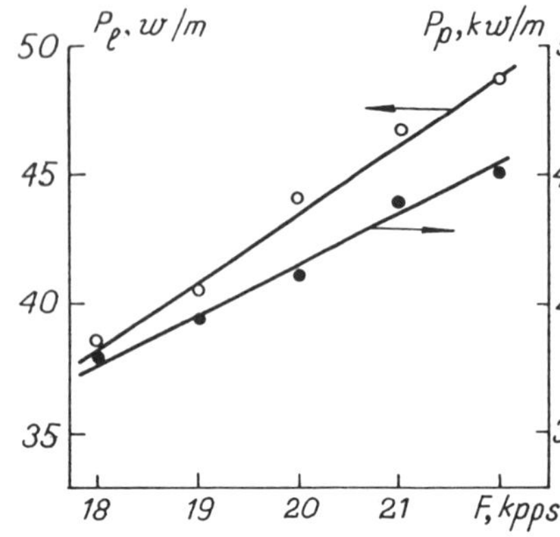

Fig.5. Lasing output P_l and pumping power P_p versus PRR at the optimal lasing conditions of CuBr + H_2 laser. d=2.7 cm.

The greatest advantage of the written procedure, contrary to the methods of double, regular or burst pulses, is the separation of N^{pp}_{ms} influence from N_{eo} during actual practice of pulse-periodic mode operating. This is evident from the fact that under changes of τ initial conditions along N_{eo} for delayed pulse are fixed, but along N^{pp}_{ms} they are changed as for a case of double pulses mode operation. But it should be noted, this is realized under conditions of operating of a concrete pulse-periodic laser.

We have exploited in investigations a power supply with transformers to pump two laser areas of 40 cm length and 27 mm in diameter, separated from each other by a distance of 10 cm. In association with low inductance tube this allows achievement of high output power. By way of illustration demonstrated in Fig. 5 is the rate-power characteristic, obtained under the condition: $C_1 = C_2 = 680$pF, rectifier voltage $U_r = 7$kV.

In Fig.6 are demonstrated typical patterns of lasing arising versus period delay between pumping pulse of areas 1 and 2. Curves 1-3 were obtained on variation of U_r from 6.25 kV up to 7.5 kV, latest corresponds the pumping energy $E_p = 0.18$ J. Pulse repetition rate was F=17 kpps and $E_p = P_r / F$.

If we operate from presumption that 38% of pumping energy will be treated for ionization [32], then to the end of a pumping pulse, electron concentration is as $N_e = 1.55 \times 10^{14}$ cm^{-3} for the curve 3 conditions. This corresponds with the deexcitation rate $k_e N_e = 2.2 \times 10^8$ s^{-1} or relaxation period $\tau_r = 0.46$ μs. Lasing initiation occurs

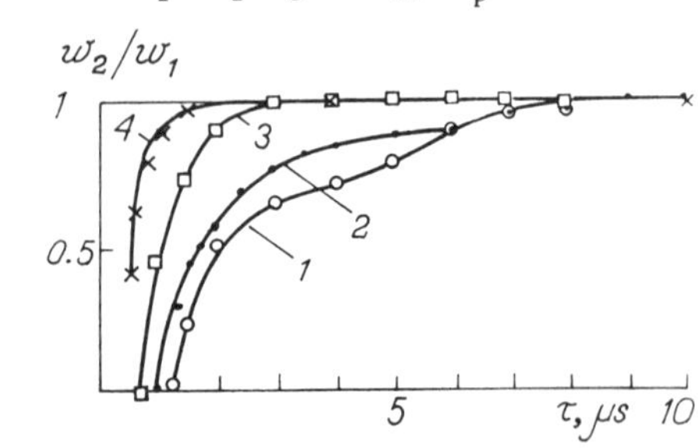

Fig.6. Lasing restoration of CuBr and e-beem Pb lasers. 1-3 -CuBr, p_{Ne}=2.9 kPa, F=17 kpps, U_r=6.25 kV (1), 6.8 kV (2), 7.5 kV (3); 4-results from [8].

at the delay period τ_d=0.7 μs and reproduction of utter lasing within 3 μs. Slower restoration for curves 2 and 3 is conditioned not only by the lesser deexcitation rate because of electron concentration reduction, but by of the broadening excitation pulse due to reduction of plasma conductivity.

Thus, when used the space-separated areas of absorption and amplification and thus is separated the influence N^{pp}_{ms} and N_{eo}, or, more precisely, the influence of N_{eo} is neutralized, reproduction of utter conditions for lasing of Cu-laser occurs within several μs as well as for Ba, Pb and Mn vapour lasers. At the moment we have not prepared to answer at what point the maximum of the curve in Fig. 5 will occur, if we are able to optimize a copper bromide laser by means of methods (a, c, d, e, see above). But these results together with data from paper [14] have not left any place for doubt that going from pulse gas discharge to an electron beams pumping methods makes possible realization of the potentialities which have been laid in metal vapour lasers on self-terminated transitions.

4. Electron - beam pumping of lasers.

Let us briefly recall the essential characteristics of the electron beams as refereed to. If we apply to the narrow gap δ cathode + net anode the voltage, which is enough for electron runaway, we will produce intense electron beam EB in the drifting space. Electron emission in this case is caused owing to intense ultraviolet illumination from drifting space. Essential data about beams generation mechanism have been published in the papers [33]-[36]. An extended computer simulation of EB generation was presented in the paper [10]. In the table (2) are presented the essential beams parameters realized at present.

TABLE 2.

Electron energy	from 2 to 70 keV	Pulse repetition rate	up to 2×10^6 Hz
Current density	up to 10^3 A/cm^2	Pulse duration	from 1.5 ns to continuous
Total current	up to 10^5 A	Accelerating gap length	from 0.2 to 5 mm
Gas concentration	up to 3×10^{18} of He	Drifting space length	from 0.5 to 500 cm

By rolling up the construction in cylindrical geometry we can obtain a well suited low inductance cell to pump lasers. At a cell diameter d=10 cm, length l=100 cm and accelerating gap $\delta \approx 0.5$ mm total pulse beam current is liable to be 50 kA at the voltage in range from 5 to 7 kV.

Researches by the double excitation pulses method have demonstrated that lasing conditions of Pb and Ba vapour lasers reproduce still more rapidly then at gas discharge pumping under conditions of neutralization of N_{eo} influence (see curve 4 for e-beam Pb laser in Fig. 6). F_{lim} under e-beam excitation is usually conditioned by the rate of electrons cooling.

Comparative investigations of a Pb vapour and Mn-vapour lasers under gas-discharge and e-beam pumping [14] have demonstrated incontestable advantages of an e-beam pumping method.

We point out three essential advantages of this pumping method.

1. Into active medium are already injected accelerated electrons. Therefore no problem to heat up them. As a result the restrictions caused by N_{eo} shall be conditioned by the two mechanisms: a) - deexcitation of resonance states; b) - heating of prepulse electrons by secondary ones and population of MS owing to this process. As a result an influence of N_{eo} under e-beam pumping displays very much later than under gas discharge pumping and have not been discovered by double pulses method. This allows to restore lasing in period from 1 to 3 μ s.

2. Energy share which have been treated to excite MS and for stepwise ionization lesser than under gas discharge excitation. This is conditioned by the other type of electron energy distribution and more rapid decreasing of MS excitation cross section.

3. Electric field is absent inside the laser cavity. Therefore effective cooling of active medium is possible owing to heat transfer to the numerous partitions placed in the cavity. As a result it is possible to use pumping power up to hundreds kilowatt without overheating of active medium.

Theory of e-beam pumping of laser on self-terminated transitions was developed in the papers [37,10]. In the figure 7, 8 are presented some from these results, which agreed with experiments [14]. As it has been seen from these figures, under conventional beams characteristics and optimal active mixture specific average output power exceed 200 mW/cm^3 for the tube diameter 10 cm. This corresponds the average

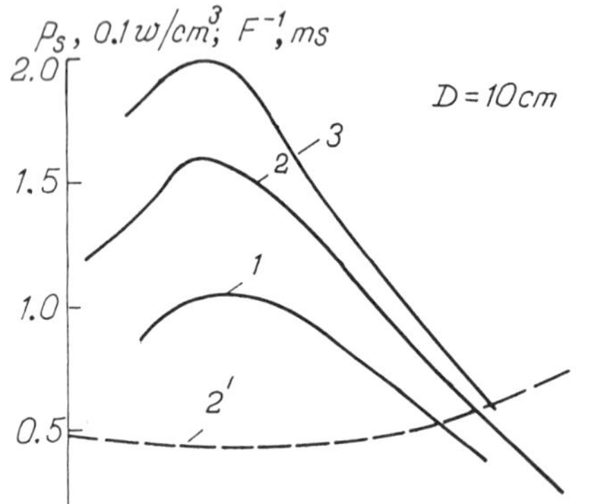

Fig.7. Lasing output (1-3) and optimal pulse repetition rate (2`) versus He concentration. $N_{Cu}=5 \times 10^{15}$ cm$^{-3}$; d=10 cm, l=100 cm, V= 7.8 l. Curves 1; 2, 2`; 3 accordingly j_b=5;10;15 A/cm^2.

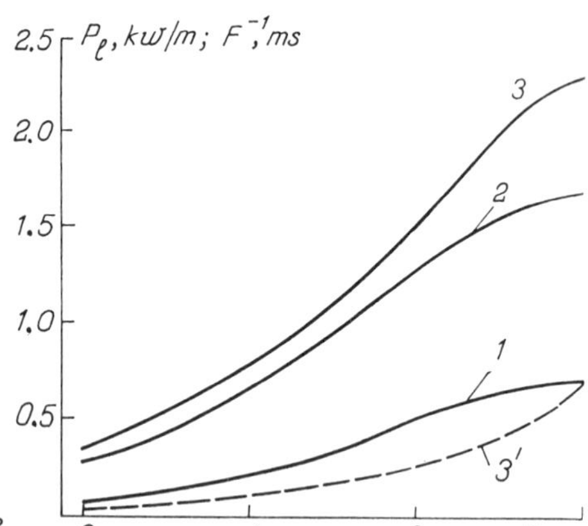

Fig.8. Lasing output (1-3) and optimal pulse repetition rate (3`) versus bore diameter. Curves 1-3, 3` accordingly: j_b = 5;10;10;10 A/cm^2; 10^{-15} N_{Cu} = 10; 5; 4; 4 cm^{-3}; 10^{-16} N_{He} =15; 7; 4; 4 cm^{-3}.

output power P_l > 1.5 kW and efficiency 5% for tube length 100 cm. Beam current for this case must be 40 kA at the pulse duration 70 ns and voltage ~ 5kV. These beam parameters have already been obtained, true, at room temperature.

It may be happened that Cu-laser will not be the best under beam excitation. Because the higher excitation energy of lower operating levels Mn-laser under the same essential conditions can have the following parameters: *eff*=(7-8)%, F_{opt}=7.5 kpps, P_l =2 kW/m.

More promising parameters of a Mn-laser are founded on the lesser thermal population of lower levels. This restriction can easily be obviated by a placement of numerous partitions in the active volume. In Fig.8 is demonstrated a suitable design of lasing tube, which capable to produce P_l >10 kW/m.

Let us briefly discuss the basic features of this laser. If we wish to produce P_l=10kW/m at the *eff*=5%, we need to inject P_p=200 kW into active volume. Because finally this power is dissipated by radiating cooling of the partitions, their temperature T_p will be defined from radiative balance between partitions and cavity wall. It can be shown that T_p-T_w=300^0, if T_w=1900°K. From the equation for the thermal conductivity we can calculate that the gas temperature of central region of any tube share will be lesser than T_c=3500°K as compared with T_c=4100°K of tube design without partitions at the output average power P_l =1 kW/m.

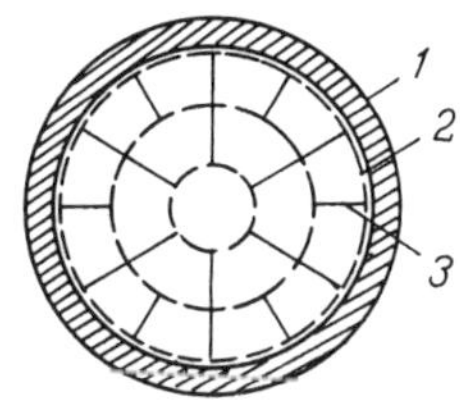

Fig.9. A sketchy design of 10 kW/m Cu e-beam laser tube. 1- BeO+Mo cathode; 1- Mo-Re net anode; 3-partition.

Thus we can conclude that the aim to obtain an average output power of metal vapour laser more than 10 kW/m is not a fantastic dream. It could be realized under electron beam pumping of tube diameter 10 cm, tube length 100 cm and 18 radial partitions placed into active volume.

5. References

1. Walter, W.T., Solimene, N., Piltch, M., Gould, G. (1966) Efficient gas discharge lasers, *IEEE J. Quant. Electr.*, **2**, 474-479.
2. Isaev, A.A., Kazaryan, M.K., Petrash, G.G. (1973) The potential for production high average power lasing in the visible spectral region. *Kvantovaya Electr.*, № **6**, 112-115.
3. Bokhan, P.A., Gerasimov, V.A. (1979) Optimization of excitation conditions in a copper vapour laser, *Kvantovaya Electr.*, **6**, 451-455.
4. Jones, D.R., Maitland, A., Little,C.E. (1994) A high-efficiency 200 W average-power copper HyBr ID laser, *IEEE J. Quant. Electr.*, **30**, 2385-2390.
5. Arlantsev, S.V., Borovich, B.L., Buchanov, V.V. and oth. (1983) Similarity relations for pulse metal vapour lasers, *Kvantovaya Electr.*, **10**, 1546-1553.
6. Batenin, V.M., Vokhmin, P.A., Klimovsky, I.I., and oth. (1981) On the multiparametric optimization of copper-vapour lasers, *Doklady Academy Nauk USSR*, **256**, 831-834.
7. Kravchenko, V.F. (1995) Similarity and simulation of the pulse gas discharge lasers, *Dissertation, Rostov.*
8. Bokhan, P.A. (1985) On the mechanism of limitation of the optimal pulse repetition rate in lasers due to self-terminating transitions in metal vapour, *Kvantovaya Electr.*, **12**, 945-952.
9. Bokhan, P.A., Sorokin, A.R. (1991) Gas laser excitation by an electron beam formed at open discharge, *Optical and Quant. Electr.*, **23**, 523-538.
10. Arlantsev, S.V., Borovich, B.L., Buchanov, V.V. and oth. (1995) On the prospects of using runaway electron beams generated in an open discharge for the pumping of metal vapour lasers, *J. Russian laser research*, **16**, 99-120.
11. Isaev, A.A., Kazaryan, M.K., Petrash, G.G. (1972) Efficient pulsed copper vapour laser, *JETP Letters*, **16**, 40-42.

12. Bokhan, P.A., Solomonov, V.I. (1973) On a lasing mechanism of a copper vapour lasers, *Kvantovaya Electr.*, **№6**, 53-57.

13. Bokhan, P.A., Gerasimov, V.A., Solomonov, V.I. and oth. (1978) On a lasing mechanism of laser action in copper vapour, *Kvantovaya Electr.*, **5**, 2162-2173.

14. Bokhan, P.A. (1986) Relaxation processes and influence of the metastable states of atoms and ions on the lasing mechanism and power characteristics of laserş, *Kvantovaya Electr.*, **13**, 1837-1846.

15. Bokhan, P.A., Silantiev, V.I., Solomonov, V.I. (1980) On a mechanism of limitation of the pulse repetition rate in a copper-vapour laser, *Kvantovaya Electr*, **7**, 1264-1269.

16. Miller, J.L., Kan, T. (1979) Metastable decay rates in a Cu-metal-vapour laser, *J. Appl.Phys.*, **50**, 3849-3851.

17. Tabata, Y., Hara, K., Ueguri, S.. (1992) Lasing characteristics of a large-bore copper vapour laser, *Proc. SPIE*, **1628**, 32-43.

18. Klimkin, V.M., Maltsev, A.N., Fadin, L.V. (1978) Investigation of a steady field of a pulse-periodic high repetition rate gas discharge, in P.A. Bokhan (eds), *Efficient metal vapour gas discharge lasers, Tomsk, p.p.* 116-132.

19. Elaev, V.F., Melchenko, V.S., Pozdeev, V.V. and oth. (1978) Influence of the radial irregularity of a gas discharge plasma on the copper vapour laser characteristics, ibid., *p.p.* 189-196.

20. Kushner, M.J., Warner, B.E. (1983) Large-bore copper-vapour lasers: kinetics and scaling issues, *J. Appl.Phys.*, **54**, 2970-2982.

21. Chang, J.J., Boley, C.D., Martinez, M.W. and oth. (1994) Beam characteristics of a large-bore copper laser with a radiatively cooled plasma, *Proc. SPIE*, **2118**, 2-8.

22. Yseki, Y., Hayashi, K., Watanabe, I. and oth. (1994) Characteristics of a large-bore copper-vapour laser with gas cooling plates, *J. Applied Phys.*, **33**, L860-L862.

23. Bokhan, P.A. (1986) Mechanism limiting the pulse repetition frequency of a barium vapour laser, *Kvantovaya Electr.*, **16**, 1041-1046.

24. Bokhan, P.A., Silantiev, V.I., Solomonov, V.I. (1979) The method of obtaining of lasing in metal vapour laser, *USSR patent*, N° 810032.

25. Vuchkov, N.K., Sabotinov, N.V., Astadjov, D.N. (1988) High-efficiency CuBr laser with interacting peaking circuits, *Optical and Quant. Electr.*, **20**, 433-438.

26. Elaev, V.F., Lyakh, G.D., Pelenkov, V.P. (1989) CuBr laser with average lasing power exceeding 100 W, *Atmospheric Optics.*, **2**, 1045-1047.

27. Bokhan, P.A., Maltsev, A.N., Silantiev, V.I. (1980) A new mechanism for limiting generation mean power of metal-vapour pulse laser, *Abstracts* IX *Conf. Quant.Electr. Nonlin. Opt. (EKON-80), Poznan,* 333-334.

28. Bokhan, P.A., Maltsev, A.N., Silantiev, V.I. (1980) News methods to improve power characteristics of the pulse periodic metal vapour lasers, *Abstracts* X *Conf. Coherent and Nonlin. Optics, Kiev,* 201-202.

29. Bokhan, P.A. (1981) Copper Vapour laser with an active volume 1.5 litre. *J.Technical Physics,* **51**, 206-209.

30. Astadjov, D.N.., Vuchkov, N.K., Sabotinov, N.V., (1988) Parametric study of the CuBr laser with hydrogen additives, *IEEE J. Quant. Electr.*, **24**, 1927-1935.

31. Bokhan, P.A., Zakrevsky D.E. (in press) On the pulse repetition rate limit of a copper vapour laser, *Kvantovaya Electr.*

32. Borovich, B.L., Yurchenko, N.I. (1984) Analysis of the exciting and relaxation kinetics of a Cu-vapour lasers with longitudinal gas discharge, *Kvantovaya Electr.*, **11**, 2081-2095.

33. Bokhan, P.A., Kolbychev, G.V. (1981) Generation of intense electron beams in a middle pressure gas, *J. Technical Physics*, **51**, (1823-1831).

34. Bokhan, P.A. (1991) Formation mechanism and producing of intense electron beams in an open discharge , *J. Technical Physics*, **61**, 61-68.

35. Bokhan, P.A., Sorokin, A.R. (1985) Open discharge, producing electron beam: mechanisms, properties and using to pump gas lasers of a medium-pressure., *J. Technical Physics*, **55**, 1823-1831.

36. Bokhan, P.A., Sorokin, A.R. (1985) Gas laser excitation by an electron beam formed at open discharge, *Optical and Quant. Electr.*, **23**, 523-538.

37. Borovich, B.L., Buchanov, V.V., Molodykh, E.I. (1984) Numerical simulation of an electron-beam pumped copper vapour laser, *Kvantovaya Electr.*, **11**, 1007-1014.

STRONTIUM AND CALCIUM ION RECOMBINATION LASERS AND THEIR APPLICATIONS

E.L.LATUSH, M.F.SEM and G.D.CHEBOTAREV

Rostov University, Physics Department
Zorge st.5, Rostov-on-Don, 344104, Russia

Different types of the Sr^+ and Ca^+ recombination lasers and operation regimes are considered together with obtained outputs and prospects for their improvements. The approach is given to their optimum scaling. Already realized and perspective applications of these lasers are discussed, such as for optical pumping of gaseous, liquid and solid active media; in spectroscopy, industry, in amplifiers and projection systems, in holography and in medicine.

1.Introduction

Laser oscillations at strontium ion (λ430.5, 416.2 nm) and calcium ion (λ373.7, 370.6 nm) lines were first reported in [1,2] where recombination pumping mechanism for them was also proposed. Even in those early studies, it became evident that these lasers were able to deliver high levels of output power at acceptable efficiencies ~ 0.1%. As for the output characteristics, the He-Sr and He-Ca lasers remain to date the best of recombination metal vapour lasers, that is due to more or less close fulfilment of requirements needed to effectively create a population inversion in the decaying plasmas [3-5].

In this paper, a review of investigations on these lasers is presented with an emphasis on our own studies of their regimes and modes of operation, and also the approaches is developed to an optimum scaling of strontium and calcium ion recombination lasers. Applications, already realized and perspective, of strontium and calcium metal vapour recombination lasers are also discussed.

2.Longitudinal-Discharge Lasers

Temperatures needed to achieve optimum strontium and calcium vapour pressures are relatively low (500-800 °C). Therefore, a self-heating regime is quite suitable for operation of these lasers and tube design can be rather simple: either with an open-air ceramic discharge tube or with a ceramic tube mounted inside silica or glass envelope [5,6].

C. E. Little and N. V. Sabotinov (eds.), Pulsed Metal Vapour Lasers, 149–160.

Since high pressures of helium, p, are necessary for optimum performance, high voltages across the tube in longitudinal Sr-He and Ca-He lasers are needed.. Typically, p.r.f. is 5-10 kHz; input power of 1-2 kW is needed for tube of 40 cm active length and 1.5 cm i.d., heat-up time for normal operation ~5-10 min. Average power output, P, is then 0.5-1.5 W, laser pulselength 200-300 ns at p = 500-600 Torr. Small-signal gain coefficient is rather considerable (0.05-0.06 cm^{-1}). Laser oscillations at weaker components of doublets (416.2 nm SrII and 370.6 nm CaII) are, as a rule, suppressed by stronger ones because of common upper laser levels. With a prism or grating resonator output power close to 60% of that for stronger component was obtained at 416.2 nm SrII line, and 20-30% - at 370.6 nm CaII line [7].

At high helium pressures 430.5 nm SrII line undergoes strong collisional broadening ($\Delta\nu \approx$ 12 GHz at p = 1 atm), while a spectral width of laser output still remains narrow (0.3-1.5 GHz), that is due to homogeneous nature of broadening [6]. It allowes to somewhat tune the laser frequency in the range of 12 GHz [8]. A self-modelocking regime under certain discharge conditions was also observed [8].

Output power characteristics of these lasers can be scaled, for example,by lengthening of a laser tube, although,because of high helium pressure and, as a consequence, high needed voltages, it is difficult to use discharge gaps wider than half a meter. Then, it makes sense to divide the discharge length into sections. By this way, we have reached the output power of 3 W with an efficiency of 0.12% from the He-Sr 430.5 nm laser tube of two sections 45 cm long each and 1.5 cm i.d. [9]. From a simular He-Ca laser tube, 1.35 W output at 373.7 nm was obtained, though an operation regime in this case was not very much optimized [7,10].

For self-heated lasers, a p.r.f. is determined,basically,by heat-transfer rate from a discharge tube and optimum vapour pressure. Then, real possibility exists to increase an average output power by enhancing the heat-transfer rate. The analysis made in [11] revealed that heat-transfer from laser tube was determined both by convective cooling (~35 %) and by thermal radiation (~65%). Therefore, heat-transfer rate can be substantially increased by blackening the outer surface of the discharge tube. That kind of experiment [11] confirmed this point allowing to increase output power by a factor of 1.6.

Considerable advantages provides a segmented discharge tube with the ceramic cylinder segments inside the tube of larger diameter. Metal pellets are placed then in gaps between the segments. With such a tube of full active length of 45 cm, P = 2 W was obtained from the He-Sr laser at p.r.f. of 6.5 kHz [7,10].

Further growth of output power can be obtained through heat-transfer enhancing by water cooling through a helium layer [11]. Thickness of the helium layer can be appropriately chosen so as to provide a needed heat fluence from a discharge tube, that is mostly determined by power and p.r.f. capacity of a driver unit. In fact, we obtained from the water-cooled Sr-He laser tube (active length of 33 cm, 1.1 cm i.d.). P = 3.9 W, P_{sp} = 130 mW/cm^3 at p.r.f. of 29 kHz [12]. These laser parameters are well compared to those of copper vapour laser and evidence of real feasibility of output powers on the order of 5-10 W from He-Sr and He-Ca lasers.

In [13-15], a regime of rarely repeated pulses was used in large-bore tubes; a Sr-He laser with active volume of 1000 cm^3 (4 cm i.d., 80 cm long) emitted laser pulses of 300 ns duration with energy of 6 mJ per pulse and peak power of 20kW

An average power 2.05 W was yielded in [16] from a He-Sr laser tube of 27 mm i.d. and 28 cm active length at p = 675 Torr, thus, output power per unit length reached 7.3 W/m. The use of slow helium flow through the tube allowed the active medium to be more uniform and clean. The flow of helium was also supplying strontium vapour into the tube, which had been generated in special separately heated zone outside the discharge length. Such a technique allowed to vary excitation current parameters and strontium vapour pressure independently and, thus, easier to obtain an optimum output characteristics. Similar technique was employed in [17] where strontium pieces had been placed in a special container preventing the strontium oxide dust from entering the helium flow. The average power in this experiment amounted 1.7 W at 430 nm line from laser tube of 15.5 mm i.d. and 70 cm active length at p = 495 Torr and p.r.f. of 6 kHz. The device showed good stability and repeatance of laser parameters.

The He-Sr laser output power enhancing by means of forced air cooling of the tube was demonstrated in [18].

A possibility was investigated of joint oscillations at recombination laser lines 430.5 nm SrII and 373.7 nm CaII in one discharge tube [7,10]. Pieces of Sr and Ca were placed in spacially seperated in length zones of the discharge tube (1.1 cm i.d., full length - 30 cm, length with Sr - 10 cm, with Ca - 20 cm). As a result, simultaneous lasing at 430.5 nm and 373.7 nm with a total power output of 0.3 W was obtained.

In addition to recombination laser lines, rather strong self-terminating electron-impact excited laser transitions exist in Sr and Ca ion and neutral spectra. Their joint oscillations over a wide range of helium pressures and p.r.f. in discharge-heated regime are possible. Total power at all lines obtained in our experiment [7,10] was 0.52 W from a laser tube of 1.1 cm i.d., active length 50 cm at p = 450 Torr, and p.r.f. = 6 kHz. The power was destributed between the lines as follows: 70% - 430.5 nm SrII,16% - 6.45 μm SrI,8% - 3.05 μm SrI, and 3% - 1.03, 1.09 μm SrII. This output power distribution can easily be varied by helium pressure and deposited electric power. Similar self-terminating laser transitions are also present in the CaI,II spectrum. Thus, combining strontium and calcium in one tube, simultaneous laser action at a number of lines ranging from 0.37 μm to 6.45 μm can be obtained.

Output power of the Sr-He and Ca-He lasers rapidly grows with the helium pressure increase, and no signs of suturation have been observed until recently up to pressure of 2 atm. So, lately [7,19], we have staged an experiment with a small-scale longitudinal-discharge tube in order to find out the pressure of helium above which output power would begin to decrease. Discharge was fired in tube of 0.8 cm i.d. and 20 cm long. The maximum power was reached at $p \approx 4$ atm, and from 1 atm up to 4 atm of helium power output increaseâ four times and specific output characteristics reached the record values for longitudinal lasers: P_{sp}(ave) = 44.1 mW·cm^{-3} , P_{sp}(peak) = 180 W·cm^{-3} , ε_{sp} = 13.8 μJ·cm^{-3} at p.r.f. of 3.2 kHz.

Out of latest, the work [20] should be mentioned, where an excitation of a small-scale strontium laser tube was performed by use of semiconductor switch. Due to careful matching of the plasma tube to a electrical discharge circuit, the current ringing was practically eliminated. As a result, rather good specific power was reached P_{sp} = 13.7 mW·cm^{-3} with total average power of 118 mW.

3.Transversely and E-beam Excited Lasers

The use of transverse discharges to excite the discussed lasers looks attractive, for it allows to work easily at high helium pressure. For the first time, laser action at 430.5 nm SrII with transverse-discharge excitation was obtained in [21] but more detailed studies on this subject were done in [22-24]. It turned out to more difficult realize a high p.r.f. regime under the transverse-discharge excitation: the p.r.f. could not exceed ~ 500 Hz [22]. Although, record values of specific and peak laser powers were reached in rarely repeated pulse regime: ε = 2.1 mJ, ε_{sp} = 53 μJ·cm^{-3}, P(peak) = 20 kW for the He-Sr laser; ε = 0.75 mJ, ε_{sp} = 35 μJ·cm^{-3}, P(peak) = 7.5 kW for the He-Ca laser. As it is seen the specific energy outputs are about an order of magnitude greater of those in longitudinally excited lasers. However, transverse-discharge lasers have more complicated tube design compared to longitudinal lasers.

Hollow-cathode discharge and so called "open" discharge [25] can also be considered as transverse ones.The disfinct feature of theirs is highly non-Maxwellian electron-energy distribution function, i.e., the presence of a group of fast electrons.

Our studies [26,27] demonstrated the possibility to employ a hollow-cathode discharge for laser action at IR lines of SrII and CaII via charge-transfer excitation by krypton ions. But lasing at 430.5 nm was not obtained then because of current densities j and helium pressures being too low. Attempts to increase j led to arcing and spoiling the discharge uniformity. In later work [28], this difficulty was overcome with the help of pre-ionization of a hollow-cathode discharge, and that made possible to obtain lasing at 430.5 nm SrII line in recombination regime, though output laser parameters were low.

Better output characteristics on 430.5 nm line were obtained in "open" discharge of coaxial type. A high-energy electron beam appears here in the gap between cylinder cathode and grid anode. At active length of 9 cm in the mixture of Sr-Kr-He, simultaneous lasing at several ion lines pumped by different mechanisms was observed [28]: self-terminating electron-impact excited λ1033, 1091 nm, charge-transfer excited by Kr^+ ions λ1087,1123 and 1201 nm and recombination pumped λ430.5 nm lines.

Far more impressive results on 430.5 nm SrII laser action in the "open" discharge were reported in [25]. From the tube of 4.5 mm i.d. and active length of 50 cm in a mixture of Sr-Kr-He, quasi-CW laser action of 20 W output power was realized under excitation with a burst of current pulses (p.r.f. in the burst ~1-3 MHz, burst duration of 1 ms). The 0.2% efficienfy of lasing with respect to power inserted into the discharge was reached.

It should be mentioned about recently obtained laser oscillations at the 430.5 nm SrII line under a pulsed microwave-pumped discharge excitation in a self-heated regime at helium pressures between 350-680 Torr [29]. Although output parameters were near treshhold, further improvements of a microwave discharge coupling to the plasma tube would probably enhance the laser output.

4. Optimum Scaling of the Sr^+ and Ca^+ Recombination Lasers

In our recent work [30], we have derived some relationships that allow to calculate optimum parameters and regimes of the excitation electrical circuit to obtain maximum output characteristics for these lasers with longitudinal plasma tubes of a given geometry provided such a regime is established experimentally for the sample (prototype) tube. This theory of an optimum scaling is based on the following assumptions that are, however, rather strongly supported by numerous experiments on the output power characterization of these lasers and their computer modelling.

1. A regime of the discharge for an oscillatory circuit which includes a storage capacitor, *C*, a plasma tube of inductance *L* and resistance *R* is close to a critical one. For the critical regime then the following is valid

$$R = 2\cdot\sqrt{\frac{L}{C}} = \frac{l}{\sigma\ S}, \qquad \sigma \sim \frac{n_e}{p\cdot\sqrt{T_e}} \tag{1}$$

where l and S are the length and cross-section of the laser tube, n_e and T_e are the density and temperature of electrons, σ is the conductivity of the discharge plasma.

2. During the current pulse, T_e obeys

$$T_e \sim E/p. \quad (2)$$

3. In optimum regime for lasing, n_e is approximately proportional to an energy input per active volume:

$$n_e \sim \mathcal{E}_{sp} = \frac{CU_c^2}{2V}, \quad (3)$$

where $U_c=El$ is an initial voltage on the storage capacitor, $V=Sl$ is a discharge volume. Out of (1)-(3), the next scaling relations can be deduced

$$U_c \sim \frac{l\ p^{1/3}}{L^{1/3}C^{1/3}}, \qquad \mathcal{E}_{sp} \sim \frac{l\ p^{2/3}C^{1/3}}{S\ L^{2/3}}. \quad (4)$$

For the self-heated lasers with pieces of lasant distributed directly inside the plasma tube, an average input power, P_{in}, is determined by a wall temperature needed to obtain the optimum vapour pressure, and is equal to the plasma tube power loss through heat-transfer, the latter is nearly proportional to the plasma tube outer diameter d_o. Assuming d_o is proportional to inner diameter, d, and remembering that the average output laser power amounts

$$P = \eta P_{in} \sim \eta l d,$$

where η is the laser efficiency coefficient, it can be found for a specific average output power

$$P_{sp} = \frac{P}{V} \sim \frac{\eta}{d}. \quad (5)$$

On the other hand, P_{sp} can be expressed through a p.r.f., f, and a specific laser pulse energy ε_{sp} as

$$P_{sp} = \varepsilon_{sp} f = \eta \mathcal{E}_{sp} f. \quad (6)$$

Using (5) and (6), we can find a scaling law for the optimum p.r.f.:

$$f \sim \frac{d\ L^{2/3}}{l\ p^{2/3}C^{1/3}}. \quad (7)$$

The experiments demonstrate that $\sigma = const$ for tubes of different dimensions. Then, out of (1), follows

$$C \sim (S/l)^2 L. \quad (8)$$

If in the discharge oscillating circuit, a prevalent part of inductance is that of the plasma tube, which can be calculated with the formula

$$L = 2l\left(\ln\frac{4l}{d} - 1\right) \sim l, \quad (9)$$

then, from (8) and (9), we can find the relation for optimum scaling of a storage capacitor value

$$C \sim S^2/l. \quad (10)$$

The scaling laws for other parameters can be also obtained [30].

Naturally, restriction borders exist for the obtained scaling relationships. In particular, an analysis of the experimental data reveals that in optimum conditions for the He-Sr longitudinal laser, ε_{sp} ranges from 2 to 10 μJ· cm^{-3} at $p <$ 760 Torr. Typically, $\eta \approx 0.1\%$ for the He-Sr laser, hence, $\mathcal{E}_{sp}$ = 2-10 mJ·cm^{-3}. The lower limit for $\mathcal{E}_{sp}$ is due to existence of critical values of n_e and a recombination pumping rate below which the inversion ceases [5]. The upper limit corresponds to $\mathcal{E}_{sp}$ when practically total double ionization of strontium atoms occurs. The excess input energy is wasted then on ionization of helium atoms, that leads to decrease of η and P, although ε_{sp} may still remain high.

As it is shown in [12,17] the limit of p.r.f. also exists due to overheating of an axis discharge zone.

We have tested experimentally the obtained relationships with tubes of various geometry. In Fig.1, some results are presented for three of them (two of He-Sr and one of He-Ca laser). Taking as starting point (prototype regime) the only very carefully optimized regime for one of the Sr tubes (circled in Fig.1), the curves for each tube were calculated with use of (4) for optimum $\mathcal{E}_{sp}$ vs. helium pressure. As can be seen from Fig.1 the experimental data are in close agreement with the calculated curves, that proves fair reliability of the above consideration.

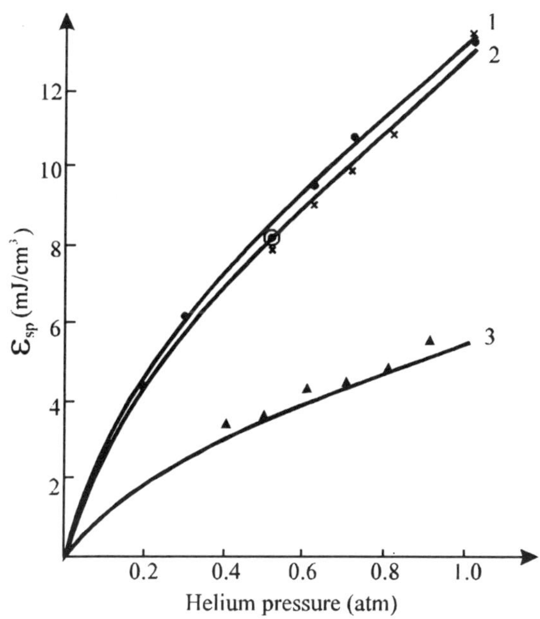

Fig.1. Dependences of optimum energy imputs on helium pressure.
(1) He-Ca laser: l=26.5 cm, d=0.7 cm, C=550 pF, L=0.713 nH
(2) He-Sr laser: l=20 cm, d=0.6 cm, C=550 pF, L=0.655nH;
(3) He-Sr laser: l=45 cm, d=1.5 cm, C=825 pF, L=0.84 nH

5. Applications

The discussed lasers emit in a short-wavelength spectral region and, therefore, are of interest for variety of applications. For instance, as optical pump sources for other laser active media in order to transform their light into other spectral regions.

In [31], we realized quasi-resonant optical pumping of thallium vapours mixed with helium by the Ca-He recombination laser resulting in laser oscillations at green 535 nm TlI line in superluminescent mode. The 373.7 nm CaII pump line falls into violet wing of the pressure-broadened thallium resonant 377.6 nm line, that is, an absorption occurs due to transition between unbounded terms of TlHe quasi-molecule. That explains the growth of a conversion efficiency with a helium pressure. In our experiment, it reached the maximum at p(He) = 4 atm and p(Tl) = 20 Torr, when at the length of thallium vapour zone of 5 cm about 90% of focused into sell pump power was absorved.The average output power of the green superluminescence amounted 8 mW at p.r.f. of 5-6 kHz. The lasing at 535 nm TlI line has a self-terminating nature because of metastable character of the lower laser level. So, actually, only leading part of the 373.7 nm pump radiation impulse (~10 ns) was effectively used, and, hence only ~ 1/10 of average pump laser power (~35 mW) was employed. With this in mind, near 30% conversion efficiency was obtained in the experiment. The marked feature was the simplicity of the experimental arrangement: the Ca-He laser radiation should be just focused into rather short (~10 cm) thallium cell without any optical resonator. In addition, sufficiently large ionization of the thallium vapour (quasi-resonant ionization) was registered, due to an efficient pumping of the thallium resonant level by a calcium vapour laser emission [32].

Another possibility to convert wavelengths of the recombination lasers with the help of gaseous media is to use optical pumping of dimer molecules. By this method, the conversion of 430.5 nm He-Sr laser radiation was achieved down to yellow-red region in Te_2 vapour cell [33]. The violet pumping light was focused through one of the mirrors of optical resonator along the axis of a heat-pipe tube filled with tellurium vapour. The lasing was observed at 17 wavelengths of a Te_2 molecule; the most intensive of them were of 642, 641.66, 631.95, 632.46 and 624.45 nm wavelengths. The pump power was 0.6 W (but only 50% of it went through the mirror) at p.r.f. of 5-6 kHz, pressure of tellurium vapour of 1.5-3 Torr at active length of 8 cm. Analysis shows that Se_2, S_2, As_2, Bi_2 ,and Sb_2 dimers might also be suitable for being pumped with calcium and strontium recombination lasers.

As a pump source for pulsed dye lasers, the He-Sr and He-Ca lasers are also of interest not only due to high-energy photons they generate, but to the relatively long pulsewidth (200-300 ns) as well, that is an order of magnitude more than that of traditional pumps, such as nitrogen and copper vapour lasers. Consequently, a dye laser output radiation of higher quality can be obtained,i.e.,with linewidth of an order of tens of MHz, that is close to monochromacy of the CW dye lasers.In fact, we have successfully tested a dye laser performance with the He-Sr recombination laser pumping [34].

The He-Sr and He-Ca lasers might occur to be good for pumping of solid-state tunable lasers. For instance, the 430.5 nm strontium laser output gets into one of the absorption bands of an alexandrite ($BeAl_2 O_4:Cr^{3+}$), which then can generate tunable light emission centered at ~ 760 nm with a tuning range of 100 nm. The wavelength conversion of the discussed lasers by means of non-linear optics is also quite possible. In particular, for second harmonic generation of the He-Sr laser output, the BBO crystal is appropriate, as it is still transparent at 215 nm [35].

The Ca-He laser emission (373.7 nm) is of special interest in semiconductor technology, for it closely coincides with absorption maxima of the most efficient photoresists. The Sr-He laser light can be employed for preparing and cleaning semiconductor substrates. The lasers may be also useful for stereolithography, stereopolymerization, laser writing, in particular, for direct laser writing of polymer optical wavegudes: their wavelengths, especially that of Ca-He laser, fit the absorption band of a corresponding polymer [36].

These lasers have good prospects in microtechnology for a thin film processing,thermal trimming of integrated circuit elements, defect repairing in IC and masks, etc. The Sr-He and Ca-He lasers can successfully compete with widely used for these purposes Nd:YAG, N_2 and eximer lasers, mainly due to more suitable wavelengths, higher quality of laser emission (compared to N_2 and excimer lasers) and much higher p.r.f., and due to simplier design and maintainance, as well.

The sufficiently narrow linewidth (0.5-1.5 GHz) and adequate coherence length (10-30 cm) together with short wavelengths of these lasers make them convenient for holography applications: in art holography, optical memory, for embossed hologram manufacturing. In the last case, the valuable advantage of the Ca-He (373.7 nm) laser is that its wavelength corresponds to maximum sensitivity of commonly used photoresists. As a result, the use of an 0.5 W Ca-He laser is equivalent to employment of an argon-ion laser delivering of more than 100 W total output power [37].

Short wavelengths of the recombination lasers are iteresting to apply in micro-analisys [38] employing method of liquid chromatography and electroforesis with a laser-excited fluorescent registration. For example, a fluorescent spectorscopy on a basis of Ca-He laser can be successfully applied in ecology for fast analysis of water pollution with polyaromatic organic molecules. They are most wide-spread toxic pollutants, whilst 373.7 nm CaII line turns out to be good for complex spectrum excitation almost all of them. Simular reasons make these lasers pertinent for laser fluorescent microscopy with a sub-micron resolution, confocal microscopes, for application in forrensics for fingerprint detecting, supplementing in this field copper and gold vapour lasers [39].

Application of a strontium laser as one of the components of a TV-image laser projection system is also possible [40]. An image-formation approach in it demands the laser p.r.f. about 15 kHz in order to match the laser operation with a standard TV-signal. That is rather attainable for a strontium laser. Thus, a blue 430.5 nm SrII line in conjunction with a green 511 nm CuI and red 628 nm AuI laser lines would provide high-quality colour image upon sufficiently large screen (~10 m^2). Although, for this, stable performance of a Sr-He laser with output powers of 5-10 W should be achieved yet.

Metal vapour lasers (mostly copper laser) are now actively used as brightness amplifiers in so called projection microscopes. The Sr-He laser has not as high small-signal gain as copper laser, but still enough for that kind of application on a basis of multipass amplifier arrangement. That was confirmed by an experiments [42-44]: magnified image of appropriate brightness, contrast and resolution was formed with the Sr-He amplifier and viewed on the screen of about 1 m^2 in size.

Various applications of strontium and calcium recombination lasers in medicine can be forecasted for UV irradiation of blood, for phototherapy treatment of neonatial jaundice, for fluorescent diagnostics and photodynamic therapy (PDT) of cancer diseases. Saying in more detail on the PDT, it should be noted that the 416.2 nm SrII laser wavelength almost exactly fits one of the absorption maxima of a gematoporfirine, the key substance used for treatment in the PDT, so that its fluorescence is effectively excited by this laser wavelength. So far, rather expensive and low-efficient Kr^{++} ion-laser is usually used for that purpose. Thus, the Sr-He laser can be a good alternative to the krypton-ion laser in cancer illness diagnostics. For treatment by the PDT, a laser irradiation into red absorption band of a gematoporfirine (610-635 nm) is normally used, usually, from a dye laser pumped by argon-ion laser. But red laser dyes can also be pumped by the Sr-He laser output. Accordingly, the medical installation based on an Sr-He laser might be much more simpler and cost-effective than employing krypton and argon-ion lasers.

The list of He-Sr and He-Ca laser applications will, by all means, be continued as soon as these simple and cost-effective sources of short-wavelength coherent radiation appear in the marketplace. Therefore, their studies, especially on prospects to further increase reliability and power outputs, are certainly useful and must be continued.

6.Acknowledgements

The authors are thankful to S.V.Gumberg and L.T.Latush for theur valuable help.

The research discribed in this publication was made possible in part by Grant No JG 1100 from the International Science Foundation and Russian Gouverment.

7.References

1. Latush, E.L. and Sem, M.F. (1973) Laser recombination transitions in CaII and SrII, *Sov. Phys.JETP* **37**, 1017-1018.
2. Latush, E.L. and Sem, M.F. (1973) Stimulated emission due to transitions in alkaline-earth metal ions, *Sov.J.Quant.Electron.* **3**, 216-219.
3. Zhukov, V.V., Latush, E.L., Kucherov, V.S., Mikhalevskii, V.S., Sem, M.F. (1977) Recombination lasers utilizing vapours of chemical elements. I,II, *Sov.J. Quant.Electron.* **7**, 704-714.
4. Sem, M.F. and Latush, E.L. (1995) Kinetics of metal ion recombination lasers. (the previous article in this volume)
5. Ivanov, I.G., Latush, E.L., Sem, M.F. (1990) *Metal Vapour Ion Lasers*, Energoatomizdat, Moscow (In Russian).
6. Zhukov, V.V., Kucherov, V.S., Latush, E.L., Sem, M.F., and Tolmachev, G.N. (1976) Strontium- and calcium-vapour recombination lasers, *Sov.Tech.Phys. Lett.* **2**, 215-216.
7. Latush, E.L., Sem, M.F., Koptev, Yu.V., Bukshpun, L.M., and Atamas', S.N. (1993) Strontium and calcium vapour recombination lasers: excitation mechanisms, operation regimes, and applications, *Proc.SPIE* **2110**, 106-127.
8. Bukshpun, L.M., Zhukov, V.V., Latush, E.L., and Sem, M.F. (1981) Frequency tuning and mode self-locking in He-Sr recombination laser, *Sov.J. Quant.Electron.* **11**, 804-805.
9. Bukshpun, L.M., Atamas', S.N., Zhukov, V.V., Latush, E.L., and Sem, M.F. (1983) He-Sr laser with 3 W output power, *Izv.Vyssh.Uchebn.Zaved., Fiz.*, No.6, 105-107 (In Russian).
10. Latush, E.L., Sem, M.F., Bukshpun, L.M., Koptev, Yu.V., Atamas', S.N. (1992) Gas-discharge recombination lasers on strontium and calcium vapours, *Optics and Spectroscopy* **72**, 672-680.
11. Bukshpun, L.M., Latush, E.L., and Sem, M.F. (1986) Role of heat-transfer in average output power increasing of the Sr and Ca vapour recombination lasers, *Teplofiz.Vys.Temp.*, **24**, 402-405 (In Russian).
12. Bukshpun, L.M., Latush, E.L., and Sem, M.F. (1988) Influence of the temperature of the active medium on the stimulated emission characteristics of an Sr-He recombination laser, *Sov.J.Quant.Electron.* **18**, 1098-1100.
13. Butler, M.S. and Piper, J.A. (1984) Optimization of excitation channels in the discharge-excited Sr^+ recombination laser, *Appl.Phys.Lett.* **45**, 707-709.
14. Butler, V.S. and Piper, J.A. (1985) Pulse energy scaling characteristics of longitudinally excited Sr discharge recombination lasers, *IEEE J.Quant. Electron.* **21**, 1563-1566.
15. Little, C.E., Piper, J.A. (1989) High-power violet Sr recombination lasers, *Proc.SRIE 1041*, 167-174.
16. Bokhan, P.A., Zakrevskii, D.A. (1991) *Sov.J.Quant.Electron.* **21**, 926-928 (p. in Russ. ed.).
17. Loveland, D.G., Orchard, D.A., Zerrouk, A.F., Webb, C.W. (1991) Design of a 1.7 W stable long-lived strontium vapour laser, *Meas.Sci.Technol.* **2**, 1083-1087.
18. Bethel, J.W., Little, C.E. (1991) Forced-air cooled strontium-ion recombination laser, *Opt.Communs.* **84**, 317-322.
19. Atamas', S.N., Latush, E.L., Sem, M.F. (1994) He-Sr recombination laser with helium pressure up to 6 atm, *J. of Russian Laser Research* **15**, 66-68.
20. Takedo, Y., Iwata, A., Uegupi, S., Fujii, K. (1994) Lasing characteristics of Sr^+ reconbination laser, *IEEE J.Quant.Electronics* **30**, 1176-1180.
21. Bogus, A.M., Dzhikiya, V.L., Chernov, A.A. (1978) *Sov.J.Quant.Electron.* **8**, 442-444 (pages in Russian ed.).

22. Brandt, M. (1984) Repetively pulsed transversely excited Sr^+ recombination laser, *IEEE J.Quant.Electron.* **20**, 1006-1007.
23. Brandt, M. (1983) Transversely excited Sr^+ recomb. laser, *Appl. Phys.Lett.* **42**, 127-129.
24. Butler, M.S., Piper, J.A. (1983) High-pressure high-current transversely excited Sr^+ recombination laser, *Appl.Phys.Lett.* **42**, 1008-1010; High-power transverse-discharge Ca^+ recombination laser, *Appl.Phys.Lett.* **43**, 823-825.
25. Bokhan, P.A., Sorokin, A.R. (1991) Gas laser excitation by an electron beam formed at open discharge, *Opt.Quant.Electron.* **23**, s523-s528.
26. Latush, E.L., Solanki, R., and Collins, G.J. (1979) CW strontium ion transitions in the infrared, *Phys.Lett.* **73A**, 387-388.
27. Zhukov, V.V., Latush, E.L., Sem, M.F., (1980) Laser action in the mixtures Sr-Kr, Ca-Kr due to charge-transfer, *Zhurn.Prikl.Spekt.* **32**, 738-740 (in Russian).
28. Koptev, Yu.V., Latush, E.L., and Sem, M.F. (1990) Lasing due to ionic transitions of strontium in a discharge with an electron beam, *Sov.J.Quant. Electron.* **20**, 346-348.
29. Bethel, J.W., Maitland, A., Beecham, P.M., Bell, T.C., Webb, S., Little, C.E. (1994) A microwave-excited strontium-ion (λ430.5 nm) recombination laser, *Opt. and Quantum Electronics* **26**, 1079-1087.
30. Chebotarev, G.D., Latush, E.L. (1995) Scaling laws of the He-Sr recombination laser. *Report at II Int.Conf. "Pulsed lasers on transitions in atoms and molecules"*, Tomsk, 27-30 March 95 (to be published).
31. Atamas', S.N., Bukshpun, L.M., Koptev, Yu.V., Latush, E.L., and Sem, M.F., (1984) Stimulated emission at 535 nm thallium line due to quasi-resonant optical pumping of the Tl-He mixture with the radiation of an Ca-He recombination laser, *Sov.J.Quant.Electron.* **14**, 229-231 (pages in Russian ed.).
32. Atamas', S.N., Latush, E.L. (1994) Quasiresonant ionization of thallium vapour by calcium-vapour laser emission, *J. of Russian Laser Research* **15**, 61-65.
33. Atamas', S.N., Koptev, Yu.V., and Latush, E.L., (1985) Laser oscillations on transitions of tellurium dimers due to optical pumping by an Sr-He recombination laser emission, *Sov.J.Quant.Electron.* **15**, 432-433 (pages in Russian ed.).
34. Zinchenko, S.P., Latush, E.L., Sem, M.F., and Tikidzhi-Khambur'yan, R.A. (1992) Dye laser pumped by an He-Sr recombination laser, *Sov.J.Quant. Electron.* **22**, 798-799.
35. Fan OiKang, Lu ZuKang, Wu Bizheng, and Hong Zhi, (1992) Generation of UV radiation of 196-228 nm with BBO crystal, *Acta Opt.Sin.* **12**, 402-405.
36. Beeson, K.W., Horn, K.A., McFarland, M., Yardley, J.T. (1991) Photochemical laser writing of polymeric optical waveguides, *Appl.Phys.Lett.* **58**, 1955-1957.
37. New Embosser Series from LiCONIX Redefines Embossed Hologram Manufacturing, Laser lines, (1991) *LiCONIX Quarterly Newsletter*, No.2, 1-2.
38. Dovichi, N.J. (1990) Laser-based microchemical analysis, *Rev.Sci. Instrum.* **61**, 3653-3657.
39. German, E.R. (1987) Forensic applications of copper laser techn. *Proc. SPIE* **737**, 28-30.
40. Messenger, H.W. (1990) Metal-vapour lasers display versatality, *Las. Foc. Wor.* **26**, 87-92.
41. Lewis, R.R., Maylor, G.A., Salkeld, N., Kearsley, A.J., and Webb, C.E. (1987) Improvements in copper vapour laser technology: new applications, *Proc. SPIE* **737**, 10-16.
42. Zinchenko, S.P., Petrash, G.G., and Sem, M.F., (1993) An active optical system with the helium-strontium amplifier, *Quantum Electron.* **23**, 677-679 (pages in Russian ed.).
43. Vlasov, D.N., Ivashkin, P.I., Isaev, A.A., Kazaryan, M.A., Kuznetsova, T.I., Chuykov, V.V. (1993) Amplification of image brightness in strontium vapour, *Physica Scripta* **48**, 461-463.
44. Vasiliev, Yu.P., Dymarevsky, Yu.D., Zemckov, K.I., Kazaryan, M.A., Medvedeva, L.V., Petrovicheva, G.A., Chuykov, V.V. (1995) A phase object in the projection system with brightness amplification, *Physica Scripta* **51**, 92-93.

THE EFFECTS OF IMPURITIES ON METAL VAPOUR LASER PERFORMANCE

D. J. W. BROWN, M. J. WITHFORD, R. P. MILDREN, J. A. PIPER
Centre for Lasers and Applications
Macquarie University, NSW 2109
Australia.

Abstract

Trace levels of volatile or metallic impurities can dramatically affect metal vapour laser performance. We report a detailed investigation of impurity effects, with emphasis on characterizing and explaining the beneficial influence of hydrogen additives on output power, repetition rate scaling, and high beam quality extraction from CVLs.

1. Introduction

The effects of trace levels of impurities on copper vapour laser (CVL) performance have been of interest since the first CVL devices were operated. The high operating temperature of elemental CVLs leads to impurity evolution from the ceramics used for vapour confinement and insulation purposes, and these impurities can have dramatic effects on CVL performance. Of most concern is the initial processing of CVL plasma tubes, and the long-term power stability of CVLs [1]. It has also been reported by many researchers that the deliberate introduction of trace levels of H_2 can significantly enhance CVL performance [2],[3].

In this paper, we review our long-running studies of impurity effects in CVLs, which have been aimed at identifying impurities and their sources, and understanding the detailed effects of these impurities on CVL operation.

2. Impurity Evolution from a Newly-Constructed CVL Plasma Tube

In order to accurately characterise impurity behaviour in a CVL, we purpose-built a CVL system incorporating an essentially "leak-free" plasma tube that could be evacuated to low base pressures (to minimise any possible ambiguity due to either leaked impurities or residual gases trapped in the insulator) and mass quadrupole analyser for rapid identification of impurities evolved into the buffer gas from the tube construction materials. The base pressure of the system we developed was $<10^{-7}$ torr at the laser head. [4].

The CVL plasma tube utilised an 18mm diameter, 750mm long alumina tube (Alsint) insulated with fibrous alumina (Zircar AL-30AAH), inside a silica

C. E. Little and N. V. Sabotinov (eds.), Pulsed Metal Vapour Lasers, 161–168.

vacuum jacket which was Conflat-sealed (via silica-glass-metal transitions) to the vacuum system. A heater was fitted to the exterior of the silica jacket to provide pre-heating or supplementary heating of the plasma tube. The CVL was excited using a standard thyratron-switched pulse-charging circuit. Using this system, in conjuction with an optical multi-channel analyser to record emission spectra, a detailed study of the evolution of impurities from a new insulator and ceramic plasma tube was undertaken [5].

A 100°C passive bake-out of the tube (no discharge) evolved a large variety of volatile contaminants, including water, nitrogen, hydrogen, carbon dioxide, oxygen, C_5H_{10}, propene and ethene/methane. Elevating the bake-out temperature caused most atmospheric contaminants to diminish, but ethene/methane, propene, C_5H_{10} and water were still present up to 800°C. In addition, new organic contaminants including toluene, xylene and benzene (presumably residue from the insulator fabrication process) were also liberated during this increase in tube temperature [5].

When the tube was heated further by continuously pulsed discharge, it was found that only free radicals (such as O-H, N-H, C-N, C-H, C-C, H-H and N-N) were detected by the mass quadrupole as larger molecules fragmented under discharge conditions. Of particular importance is the tendency for hydrogen-bearing molecules (ie water, propene) to dissociate to produce free H_2 (which was verified by re-introduction of these contaminants in a separate experiment). Metallic impurities such as Na, Al, Ca, Mg, Mn and Zn were also detected by optical spectroscopy as the tube temperature was elevated towards lasing range.

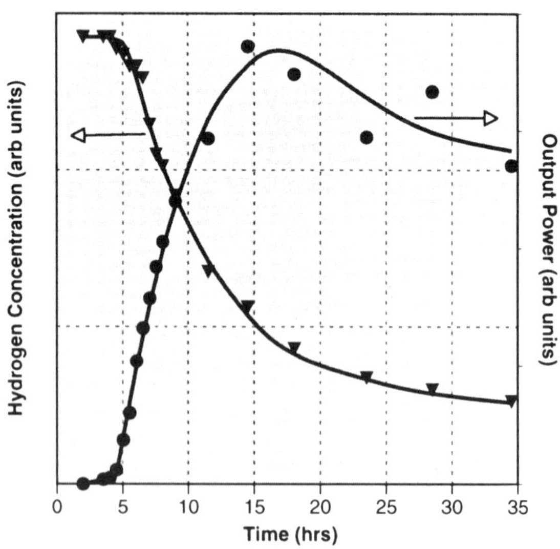

Figure 1: *CVL output power and H_2 concentration during tube processing*

As the tube was processed, the output power increased to a peak value and then decreased somewhat to a lower steady-state value. Analysis of impurity concentrations revealed that the output power was correlated with the H_2

concentration in the plasma tube, as shown in figure 1. There was no correlation of output power with metallic impurity concentration in this CVL, which varied over a time scale of hundreds of hours. It was therefore verified (as discussed in [2],[3]) that the presence of small (few %) quantities of hydrogen in the neon buffer gas is beneficial to CVL performance.

3. Hydrogen effects in repetition-rate scaling and aperture scaling of CVLs

Having identified that hydrogen additive can improve CVL output power, we undertook a detailed series of measurements examining CVL performance when a controlled amount of H_2 was re-introduced into a clean CVL. Significant H_2-induced effects were observed in both repetition-rate scaling [6] and aperture scaling [7] of the CVL, and are described more fully below.

The degree to which H_2 additive improved CVL output power was found to be strongly dependent on the pulse repetition frequency (prf). Figure 2(a) shows the variation in CVL output power as a function of hydrogen add-gas percentage (in 25 torr Ne) for a range of repetition rates. Note that the CVL was operating with a plane-plane resonator for these measurements.

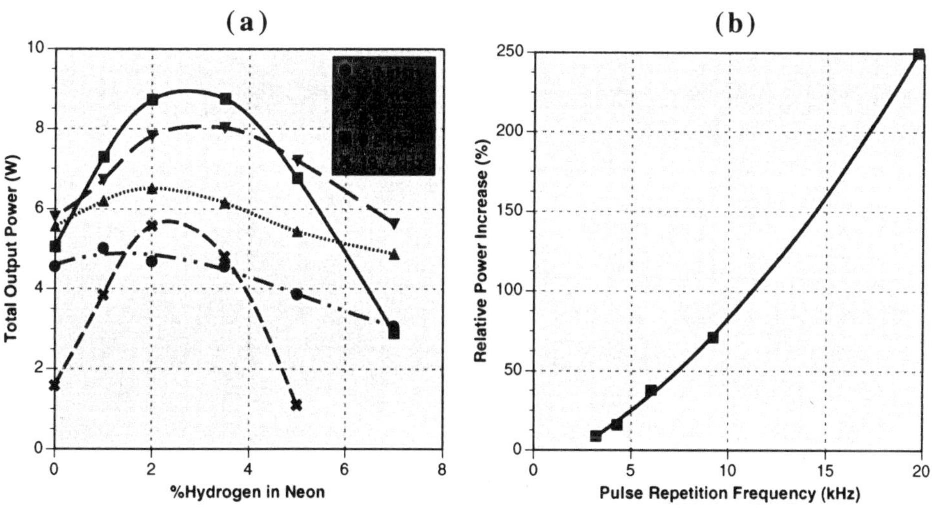

Figure 2. *(a) Output powervariation with H_2 percentage for a range of prf's.*
(b) Relative output power increase as a function of prf.

At the lowest prf (3.3kHz), adding H_2 only produced a marginal improvement in CVL power (~9%). However, as the prf was increased, adding H_2 produced progressively larger percentage improvements over the output power obtained in pure Ne buffer gas, as shown in figure 2(b). At the highest repetition rate used, a 250% increase in power (from 1.6W to 5.6W) was observed. The optimum H_2 admix percentage (ie at which maximum output power is obtained) increases slightly from ~1-2% at low repetition rates to ~2-3% at elevated prf's. Above the optimum H_2 concentration CVL output power falls, however the CVL becomes more intolerant of *excess* H_2 as the repetition rate is increased. There is also some shift in the optimum repetition rate when hydrogen additives are employed. For this CVL, the maximum output power under clean (no H_2) conditions of 5.8W is obtained at a prf of 6kHz; however when H_2 is added 8.7W is obtained at 9.2kHz for a 2% H_2 admix [6].

Measurements of the output power behaviour of three CVLs of different diameter (the 18mm tube described above, a 25mm diameter CVL, and a 40mm diameter CVL) as a function of H_2 add-gas concentration are shown in figure 3. All three CVLs were operated at the same repetition rate (4.3kHz) and the same neon buffer gas pressure (40 torr), and utilised plane-plane resonators. Approximately the same relative power improvement over pure Ne buffer gas operation (~20%) was observed in all three CVLs, and the optimum H_2 admix percentage was in the 1-2% range for all three CVLs [7].

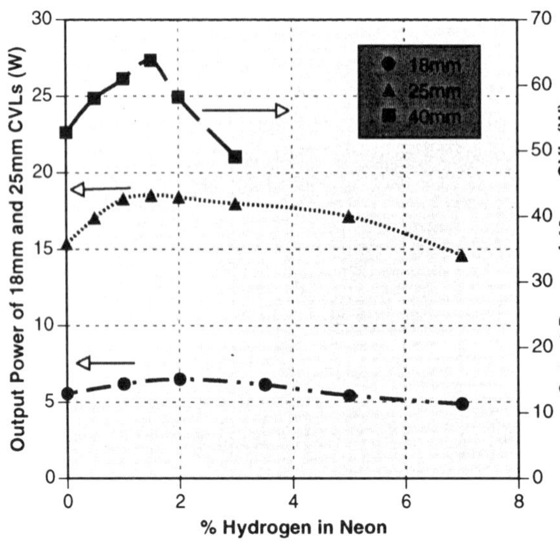

Figure 3. *Performance of CVLsof different apertures as a function H_2 percentage.*

The observed output power improvements in all above cases are due to the increase in the instantaneous power of the latter stages of the output pulse, and also some slight extension of the overall pulse duration. These effects are generally more significant near the tube axis, which leads to some filling-in of the spatial output profile, improving the beam homogeneity [6].

There are a number of other observable changes in the CVL characteristics evident when operating with H_2 additive. For instance, the peak cathode voltage increases and the peak discharge current decreases when H_2 is added, as also observed by other workers. The power deposition into the plasma tube also increases (ie improved impedance matching), as evidenced by slight increases in tube wall temperature and decrease in power loss in the thyratron. Note that we have conducted a thorough investigation of impedance matching effects, and have found that improved impedance matching only makes a relatively small contribution to observed CVL power improvements [4].

4. Hydrogen Effects in High Beam Quality Generation from CVLs

We have also shown that there are several beneficial effects of H_2 additive on high beam quality (ie unstable resonator) CVL operation, including improved pulse-average beam quality and higher yield of low-divergence output power [8]. For example, the low-divergence (or non-ASE) output of the 18mm CVL operating with M=16 positive-branch unstable resonator at a prf of 9.2kHz increased from 5.3W (clean) to 9.5W when 3-5% of H_2 was added to the buffer gas. Further, the pulse-average far-field intensity distribution (figure 4) improves when H_2 is added, with a higher concentration of power within a smaller range of divergence.

The percentage improvement in high beam quality power extraction progressively increases with repetition rate. For elevated prf (~20kHz) operation, non-ASE output power improvements of ~3.9× have been observed, along with significant improvements in the (pulse-average) beam quality.

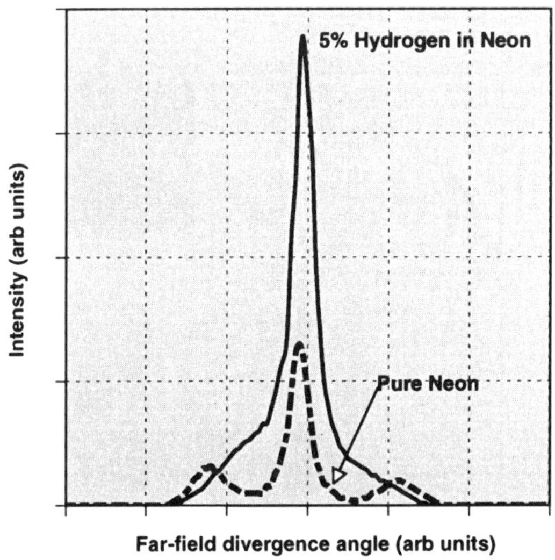

Figure 4. *Pulse-average far-field intensity distribution of CVL output when utilising an M=16 unstable resonator.*

The optimum H_2 concentration is found to be slightly higher for unstable resonator operation (3-5%) than for plane-plane cavity operation (1-3%). This is because improved high-beam-quality power extraction with H_2 additive is not only a result of the improved gain behaviour (ie improved energy extraction from the latter portion of the output pulse, as evidenced in the plane-plane cavity behaviour) but also improvements in the spatial uniformity of the ASE seed [9] which optimise at higher H_2 partial pressures. In fact, it is this improvement in the seed's spatial profile that eliminates the annularity apparent in the far-field profile for clean CVL operation (figure 4).

5. Mechanism of Improved CVL Performance with Hydrogen Additive

There has been much debate about the detailed kinetic effects that lead to the observed enhancements in CVL performance when utilising H_2 additive. We have conducted a wide range of experiments confirming that the dominant H_2-induced effect leading to improved CVL performance is enhanced electron cooling in the afterglow, via an increase in the elastic cooling rate for electrons. This promotes more rapid three-body recombination of electrons and ion species, which provides two benefits directly linked to improved CVL performance. Firstly, improved recombination reduces the depletion of metal atoms from the axial region by "ion pumping" (radial cataphoresis); secondly, reduced pre-pulse electron densities mean lower pre-pulse tube conductivity and hence higher peak cathode voltages (higher electric fields) across the plasma tube, as well as reducing any electric field penetration delay due to the skin effect. In this paper we present the results of two experiments (selected from a much larger body of evidence) that provide strong evidence supporting the "enhanced afterglow electron cooling" mechanism.

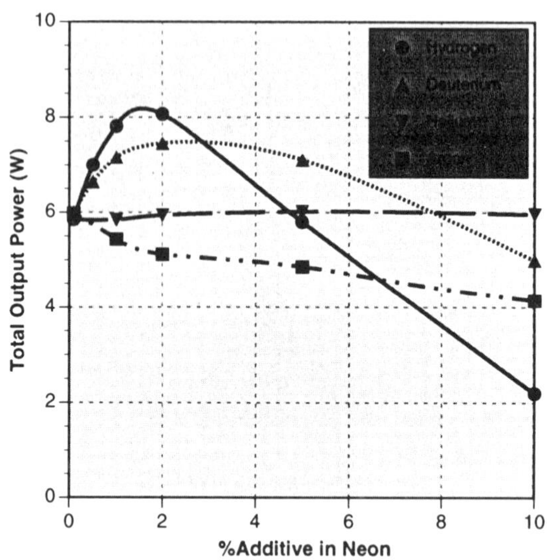

Figure 5. *CVL output power variation with different admix gases.*

Figure 5 shows the variation of CVL output power (for the 18mm UHV CVL) when different gases (H_2, D_2, He and Ar) are admixed with pure Ne buffer gas (maintained at 25 torr) at a fixed prf of 6kHz and fixed input power [4]. The optimum hydrogen additive of ~1.5–2% produced the largest increase in output power (~2W), while twice as much ~D_2 additive (~3–4%) produced a smaller increase in CVL output power and a He additive of ~10% produced a marginal improvement in CVL performance. Addition of argon reduced CVL power. The ratio of optimum admix percentages (ie H_2:D_2:He is 1:2:5) is similar to the ratio of the particle mass (M) divided by the momentum transfer collision cross-section σ^{mom} (1:2:4), suggesting that elastic cooling processes (rate $\propto$ σ^{mom}/M) are involved in improving CVL performance. Further, additional experiments showed that adding H_2, D_2 and He produced similar macroscopic changes in the electrical characteristics of the CVL, which we attribute to enhanced interpulse plasma relaxation (driven by enhanced electron cooling). In particular, similar increases in peak cathode voltage, and decreases in peak tube current and interpulse tube conductivity were observed when twice as much D_2 (or 5 times as much He) was added than H_2. It should be noted that great care was taken to eliminate uncertainty regarding the influence of improved impedance matching in these experiments [4].

More recently, we have directly investigated the effects of adding H_2 by measuring (with spatial and temporal resolution) the population densities in a barium vapour laser (BVL) operating with burst-mode excitation [10]. The measured interpulse variation of Ba and Ba^+ ground state densities, during the afterglow of a *single* excitation pulse (ie first pulse of burst), is shown in figure 6(a) for both pure Ne buffer gas and an 8% H_2 in neon mixture. The inferred interpulse Boltzmann temperatures for the two buffer gas conditions, determined from the population densities of two closely-spaced excited states, are shown in figure 6(b).

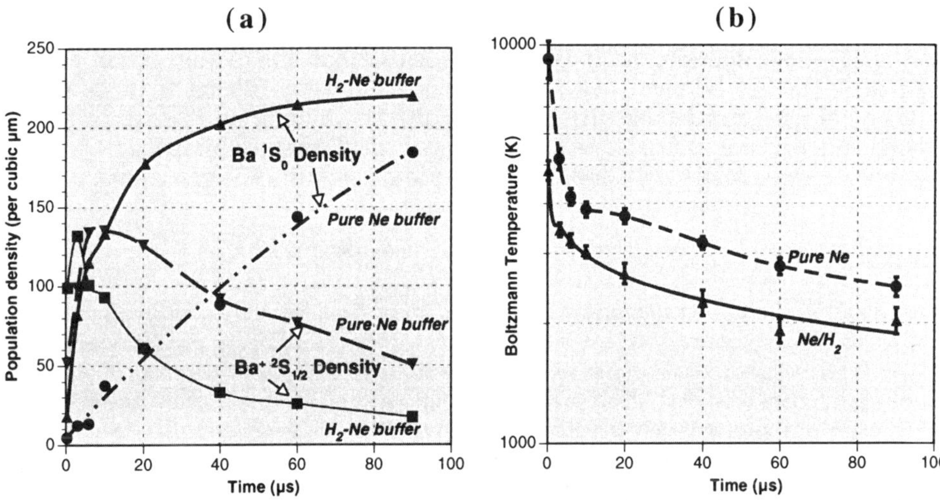

Figure 6. *(a) Population densities during the afterglow (single-shot operation) (b) Boltzmann (electron) temperatures in the afterglow.*

Despite the pre-pulse conditions and excitation pulse characteristics being essentially identical for the two buffer gas compositions, there are significant differences in the relaxation of the plasma during the afterglow due to the hydrogen additive. In particular, barium ion loss (through recombination) and barium atom ground-state recovery is much faster in the presence of H_2. The inferred Boltzmann temperature ($\approx T_e$) shows an initial rapid fall followed by a much slower decay for both buffer gas compositions. However, electron cooling is clearly enhanced in the presence of H_2, as T_e initially falls faster than in pure Ne, and remains lower for the remainder of the afterglow. It is this enhanced electron cooling that drives the increased rate of three-body recombination apparent in the measured barium ion densities.

Additional density measurements performed late in the burst [10] show how the differences in afterglow relaxation induced by H_2 accumulate to dramatically alter the pre-pulse spatial profile of the barium ground-state density, principally by affecting the relative importance of "ion-pumping" in establishing the steady-state density profile. These significant differences in the character of the BVL plasma under these conditions lead to the output energy for H_2:Ne buffer gas being 65% higher than for pure Ne.

Questions still remain regarding which hydrogen species are facilitating electron cooling in the afterglow. In addition to H_2, other hydrogen species such as H, H^+ and H^- may also be important in augmenting electron cooling. Further experimentation and computer modelling is required to ascertain which of these species are the most important in enhancing electron cooling.

6. Conclusion

Detailed studies of the evolution of impurities from a CVL plasma tube showed that hydrogen or hydrogen donors can enhance CVL performance above that obtained using pure Ne buffer gas. Exhaustive studies of re-introduction of H_2 into clean CVLs showed that H_2 additives improve repetition rate scaling and aperture scaling of CVLs, as well as enhancing extraction of high beam quality output when unstable resonators are utilised. A wide range of experiments confirm that the beneficial effects of hydrogen arise from the enhancement in the total electron cooling rate in the afterglow because of the improvement in elastic collision cooling with the light buffer gas additive.

7. References

1. Karras T W (1980) Contaminants in metal vapor lasers *Proc. Int. Conf. Lasers '80* 168-176
2. Bokhan P.A., Silant'ev V.I., and Solomonov V.I. (1980) Mechanism for limiting the repetition frequency of pulses from a copper vapor laser *Sov. J. Quant. Electron.* **10** 724-727
3. Huang Z., Namba K., and Shimizu F. (1986) Influence of molecular gases on the output characteristics of a copper vapor laser *Jpn J. Appl. Phys.* **25** 1677-1679
4. Withford M.J., Brown, D.J.W., and Piper J.A. (1994) Investigation of the effects of hydrogen and deuterium on copper vapour laser performance *Opt. Commun.* **110** 699-707
5. Withford M.J., Brown, D.J.W., and Piper J.A. (in press) Investigation of the evolution of trace impurities from a newly-constructed copper vapour laser *J. Phys. D: Appl. Phys*
6. Withford M.J., Brown, D. J. W., and Piper J.A. (1994) Effects of H_2 buffer gas additive on repetition rate scaling of a copper vapour laser *Opt. Quant. Electron.* **26** 1089-1100
7. Withford M.J., Brown, D.J.W., and Piper J.A. (submitted) Optimisation of H_2-Ne buffer gas mixtures for copper vapour lasers *IEEE J. Quantum Electron.*
8. Withford M.J., Brown, D.J.W., Coutts D.W., and Piper J.A., (1995) Increased Efficiency of High-Beam Quality Extraction from a Copper Vapor Laser with H_2-Ne admixtures *IEEE J. Quantum Electron.* **QE-31** 898–902
9. Coutts D.W. and Brown D.J.W. (1995) Formation of output in copper vapor lasers *Appl. Opt.* **34** 1502-1512
10. Mildren R.P., Brown D.J.W., Carman R.J. and Piper J.A. (1995) The effect of Hydrogen Additive on Population Densities in the Afterglow of Barium Vapour Lasers *Opt. Commun.* **120** 112-120

INFLUENCE OF HYDROGEN ON THE KINETICS OF COPPER BROMIDE LASERS

DIMO N. ASTADJOV
Department of Metal Vapour Lasers
Institute of Solid State Physics
Tzarigradsko Chaussee 72
Sofia 1784, Bulgaria

1. Introduction

The *positive* aspect of hydrogen influence on the performance of copper vapour lasers was finally established in 1985. The times of 'to be or not to be' for hydrogen had passed. That year we *for first time* reported an impressive doubling (but not a marginal 'within a few percent' increase) in the power and efficiency of the CuBr vapour laser as hydrogen of a few percent had been added to the neon buffer gas [1].

The next decade has been devoted not only to attempts to elucidate the physical nature of the hydrogen effect but yet to win over the scepticism about the positiveness of the effect at all. However it took years more people to come to trust those results.

Now the hydrogen effect is a *recognized* fact in metal vapour lasers (Cu, Au, Ba, etc.). It also gave birth to a new generation - the HYBrID lasers. Moreover it challenges most of experimental and simulation issues reported so far.

In this paper I give a recapitulation of the major results of our study of the hydrogen effect with various experimental and calculation techniques during the last decade of years. The majority of experiments and calculatuons were done under the typical conditions for long-term optimum operation of the CuBr laser : 1.2 kW-input power and 15.5 kHz- repetition frequency. Two identical silica sealed-off tubes of 50 cm -electrode separation and 2 cm-inside diameter (of the confining thin rings) were employed; one of them was with pure neon of 15 torr and the other was with a gas mixture of 15 torr neon and 0.3 torr hydrogen.

2. Time and radial measurements of Cu population densities by absorption spectroscopy [2, 3]

Figure1 presents the time evolution of the ground $^2S_{1/2}$, resonance $^2P_{3/2}$ and metastable $^2D_{5/2}$ levels of Cu at the tube axis. The addition of hydrogen slows down the excitation rates of the laser levels. As a result the depopulation of the ground state is slower too.

C. E. Little and N. V. Sabotinov (eds.), Pulsed Metal Vapour Lasers, 169–174.

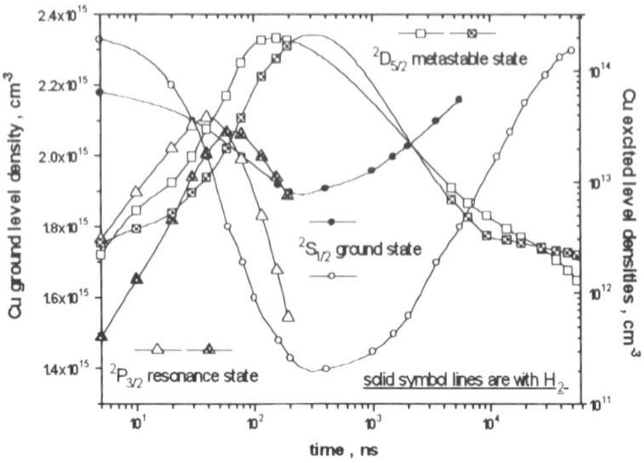

Figure 1

Its depletion is twice smaller and occurs some later. The maxima of the resonance and metastable populations are actually the same with or without hydrogen.

The recovery times for all measured states are much shorter with hydrogen e.g. the time constant for ground state repopulation is only 0.6 μs compared with 14 μs with pure neon buffer gas. The decay of $^2D_{5/2}$ metastables initially is faster with hydrogen (2-3 μs) than without it (5-6 μs). In the middle of the interpulse period the situation reverses. The metastables with hydrogen 'hang' with time constant of about 200 μs while the metastables with pure neon buffer go down with time constant of 55 μs.

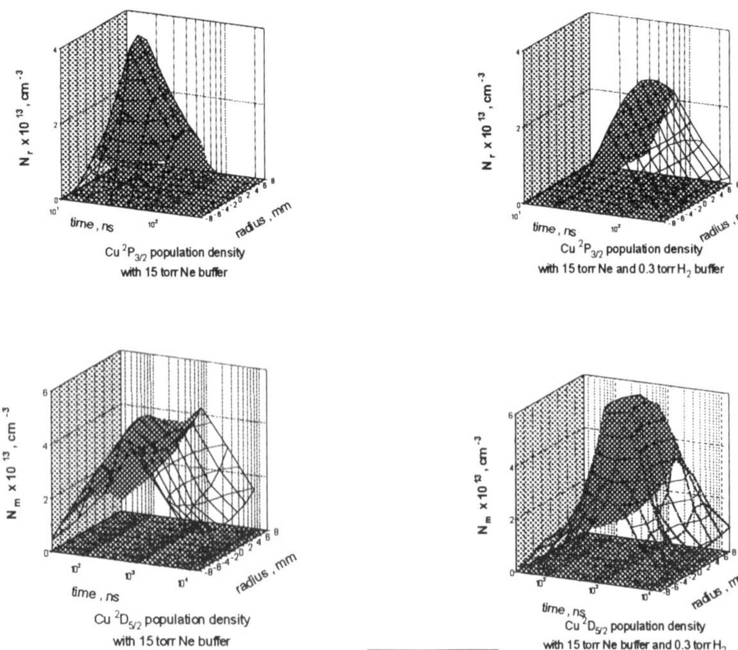

Figure 2

Figure 2 is an illustration of the time-spatial behaviour of Cu resonance$^2P_{3/2}$ and metastable $^2D_{5/2}$ levels. The $^2P_{3/2}$ radial profile with hydrogen is obviously flatter than in the case without it. However, the evolution of the $^2D_{5/2}$ metastables is more complicated. In both cases, there are continuous complex changes in the shape of radial profile during the relaxation period of laser plasma.

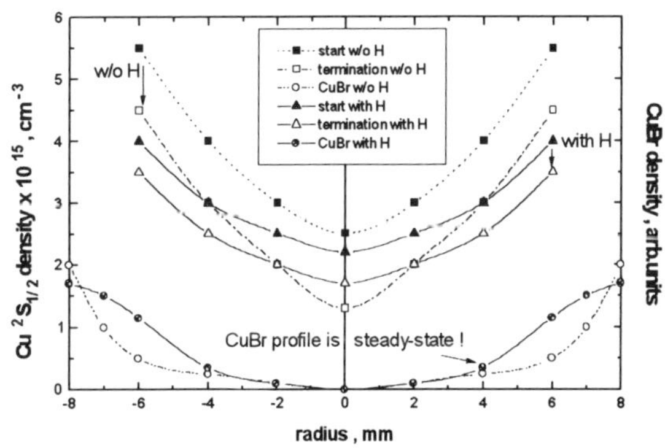

Figure 3

In *Figure 3* are shown the radial profiles of Cu ground $^2S_{1/2}$ and CuBr ground levels. The radial distribution of the $^2S_{1/2}$ ground state of Cu is also flatter with hydrogen than without it. The ground state density of CuBr molecules was measured by absorption of 434.2 nm and 432.2 nm corresponding to the lowest vibronic transitions of X-B and X-C bands. The profiles, in both cases, are actually steady-state in time. CuBr was detected only in the periphery of the discharge volume.

3. Calculation of electron density via momentum transfer cross sections and estimation of the power input channels during the excitation pulse [4].

The time dependence of electron density in the CuBr laser during the excitation pulse is plotted in *Figure 4*. .With pure neon buffer gas the initial electron density (conditioned to be at a current of 10 A) is $2.7x10^{12}$ cm^{-3} while with 0.3torr-hydrogen additive it is $8.6x10^{11}$ cm^{-3}. This is an indication that the *prepulse* electron density with 0.3torr-hydrogen additive is more than three times (at least) lower than with pure neon buffer gas. However after about 70 ns the electron density is already the same, which means higher rate of electron production in the case with 0.3torr-hydrogen additive. The measured values of electron production are 0.04 ns^{-1} and 0.06 ns^{-1} , respectively. If negative hydrogen ions

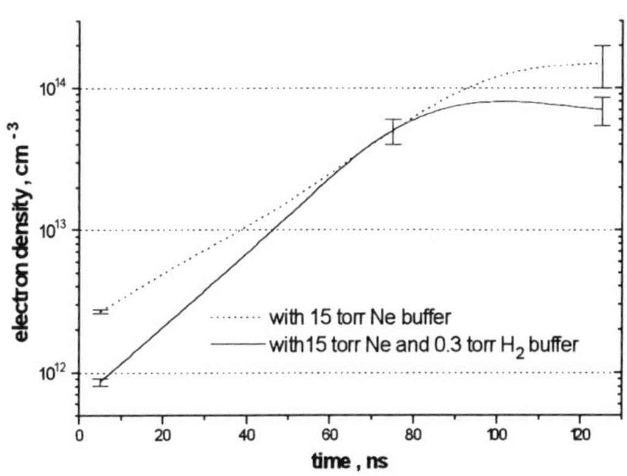

Figure 4

(< 10^{14} cm^{-3}) are supposed to be the only electron donor during the excitation pulse the electron production rate should be 0.05 ns^{-1}.

Estimations of the power input channels in various gas/vapour mixtures can give us an *idea* of the ways in which electrons lose their energy (for instance, see [5]). At the initiation of pulsed gas discharge when populations of excited states are very low they can not be taken into account.

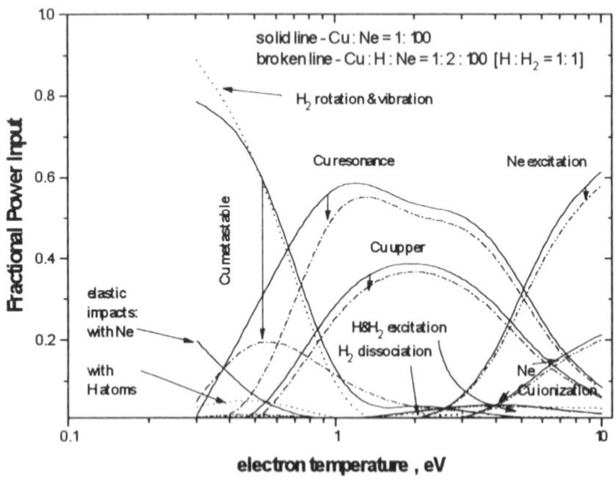

Figure 5

Figure 5 shows the changes in the power input balance in a mixture of Cu with two different buffer gases. The first case is pure Ne of two-order higher density than Cu. In the second case 2% H_2 is added to the Cu-Ne mixture. As can be seen the greatest changes are in Cu metastable state contribution to electron energy dissipation at T_e<1eV. It is noteworthy that rotational and vibrational excitation of H_2 is a substitute for Cu metastable state in the second case. The part of elastic collisions is low. With pure Ne buffer gas the highest value is 0.2 while with the 2% H_2 mixture it is only about 0.05. If D_2 is substituted for H_2 , the power input balance does not alter too much (*cf.* [5] as well).

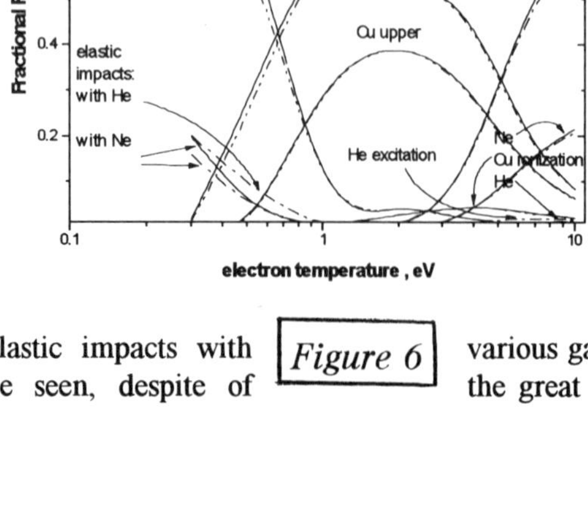

Figure 6

Figure 6 is like *Figure 5* but now 2% H_2 are replaced by 10% He. The changes in the power input balance with 10% He are not so great as with 2% H_2 . However, the contribution of elastic impacts is larger than in the case of 2% H_2 admixture (*cf. Figure 5* or next *Figure 7*).

Figure 7 depicts the *total* (i.e. the sum of contributions of all components due to elastic impacts) input fraction of elastic impacts with various gas admixtures to Ne buffer gas. As can be seen, despite of the great spread of the 2%-H_2 lines due to the considered range of H_2 dissociation (2 orders), the contributions of 10%-He (even 5%-He) admixtures are comparable or higher than the contribution of 2%-H_2 admixture.

Thus it seems implausible that hydrogen effect is only (or even mainly) in the increase of elastic impacts due to presence of the light hydrogen atoms and molecules.

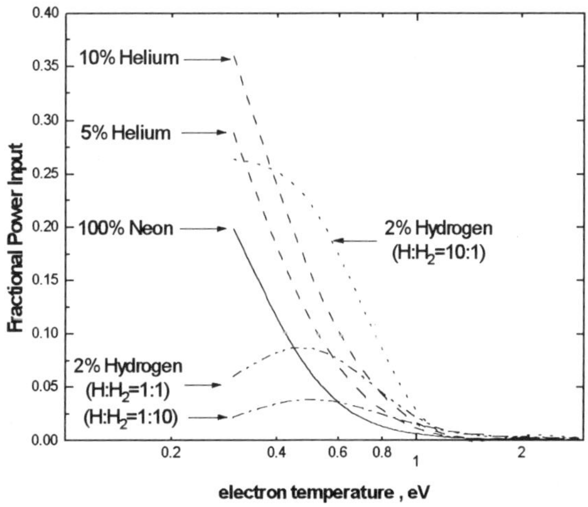

Figure 7

4. Summary

The most obvious result of hydrogen additives is the decrease of the prepulse electron density which is experimentally found in all the family of copper vapour lasers *viz.* the metallic CVL, the CVL utilizing copper halides and the Cu-HYBriD lasers. The much lower depletion of the $^2S_{1/2}$ ground state of Cu during the excitation pulse followed by its more rapid recovery (measured in the CuBr laser [3]) is an implication of reduced Cu ionization and faster return of plasma parameters to their initial prepulse values.

These experimental and calculation results (as well as reported by other researchers not mentioned here because of paper limitations) suggest that the positive influence of hydrogen on laser characteristics of number of metal vapour lasers *probably* does not allow an offhand explanation *e.g.* exaggerated role of elastic collisions or *proposed* production of negative hydrogen ions or even more complex formations, *etc.* It's quite possible that its mechanism is not absolutely the same in that diversity of lasers *viz.* with pure metal or with metal halides, or with a flow of hydrogen halide. It is also likely the domination of one or another process with hydrogen participation over the rest of processes. Further accumulation of experimental and computer-modeling results should shed more light upon that problem.

5. References

1. Astadjov, D.N, Sabotinov, N.V., and Vuchkov, N.K. (1985) Effect of hydrogen on CuBr laser power and efficiency, *Optics Communication* **56**, 279-282.
2. Astadjov, D.N, Vuchkov, N.K., Isaev, A.A., Petrash, G.G, Ponomarev, I.V., and Sabotinov, N.V (1987) Relaxation of metastable copper atoms in CuBr vapour laser emitting regular pulses, *Kvantovaya Elektronika* **14**, 396-399.
3. Astadjov, D.N, Isaev, A.A., Petrash, G.G, Ponomarev, I.V., Sabotinov, N.V, and Vuchkov, N.K. (1992), Temporal and radial evolution of the populations of CuI levels in the CuBr vapour laser, *IEEE Journal of Quantum Electronics* **28**, 1966-1969.
4. Astadjov, D.N, Vuchkov, N.K., and Sabotinov, N.V (1988), Parametric study of the CuBr laser with hydrogen additives, *IEEE Journal of Quantum Electronics* **24**, 1927-1935.
5. Engelhardt , A.G. and Phelps, A.V. (1963), Elastic and inelastic collision cross sections in hydrogen and deuterium from transport coefficients, *Physical Review* **131**, 2115-2128.

MVL PARAMETER MANAGEMENT THROUGH ELECTRON PLASMA COMPONENT

A.N. SOLDATOV

Tomsk State University
Tomsk 634050, Russia

The management of the MVL energy and optical characteristics and time behaviour by varying the parameters of the electron plasma component is discussed. The experimental results obtained from a laser with controlled ionization of the active medium are reported. An improvement of the laser efficiency and control over the lasing pulse duration are shown to be feasible.

1. Introduction

Self-terminating pulsed lasers on atomic transitions in metal vapour lasers offer a favourable combination of high pulse and average power, high pulse repetition rate (PRR), etc. With the classical self-heated discharge excitation, however, there is no way of purposeful variation of the excitation pulse (EP) shape, realization of high laser efficiency and expeditious management of lasing parameters [1, 2].

Comprehensive research into the self-heated discharge dynamics and the electrokinetic and optical characteristics of the repetitively pulsed discharge (RPD) plasma in metal vapour lasers (MVL) has shown that the major parameters capable of fast manipulation of lasing characteristics are the electron temperature T_e and the electron density N_e [2–5]. Two PRD phases, i.e. EP and pulse delay time (PDT) are interrelated, with the highest attainable lasing power W and PRR f depending on T_e and N_e during EP and at PDT. High degree of ionization of the active medium (30–80% of metal vapour per EP) necessitates a long relaxation time and sets an upper limit on f and W, while high input energies cause the laser efficiency η and W to decrease substantially.

C. E. Little and N. V. Sabotinov (eds.), Pulsed Metal Vapour Lasers, 175–182.

In this paper the MVL output characteristics are shown to be manageable through controlled ionization during EP. The problem of the laser output management by varying the plasma parameters at PDT has been discussed at length in [5].

2. Controlled Ionization Experiments

A schematic of the MVL excitation circuit in question is shown in Fig. 1. The bore of gas discharge tube 3 (GDT) is a BeO tube 170 mm in length and 6 mm in diameter. The neon buffer gas temperature P_{Ne} = 6.7 kPa. A combination switch makes possible the current cutoff at a level of 20 A. The switch is made up of thyratron 1 in series with tacitron 2. In the initial state the tacitron is open. When the thyratron is triggered a negative pulse is applied to the tacitron using a controlled delay line to close the tacitron. Unlike the self-heated discharge excitation (regime 1), this admits of the EP shape management and of the cutoff of the energy delivered to the laser tube upon cessation of the lasing pulse, thus realizing the controlled ionization mode (regime 3). This design also permits of the energy "slicing" mode (regime 2) where within 200–300 ns after the current cutoff the remainder of the energy stored in the capacitance is delivered to GDT to maintain thermal conditions. In the latter case the first peak of EP and the entire PRD correspond to regimes 3 and 1, respectively.

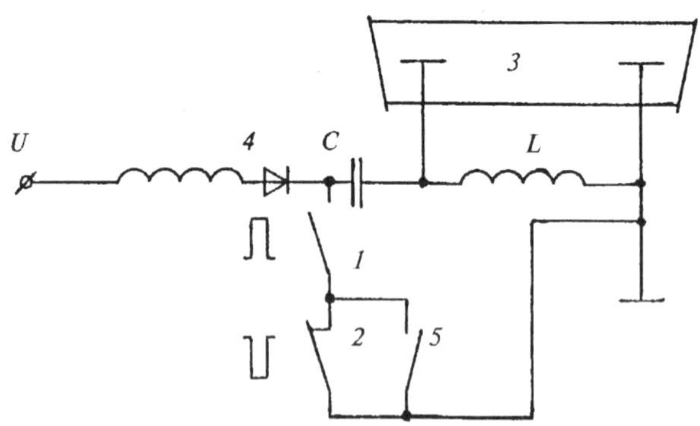

Figure 1. Schematic of MVL excitation circuit: 1 and 2 - switches, 3 - GDT, 4 - resonance charging circuit, c - storage capacitance, *L* - charging inductance, and 5 - vacuum switch.

2.1. CVL EFFICIENCY IN CONTROLLED IONIZATION REGIME

Importantly, one point appears to be clearly established - in regime 3 the input energy to the discharge is decreased drastically without loss in the lasing power. In the case where the input energy to the active medium is entirely cut off upon cessation of the lasing pulse the copper vapour laser (CVL) efficiency is 9%. In regime 3 spontaneous emission from the upper states of Cu atoms and Cu and Ne buffer gas ions is much lower. The higher the radiative level the weaker its excitation. A lower degree of ionization of Cu atoms can be judged both by the discharge current pulse amplitude and by intensity variations of the emission lines from ionic species of Cu. Recombination becomes appreciable within 10^{-6}–10^{-5} s after EP.

In regime 1 the step-by-step excitation and ionization act to decrease the population of the upper $4p$ laser levels. The role of these processes at high concentrations of Cu atoms N_{Cu} has been pointed out in [7]. In the measurements reported here the explicit effect of these processes is observed at $N_{Cu} = 10^{15}$ cm^{-3}. Furthermore, following the lasing pulse the electron density increases considerably and the plasma deactivation and deionization processes slow down at PDT. In regime 3 the detrimental effect of the step processes is reduced substantially.

2.2. DISCHARGE AND LASER EMISSION UNIFORMITY

We have compared the transverse distribution of the laser emission and spontaneous emission intensity from the discharge in regimes 1 and 3 (see Fig. 4). Higher energy delivered to the discharge in regime 1 gives rise to more severe radial inhomogeneities in the discharge and in laser emission as compared to regime 3.

It should be borne in mind that, in regime 3, the voltage across the discharge gap and consequently T_e during EP are kept constant. What happens in this case is not a mere decrease in the input energy to the discharge; rather it is time selectivity of excitation of resonance levels, otherwise a decrease in the lasing power would have been observed.

3. Kinetics of Controlled Ionization Regime

The discharge parameters corresponding to regime 3 (see Fig. 2) and the thermodynamic characteristics of plasma (buffer gas pressure P_{Ne}, copper vapour pressure P_{Cu} and gas temperature T_g) have been used to assess the role of primary processes at work during EP. The prepulse N_e and T_e have been determined from the plasma conductivity to give $N_e^{(0)} = 2{\cdot}10^{12}$ cm^{-3} and $T_e = T_g = 0.25$ eV.

The respective buffer gas and copper vapour concentrations are $N_{Ne} = 3{\cdot}10^{17}$ cm^{-3} and $N_{Cu} = 2{\cdot}10^{15}$ cm^{-3}. The excitation and ionization rates have been derived by the quasi-classical theory formulae [9]. A set of equations used in the theoretical model described in [10, 11] has been solved numerically.

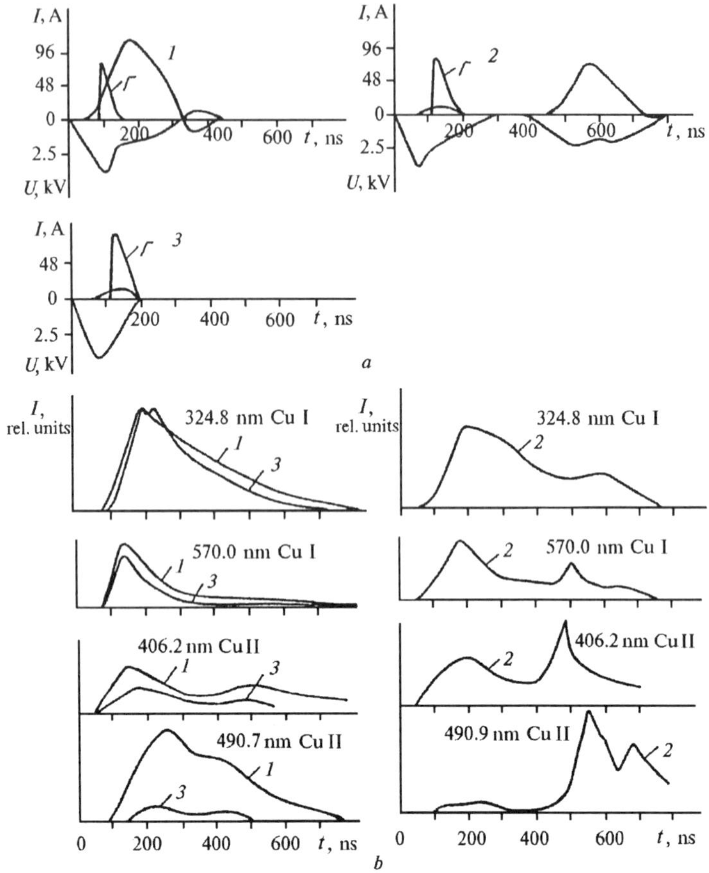

Figure 2. CVL modes of operation: a) oscilloscope traces of current $I(t)$, voltage $U(t)$ and lasing $\rho(t)$ pulses: 1 - self-heated discharge, 2 - input energy slicing regime, 3 - controlled ionization discharge and b) oscilloscope traces of spontaneous emission $I_{sp}(t)$.

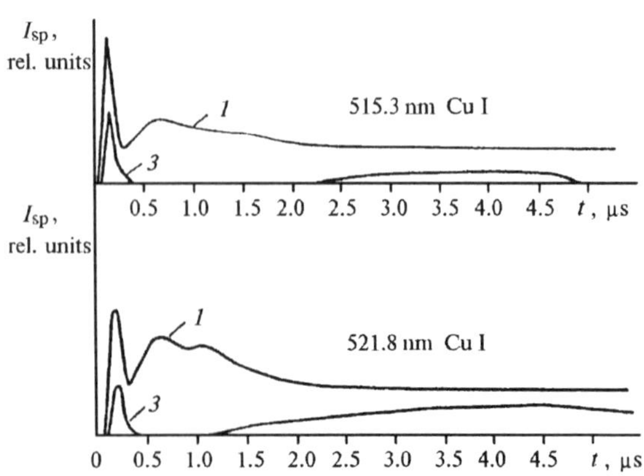

Figure 3. Time behaviour of spontaneous emission in regimes 1 and 3.

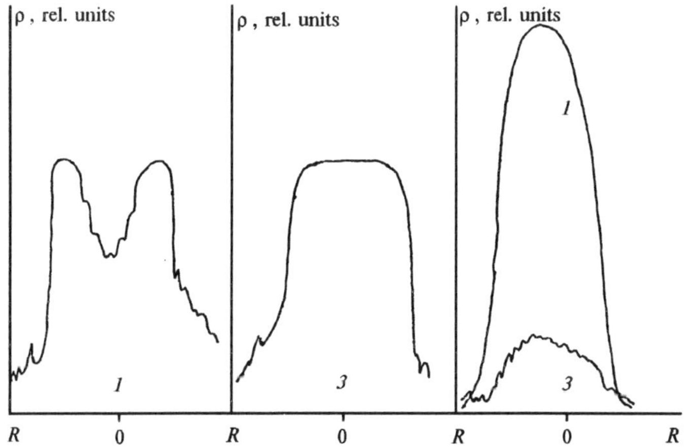

Figure 4. Radial distribution of lasing power (ρ/r) and spontaneous emission intensity (I_{sp}/r) in regimes 1 and 3.

Figure 5 illustrates calculated time dependence of N_e and N_{Cu} at laser levels as well as of T_e and lasing energy. It is evident that N_e with a conventional self-heated discharge is much higher than in regime 3. The degree of ionization is 75 and 5%, respectively, within 160 ns after the onset of EP. This accounts for the lower intensity of ionic and atomic lines from high-excitation states in regime 3 (see Figs. 2 and 3).

As indicated by our experimental and calculated data both in regime 1 and in regime 3 the lasing energy per pulse is practically the same. Consequently, regime 3 provides higher laser efficiency. Table 1 gives calculated power W_d delivered to the discharge, laser output power W_l and efficiency η for the discharge conditions of Fig. 5.

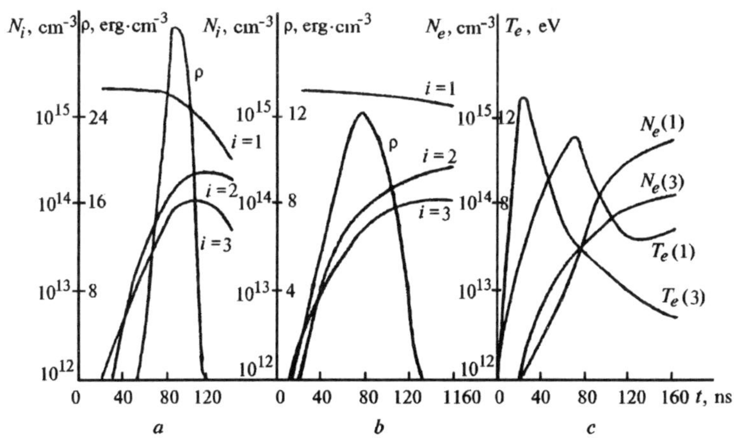

Figure 5. Calculated time dependence of CVL plasma parameters: a) regime 1, b) regime 3, c) N_e and T_e vs time for regimes 1 and 3.

TABLE 1.

Parameters	Regime	
	Self-heated discharge	Controlled ionization
W_d	330 W	30 W
W_l	4 W	2.9 W
η	1.2%	9%

4. Forming Diffraction Beam Divergence and Management of Lasing Pulse Shape and Duration

A CVL limitation is large beam divergence associated primarily with a short lasing pulse duration of about 10–20 ns. Under typical experimental conditions the beam divergence in a plane-parallel resonator laser is two orders of magnitude greater than the diffraction divergence, while in an unstable resonator laser it is an order of magnitude greater than the diffraction value. With the magnification $M = 200$ the ratio W_{dif}/W_{total} is 55% [12] (where W_{dif} is the average power for the diffraction beam divergence and W_{total} is the total average power from CVL).

Thus with typical CVL excitation modes attempts at efficient forming of the diffraction beam divergence with low energy losses prove to be futile. The beam divergence can be decreased by increasing the lasing pulse duration. Regime 3 results in longer lasing pulses. A combination of regimes 1 and 3 presents a control means of the lasing pulse shape and duration. To provide a substantial improvement of the conversion coefficient of laser emission into the diffraction-divergence beam it is essential that EP is shaped in such a way as to produce the initial part of the lasing pulse in the form of a shelf with a minimum amplitude and a duration equal to the formation time of the diffraction-divergence beam. In this case the laser energy expended for the formation of the diffraction-divergence beam is minimum, which provides the highest attainable ratio W_{dif}/W_{total}.

Let us revert to the diagram of Fig. 1. When switch 5 is closed a conventional power circuit of MVL is in operation (regime 1). For zero bias (switch 5 is open) the tacitron is open and regime 1 is realized. When the bias across the tacitron grid is increased the discharge current through the active medium is limited before and during the lasing pulse. This allows optimum voltage across GDT to be maintained for a longer time and the lasing pulse shape and duration to be varied (regime 4).

The design in question employs GDT with the active length $L = 100$ cm and inner diameter $D = 2$ cm. Using zero bias across tacitron 2 brings no changes as compared to regime 1 (Fig. 6a and b). Increasing the bias leads to controlled voltage and lasing pulse shape and duration (Fig. 6c and d). As the bias across the tacitron grid is further increased the lasing pulse shape gets distorted to the extent that the pulse splits up into two pulses or more.

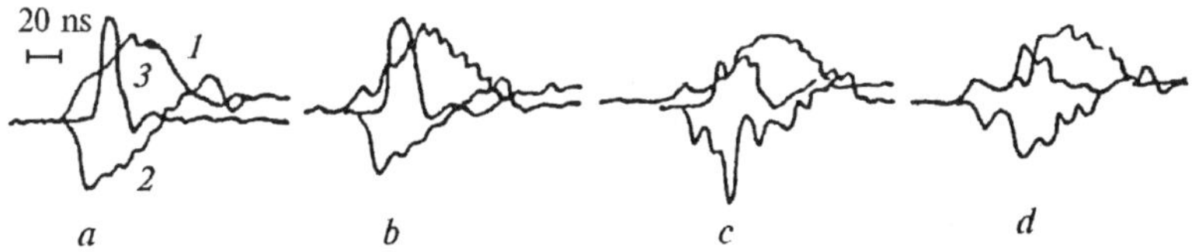

Figure 6. Oscilloscope traces of current (1), voltage (2), and laser oscillation (3) in regime 1 (a) and regime 2 at the bias voltage across tacitron grid $U_g = 0$ (b), 190 V (c), and 210 V (d).

We have examined characteristics of CVL with an unstable resonator ($L = 162$ cm and $M = 30$). The calculated formation time for the diffraction divergence beam in this resonator is 42 ns [13]. When the laser is operated in regime 1 the lasing pulse exhibits characteristic spiking of 70 ns duration at its base and the major portion of the pulse energy is released within the first 40 ns (Fig. 7a). Increasing the bias voltage across tacitron grid 2 in regime 4 decreases the laser energy at the onset of the lasing pulse, whereupon the major portion of the lasing energy is released (Fig. 7b, c). Due to this effect the amount of energy produced in the diffraction-divergence beam is increased (curve 1 in Fig. 8).

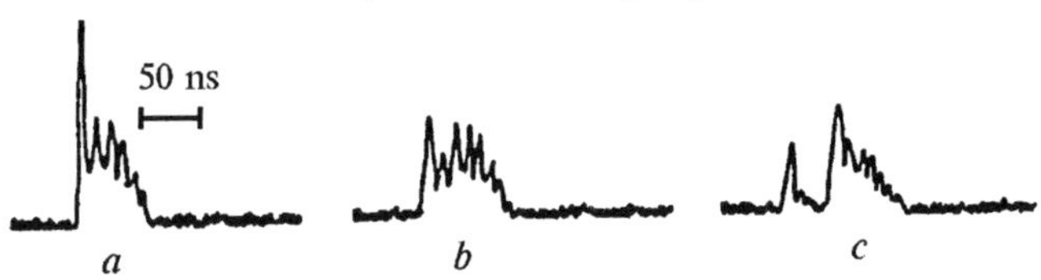

Figure 7. Oscilloscope traces of lasing pulses in regime 1 (a) and regime 2 at $U_g = 180$ V (b) and 210 V (c).

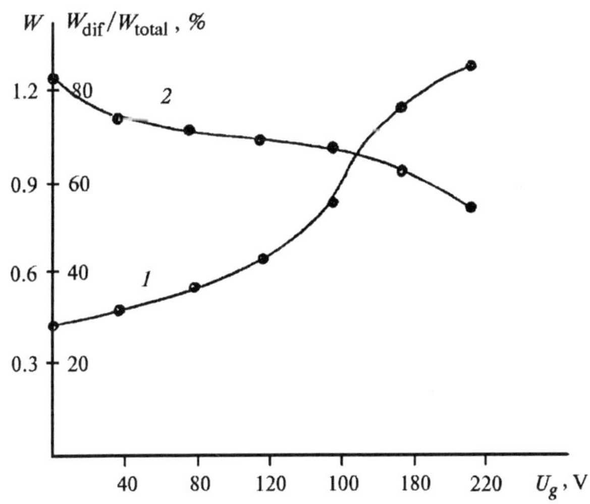

Figure 8. Variation of W_{dif}/W_{total} (1) and lasing power (2) with U_g.

Increasing the bias voltage across tacitron grid 2 from 0 to 240 V increases the ratio W_{dif}/W_{total} from 28 to 83% (curve 2 in Fig. 8). Despite the fact that the total average power decreases by 22% at constant average input energy the average power in the diffraction-divergence beam is more than doubled.

5. Concluding Remarks

MVL operation in the controlled ionization regime has been evaluated. The results obtained point to the feasibility of purposeful variation of the excitation pulse shape and expeditious management of the output laser parameters. It has been found experimentally that the laser efficiency based on the input energy to the gas discharge tube is 9% in the controlled ionization regime. Effective control over the CVL pulse duration and beam divergence is shown to be practicable. A twofold increase in the share of radiation in the CVL diffraction divergence beam has been achieved.

6. References

1. Isaev, A.A., Kazaryan, M.A., and Petrash, G.G. (1972) Efficient pulsed copper vapor laser with high average output power, *Pribory i Tekhnika Eksperimenta* **16**, 1, 40–42.
2. Soldatov, A.N. and Solomonov, V.I. (1985) Self-Terminating Gas Discharge Metal Vapor Lasers, *Nauka, Novosibirsk.*
3. Elaev, V.F., Soldatov, A.N., and Sukhanov, V.B. (1981) Determination of electron temperature and copper vapor laser afterglow, *Teplofizika Vysokikh Temperatur* **19**, 426–428.
4. Soldatov, A.N. and Fedorov, V.F. (1982) Repetitively pulsed copper vapor laser operating at maximum pulse repetition rates, *Proceedings of International Conference on Lasers' 82*, STS Press McLean, New Orleans, 478–485.
5. Soldatov, A.N. (1993), Physics and technology of copper-vapor lasers with controlled parameters, *Atmospheric and Oceanic Optics* **6**, 6, 383–387.
6. Burlakov, V.D., Gorbunova, T.M., Soldatov, A.N., et al. (1984) The use of Rozhdestvensky's hook technique for studying metal vapor laser, *Izvestiya Vysshikh Uchennykh Zavedenii, Fizika,* Deposited Paper N2856-84, 1–24.
7. Demkin, V.P., Soldatov, A.N., and Yudin, N.A. (1993) Efficiency of the copper vapor laser, *Atmospheric and Oceanic Optics* **6**, 6, 388–391.
8. Eletsky, A.V., Zemtsov, Yu.K., Rodin, A.V., et al. (1975) Optimum characteristics of high pressure metal vapor lasers, *Doklady Akademii Nauk SSSR* **220**, 318–321.
9. Mnatsakanyan, A.Kh., Naidis, G.V., and Shternov, N.P. (1978) Electron energy distribution in copper vapor - neon/helium mixtures, *Kvantovaya Elektronika* [Sov. J. Quantum Electron.] **5**, 597–602.
10. Vainshtein, L.A., Sobel'man, I.I., and Yukov, E.A. (1973) Electronic Excitation Cross Section For Atoms And Ions, *Nauka, Moscow.*
11. Demkin, V.P., Kupchinsky, N.L., Soldatov, A.N., and Fedorov, V.F. (1988) Calculation of copper vapor laser efficiency, *Tenth All-Union Conference on Electron and Atomic Collision Physics*, Digest of Technical Papers, Part 2, Uzhgorod, 193.
12. Zubov, V.V., Lyabin, N.I., Mishin, V.I., et al. (1983) Investigation of copper vapor laser with long service life and improved parameters of excitation pulse, *Kvantovaya Elektronika* [Sov. J. Quantum Electron] **10**, 1908–1910.
13. Pasmanik, G.A., Zemskov, K.I., Kazaryan, M.A., et al. (1988) Optical systems with brightness intensifiers, *Applied Physics Institute*, Gorkii.

NOVEL CIRCUITS FOR THE EXCITATION OF METAL VAPOUR LASERS

NIKOLAI K. VUCHKOV
Department of Metal Vapor Lasers,
Institute of Solid State Physics, Sofia 1784, Bulgaria

I. Introduction

The self-terminating metal vapour lasers are now a source of high power laser radiation. Amongst the most attractive characteristics of these lasers as high average and pulse power, is their high efficiency of electric-into-optical energy conversion. So the reproducible steady-state laser operation at high efficiency is an important problem in self-terminating lasers development.

Now, there are two principal methods to increase the efficiency of the most famous of these lasers - the copper vapour laser. The first is by changing of the gas mixture - addition of hydrogen or by HBr flow at large volumes of the active medium [1,2,3]. The second method is by governing of the excitation electrical pulse by means of new excitation circuits.

In our laboratory we have proposed and developed two novel excitation circuits for CuBr vapour laser. In this paper we report the main properties of the novel circuits and their influence on the CuBr laser performance.

2. Electrical circuits

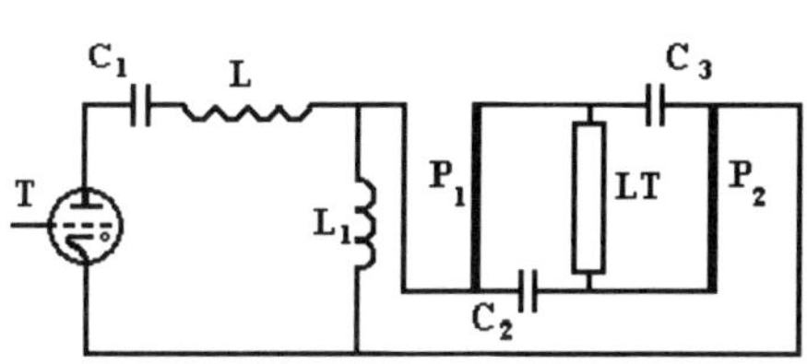

Fig.1. Schematic diagram of the IPC circuitry: C_1, storage capacitor; C_2 and C_3, peak capacitors; P_1 and P_2, metal plates; L and L_1, inductors; LT, laser tube; T, thyratron.

An the ordinary circuit (OC) used for excitation of pulsed metal vapour lasers, the triggering of the thyratron initiates the discharge of the energy storage capacitor through the peaking capacitor, resulting in the discharge of both capacitors through the laser tube.

The first novel circuit involves two interacting peaking circuits (IPC) [4]. A schematic diagram of this circuit is shown in Fig.1.

This work was supported by Ministry of Education and Science, Contract # 85 / 1991 and # 217 / 1992.

C. E. Little and N. V. Sabotinov (eds.), Pulsed Metal Vapour Lasers, 183–188.

Each interacting peaking circuit consists of a peak capacitor (C_2 or C_3), a metal plate (P_1 or P_2) located close to the laser tube between the electrodes and the laser tube itself, which is a common component.

A comparative investigations between OC and IPC excited CuBr laser have been carried out. It is established that the optimal with respect to maximal output power and laser efficiency concentration of the CuBr vapour, the pressure of the buffer gas Ne and H_2 and the pulse repetition rate are equal for both excitation. It is found that the IPC excitation leads to 40 % increasing of the laser efficiency compare with the OC excitation.

In spite of these results, we consider the IPC excitation as an improved version of the OC excitation. When OC or IPC excitation is applied to the CuBr laser and the input power is constant, it is not possible to change essentially or to govern intentionally the wave form of the voltage and the current. Sometimes this is of great importance, e.g. for lasers operating on transitions from short-living levels, a steep rising front is required, or for recombination lasers, current pulses with rapidly falling front are needed.

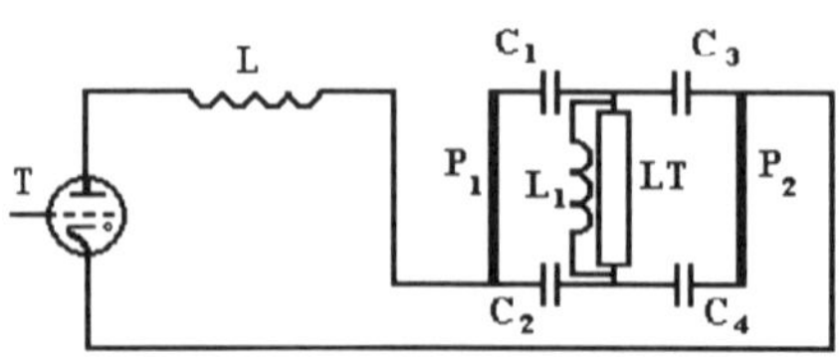

fig.2. Schematic diagram of the IC circuitry: C_1 to C_4, capacitors; P_1 and P_2, metal plates; L and L_1, inductors; LT, laser tube; T, thyratron.

The second new circuit, which meets these requirements is shown in Fig.2 [5]. The circuit comprises two interacting circuits (IC) each consisting of a metal plate that is connected to the electrodes by two capacitors. An essential feature of this circuit is that there is no separate storage capacitor and neither of the electrodes is grounded, but is under potential to ground. In the OC and IPC excited lasers the cathode of the tube is heated to a significantly higher temperature than the earthed electrode (the anode). In the IC excited laser both electrodes are heated near to the same temperature.

The use of the IC excitation leads to an significant increase of the average power and laser efficiency compare with the OC and IPC excitation. Below some applications of the IC excitations are given.

3. Influence of the excitation circuits on the CuBr laser performance

A comparative study of the output characteristics, electric parameters and main kinetic plasma processes concerning the CuBr laser performance during the excitation pulse with the four electric circuits: WPC (without peaking capacitor); OC (ordinary circuit, i.e. with a peaking capacitor); IPC (interacting peaking circuits) and IC (interacting circuits), for CuBr vapor laser excitation has been carried out [6].

The experiments were held with a laser tube made of fused silica with discharge confining rings with inside diameter of 2 cm and electrode separation of 50 cm; three side arm reservoirs containing the lasant CuBr. As a buffer gas was used a mixture of 15 torr Ne and 0.3 torr H_2.The pulse repetition rates is 21 kHz. The experiment was carried out at 1.2 kW power from a high voltage rectifier.

Discharge circuits simulation was undertaken to explain the difference in efficiencies of the four electric circuit employed in CuBr laser excitation. The numerical modeling of the electric circuits which also includes kinetic processes in the discharge plasma, could clear up the relation between the features of the electric circuit and the plasma parameters, and thus the properties of laser emission. The sets of equations and the conditions at which they are solved are given in [6].

Table 1. C_1 to C_4 - hot-value capacitances; U_0 - starting voltage of C_1 (in IC, of the capacitor bank); P - average laser power; η - practical laser efficiency; U_R^{max}, I_R^{max} and W^{max} - amplitudes of the voltage, current and power pulse, respectively; E_0 - electric energy input into the discharge; E_1 - electric energy input till the end of laser pulse; $\Delta I_R/\Delta t$ - tube current rise rate; $\eta_{ph} = E_g/E_1$, where E_g is laser pulse energy. Experiment and calculation are abridged as exp and cal.

circuits		WPC	OC	IPC	IC1	IC2
C_1, pF		1045	1045	1045	1145	950
C_2, pF		-	400	215	215	215
C_3, pF		-	-	215	215	215
C_4, pF		-	-	-	3140	3140
U_0, kV		10.3	10.3	10.3	10.3	11.3
P, W	exp	6.25	10.25	14.50	16.50	17.75
	cal	4.7	10.25	14.50	16.12	18.09
η, %	exp	0.52	0.85	1.19	1.37	1.46
	cal	0.40	0.85	1.19	1.34	1.51
U_R^{max}, kV	exp	9.20	8.20	9.80	9.90	10.10
	cal	8.64	8.64	9.74	9.87	10.30
I_R^{max}, A	exp	101	124	155	171	185
	cal	108	129	153	167	183
W^{max}, kW	exp	290	553	908	980	985
	cal	303	618	870	962	1090
E_0,mJ	exp	37	41	43	43	43
	cal	30	34	38	38	41
E_1, mJ	exp	28	33	36	38	39
	cal	17	24	29	31	33
$\Delta I_R/\Delta t$,A/ns	exp	1.2	2.2	3.2	3.6	4.0
	cal	1.0	1.9	2.6	3.1	3.4
η_{ph}, %	exp	1.06	1.43	1.92	2.07	2.17
	cal	1.31	2.01	2.37	2.50	2.60

The real (hot) values of the capacitors forming the respective electric circuits are given in Table 1. The IC CuBr laser was investigated in two versions - IC1 and IC2. With IC1 the CuBr laser efficiency is less than that of IC2 but it meets the requirement of starting voltage U_0 = 10.3kV of the capacitor bank to be equal to the starting voltage of the storage capacitors in the other electric circuits. For IC2 the starting voltage is U_0 = 11.3kV. The average output power (calculated and experimentally measured), P and the efficiency (experimental and calculated), η of the CuBr laser with different excitation circuits are given in Table I as well.

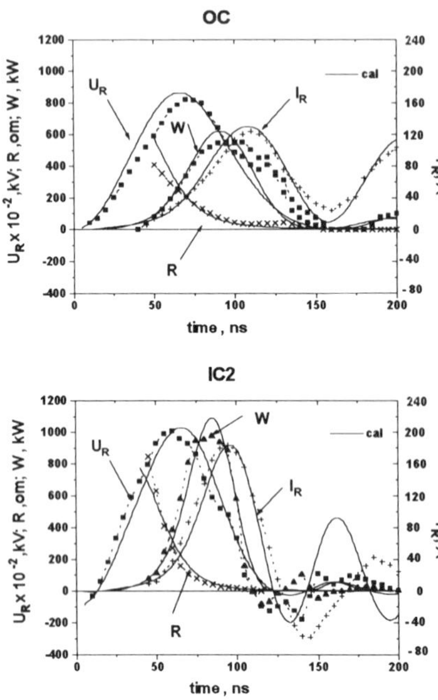

Fig.3.Wave forms of plasma resistance, R, tube voltage, U_R, electric power, W and tube current, I_R for OC; and IC2 excitation. The experimental curves are dashed and the calculated curves are solid.

. Fig.3 depicts the wave forms of the active voltage, U_R , the time dependent tube resistance, R, the electric power pulse, W and the tube current I_R only for two circuits. As can be seen, there is very good coincidence between experimental and modeling results, especially in the first half-waves.

The results given in Fig.3 and Table 1 can be summarized as follows:

(a).The amplitude of the tube current, I_R and the amplitude of the pulsed power W increases from WPC to IC2. The rise rate and fall rate of the first half-wave of I_R and W increase too.

(b).The pulse energy input into the gas discharge, E_0 and the pulse energy input till the end of the laser pulse, E_1 increase from WPC to IC2.

The electric circuit modeling shows that the improvement of the electrical and energetic parameters of the CuBr laser from WPC to IC2 is due to the appearance of additional terms in the equations which describe the operation of the IPC and IC, compared to WPC and OC. These new terms represent specific interactions between different parts of the electric circuit of IPC and IC-excited lasers, leading to an increase of I_R, W, E_1 and E_0.

On the base of the model ,we have simulated the time behavior during the excitation pulse of populations of the main Cu I states responsible for the laser oscillation. The electron density and the electron temperature were calculated too.

The comparative investigation of the five excitation circuits shows that the laser output depends on the amplitudes and alteration rates of the excitation pulses. A very good correlation is found between the laser power and the quantity $S=(W^{max}\ \Delta I_{R}/\Delta t)^{1/2}$, where W is the maximum of electric power input into the discharge and $\Delta I/\Delta t$ is the rise rate of current front through the laser tube. The correlation of laser power P and the quantity S is plotted in Fig.4.

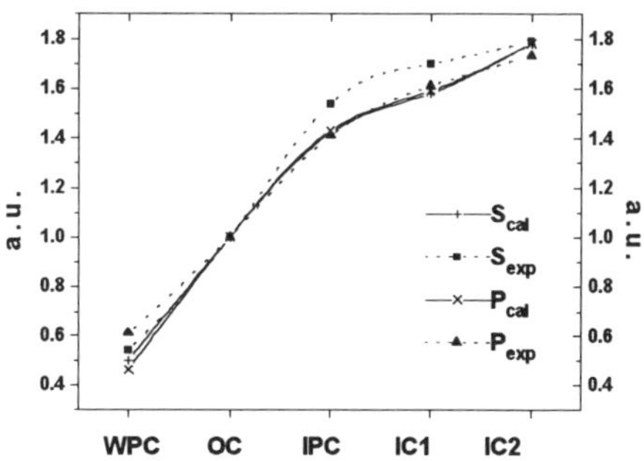

Fig.4.Magnitude of average laser power, P and $S=(W^{max}\Delta I_R/\Delta t)^{1/2}$ (each normalized to OC and trend shown) with different electric circuits.

Laser efficiency can be expressed as a function of some relatively independent factors which concern the operation of different parts of the laser device as a whole and present in steps the process of conversion of electrical energy into the optical energy of the laser beam. The first factor is the ratio E_0/E_C, where E_0 is the pulse energy dissipated into the discharge and E_C is the electric energy stored in the capacitor bank or storage capacitor. The second factor is the energy fraction E_1 of E_0 which is deposited into the discharge plasma till the end of laser pulse. The third factor is the physical laser efficiency $\eta_{ph} = E_g/E_1$, where E_g is the energy of laser pulse. So the practical (i.e. overall) laser efficiency is a product of these three factors. As can be seen in Table 1 all the factors increase from WPC to IC2 but mostly η_{ph} which means that the increase of E_g is not only due to the increase of E_1 and E_0 but mainly to the improvement of conversion of the electric energy E_1 into laser energy E_g at the final step. The last implies that IPC and IC excitation of the CuBr laser causes significant changes in the kinetic processes responsible for the laser oscillation.

The improved laser performance from WPC to IC2 excitation is attributed to the following reasons:

1.Efficiency of the pulsed power supply i.e. E_0/E_C increases.

2.The electromagnetic interaction between different parts of the excitation circuit (especially, within IPC and IC) leads to an increase in I_R (the tube current), in W (the electric pulsed power dissipated into the discharge), in E_1 (the electric energy deposited into the discharge plasma till the end of laser pulse) and in E_0 (the pulse energy dissipated into the discharge). It is experimentally measured and theoretically ascertained by the solution of circuit equations.

3.The rise of E_1 aids to the increase of E_g (the laser pulse energy).

4.The increase in E_g is principally determined by the rise of the full number of transitions from ground to resonance state. Laser oscillation is terminated for the decrease of pumping rate and the increase of depopulation rate of resonance state.

5.The increase of η_{ph} is a result of the increase of the efficiency of upper laser level pumping from ground state and the decrease of the energy losses in other channels starting from ground state.

4. Application of the IC excitation

In our laboratory some high output characteristics are obtain with the IC excitation CuBr laser. Two types of laser tubes are investigated - with small and middle volume.

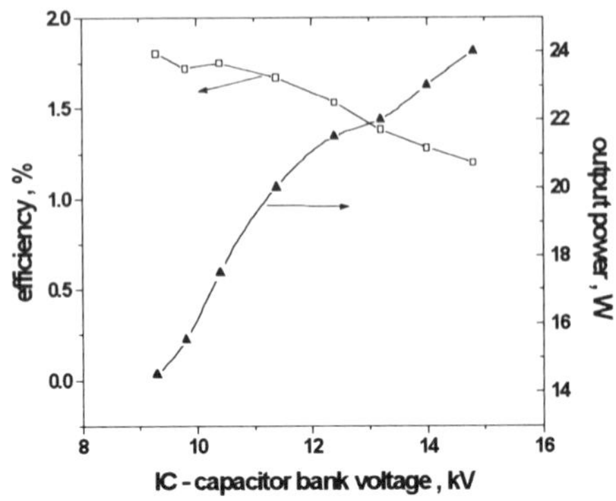

Fig.5 Dependence of the average output power and laser efficiency on the capacitor bank voltage; laser tube with electrode separation of 50 cm and discharge-confining rings with an inside diameter of 2 cm.

The first laser tube is with discharge-confining rings with an inside diameter of 2 cm, electrode separation of 50 cm and three sidearm reservoirs containing the lasant CuBr. With this tube (Fig.5) a maximal average output power of 24 W and maximal laser efficiency of 1.8 % are obtain (Fig.5) (for comparison - with OC excitation the maximal efficiency is 1.1 %).

In the second laser tube the thin (1.5 mm) diaphragms are of 4.5 mm aperture, which defined the active zone [7]. The distance between the diaphragms located close to each electrode is 30 cm. Thus the active volume of the tube by construction is 4.77 cm^3. The highest laser power of 6.7 W is obtained, which yield 1.4 W/cm^3 specific average power.

IC excitation is used very successfully in the University of St. Andrews, UK by C. Little and coworkers for excitation of a copper HyBrID laser. Recently they reported for a laser, which produced an average output power of 9.5 W from an active volume of 4.77 cm^3 corresponding to a specific output power of 2 W/cm^3 [8].With large bore tube - 6 cm and active length of 200 cm they obtain an average power of 201 W at 1.9 % efficiency. For an output power of 120 W, the efficiency reached 3.2 % [9].

5 References

1. Astadjov D.N., Vuchkov N.K and Sabotinov N.V. (1988) Parametric study of the CuBr laser with hydroge additives, IEEE J. Quantum Electronics .24, .1927-1935,
2. Jones D.R ,.Sabotinov N.V,.Maitland A, .Little C.E. (1992) A high-power high-efficiency Cu-Ne-HBr (1510.6, 578,2 nm) laser, Optics Commun.,94,.289-299.
3..Elaev V.F., .Lyakh G.D., .Pelenkov V.P. (1989) CuBr laser with average lasing power exceeding 100 W, Atm.opt., .2, 1045-1047.
4. Vuchkov N.K., Astadjov D.N, Sabotinov N.V (1988) High-efficiency CuBr laser with interacting peaking circuits, Optical and Quantum Electronics, 20, 433-438.
5 Vuchkov N.K., Astadjov D.N, Sabotinov N.V. (1991) A new circuit for CuBr laser excitation, Optical and Quantum Electronics, 23, S549-S553.
6. Vuchkov N.V. Astadjov D.N. Sabotinov N.V. (1994) Influence of the excitation circuits on the CuBr laser performance, IEEE J.Quant.Electr., 30, 750-758.
7. Astadjov D.N., Dimitrov K.D., Littel C.E., Sabotinov N.V and Vuchkov N.K (1994) A CuBr laser with 1.4 W/cm^3 Average output power, IEEE J.Quant.Electr., 30, 1358-1360.
8. Sabotinov N.V., Kerboom A.F., Jones D.R., Maitland A., Little C.E. (1995) A copper HyBrID laser with 2 W/cm^3 specific average output power, IEEE J.Quant.Electr., 31, 747-753.
9. Jones D.R., Maitland A., Little C.E (1994) A high-efficiency 200 W average power copper HyBrID laser, IEEE J.Quant.Electr., 30, 2385-2390.

NOVEL LOW-TEMPERATURE PULSED METAL-VAPOUR LASERS

The High-Voltage Segmented Hollow-Cathode Metal Ion Laser

R.C. TOBIN, K.A. PEARD, B. LEIGH AND G. BODE
Department of Physics, Monash University
Clayton, Victoria 3168, Australia

K. RÓZSA, Z. DONKÓ AND L. SZALAI
Research Institute for Solid State Physics
H-1525 Budapest, POB. 49, Hungary

Abstract Hollow-cathode discharge sputtering allows the immediate production of metal vapour together with efficient charge-transfer pumping of ultraviolet (~ 220 - 290 nm) metal ion laser transitions, most notably those for copper, silver and gold. This paper discusses a novel high-voltage hollow-cathode discharge with increased efficiency and gain to facilitate the development of compact pulsed lasers for applications to biomedical instruments.

1. Introduction

There is a considerable incentive to develop a low-cost, compact laser emitting directly on wavelengths in the ultraviolet 200 to 300 nm range at a milliwatt power level. The laser would be useful as an economic light source for instruments for such diverse applications as biopolymer monitoring in electrophoretic capillary channels with a sensitivity approaching the femtomole level, the high-sensitivity detection of pollutants in the gas phase, cell surgery, etc. The envisioned applications require a repetitively pulsed laser operating at a low duty cycle. The existing devices, such as the Optical Parametric Oscillator and the Argon Ion Laser with intracavity Frequency Doubling are expensive. In the longer term, the most significant competitive laser device may prove to be the semiconductor pumped solid state laser with frequency quadrupling. The purpose of this paper is to show that, using a novel high-voltage segmented hollow-cathode discharge, the known He-Au^+, He-Ag^+ and Ne-Cu^+ charge transfer pumped metal ion transitions offer good prospects for the development of the above laser. Since the discharge scheme derives from the conventional hollow-cathode discharge, we begin with a review of the most relevant physics of hollow-cathode metal ion lasers.

C. E. Little and N. V. Sabotinov (eds.), Pulsed Metal Vapour Lasers, 189–194.

2. Hollow-Cathode Sputtering Lasers

The early studies of charge transfer pumped metal ion lasers used a conventional hollow cathode discharge to produce the metal vapour by discharge sputtering and to generate the rare gas ions necessary to excite the metal ion laser lines, for example [1 - 3]. The process of thermal charge transfer is represented by the following equation:

$$B^{+} + M \rightarrow B + M^{+*} + \Delta E$$

where B and B^+ denote a buffer gas atom and ground state ion respectively and M and M^{+*} represent a ground state metal atom and ion in an excited level respectively. Thus the charge transfer pumping rate is proportional to $[B^+][M]$, where $[B^+]$ and [M] denote the concentration of the buffer gas ions and ground state metal atoms respectively. The cross section for charge transfer excitation is a maximum when the excited metal ion level lies at $\Delta E \approx 0.2$ eV below that of the ground state buffer gas ion [4]. There are two main advantages of the HC discharge. Firstly, a high degree of ionization is produced in the negative glow region due to the oscillation of the beam electrons in the potential well between opposed faces of the cathode [5]. Secondly, a high density of rare gas ions can be maintained in the rare gas/metal vapour mixture, despite the high concentration of metal atoms [6]. The conventional hollow-cathode discharge is characterised by a low running voltage ($\sim$ 300 V) and a low slope resistance ($\sim$ 10 Ω). Consequently, a Ne-Cu^+ 249-260 nm laser of 15 cm active length requires a high operating current density, J, of about 500 mA cm^{-2} [3].

A one dimensional model based on spatially-averaged continuity equations for the $[B^+]$, $[M^+]$ and [M] has been applied to a Ne-Cu^+ slotted hollow-cathode [7,8] and to a He, Ar-Cu cylindrical hollow-cathode [9] sputtering discharge. It was shown that the $[Ne^+]$ saturates at a low current density ($\sim$ 10 mA cm^{-2}), due to the balance between the rate of generation of the ions and their rate of loss by the dominant process of thermal charge transfer to the metal atoms [7]. Further, although the sputtering process at low current densities is due mainly to rare gas ions, there is a dramatic transition to metal ion dominated sputtering as the current density is raised. This transition, which occurs for J $\sim$ 10 mA cm^{-2}, exhibits a rapid rise in [Cu] by two orders of magnitude. This latter effect was also found in the study of reference [9]. Reference [8] describes an interesting method for deriving estimates of [Cu] and the (otherwise difficult to obtain) $[Ne^+]$ and $[Cu^+]$ from a measurement of the dependence of the intensity of the Ne-Cu^+ 260 nm fluorescence on J and the neon pressure. The high threshold current for the HC laser is a major obstacle to the development of a compact sputtering-based ultraviolet laser pumped by charge transfer. The underlying technical issues of vapour containment, cathode erosion and deposition of metal vapour on insulators and internal mirrors, deterioration of the mirrors under exposure to deep UV from the discharge and depletion of the rare gas atoms by condensing metal vapour, directly limit the tube lifetime. The lifetime issues for thermally loaded hollow-cathode cw cadmium and zinc ion lasers have been addressed comprehensively [11] and a lifetime of 1000 hours has been demonstrated for the He-Cd system operated at a current density of about 600 mA cm^{-2}. In contrast, as yet only a few studies have been made of the factors limiting the lifetime of a sputtering-based laser [12-15] and these have not examined sealed off operation. However, it has been shown that the rate of erosion of a helical hollow-cathode is proportional to $J^{5/2}$ and a 1000 hour cathode lifetime predicted for J = 200 mA cm^{-2} when continuously pulsed at 50 p/s [15]. This result indicates that the tube lifetime could be extended significantly by reducing the threshold current density. The key to raising the efficiency of the discharge is to modify the hollow-cathode to give a higher voltage and slope resistance, whilst retaining the oscillation of the beam electrons. This may be achieved by increasing the rate of loss of ions and electrons at a boundary of the negative glow. For a given value of J, this increased loss is compensated by an increase in the cathode fall which increases the rate of ion production in the negative glow. The higher discharge voltage results in a more favourable electron energy distribution which has an increased concentration of fast electrons and enhances the sputtering due to the increased energy of the ions impinging on the cathode surface. An additional benefit is that the higher slope resistance increases the stability of the discharge against the formation of arcs.

The first demonstration of this principle used a hollow anode cathode (HAC) discharge [16]. A cylindrical array of anode rods was placed concentrically within and close to a cylindrical hollow-cathode to give a partially obstructed discharge. However, despite the innovation of a slotted thin-walled tubular anode [17], the anode of the HAC acts as a sink for sputtered metal vapour and restricts the usable volume within the hollow-cathode. In an alternative design, the helical hollow-cathode (HHC) was employed [18] and used to pump the $He-Cu^+$ 781 nm [19] and the $Ne-Cu^+$ 248-270 nm [20] lines. In this device, the voltage for a given current is raised by increasing the pitch of a helical coil cathode. For a current density similar to that used in a HC laser discharge, the voltages for the HAC and HHC devices are in the range 500 to 1500 V, depending on the geometry and the buffer gas and its pressure. The performances of the HAC and HHC lasers were shown to be superior to that obtained using the HC discharge. The efficiency of the discharge is increased because the buffer gas ion concentration $[B^+]$ saturates at a higher level and because the metal atom concentration [M] rises with current due to both the increase in the sputtering ion flux and in the ion energy (hence a higher sputtering yield).

3. Segmented Hollow-Cathode Laser

The segmented hollow-cathode (SHC) is a high-voltage hollow-cathode [20] which avoids interposing anodes or other surfaces within the cathode. The anodes and cathodes are arranged around a common cylindrical surface which defines the bore of the discharge tube. Although the anode surfaces act as sinks for the metal vapour, the potential exists for commutating the electrodes. In addition this geometry and structure are compatible with the assembly of a discharge tube using metal ceramic seals. The discharge voltage for a given current density is raised compared with that for a conventional HC because the adjacent anode surfaces introduce an increased localised loss of ions and electrons from the negative glow [21].

A parametric study of a SHC $He\text{-}Cu^+$ 781 nm laser has been made [20] and a lower threshold and a superior scaling of the output demonstrated compared with the HC and the HAC devices [22]. A gain of about 85 % m^{-1} is obtained for a linear current density of 480 mA cm^{-1}. For a fixed linear current density, the gain is approximately doubled when the bore diameter is reduced from 4 to 2 mm [23]. This result is highly significant for the following reasons. First, a reduced diameter is conducive to restricting operation to the TEM_{00} mode. Second, a given current density (mA cm^{-2}) requires a reduced current. Finally, the higher gain at small diameters allows the use of a shorter active length. A further increase in the gain is obtained by reducing the ratio of the cathode to anode areas [24].

A gain in excess of 50 % m^{-1} on the $He\text{-}Au^+$ 282 nm transition is obtained for a linear current density of 250 mA cm^{-1}. For a given discharge current, the optimum output power and gain of the laser are obtained for a discharge length for which the linear current density is approximately 65 mA cm^{-1}. A threshold current of 280 mA was obtained for a 5 cm active length and bore diameter 4 mm with external mirrors each of transmission of 0.1 % [25]. Both the threshold current and the threshold power are significantly lower than the respective values for a conventional HC laser of the same active length and mirrors of the same transmission [3]. The results indicate the striking result that, if the laser was operated with internal mirrors and repetitively pulsed (~ 1000 p/s, duty cycle of 0.1), a peak power of 0.7 mW could be obtained for a mean input power of 7 W. The current density is ~ 50 mA cm^{-2}, so that the laser could provide a useful ultraviolet beam under conditions which reduce the cathode erosion [26].

The highest gain achieved so far on the $Ne\text{-}Cu^+$ 270 nm line is 12 % m^{-1} at a linear current density of about 400 mA cm^{-1}, which compares very favourably with the values of 3.6 % m^{-1} [2] and 7 % m^{-1} (combined lines) [27] obtained by other workers for HC lasers using much higher current densities of 1.3 and 0.83 A m^{-1} respectively. It has been projected that a tube of active length 10 cm with internal mirrors, repetitively pulsed at a duty cycle of 0.01, could give a peak output power of 1.7 mW for a mean power input of 5 W. The corresponding current density is about 160 mA cm^{-2} [26].

4. Conclusions

The high-voltage segmented hollow-cathode discharge is more efficient for pumping a metal ion laser transition by charge-transfer than its predecessors the hollow-cathode, hollow anode-cathode and helical hollow-cathode systems. The segmented hollow-cathode allows a reduction in the operating current density for a short active length, which is a necessary requirement for a compact tube with an extended lifetime. The results obtained with external mirrors indicate that a repetitively pulsed ultraviolet laser, with an active length of 10 centimetres and internal mirrors, could provide milliwatt level peak power for a mean electrical input of less than 10 Watts.

5. References

1. Gersternberger,D.C.,Solanki,R. and Collins, G. J. (1980) Hollow cathode metal ion lasers, *IEEE J. Quant. Electron.* **QE-16**,820-834.
2. Solanki, R., Fairbank Jr, W. M., and Collins, G. J. (1980) Multiwatt operation of CuII and AuII hollow cathode lasers, *IEEE J. Quant. Electron.* **QE-16**, 1292-1294.
3. Jain, K. and Newton, S. A. (1981) Operating characteristics of UV and IR hollow-cathode silver, gold and copper ion lasers, *Appl. Phys.* B **26**, 43-48.
4. Turner-Smith, A. R., Green, J. M. and Webb, C.E. (1973) Charge transfer into excited states in thermal energy collisions , *J. Phys. B: Atom. Molec. Phys*. **6**, 114-130.
5. Helm, H. (1972) Experimenteller nachweis pendel-effektes in einer zylindrischen niederdruck-hohlkathoden-entladung in argon, *Z. Naturforsch.* **27a**, 1812-1820.
6. Piper, J. A., Collins, G. J. and Webb, C. E. (1972) CW laser oscillation in singly ionized iodine, *Appl. Phys. Lett.* **21**, 203-205.
7. Warner, B. E., Persson, K. B. and Collins, G. J. (1979) Metal-vapour production by sputtering in a hollow-cathode discharge: Theory and experiment, *J. Appl. Phys.* **50**, 5694-5703.
8. Koch, H. and Eichler, H. (1983) Particle densities in high current hollow cathode discharges, *J. Appl. Phys.* **54**, 4939-4946.
9. Rae, S. C. and Tobin, R. C. (1988) Sputtering in the presence of a rapid gas flow, *J. Appl. Phys.* **64**, 1418-1421.
10. Tsuda, T. and Piper, J. A. (1989) Practical small-scale hollow-cathode cw metal ion lasers, *J. Phys. E: Sci. Instrum.* **22**, 462-465.
11. Auschwitz, B., Eichler, H. J. and Wittwer, W. (1980) Extension of the operating period of an UV CuII laser by admixture of argon, *Appl. Phys. Lett.* **36**, 804-805.
12. Yang, M. (1983) A cw copper hollow cathode laser in the 260 nm region, *Appl. Phys. B* **32**, 127-135.

13. Eichler, H. J., Hamisch, J. and Song, S. (1987) Output power and operation period of the CuII-laser in sherical and circularly slotted hollow cathodes, *Opt. Comm.* **61**, 61-65.
14. Yu, J. and Macdonald, R. (1989) Erosion of helical cathodes for copper ion lasers, *J. Phys. D: Appl. Phys.* **22**, 1627-1629.
15. Rózsa, K., Jánossy, M., Csillag, L. and Bergou, J. (1977) CW CuII laser in a hollow anode-cathode discharge, *Opt. Comm.* **23**, 162-164.
16. Zhang, Z., Perry, N. D. and Tobin, R. C. (1992) Gain measurements at 780.8 nm in a copper hollow anode cathode laser, *J. Appl. Phys.* **71**, 64-69.
17. Grozeva, M. and Sabotinov, N. (1982) Coil hollow cathode for metal ion lasers, *Opt. Comm.* **41**, 57-58.
18. Koch, H. (1983) CuII laser with a helical hollow cathode discharge, *J. Phys. E: Sci. Instrum.* **16**, 122-123.
19. Eichler, H. J., Hamisch, J., MacDonald, R., Sollinger, M. and Chen, Yuging (1987) UV-CuII-laser with helical cathodes, *J. Appl. Phys.* **61**, 2069-2071.
20. Peard, K. A., Rózsa, K. and Tobin, R. C. (1994) Parametric study of a high-voltage hollow-cathode infrared copper-ion laser, *J. Phys. D: Appl. Phys.* **27**, 219-227.
21. Donkó, Z., Rózsa, K. and Tobin, R. C. (in press 1995) Monte Carlo simulation of the electrons' motion in a segmented hollow cathode discharge, *J. Phys D: Appl. Phys.*
22. Peard, K. A., Donkó, Z., Rózsa,K., Szalai, L. and Tobin, R. C. (1994) Comparison of Cu-II 781 nm lasers using high-voltage hollow-cathode and hollow-anode-cathode discharges, *IEEE J. Quant. Electron.* **QE-30**, 2157-2165.
23. Szalai, L., Rózsa, K., Donkó, Z. and Tobin, R. C. (1995) Dependence of gain and output power for CuII 780.8 nm transition on the diameter of a segmented hollow cathode discharge, *IEEE J. Quant. Electron.* **31**, 1549-1553.
24. Leigh, B., Robilliard, F. R., Tobin, R. C., Rózsa, K. and Donkó (1995) Dependence of TEM_{00} gain on cathode area in a metal ion laser, *Proceedings of 10th Conference of the Australian Optical Society Brisbane,* 103.
25. Peard, K. A., Tobin, R. C., Rózsa, K. and Donkó, Z. (1994) A high-voltage hollow-cathode AuII 282 nm laser, *IEEE J. Quant. Electron.* **QE-30**, 1181-1186.
26. Tobin, R. C., Peard, K. A., Bode, G., Rózsa, K., Donkó, Z. and Szalai, L. (in press 1995) High-gain hollow-cathode metal ion lasers for the uv and vuv, *IEEE J. of Special Topics in Quant. Electron.*
27. Eichler, H. J., Koch, H., Molt, R. and Qui, J. L. (1981) Optimization of the UV CuII laser, *Appl. Phys. B*, **26**, 49-56.

MEASUREMENT OF "HOT SPOTS" IN COPPER VAPOUR LASERS

M. AMIT
R&D Department, Nuclear Research Centre – Negev,
PO Box 9001, Beer Sheva, Israel

Abstract

We have measured the radial dependence of the gain in a copper vapour laser amplifier as a function of medium gas temperature and gas flow. The gain was found to change over a period of tens of minutes. In particular, regions of increased beam intensity (hot spots) appear, migrate and disappear with time. The hot spots appear to be responsible for changes in the beam divergence.

1. Introduction

In a copper vapour laser system the first component is always the master oscillator, followed by at least one amplifier. In this disposition the oscillator is considered to provide the beam quality (BQ), while the attached amplifiers are used to achieve the desired power. In a typical type of such a laser system, in addition to the main oscillator and amplifiers, there are also adjoining optics in between them.

The beam quality is usually measured after the master oscillator and/or after the last amplifier. It has to be noted here that it is not straightforward to evaluate the origin of BQ degradation along the laser chain according to the above mentioned measurements. BQ degradation may be caused by either changes in the amplifier characteristics, time dependent jitters, or optical stability of the coupled optics.

In this paper, I would like to suggest a simple way to locate the main causes of BQ degradation.

2. Experimental and Results

In order to study BQ degradation along the laser chain, we constructed a copper vapour laser MOPA (master oscillator power amplifier). All heads were 30 mm inner diameter and 800 mm effective length. Each head was of 20 W power as a single stable oscillator.

Radial total gain measurements of both the output beams from the oscillator and from both amplifiers, showed instability of two kinds: (I) time dependent gain fluctuations, of a typical 10–30 minutes "period", were of 20–30% (long

C. E. Little and N. V. Sabotinov (eds.), Pulsed Metal Vapour Lasers, 195–199.

term); (II) short-time period fluctuations (seconds), the investigation of which is beyond the scope of this work.

We assume that the main causes for long-term BQ degradation is due to a change in temperature uniformity of the laser tubes. As a result of this assumption, we decided to measure the radial amplifier spontaneous emission (ASE) for each laser head separately. This was carried out in four temperature conditions: (I) optimal conditions, (II) cooled laser, (III) warmed laser, and (IV) hot laser.

In Figures 1 and 2 optimal conditions are presented, and it can be noticed easily that the graphs are considerably smooth as expected. The cooled laser (Figure 3) showed a similar behaviour, and even better in terms of the smoothness (it is obvious that cooling the laser results in lowered power). Figures 4 and 5 show the warmed and hot lasers, respectively, in which outputs are distorted and no longer have radial uniformity. These measurements were carried out at least 20 times in heating-cooling cycles, and loss of uniformity always appeared in the warmed/hot experiments but in a different radial distortion. The same results were obtained when total gains, and small signal gains, of the components were measured. Such measurements are usually tedious and time consuming.

3. Discussion

Gain stability origin of the beam quality fluctuations was measured by us and by others usually on a single laser head. In a chain disposition it is almost impossible to measure gain stability and hence the beam quality fluctuation of each chain component.

By getting the same results in ASE measurements as in gain stability, it is justified to construct a relatively simple set-up (ASE measurements) in order to evaluate beam quality fluctuations. In the laser pulse sequence it is possible to block the radiation for a few seconds and to accomplish the measurements of the radial ASE distribution.

4. Conclusion

The main source of BQ degradation, according to this study, is a loss of uniformity as a result of tube temperature fluctuations. Non-uniformity of the radial gain is 20–30%. It is possible to measure BQ degradation by means of measuring each ASE component.

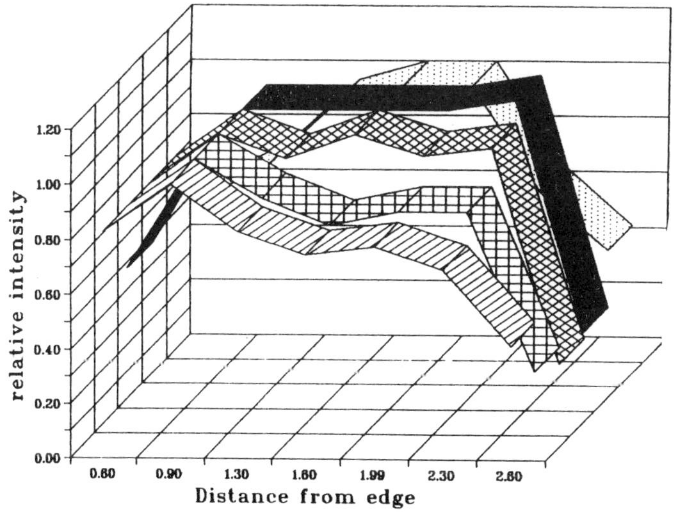

Figure 1. Radial ASE in CVL optimum condition. Green line. High voltage 7 kV, pressure 140 mbar, current 360 mA.

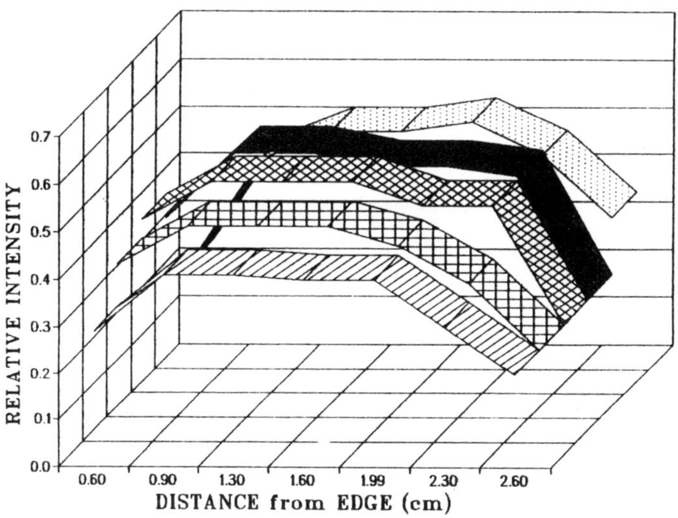

Figure 2. Radial ASE in CVL optimum condition. Yellow line. High voltage 7 kV, pressure 140 mbar, current 360 mA.

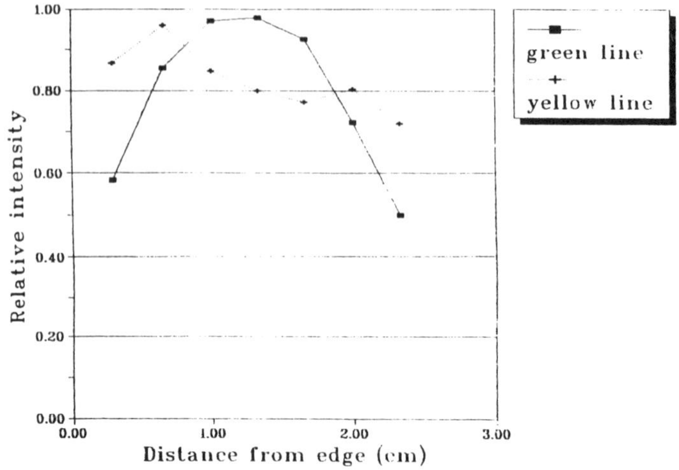

Figure 3. Radial ASE in CVL "cool laser". High voltage 6.7 kV, pressure 160 mbar, current 330 mA.

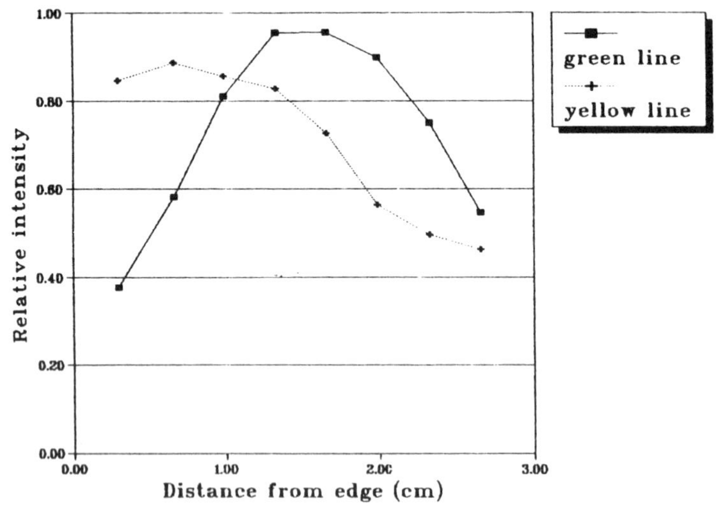

Figure 4. Radial ASE in CVL "warm laser". High voltage 7.1 kV, pressure 160 mbar, current 430 mA.

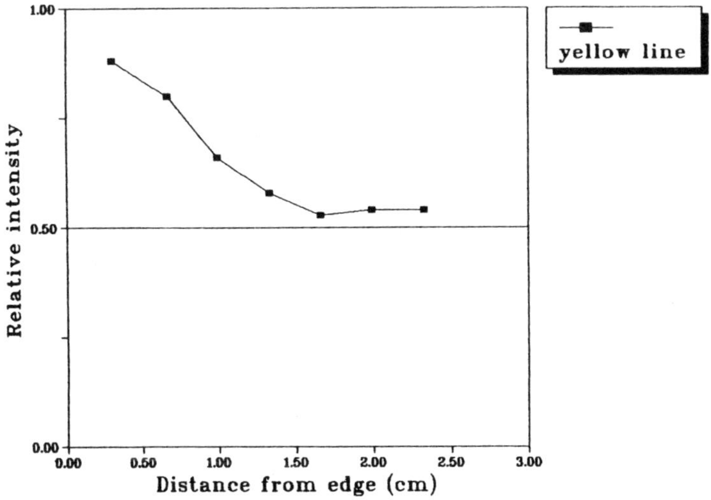

Figure 5. Radial ASE in CVL "hot laser". High voltage 7.4 kV, pressure 160 mbar, current 460 mA.

4. Modelling Metal Vapour Lasers

COMPUTER MODELLING OF LONGITUDINALLY EXCITED ELEMENTAL COPPER VAPOUR LASERS

R.J. CARMAN
Centre for Lasers and Applications
Macquarie University, Sydney
NSW 2109, Australia

1. Introduction

As an integral part of the steady development of copper vapour lasers over the past 30 years, computer modelling has been used to provide an insight into complex issues relating to the plasma kinetics and evolution of the optical laser fields. For example, calculations of the behaviour of the electron temperature and the flux rates of various processes are of critical importance when interpreting kinetic data obtained from experiments. However, these parameters are virtually impossible to measure experimentally due to the rapid temporal variation of the plasma variables. Numerical modelling has been employed to unravel the complicated and highly-coupled multi-species kinetics which evolves both spatially and temporally through the excitation (lasing) phase and interpulse relaxation period. Over the history of CVL development, there has been a steady increase in sophistication and reliability of the numerical models. This has been due primarily to the improved accuracy of the electron-impact impact cross-sections for copper, the availability of spatially and temporally resolved species population measurements for comparison with model predictions [1-4], and a steady increase in computing power. In this paper, the current status of computing modelling of CVL's is presented. This includes a summary of the currently available cross-sections, and a discussion of the published models for the kinetics and the optical laser fields.

2. Electron-impact collision cross-sections

The reliability of any computer model in simulating CVL behaviour is limited fundamentally by the accuracy of the available electron-impact collision cross-

C. E. Little and N. V. Sabotinov (eds.), Pulsed Metal Vapour Lasers, 203–214.

sections and optical transition probabilities, particularly those relating to the upper and lower laser levels. Early models [5-7] tended to use e-impact cross-sections calculated using classical methods [8]. To date, the most reliable and comprehensive set of cross-sections for transitions in Cu I are those from 4-state [9] and 10-state [10,11] close-coupling calculations based on R-matrix theory covering the energy range E=0-20eV. The 10-state calculations give cross-sections for excitation from the ground state $(3d^{10}4s)^2S$ to the upper $(3d^{10}4p)^2P$ and lower $(3d^94s^2)^2D$ laser levels, to the quadruplet $(3d^94s4p)^4P^4D^4F$ levels, and high-lying $(3d^{10}4d)^2D$ and $(3d^94s4p)^2P^2D^2F$ levels, in addition to excitation between the excited states, and elastic e-Cu scattering. These calculated cross-sections are well converged for transitions among the ground, lower laser and upper laser levels [12], and the cross-sections for excitation of the laser levels are in very close agreement with the latest experimental measurements [13,14] which are well resolved over the energy range E=0-10eV, corresponding to the typical electron energies found in CVL discharges (figure 1). The large resonance structure for excitation of the metastable lower laser levels (E=2-4eV) [9,13] is particularly important in CVL kinetics since it gives rise to strong electron-collisional coupling between the metastable and ground states [15]. The satisfactory agreement that now exists between experimental data and theory concerning electronic excitation of the laser levels is encouraging from the point of view of determining accurate rate-coefficients for inclusion in kinetic modelling studies, and may be contrasted with the scarcity of data for the range E=0-10eV apparent in an earlier compilation of cross-sections [16]. A cross-section for ionisation of the atomic ground state is available in a compilation of experimental data [17] (E=0-200eV). Excitation cross-sections of the low lying $3d^94s$ and $3d^94p$ configurations in Cu II from the $3d^{10}$ ionic ground state have been evaluated using a 41-state R-Matrix close-coupling calculation for E=0-16eV [18]. A detailed review of the transition probabilities for the principal optical lines in Cu I was reported in [19] whilst transition probabilities for selected Cu II lines were given in [20,21].

3. Kinetics studies

The first theoretical analysis of kinetics in a CVL, reported in 1967 [5], involved a calculation of the electron collisional pump rates to the laser levels for the 510nm line and the associated inversion density, for typical values of the plasma tube current density. Subsequently, more detailed modelling of the discharge kinetics has been based on time-dependent rate-equation analyses involving typically 10-20 plasma variables, including the thermodynamic parameters, population densities of various atomic and ionic species and the

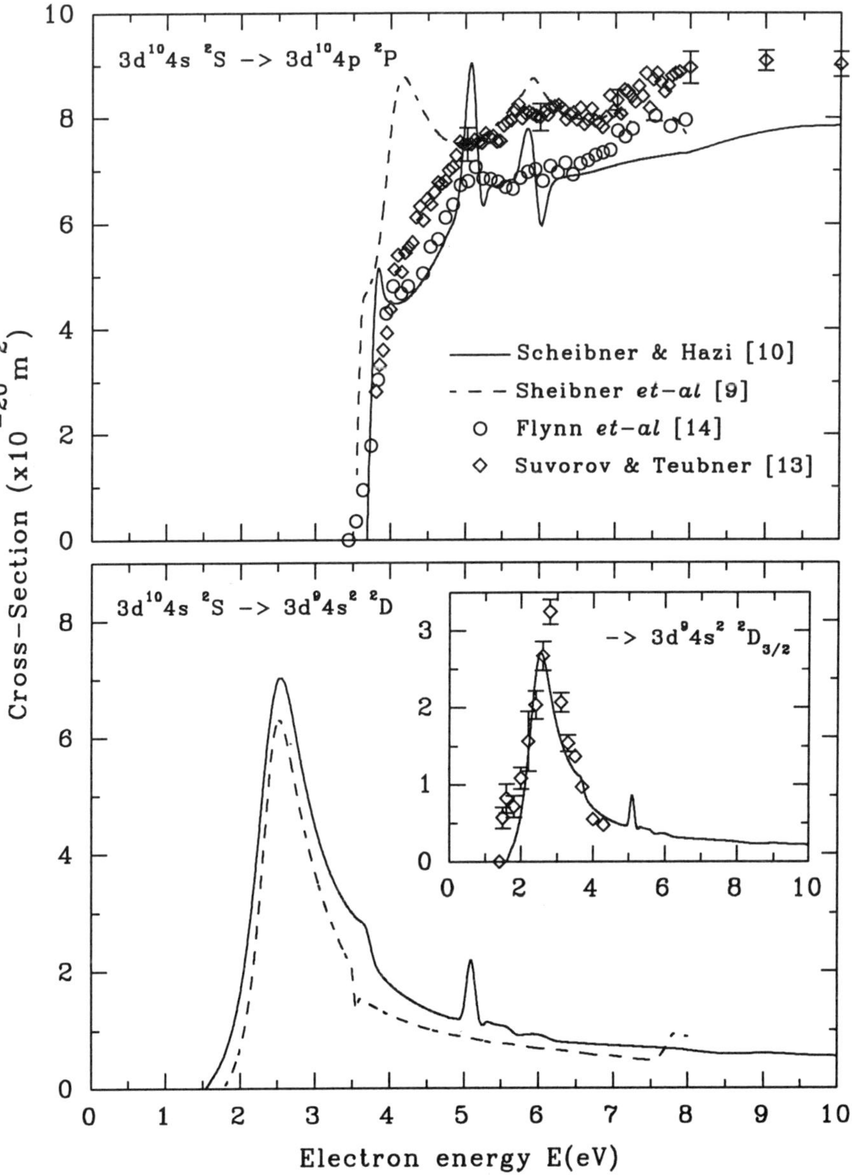

Figure 1. Electron impact excitation cross-sections for the atomic ground state of copper. The experimental data of [13] is uncalibrated and has been normalised to the 10-state calculations [10] at E=2.6eV.

laser field intensities. Specialised "stiff" integration methods have been generally required to solve the set of highly-coupled ordinary differential equations used to describe the system, owing to the markedly different time constants and magnitudes of the various kinetic processes. Due to limitations in computing power, early models [7,22-24] treated the active medium as spatially homogeneous, and calculations were carried out only during the excitation phase of the discharge cycle encompassing the lasing period. A set of initial conditions at time zero for the plasma variables were usually selected on the basis of estimated steady-state values. These "single-shot" models provided an insight into the relative importance of the various parasitic processes that limit the energy conversion efficiency for typical laser operating conditions, and demonstrated general qualitative agreement with experimental data. The interaction of the external electrical circuit with the laser discharge impedance, including the cold end zones containing pure buffer gas, were features also incorporated as part of the rate equation analysis in some models [7,22,23]. Analyses of both the excitation and relaxation phase kinetics were also undertaken [6,25,26]. The strong dependency of the calculated kinetics and lasing behaviour on the user specified initial conditions in single-shot models was nevertheless a limiting factor in their reliability and wider application in the interpretation of kinetic data.

A landmark development in CVL modelling was the development of the first "self-consistent" code [27] where the behaviour of the plasma is simulated through multiple excitation-relaxation cycles until a convergence is reached for the shot to shot variation of the predicted results. Such codes are particularly demanding in terms of computing power, but are able to simulate the complete temporal evolution of the discharge kinetics and the behaviour of the external electrical circuit with a relatively small number of easily determined macroscopic input parameters dependent on the operating conditions of the laser. Thus, it becomes possible to unravel the complex inter-dependence between the kinetics during the excitation (lasing) phase and the afterglow period. Kinetics issues relating to average power scaling of CVL's were investigated in [27] as a function of the laser operating conditions (eg. prf, tube temperature, storage capacitance and charging voltage, etc.) for narrow bore devices. Factors that were found to strongly affect the laser performance were the peak electron temperature and the pre-pulse lower laser level population. An improved version of the model reported in [27], modified to include radial dependence of the microscopic plasma variables and the electric field, was later used to investigate the average-power scaling issues for large diameter CVL's ($60<\phi<120$mm) [28]. Close agreement with experimental hook population measurements [29] for the laser levels was demonstrated, and the principal considerations in scaling to large bore CVL's were shown to be volumetric gas heating and non-uniform electric field penetration due to the plasma skin effect. The interplay between the spatio-temporal evolution of the electric field and the discharge tube impedance has recently been modelled in detail for a large bore CVL using both analytical and numerical methods [30,31]. As a result of these studies, it is now generally accepted that the various features of the CVL, such as the discharge kinetics, electric field penetration and external circuit, are highly inter-dependent and cannot be modelled in isolation.

For reasons of computational economy, an almost universal assumption in CVL models is that the electron energy distribution function (EEDF) has a Maxwellian profile characterised by a single electron temperature. More precise calculations of the EEDF for the plasma conditions in a CVL discharge have been performed by solving the time-dependent Boltzmann equation including electron-electron collisions [32,33]. The results suggest that any departure of the EEDF from a pure Maxwellian at low energies (E=0-10eV) is probably fairly small (~20%). Larger deviations are expected at higher energies (E~20eV) due to reduced electron-electron thermalisation rate, resulting in a significant overestimate (typically x10) of the buffer gas excitation rates if a Maxwellian is assumed [28,34]. In a recent self-consistent model of CVL kinetics [15] incorporating the 10-state e-impact cross-sections given in [10], the EEDF was approximated by a bi-Maxwellian profile with two electron temperatures [35] in order to calculate the excitation rates of the buffer gas more accurately. The model calculated the spatio-temporal evolution of eighteen plasma variables including the thermodynamic parameters, and is probably the most comprehensive model of CVL kinetics published to date. The predicted results were in close agreement with spatially resolved hook population measurements [1] for a small bore device (figure 2). The kinetics issues arising from this study were also in broad agreement with those reported in an earlier self-consistent model [28].

The steady increase in sophistication of the computer models describing CVL behaviour, coupled with the availability of accurate cross-sections, has allowed comparisons to be made between theory and experimental data with good quantitative agreement in most key areas. On this basis, it seems reasonable to suggest that the kinetics of the copper laser are now moderately well understood, particularly for devices operating at or near the condition for optimum power output. Further modelling studies are required, however, to fully explain kinetics issues for the wide range of non-optimum conditions corresponding to the observed parametric behaviour of CVL's. Certain aspects of the kinetics in elemental devices operating under non-optimum conditions have yet to be fully resolved. For example, the relative influence of the pre-pulse electron and lower laser level densities on laser performance is still the subject of some debate. The complex inter-dependence of the pre-pulse

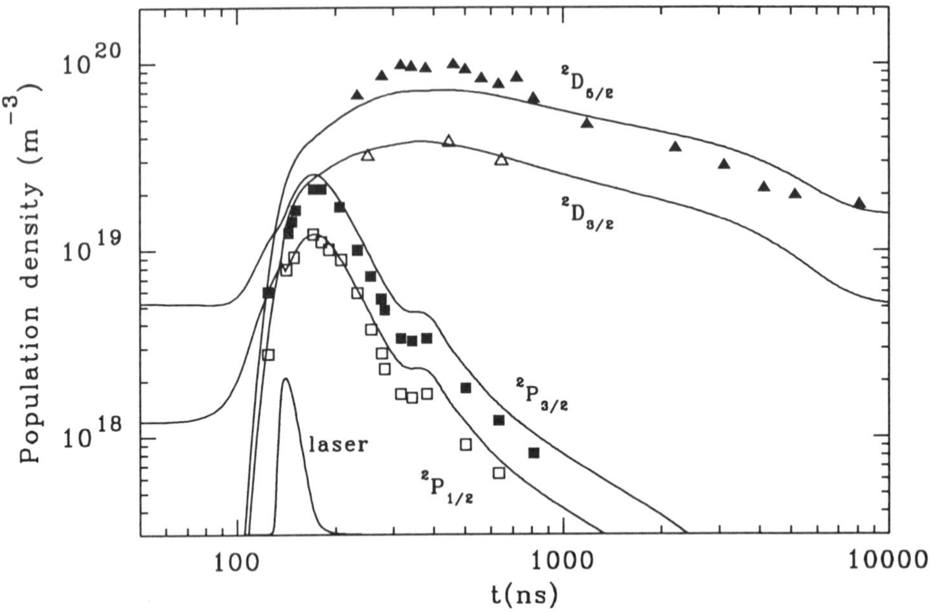

Figure 2. Temporal evolution of the populations densities of the upper and lower laser levels on the tube axis. The experimental points are from [1].

populations densities on key plasma parameters such as the electron temperature and electron density makes the analysis of even detailed experimental data non-trivial. The role of radial cataphoresis, which gives rise to severe radial depletion of the metal ground state density when the interpulse period is short [36], may have underestimated in previous studies. Analysis of the kinetics for CVL's with different plasma tube dimensions operating under optimum conditions for output power or efficiency would also be helpful in understanding the average-power scaling issues on a fundamental level.

4. Intracavity optical laser fields

The spatio-temporal evolution of the intracavity optical fields in the active medium of a CVL is of major importance when selecting external resonator configurations to generate the high beam quality output required for a variety of non-linear frequency conversion and micromaching applications. In addition, the formation of the beam output by the amplification of an initial "seed" of spontaneous emission, and the relationship between the gain and the underlying discharge kinetics, are phenomena that are intrinsically interesting. Laser action within the active medium in a CVL is characterised by a large small-signal gain coefficient (γ_o=10-20 m^{-1}), large Fresnel number/large active volume, and short pulse duration (<100ns) which precludes the establishment of normal cavity modes within the resonator [37]. Experimental measurements of the optical gain and laser field intensities, which give results with temporal, radial and longitudinal resolution, are complicated by the inaccessibility of the active medium, the short gain duration, and gain saturation by the probe. Thus, modelling studies are particularly valuable in gaining an insight into the state of the active medium during the establishment of optical gain, and thereafter the growth of the laser optical fields into the gain-saturation regime.

There have been a number of theoretical studies on the influence of the external unstable resonator configuration in determining beam quality [37-39]. These studies have the advantage of not requiring a detailed knowledge of the active medium and discharge kinetics, and require only moderate computer power. Ray-matrix calculations of the spatio-temporal evolution of the spatial-coherence/divergence of the beam as it propagates inside the cavity or through an equivalent "unfolded" resonator, have been shown to be in good qualitative agreement with experimental results [37]. Modelling studies of the full spatio-temporal evolution of the intracavity fields in the presence of an active medium, including the gain saturation characteristics, remain scarce despite the fact that detailed experimental data exists for comparison [38,40]. This can be attributed in part to the requirement for substantial computing power and memory storage to simulate the evolution of multiple photon avalanches with good spatio-temporal resolution (~10-30ps) concurrently with the discharge kinetics. However, such studies would be particularly helpful in understanding the observed output beam characteristics, synchronisation issues relating to injection seeding in oscillator/amplifier configurations, and the chaotic behaviour observed in the diffraction limited output when using high magnification unstable resonators [41].

In modelling the interaction of the intracavity fields and the laser level populations in CVL's, quantum coherency effects [42] have generally been neglected and rate-equation analyses employed to evaluate the field intensity. This assumption is valid if the phase-destroying collision frequency mainly due to pressure broadening is sufficiently large [6]. A solution to the scalar wave equation for the intracavity optical field in a copper halide laser has been reported in [43] taking into account coupling between the field and the laser level populations. The spatio-temporal characteristics of the optical field and gain at 510nm in a conventional CVL has also been modelled using a Monte-Carlo technique [44,45], including the angular distribution of the emitted radiation and losses at the plasma tube walls due to non-axial rays. A similar but more comprehensive study was recently reported [46] based on an analysis of the propagation of photon flux components at 510nm and 578nm inside the active medium and optical resonator. The model was based on a rate-equation analysis of the discharge kinetics, taking into account the hyperfine structure of the laser lineshape functions and beam divergence losses. The results from two modelling studies in [44,46] have established that marked longitudinal variations of the gain and intracavity field intensities exist in CVL's during the lasing period as gain saturation is sets in.

5. Electrodes

As far as the author is aware, no detailed modelled studies describing the physics of the interaction between the electrodes and the bulk plasma in CVL's have been published to date. Experimentally, electrode designs and the choice of materials vary widely, and are usually selected on the basis of their compatibility with the high temperature environment adjacent to the plasma tube ends and lifetime considerations. Cathode designs also often incorporate re-entrant geometries to take advantage of the hollow-cathode effect [47], or narrow edge profiles to promote field emission [47], but the relative importance of the various secondary electron-emission processes remains unclear. Information is also lacking concerning the formation phase of the space-charge sheaths separating each electrode from the bulk plasma, the expected degree of distortion of the electric field, and the overall impedance of these zones compared with the bulk plasma. Discharge instabilities originating from plasma filamentation at the cathode surface are also commonly observed in CVL's [48] degrading the overall laser performance.

A number of modelling studies have been reported concerning the spatio-temporal evolution of the plasma near the electrodes in transverse discharge-excited excimer lasers [49-52]. Since the electrical excitation phase occurs on a timescale not too dissimilar to that in CVL's (~100ns), many of the underlying features of sheath formation will be common to both systems. Using a one-dimensional hybrid model for the cathode sheath and a zero-dimensional model in the bulk plasma of a high pressure XeCl laser [49], it was shown that the sheath formation time is typically 40-50ns corresponding to a transition from purely capacitive to resistive behaviour. Secondary electron emission at the cathode was of critical importance during the formation phase with low values of the emission coefficient giving rise to discharge instabilities. Substantial heating of the cathode due to ion bombardment also occurred giving rise to the development of shock waves in the afterglow period. Large electric fields (10^7-10^8 $V.m^{-1}$) were also predicted to occur in the cathode sheath [49,50], giving rise to the possibility of field emission from microscopic inhomogeneities at the cathode surface. Using the modelling techniques developed for the excimer laser electrode studies, it should be possible to generate similar results pertaining to the CVL electrodes, with a 2-D code [51] allowing for the introduction of various electrode shapes.

6. Electric circuit modelling

As part of the rate-equation analysis of CVL kinetics, the behaviour of the external electrical circuit and thyratron switch is almost always included to accurately determine the voltage applied to the laser head [7,15,23,27,28]. The representation of the passive components in a circuit is relatively straight forward, however, the time-dependent impedance of the thyratron is more complex since it cannot be treated in complete isolation to the rest of the circuit. The energy loss mechanisms in thyratrons in the commutation phase have been discussed in detail elsewhere [53]. For the purposes on CVL modelling, the thyratron "switch-on" characteristic is usually modelled by assuming an exponentially falling resistance with a time-constant τ~5ns [15,28,54]. An additional resistive component was included in [15], based on the energy deposition in the conducting plasma of the thyratron, to simulate losses during the high-current conduction phase in accordance with experimental I-V characteristics [55]. A detailed and flexible model for CVL circuits has been reported recently based on analysis using a SPICE simulator package [56]. The laser discharge was approximated as a transmission line with time-varying impedance elements. Factors contributing to the power deposition and stress loading on the thyratron were discussed in relation to the fast risetimes desired for the excitation pulses to the laser. Finally, there are a handful of reports on modelling studies for non-linear circuit elements such as saturable inductors used for magnetic pulse compression [57-59]. Close agreement has been demonstrated between experimental results and theory, and the simulations serve as a useful guide in understanding circuit behaviour and design issues.

7. Conclusion

Computer modelling of the kinetics and lasing characteristics of copper vapour lasers is now moderately well advanced. The available electron impact excitation and ionisation cross-sections for atomic copper are probably the most comprehensive and reliable for any metal-vapour laser system. The availability of detailed experimental data for CVL's operating under optimum conditions has made it possible to make detailed comparisons with the calculated results, and favourable agreement has been demonstrated in most key areas. Further modelling studies are still required to investigate kinetics issues for devices operating under non-optimum conditions, and to provide a deeper understanding of the average-power scaling issues.

Modelling of the optical laser fields in CVL's is less well advanced than kinetics, but is important from the point of view of understanding the spatio-temporal evolution of the fields when aiming to produce high beam quality output. Models that simulate the evolution of the fields in the presence of the active medium are particularly demanding on computing time and memory storage which are limiting factors at this time with regard to the level of detail that may can be included. Modelling studies of the behaviour of the plasma at the electrodes in CVL's are virtually non-existent, but a number of studies concerning analysis of the external circuit and thyratron have been reported.

8. Acknowledgements

The author thanks K. Scheibner and P. Teubner for generously providing unpublished cross-section data, and the support of an Australian Research Fellowship from the Australian Research Council.

9. References

1. Brown, D.J.W. (1988) *Experimental study of excited state densities in a copper vapour laser*, PhD dissertation, Univ. New England, Armidale, Australia.
2. Hogan, G.H. (1993) *A study of the kinetics of copper vapour lasers*, PhD dissertation, Oxford Univ., UK.
3. Isaev, A.A. and Petrash, G.G. (1993) Kinetics of excitation and physical processes in active media of copper-vapor and copper-bromide-vapor lasers, *Proc. SPIE*, **2110**, 2-45.
4. Smilanski, I. (1979) Copper hooks-Investigation of the copper vapour laser kinetics, *Proc. Int. Conf. Lasers '79*, Orlando, Florida, 327-334, STS Press, McLean Virginia.
5. Leonard, D.A. (1967) A theoretical description of the 5106A pulsed copper laser, *IEEE. J. Quantum Electron.,* **QE-3**, 380-381.
6. Harsted, K.G. (1980) Computer simulated rate processes in copper vapor lasers, *IEEE. J. Quantum Electron.*, **QE-16**, 550-558.
7. Walter, W.T., Solimene, N. and Kull, G.M. (1981) Stepwise excitation - a limiting process in pulsed gas lasers, *Proc. Int. Conf. Lasers '81*, 853-865.
8. Gryzinski, M (1965) Classical theory of atomic collisions. I. Theory of inelastic collisions, *Phys. Rev.*, **138**, 336-358.
9. Scheibner, K.F., Hazi, A.U. and Henry, R.J.W. (1987) Electron-impact excitation cross-sections for transitions in atomic copper, *Phys. Rev. A*, **35**, no 11, 4869-4872.
10. Scheibner, K.F. and Hazi, A.U., to be published.
11. Scheibner, K.F. and Hazi, A.U. (1988) Photo-detachment cross-section of Cu^-, *Phys. Rev. A*, **38**, no 1, 539-542.

12. Hazi, A.U. Private communication.
13. Suvorov, V. and Teubner, P.J.O., to be published.
14. Flynn, C., Wei, Z. and Stumpf, B. (1993) Excitation of the copper resonance lines by low-energy electrons, *Phys. Rev. A.*, **48**, no 2, 1239-1242.
15. Carman, R.J., Brown, D.J.W. and Piper, J.A. (1994) A self-consistent model for the discharge kinetics in a high-repetition rate copper-vapour laser, *IEEE. J. Quantum Electron.*, **QE-30**, no 8, 1876-1895.
16. Winter, N.W. and Hazi, A.U. (1982) *Review of electron impact excitation cross-sections for copper atom*, Lawrence Livermore Lab., Rep. UCID 19 314.
17. Freund, R.S., Wetzel, R.C., Shul, R.J. and Hayes, T.R. (1990) Cross-section measurements for electron impact ionisation of atoms, *Phys. Rev. A*, **41**, 3575-3595.
18. Griffin, D.C. and Pindzola, M.S. (1995) Electron-impact excitation and ionisation of Cu^+, *J. Phys.B.*, in press.
19. Bielski, A. (1975) A critical survey of atomic transition probabilities for Cu I, *J. Quantum Spectros. Radiat. Transfer*, **15**, 463-472.
20. Corliss, C.H. and Bozman, W.R. (1962) *Experimental transition probabilities for spectral lines of seventy elements*, National Bureau of Standards Monograph 53, Washington, 81.
21. Froese Fischer, C. and Glass, R. (1980) The effect of correlation in the $3d^n$ core on the resonance transition in Cu II, *Physica Scripta*, **21**, 525-526.
22. Kull, G.M., Solimene, N. and Walter, W.T. (1979) Modelling of pulsed gas lasers and comparison with experimental results for the copper vapor laser, *Proc. Int. Conf. Lasers '79*, Orlando, Florida, STS.
23. Miller, J.L. (1979) *Copper vapor laser kinetics model*, Lawrence Livermore Lab Rep. UC-18182.
24. Bokhan, P.A. and Gerasimov, V.A., Solominov, V.I. and Skcheglov, V.B. (1978) Stimulated emission mechanisms of a copper vapor laser, *Sov. J. Quantum Electron.*, **8**, 1220-1227.
25. Harsted, K.G. (1983) Interpulse kinetics in copper and copper halide lasers, *IEEE. J. Quantum Electron.*, **QE-19**, no 1, 88-91.
26. Borovich, B.L. and Yurnchenko, N.I. (1985) Analysis of the excitation and relaxation kinetics in a copper vapor laser excited by a longitudinal discharge, *Sov. J. Quantum Electron.*, **14**, no 10, 1391-1400.
27. Kushner, M.J. (1981) A self-consistent model for high-repetition rate copper vapor lasers, *IEEE J. Quantum Electron.*, **QE-17**, 1555-1565.
28. Kushner, M.J. and Warner, B.E. (1983) Large-bore copper-vapor lasers: Kinetics and scaling issues, *J. Appl. Phys.*, **54**, no 6, 2970-2982.
29. Warner, B.E. and Kushner, M.J. (1981) Controlling kinetic parameters of 100W large bore copper vapor lasers, *Proc. Int. Conf. Lasers '81*, New Orleans, 845-852.
30. Blau, P. (1994) Analysis of the impedance of a coaxial, large-bore copper-vapor laser, *IEEE. J. Quantum Electron.*, **QE-30**, 763-769.
31. Blau, P. (1995) Spatial and temporal evolution of the electric field in a longitudinaly electric discharge pumped gas laser: Application to a large bore copper vapor laser, *J. Appl. Phys.*, **77**, 2273-2278.
32. Borgard, J.M., Goossens, J.P., Lemaire, P. and Valrho, C.E. (1994) Impact of the electron distribution function on the copper vapor laser kinetic, *Proc. Conf. Lasers and Electro-Optics Europe*, Amsterdam, technical digest, CWB7.
33. Lewis, R.R. (1985) *Mechanisms of copper vapour lasers*, PhD dissertation, Oxford University, UK.
34. Morgan, W.L. and Vriens, L. (1980) Two electron group model and Boltzmann calculations for low pressure gas discharges, *J. Appl. Phys.*, **51**, 5300-5306.

35. Carman, R.J. (1991) A time-dependent two electron group model for a discharge excited He-Sr recombination laser, *J. Phys.D.*, **24**, 1803-1810.
36. Mildren, R.P., Brown, D.J.W., Carman, R.J. and Piper, J.A. (1995) The effect of hydrogen additive on population densities in the afterglow of barium vapour lasers, *Opt. Comm.*, **120**, 112-120
37. Coutts, D.W. (1995) Time-resolved beam divergence from a copper vapour laser with unstable resonator, *IEEE. J. Quantum Electron.*, **QE-31**, 330.
38. Coutts, D.W. (1991) *Beam quality and frequency conversion of copper laser output*, PhD dissertation, Macquarie University, Sydney, Australia.
39. Eggleston, J.M. (1988) Theory of output beam divergence in pulsed unstable resonators, *IEEE. J. Quantum Electron.*, **QE-24**, 1302.
40. Coutts, D.W. and Brown, D.J.W. (1995) Formation of output in copper vapour lasers, *Appl. Optics*, **34**, 1502-1512.
41. Coutts, D.W. and Brown, D.J.W., (1995) Chaotic properties of copper-vapor-laser output pulses, *Proc. Int. Conf. Lasers and Electro-Optics*, Baltimore, technical digest, **15**, CThG3.
42. McIntyre, I.A. and Rhodes, C.K. (1991) High power ultrafast excimer lasers, *J. Appl. Phys.*, **69**, R1-R19.
43. Kushner, M.J. (1979) *Electronic and kinetic processes in the Cu/CuCl double pulse laser*, PhD dissertation, California Institute of Technolgy, Pasadena, CA.
44. Buchanov, V.V., Molodykh, E.I. and Yurnchenko, N.I. (1983) Calculation of the dynamics of the radiation divergence and losses in nonaxial copper-vapor laser beams, *Sov. J. Quantum Electron.*, **13**, 1022.
45. Borovich, B.L. and Yurnchenko, N.I. (1985) Efficiency of coupling out radiation from a resonator of a laser emitting short pulses, *Sov. J. Quantum Electron.*, **15**, no 7, 912-918.
46. Carman, R.J. (1995) Modelling of the intracavity optical fields in a copper vapour laser, *Opt. Comm.*, **119**, 415-423.
47. Chapman, B. (1980) *Glow Discharge Processes*, Wiley, NY.
48. Brown, D.J.W., private communication.
49. Belasri, A., Boeuf, J.P. and Pitchford, L.C. (1993) Cathode sheath formation in a discharge-sustained Xe-Cl laser, *J. Appl. Phys.*, **74**, no 3, 1553-1567.
50. Akashi, H., Sakai, Y. and Tagashira, H. (1994) Modelling of a self-sustained discharge-excited ArF excimer laser, *J. Phys. D.*, **27**, 1097-1106.
51. Simon, G. and Botticher, W. (1994) Two-dimensional model of the ignition phase of high-pressure glow discharges, *J. Appl. Phys.*, **76**, 5036-5046.
52. Kulikovsky, A.A. (1993) Nonlinear expansion of the cathode region in atmospheric pressure glow discharge, *J. Phys. D.*, **26**, 431-435.
53. Turnquist, D. (1979) Thyratron development for high repetition rate gas lasers, *Proc. Int. Conf. Lasers '79*, Orlando, Florida, 378-387, STS Press, McLean Virginia.
54. Turnquist, D., Caristi, R., Friedman, S., Merz, S., Plante, R. and Reinhardt, N. (1980) New hydrogen thyratrons for advanced high power switching, *IEEE. Trans. on Plasma Sci.*, **PS-8**, 185.
55. Withford, M.J., unpublished data.
56. Buckley, J.M. and Maitland, A. (1992) The electrical simulation of a gas discharge excited copper laser, *IEEE. Trans. Plasma Sci.*, **20**, no 6, 979-987.
57. Boley, C.H. (1993) *Computational model of a magnetic modulator for copper lasers*, Lawrence Livermore Lab, report UCRL-JC-111938.
58. Nehmadi, M., Kramer, Z., Ifrah, Y. and Miron, E. (1989) Mangentic pulse compression for a copper vapour laser, *J. Phys. D.*, **22**, 29-34.
59. Cook, E.G. (1991) High average power magnetic modulator for copper lasers, *Proc. Eighth IEEE. Int. Pulsed Power Conf.*, San-Diego, CA, 537.

IMPEDANCE MATCHING AND ELECTRIC FIELD PENETRATION IN METAL VAPOUR LASERS

PINHAS BLAU, Ph.D.
Laser Dept., Nuclear Research Centre - Negev
P.O.Box 9001, Beer-Sheva, Israel

1. Introduction

In metal vapour lasers (MVL), the lasing atoms are excited by collision with free electrons. The electrons gain their energy from the electric field (EF) within the plasma. High electron energy (in copper vapour laser, about 10 eV) is required in order to create maximal population inversion. This requires strong EF in the plasma (about 100 V/cm for copper vapour laser under normal operation conditions). Ideally the EF should be switched on with as short as possible risetime because gradual rise of the electron energy causes supplement excitation of the lower laser levels. The risetime and strength of the EF depend on the excitation electric pulse, but also on the geometry of the laser tube and on the plasma conductivity. The amplitude and risetime of the electric pulse are limited by the parameters of the excitation circuit, namely, in the conventional circuit, the rating of the thyratron. While working in the maximal thyratron ratings, the value of the capacitor is adjusted in order to deliver the required input energy/power into the laser head. Under these constrain there is poor impedance matching between the circuit and the laser head (this is demonstrated by the amount of reflected power) and apparently the EF in the plasma is not optimal. An optimization of the tube geometry is required in order to enhance the EF. The plasma conductivity, that is determined mainly by the residual electron density, affects the EF and the laser performance dramatically, as will be demonstrated below, but is much more complicated to control.

Two technologies that were adapted to MVL excitation circuits remove the circuit limitation. One is the magnetic compression, that enables shortening the risetime, and the other is pulse transformation, that enables increasing the amplitude. Thus the capacitance can be determined independently. Yet an optimization of all these parameters and investigation of the correlation between them, the tube and plasma parameters and the resulted EF is required.

Most of the published theoretical work on the excitation of MVL over simplify the issue of laser head impedance. The issue of the EF inside the plasma, including radial penetration effect, was dealt with in some kinetic models of the laser. But this

C. E. Little and N. V. Sabotinov (eds.), Pulsed Metal Vapour Lasers, 215–220.

approach was directed to investigation of the laser kinetics, and is too time-consuming to be an efficient working tool for excitation circuit development.

We developed a model that links the excitation electric pulse, the circuit capacitance, the tube geometry and the plasma parameters to the EF in the laser. It provides the spatial and temporal evolution of the EF and the current in the laser, thus enables evaluation of the impedance matching between the given circuit and the laser head.

The application of this model to our 100 W, 170 cm long, 8 cm diameter copper vapour laser is presented. The model also revealed some new directions to improve the impedance matching.

2. The theoretical model

The main assumption in the model is that the plasma conductivity is not changing significantly during the excitation pulse. The main agent for conductivity change is increase in electron energy and electron density as a result of the pulse. These changes influence the conductivity in opposite directions so they partially cancel one another. They also take place gradually, and practically the EF encounters the prepulse conditions.

We estimated the radial dependent plasma conductivity from the available values of plasma parameters. The full mathematical elaboration of the model is given elsewhere [1], thus we will present only the main stages.

2. 1. ELECTRIC FIELD PENETRATION

The electric field in the plasma obeys the diffusion equation:

$$\nabla^2 E = \mu_0 \frac{\partial J}{\partial t} = \mu_0 \frac{\partial(\sigma E)}{\partial t} \tag{1}$$

where E is the electric field, J the current density and σ is the plasma conductivity. This is obtained directly from Maxwell's equations by neglecting the displacement current and assuming plasma quasi-neutrality. In cylindrical coordinates, assuming azimuthal symmetry and constant conductivity, Eq. 1 becomes

$$\frac{\partial^2 E}{\partial r^2} + \frac{1}{r}\frac{\partial E}{\partial r} + \frac{\partial^2 E}{\partial z^2} = \mu_0 \sigma \frac{\partial E}{\partial t} \tag{2}$$

An analytical solution exists for uniform conductivity, $\sigma \neq \sigma(r)$, which is the case in conventional transmission lines. The basic solution is given by

$$E_{\omega,k}(r,z,t) = E_{\omega,k}\mathbf{J}_0(r/\delta)\exp\left[i(\omega t + kz)\right], \tag{3}$$

where $E_{\omega,k}$ is a complex constant, $\mathbf{J}_0(x)$ is the Bessel function of the first kind of order 0, and δ is the skin depth:

$$\delta = \left(i\omega\sigma\mu_0 + k^2\right)^{-\frac{1}{2}} \tag{4}$$

The general solution is a superposition of basic solutions with various ω and k: $E = \sum_{\omega,k} E_{\omega,k}(r,z,t)$, where ω and k are determined by the boundary conditions.

From Eq. 3, the basic solution for the total current is obtained by integration over the tube cross-section:

$$I_{\omega,k}(z,t) = \int_0^R 2\pi E_{\omega,k}(r,z,t)\sigma r\, dr = 2\pi R\delta\sigma E_{\omega,k}\mathbf{J}_1(R/\delta)\exp[i(\omega t + kz)] \tag{5}$$

where R is the tube radius and $\mathbf{J}_1(x)$ is the Bessel function of order 1.

In our case, the conductivity is not uniform, because the gas and electron densities are not uniform. We developed a semianalytical approach that considers the radial-dependent conductivity. The solution for the radial direction is obtained numerically by stepwise integration:

$$R_\omega(r) = i\omega\mu_0 \int_0^r \frac{1}{r'}\left[\int_0^{r'} r''\sigma(r'')R_\omega(r'')dr''\right]dr' . \tag{6}$$

and the electric field is:

$$E_{\omega,k}(r,z,t) = E_{\omega,k}R_\omega(r)\exp[i(\omega t + kz)] . \tag{7}$$

The component of the current at frequency ω is:

$$I_{\omega,k}(z,t) = \int_0^R E_{\omega,k}R(r)\sigma(r)2\pi r\, dr\, \exp[i(\omega t + kz)] . \tag{8}$$

2. 2. BOUNDARY CONDITIONS

To get the boundary conditions we use the integral form of Maxwell's equations, and we get:

$$\int_0^L E(R,z,t)\, dz - V(t) = -\frac{\partial}{\partial t}\int_R^{R_S}\left(\int_0^L \frac{\mu_0 I(z,t)}{2\pi r}dz\right)dr , \tag{9}$$

where V is the applied voltage, $E(R,z,t)$ the electric field near the tube wall, $I(z,t)$ is the laser current, L the tube length, R is the tube radius and R_S is the radius of the metal coaxial sleeve.

We have to consider both the incoming and reflected electromagnetic wave. The boundary condition for the laser edge is a short circuit, thus $I_+ = I_-$. Therefore the axial dependence of the EF and the current should be $\frac{1}{2}\left(e^{ikz} + e^{-ikz}\right) = \cos(kz)$. Eq. 9 thus becomes:

$$V_{\omega} = E_{\omega,k}(R)\frac{\sin(kL)}{k} + i\omega\frac{\mu_0}{2\pi}\ln\left(\frac{R_S}{R}\right)\frac{\sin(kL)}{k}I_{\omega,k} \tag{10}$$

k represents the longitudinal evolution of the EF and is an outcome of the parasitic capacitance of the tube. It is given by [1]:

$$k_{\omega} = \left[-i\omega C\left(R_{\omega}(R)\left(2\pi\int_0^R R_{\omega}(r)\sigma(r)r\,dr\right)^{-1} + i\omega\frac{\mu_0}{2\pi}\ln\left(\frac{R_S}{R}\right)\right)\right]^{1/2} \tag{11}$$

By the above equations one can calculate the spatial and temporal evolution of the EF for a given excitation pulse [2], or calculate the impedance of the laser head [3], Z_{ω}, that is the ratio between the applied voltage and the current. Z_{ω} includes the tube inductance and capacitance. Z_{ω} can be combined with a model of the electric circuit within a full electric model of the laser.

2. 3. LONGITUDINAL EFFECT

Longitudinal evolution of the EF is known to exist in transmission lines but was not previously considered in longitudinally excited, coaxial lasers. Indeed in some cases the longitudinal time delay can be neglected, but the effect on the impedance matching is more significant. For long tubes or when the tube capacitance is high the temporal effect should also be considered. We suggest that some of the impedance matching techniques of transmission lines can be adapted to our case.

3. Application to 100 W copper vapour laser

A test case for the model was our 100 W, 170 cm long, 8 cm diameter laser. We have calculated the laser current using the measured excitation voltage. The plasma conductivity was calculated using plasma parameters that we measured on the same laser [4]. We compared the calculated current to the measured one (Figure 1). The good agreement substantiates the assumption of the model.

We calculated the EF in the laser (Figure 2), and we can separate the skin-effect and the effect of the poor impedance matching. The ‘external’ EF is 130 V/cm. The EF in the tube near the wall is 59 V/cm (impedance mismatch) and the EF on the axis is only 28 V/cm (skin-effect). This EF is evidently insufficient for proper excitation of the laser. A solution of the Boltzman equation for the electron in these plasma parameters [5] obtained an electron energy of 4.2 eV, which is just a little above the minimal that is required for population inversion. The specific power of this large-bore laser is indeed low in comparison to small-bore copper vapour laser. We obtained a delay of 40 nsec between the peak EF near the tube and the peak on axis.

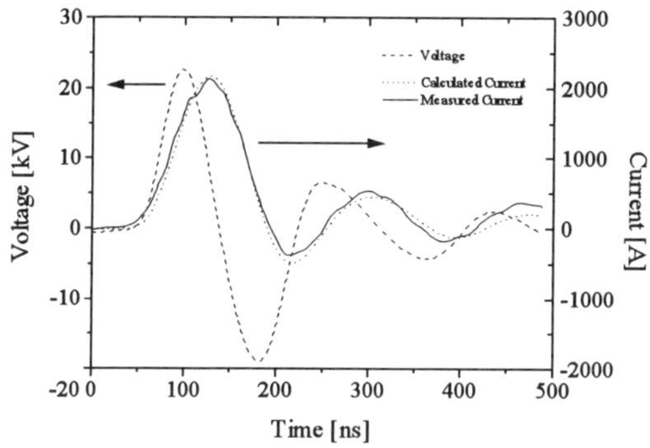

Figure 1. Calculated and measured laser current for the given laser voltage.

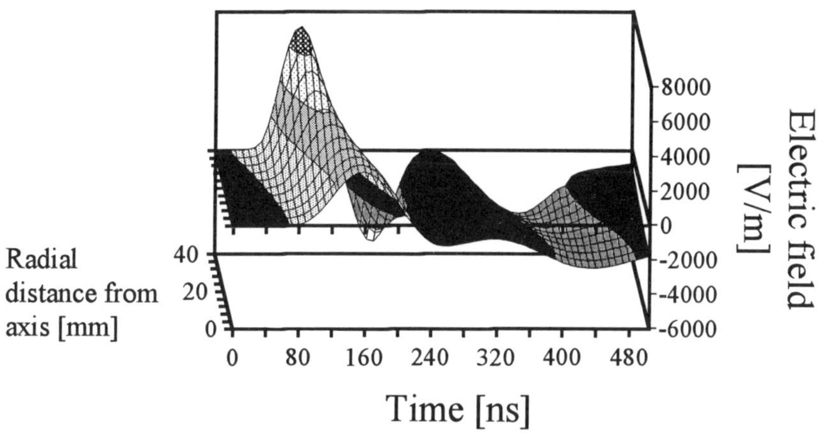

Figure 2. Radial and temporal evolution of the electric field.

4. The effect of the conductivity

To demonstrate the effect of the high plasma conductivity on the internal EF we calculated the peak EF as a function of plasma conductivity, assuming the same radial dependence of the conductivity (Figure 3). As one can see, for very low conductivity, the internal EF approaches the external EF, as a result of the improved impedance matching and EF penetration.

5. Optimization of the capacitance

The longitudinal effect is a result of the tube capacitance. The longitudinal delay in our laser is only 14 nsec but the effect on impedance matching is significant. To demonstrate this we have calculated the peak EF and peak current as a function of the metal sleeve radius (Figure 4). As the sleeve radius decreases the tube inductance decreases but the capacitance increases, and below certain value of sleeve radius, it destroys the impedance matching completely. The optimal value of sleeve diameter depends on the tube and plasma parameters, but also on the circuit parameters, in particular on the value of the capacitor in the circuit [6].

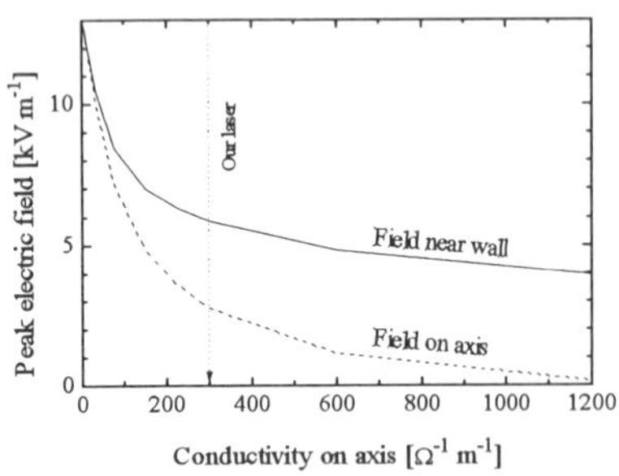

Figure 3. Conductivity dependent EF.

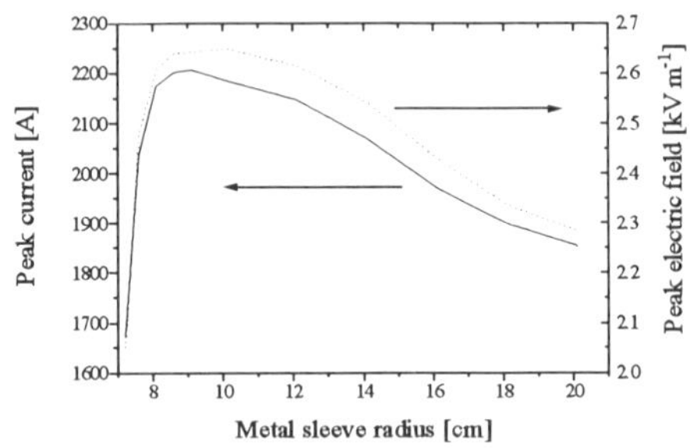

Figure 4. Sleeve diameter dependent EF.

6. Discussion and conclusion

The specific output power of a large-bore copper vapour laser is much lower than that of a small-bore one. The reason is that the high conductivity and high inductance of large-bore lasers prohibit good impedance matching and prevent EF penetration. Our model enables optimization of the tube geometry and improvement of the impedance matching between the excitation circuit and the laser head, leading to an increase in the EF in the plasma and the specific laser power.

7. References

1. Blau, P. (1994) *Investigation of Copper Vapor Laser Plasma Parameters and Electric Discharge*, Ph.D. thesis, Ben-Gurion University of the Negev, Israel, NRCN-619.
2. Blau, P. (1995) Spatial and temporal evolution of the electric field in a longitudinally electric-discharge-pumped gas lasers: Application to a large-bore copper vapor laser, *J. Appl. Phys.* **77**, 2273-2278.
3. Blau, P. (1994) Analysis of the impedance of a coaxial, large-bore copper-vapor laser, *IEEE J. Quantum. Electron.* **30**, 763-769.
4. Blau, P., Smilanski, I., Gabay, S., and Rosenwaks, S. (1993) Radially and temporally resolved measurements of electron density and gas temperature in a copper-vapor laser, in *Conference on Lasers and Electro-Optics 93*, May 2-7, Baltimore, Maryland, *OSA Technical Digest Series,* vol. 11, 456-458.
5. Blau, P. (1994) Power limitation of a large-bore copper-vapor laser due to inadequate penetration of the electric field, in *IEEE Digest of Conference on Lasers and Electro-Optics*, Europe **94TH0614-8**, 148.
6. Smilanski, I. and Blau. P. (1995) Optimization of the characteristic capacitance of the excitation circuit for improved copper vapor laser performance, unpublished.

PHYSICAL CONSEQUENCES OF THE RAPID DROP OF THE ELECTRON TEMPERATURE IN COPPER VAPOUR LASER (CVL)

J. MAURY, P. LEMAIRE, J.P. GOOSSENS, E. LE GUYADEC, J.M.BORGARD
CEA Pierrelatte/ DTE/SLC
BP 111 26702 PIERRELATTE CEDEX, FRANCE

1. Abstract

Thomson scattering measurements had shown a rapid drop of the electron temperature in CVL. The main consequence is a three body recombination governed solely by the electron density during the post discharge phase. Hence the low laser levels are populated through a very slow stepwise radiative desexcitation, for small and medium bore size CVL. Prior calculations involving Electron Distribution Function (EDF) and kinetic are in good agreement with observations.

2. Electron temperature measurements

The electron energy is of major importance for performing an efficient population inversion in CVL. We have studied in both a theoretical and an experimental way the EDF. The choice of Thomson scattering measurement comes from the use of a CVL commercial version excluding disturbing modification. From those experiment we are able to have information on the electron distribution along the tube axis [1].
The laser studied is a 40 mm bore tube, 1.5 m long. It delivers up to 20 W at a repetition rate of 5000 Hz, for a buffer gas pressure of 20 mbars. Electrical circuit is a classical resonant one. The experimental set up is shown in figure 1. The injected YAG laser delivers 0.5 J @ 532 nm at a 20Hz repetition rate. The spectrograph is a double monochromator from Jobin-Yvon which allows a rejection up to 10^{-13} to

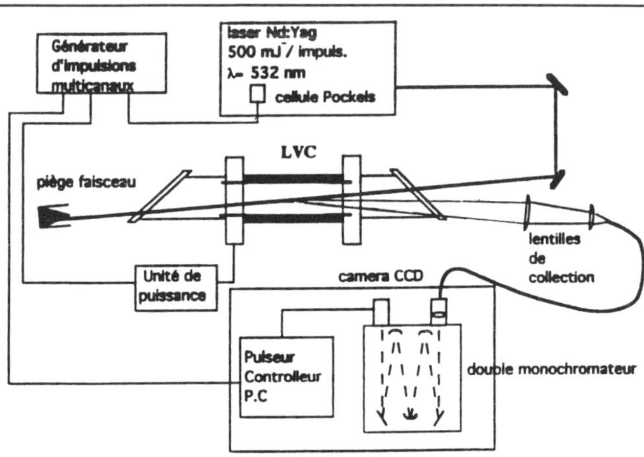

Figure 1. Experimental set up for Thomson backscattering acquisition (details in ref 1).

C. E. Little and N. V. Sabotinov (eds.), Pulsed Metal Vapour Lasers, 221–226.

remove direct scattering. A low noise intensified CCD camera is set at the rear of the spectrograph. Results and errors bars are given in § 4 . In the discharge phase, when the Thomson spectrum is wide, accumulation time was very long and it limits the temperature measurement accuracy. Improvements are in progress.

3. Electron temperature calculations

3.1 MATHEMATICAL MODEL

Results on temperature given by a self consistent kinetic model for discharge and post-discharge in a copper vapour laser are presented within a non stationary treatment of the Boltzman equation of electrons [2]. The model involved two coupled codes, one for the kinetic of excited atomic species, the other for the description of the Electron Distribution Function (EDF).

3.1.1 *Species and rates*

A total of 16 atomic species are considered (12 for copper and 4 for the buffer gas -neon herein-) and 92 electron-heavy-body collisional or radiative processes. The non elastic collisional processes include single-step excitation or ionisation, stepwise ionisation via excited atomic states, three-body recombination, heavy-body Penning collisions and the reverses processes from detailed balance.

3.1.2 *Kinetic*

The kinetic code used is a classic model, solved by a Runge-Kutta type algorithm. Longitudinal inhomogeneity effects, as well as radial effects will not be studied in the present paper, excepted a radialy averaged electronic and heavy species thermal conductions. The atomic species temperatures are taken equal.

3.1.3 *Boltzmann equation for electrons*

Models with bi or trimaxwellian distribution function (which means that fast electrons are included in a separate maxwellian distribution) are not efficient enough to characterise the plasma kinetic. A too high energy exchange between fast electrons and copper atoms through inelastic collisions, strongly perturbates the high energy tail of the EDF. It is necessary to solve a model with the whole EDF.

The non stationary Boltzman code used here describes the EDF through energetic classes discrimination. It is solved with an iterative technique, to avoid matrix inversion, and to allow a large grid size using more than 1200 nodes. This high energy resolution is necessary because of quick variations of most electron collisions cross-sections and their important energy gap (as far as 21.4 eV for neon ionisation) as well as to provide a sufficient accuracy in the treatment of the non-linear electron-electron interaction and also in transport coefficients calculation in the weakly ionised gases approximation.

The non stationary treatment of this equation includes the previous superelastic, or inelastic as well as electron-electron collisions. It is based upon Phelps's [3]conventional two-term approximation, and is also applied to weakly ionised gases.

The algorithm used to describe the variations of the EDF is based on the formulation of Rockwood [4]. Ionisation processes are simply treated as inelastic processes except that in the algorithm we suppose the creation of one electron of null energy at the corresponding rate.
To solve the Bolztmann equation we truncate our time domain in small time intervals where the electric field is assume to be quasi constant, that is $E(t)=E_0+E_1(t)$ with $E_1(t) << E_0$.

Finally it is necessary to solve the following kind of equation :

$$\frac{\partial n}{\partial t}=A.n+B_{(t)}.n+C_{(n(t))}+D_{(t)}$$

where n is the energy resolved density, A is the matrix for inelastic and superelastic processes and energy gain from the electric field E_0, B(t) the matrix for energy gain from electric field $E_1(t)$ and recombination as well as ionisation processes, C the matrix for electron-electron collisions and D the matrix for external sources.
Following an approach similar to Morgan and Rockwood [4] method, we used both half implicit and half explicit algorithm.

Electro-neutrality is used for codes coupling

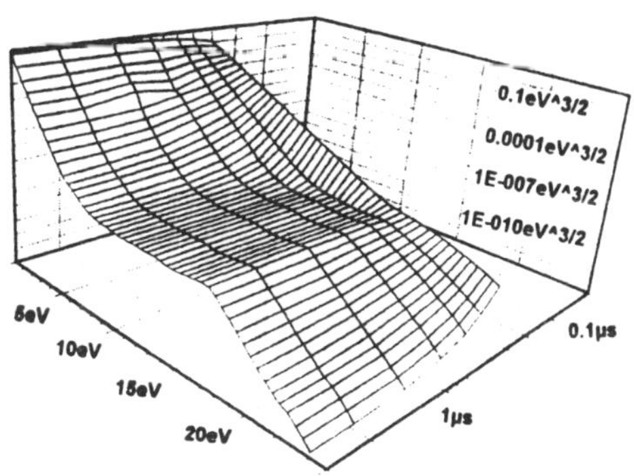

Figure 2. CVL calculated electron distribution function versus time and energy

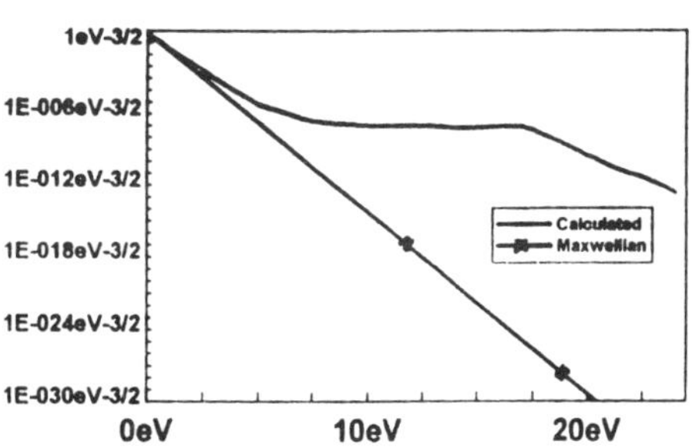

Figure 3. Calculated CVL electron distribution function versus energy for t=3µs.

3.1.4 *Results*

Reaction rate calculations with a maxwellian distribution usually differs from more than 20% with the model described above, in the discharge period, the first 600 ns and in the early postdischarge[5] up to 3 microseconds, [this work]. This approach was first used to characterise quantitatively the laser time behaviour during the discharge phase [5]. The EDF is closer to a Druyvesteyn than a Maxwellian during the first 0.6µs.
We observed on Figure 2 the calculated EDF resolved temporally and in energy for the CVL describe

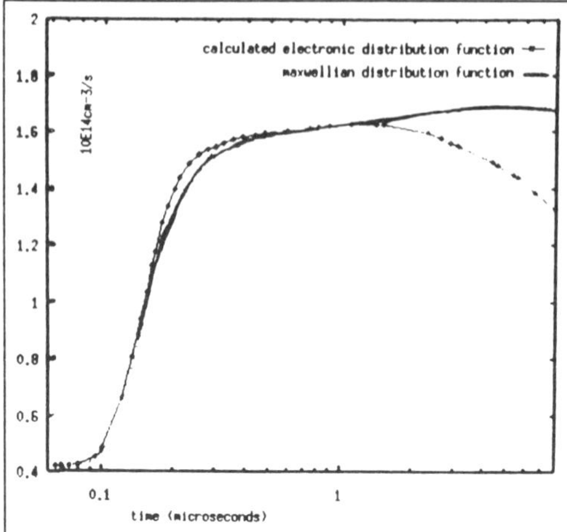

Figure 4. Calculated CVL plasma electron density versus time.(Maxwellian calculation is shown)

above. A strong modification of the tail of the EDF could be seen after 0.6 microsecond. We may also observe on that figure a local maximum at 16.67 eV which corresponds to the neon metastable level. The impact of the EDF is less important on the discharge period than on the next one, the early postdischarge. The computed EDF versus energy is shown in Figure 3, for t=3 μs, the Maxwellian EDF is added for comparison (The Maxwellian EDF is given by a strait line in the figure scale. The slope is proportional to the electron temperature).

The calculated electron density, the EDF integral (Figure 4), drops much faster in the post-discharge than those calculated with a maxwellian EDF. In the early post-discharge period, we could expect that electron-electron collisions tend to drive the EDF toward a Maxwellian one. But these collisions affect partially only the slowest electrons. Fast electrons are mainly affected by inelastic processes with atomic copper.

4. Comparison and consequences on the CVL kinetic.

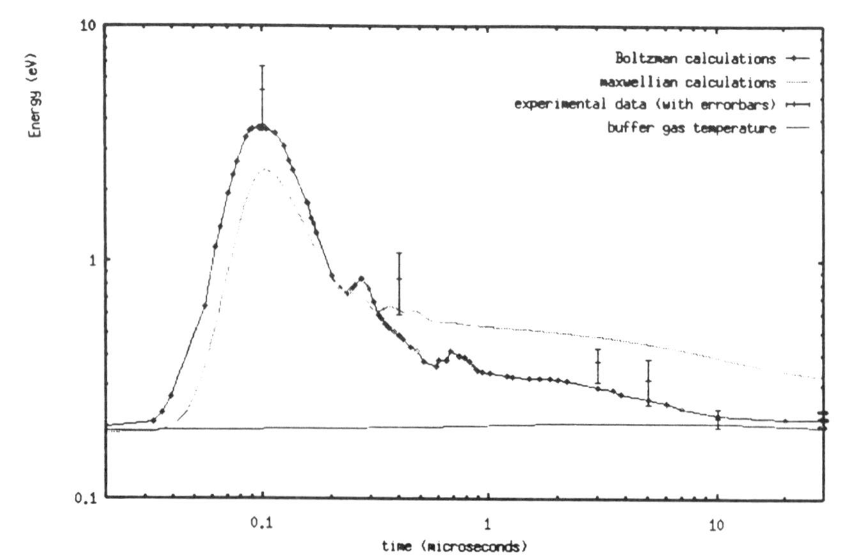

Figure 5. CVL electron temperature calculation versus time.(with a calculated EDF -dotted solid line-; Mawellian EDF -dashed line- experimental -dots-.

The electronic temperature given by $2/3 \int_0^{+\infty} f(E)\, E^{3/2}\, dE$ is shown in figure 5, (f(E) is the EDF in energy). It drops faster both in the calculations and experimentally than for Maxwellian EDF. Close to 10 microseconds the electron temperature reaches the gas temperature. This result is confirmed experimentally. It is not the case for maxwellian calculations.

For the fast electrons, inelastic collisions with heavy species is the major phenomenon for energy losses and it is more efficient than elastic transfers intended for slow electrons. The fast electron proportion is higher with the calculated EDF in the early post discharge than with the Maxwellian EDF. The rapid drop of the electronic temperature involves a more important three-body recombination in the early post-discharge which means that the maximum of electronic density is reached earlier as confirmed experimentally [6,7].

After a few microseconds, the three body recombination is governed solely by the electron density. High lying levels are involved in the three body recombination process. Figure 6 shows the time variation of relative intensities of some copper lines (the same behaviour is observed for all the buffer gases tested [1]). In the late post discharge (>5 μs) , the intensities are proportional to the population of the level (no trapping). During the next few microseconds, they are rising again due to the three body recombination. High lying levels stay excited because of their small radiative transition probabilities. They play a buffer role. In such case, the radiative decay of the high lying levels is one of the major source for the copper lower laser metastable population in the post discharge for 30 to 50 mm CVL bore diameter, after few microseconds. Two other sources could contribute to the metastable excitation in post discharge. The first one is the direct electron excitation, electron energy comes from a previous superelastic collision, and the second one is thermal excitation, like in large bore CVL.

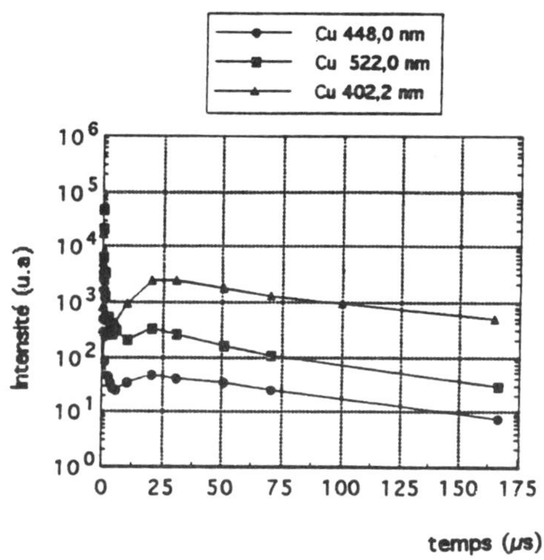

Figure 6. Time variation of the relative intensity of some lines from neutral copper excited state.

We verify that the desexcitation of the metastable is exponential in time during the post discharge [8] with the same slope as the other lines, independently of the electron density.

5. Conclusion

We have developed a kinetic code for CVL which take into account a calculated electron distribution function. The CVL kinetic behaviour computer modelling is limited primarily by the accuracy of the available data (cross-sections, optical transition probabilities), and by the small number of atomic levels involved. Afterwards, Thomson backscattering experiments have shown a rapid drop of the electron temperature in accordance with the code. High lying levels time evolution and the previous analysis permit us to conclude that lower laser population in post discharge come essentially from radiative decay of such levels which are populated by three body recombination independently of the electron temperature.

6. References

1. Goossens J.P., (1995) *Thesis* from Grenoble 1 Université.
2. Morgan W.L. and Penetrante B.M. ,(1990) *Computer Physics Com. 58*, 127
3. Frost L.S.,and Phelps A.V. ,(1962) *Physical Review 127*, 1621
4. Rockwood S.D. , (1973) *Physical Review A* 158, 70
5. Goossens J.P. et al. (1994) *Cleo Europe -Amsterdam-*
6. Hogan G., (1993) *PhD Thesis* at the University of Oxford
7. Le Guyadec E., Chamouard C.,Goossens J.P.,Gilbert D.,Lemaire P. (1992) *Optic Communication 100*, pp 461,466
8. Chamouard C. (1993) *Thesis* from Orsay Université n°2582.

5. Optical Resonators and Beam Quality

BEAM QUALITY ISSUES IN CVL APPLICATIONS

R. SALIMBENI
Istituto di Elettronica Quantistica del CNR
Via Panciatichi 56/30, 50127 Florence, Italy

1. Introduction

Despite of the long development age of CVL, only recently this class of lasers has escaped the classified cage of national energy related programs to merge into the wide interest of material processing. The CVL emission parameters have been actually found to meet the conditions for very high precision machining of many kinds of materials [1,2,3]. One crucial condition is the beam quality, to operate in material processing with a highly competitive role in respect with the other class of lasers. In facts infrared lasers have a well established utilisation in coarse cutting and drilling of materials. Shorter wavelength lasers, such as UV excimer lasers, are being under study to overcome the focusing capability and the energy coupling limits of the IR lasers, with an extensive set of Projects devoted to the development of powerful and reliable laser sources to be employed in material processing with a higher standard of precision. Nevertheless the excitation parameters and the halogen compound gas mixture imply specific solutions which rise the cost of the emitted photon to the top of the various laser classes proposed for the industry.

CVL which have demonstrated high reliability at considerable output fluxes could meet the parameters for high precision machining in competition of excimer lasers.

The two different laser radiation interacts by means of different photophysical effects:

- the excimer laser is especially selected to provide high precision removal by photoablation of material with high absorption coefficient in the UV (organic, polymer, semiconductors), provided that an energy threshold is homogeneously overcame on the irradiated area,
- the CVL can induce material removal with a relatively reduced heat affected zone, only if in a specific point of focused irradiation the power density triggers a complex combination of effects.

To some extent the CVL interaction can be schematically defined as a train of microscopic explosions, every one driven by a fast photothermal effect.

To explain this definition we rely on several reported observations concerning CVL application in machining of both absorbing and transparent materials, which are synthetically described in the following paragraph. In this introduction let speculate

C. E. Little and N. V. Sabotinov (eds.), Pulsed Metal Vapour Lasers, 229–240.

only on the obvious conclusion about the need to interact with the maximum power density to provide such a fast photothermal effect in optimum conditions of both thermal and inertial confinements in such a way to achieve the maximum rate of rise of temperature in the interacting volume.

By these considerations the beam quality issues related to CVL applications become essentially two: why such a quality is needed, and how a high quality beam can be achieved. The following paragraphs will give some answers to these issues.

2. Beam quality dependent processes

As it is well known the ablation can be completely described by a etching depth vs. energy density or fluence diagram, as is shown in Fig. 1 [5]. The characteristic curve starts as soon as a specific threshold is overcame, due to the energy requirement for direct material vaporisation. With excimer lasers in the UV the absorption coefficient is quite high for most of the organic materials, so also the threshold fluence is reduced to relatively low levels (0.01 J/cm^2 at 193 nm, 0.1 J/cm^2 at 249 nm and 1 J/cm^2 at 308 nm). This feature allows to etch very fine details by irradiating the material through a suitable mask with a beam with little or no quality requirement.

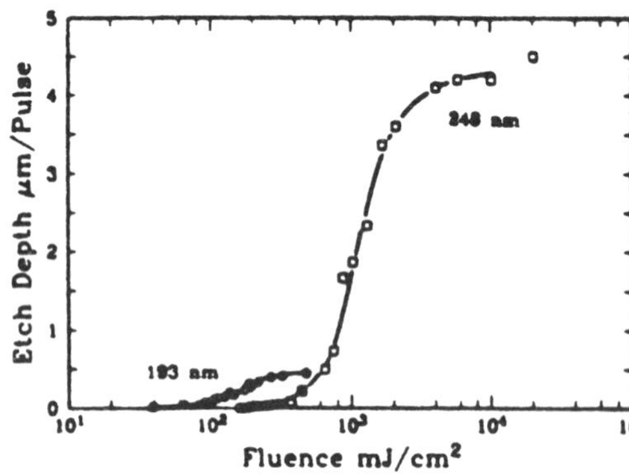

Figure 1.

Differently from the ablation with excimer lasers, CVL have shown the possibility to scribe very precise holes and cuts by means of a completely different combination of photophysical effects. Both in metals and dielectrics the wall quality was found extremely good and with a very narrow heat affected zone lying behind. The depth could easily reach 100 times the average width in blind or open holes. These general features evenly shown by very different kinds of materials are also accompanied by a similar morphology of the hole. The hole starts with an initial irregular crater, deepens regularly with slight conicity and smooth walls, and finally ends with a constant shape apex. These characteristics of the CVL machining properties will be more extensively described elsewhere, so the description can be limited to the need to support some of the following considerations.

First of all it is clear that CVL operating parameters provide an inherently repetition rate of several kHz and a single pulse energy limited to the mJ range in medium size lasers. The diagram of Fig. 2 describes the hole depth as a function of the pulse number. Considering that the reported results have been achieved with a few watts laser it remains clear that the threshold parameter is neither the energy density nor the average

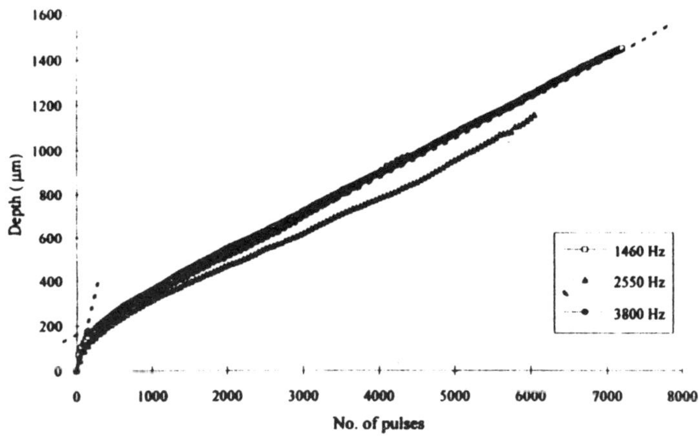

Figure 2.

power density. Furthermore about the energy absorption it is interesting to consider how the energy is transferred in the case of transparent materials such as glasses and crystals. The answers to the threshold and energy deposition issues are favoured by the common observation of the experimenters about the critic start of the drilling process only in close proximity of the beam waist of a converging lens. In other words only if the material is irradiated with a certain amount of power density the drilling is triggered for all the kinds of materials. This common dependence observed in the tests on glasses indicates a threshold parameter close to the range of 10 GW/cm^2 which is the damage threshold for optical breakdown in glasses. This recall of such a high power density threshold observed in not absorbing materials is also related to the similar observations reported for highly absorbing materials such as metals or dielectrics. In facts besides the quantitatively lower threshold for metals, the general interpretation of the complex combination of phenomena [6] can be valid as well:

1) plasma generation
2) fast temperature rise at the surface of the well apex
3) expansion in a rifle-butt like geometry
4) removal of the vaporised volume
5) advancement of the hole apex

The beam quality plays an important role in this sequence. First in determining the condition for a power density dependent threshold, second in providing the matching conditions of the laser beam to the optical guide formed by the hole. To substantiate this statement let consider the relevant depth reached by the drilling process in comparison with the Raleigh range set at the beam waist, and the small conicity of the hole channel. In our measurements the hole depth was up to 3-4 times the Raleigh range and the conicity was around 1%. These values are explained in terms of guiding effects induced by the high quality hole walls, which propagate the radiation with limited losses and intensity lowering. An optimum coupling condition asks for a specific relationship between the H_{11} Hermitian describing the lowest order mode which can propagate along the leaky capillary guide, and the supposed TEM_{00} mode of the laser beam [7]. Assuming that the intensity distribution in the beam waist will be gaussian the following relation applies

$$3W_o = 2a \tag{1}$$

with W_o the waist radius, a the guide radius.

In this condition the power coupling is almost entirely bound to the H_{11} mode with relatively low propagation losses. In the hole however several imperfections limit the ideal straightness of the guide: the beam pointing stability affects the centring of the input beam on the hole entrance, causing deviances from the axis which appear in some cases as slightly curved, if the deviance is relatively small, or tortuous if the deviance is appreciable and reflects itself periodically.

As a preliminary conclusion a very high quality beam is actually required for CVL high precision machining, in terms of minimum divergence, down to the diffraction limit, and high beam pointing stability. Because of the essentially thermal nature of the processes involved no requirement is demanded about the spectral composition of the laser output.

The following paragraph will explain how such beam quality can be achieved.

3. CVL resonators

3.1. GENERAL FEATURES

Due to the intrinsic characteristics of the laser transition CVLs have a quite high gain coefficient, typically in the order 0.1-1 cm^{-1}, for the two lines at 510 and 578 nm.

The atoms density in the order of 10^{15}, considered an upper limit before the formation of dimers, supports a high excitation current and determine a relevant flux of photons being emitted because of the high quantum efficiency of the transition. The metastable nature of the lower state extinguishes the laser emission which is allowed to take place only for a few tens of nanoseconds. For typical dimensions in the order of a meter of amplification length, such high gain coefficient induces a superradiant emission, with the self amplification up to gain saturation of the emitted fluorescence (ASE). In this conditions a large fraction of the available energy stored in the active medium is extracted as noise, with large divergence and no coherence. The control on these parameters which can be imposed by the presence of an optical resonator, has to compete with this noise.

This has been frequently observed as well discriminable components of the pulse waveform, with an initial strong superradiance onset which remains progressively quenched by the photons flux rebounding from one mirror to the other. In facts with resonator length in the order of a meter only a few passes are allowed within the gain duration. The effect of saturation of the gain by the oncoming oscillation depletes the initial noise, and raises the beam quality. If the tube diameter and the mirrors pupils are d, and the mirrors separation is D, the very first superradiant emission develops with a full angle divergence of the two-pass output equal to d/2D, which for D=100 cm and d=2 cm gives θ= 10 mrad. Because the small signal gain saturates after a few tens of cm of amplification, the emission of the laser in the first round trip duration cannot be different than this low quality ASE for any kind of resonators.

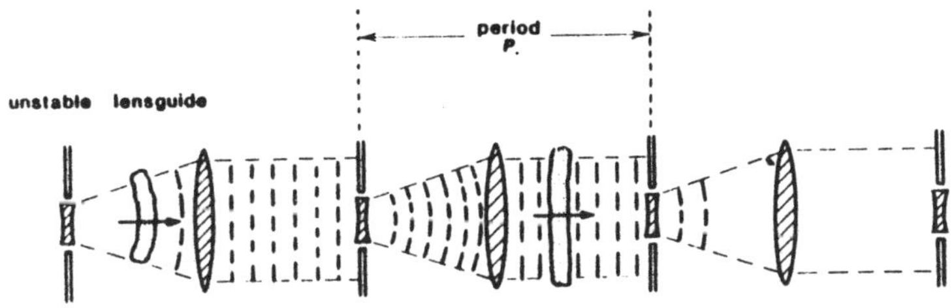

Figure 3.

A simple representation can be given in geometrical optics, as shown in Fig. 3. After the first round trip the mirror reflection propagating along the amplifier suffers another geometrical selection, reducing pass after pass its own divergence. For these reasons even the unstable resonator represented by the positive and negative lenses of Fig. 3, determines a pulsed emission with quite different intrinsic quality between the risetime and the falltime.

About the expected quality of the beam the absolute reference is the diffraction of the beam itself. For gaussian beam profile some properties do apply: the Fourier transform of a gaussian beam as determined in the focal plane of a lens is gaussian, the far-field can be approximated as spherical, if the diameter in the focal plane is $2W_o$ the divergence for diffraction is θ_D=1.27 $\lambda/2W_o$. Typical intensity distribution achieved by unstable resonators can be far from gaussian, and the relations have to be modified for the evaluation of the divergence limit set by the diffraction.

3.2. UNSTABLE RESONATORS

As it is known to implement the output characteristics of high gain short pulse lasers several types of unstable resonators are usually employed. This situation applies similarly in nitrogen lasers, excimer lasers, TEA CO_2, Dye lasers and also for CVLs. As is known the stability of a laser resonator can be described by a point in a diagram, with coordinates G_1 and G_2, respectively equal to $1-L/R_1$ and $1-L/R_2$. The centre of the axis represents the confocal resonator, while the two hyperbolas $G_1G_2=1$ are the limits for the resonator stability. The points located out of the dashed regions of Fig. 4 represent unstable solutions for the radius of curvature choice. A high magnification coefficient $M=R_1/R_2$ is considered a crucial condition to drive the choice of a highly selective solution which despite of the short duration of the gain, has a mode volume involving a major fraction of the excited medium volume.

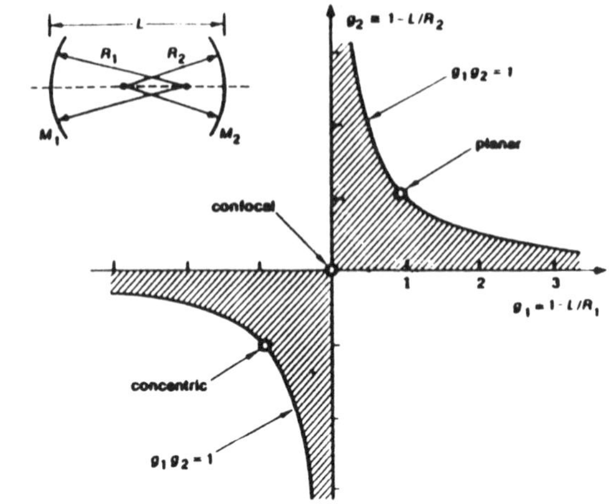

Figure 4.

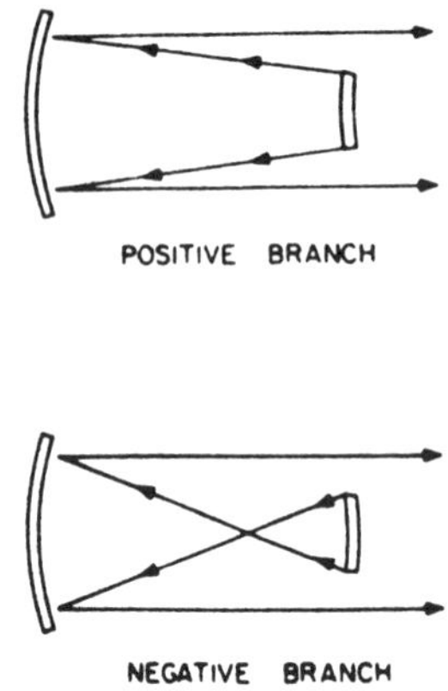

Figure 5.

Typically the unstable resonators are classified in two groups: positive and negative branch resonators, according to the relative position in the stability diagram. Fig. 5 shows the two optical schemes. The PBUR has $R_1>0$, $R_2<0$, $/R_2/<L$, hence (G_1,G_2) is in the positive quadrant of the diagram. The NBUR has $R_1>0$, $R_2>0$, $/R_2/<L$, so the point (G_1,G_2) falls under the abscissa in the negative 4th quadrant. Generally speaking an effective transverse mode selection takes place if there is some discrimination in the losses among the lowest order transverse modes. Between the others, diffraction losses represent the discriminating factor if they are significant in respect to others. For the description of the loss per bounce in an unstable resonator we can consider from a geometrical or spherical wave analysis [8] Loss= $1-(1/M)^2$. More detailed evaluations can be derived by solving the Huygens integral equation

$$\gamma_n\, U_n(x,y) = K(x,y,x',y')\, U_n(x',y')\, dx'\, dy' \tag{2}$$

where $K(x,y,x',y')$ is the Huygens kernel that propagates the complex optical wave amplitude $U(x,y)$, and the $U_n(x,y)$ eigenfunctions represent the transverse modes of the unstable resonator. If the parameter γ_n are the eigenvalues, the relative losses for the *n*th eigenmode are given by

$$\text{Loss} = 1-(\gamma_n)^2 \tag{3}$$

The Fig. 6 shows the results for the two approaches on a diagram with the relative losses vs. the equivalent Fresnel number, which confirm a stabilisation of the losses with the mirror diameter. Neq is in fact given by

$$N_{eq} = [(M-1)/2M^2]\; a^2/L\lambda \tag{4}$$

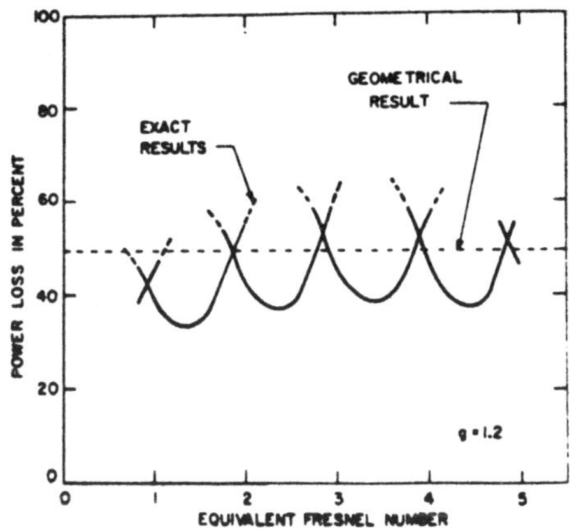

$$N_{eq} = \frac{M-1}{2M^2} N$$

Figure 6.

where the factor $a^2/L\lambda$ is the Fresnel number. As is shown the losses for the lowest order eigenmode can be significantly less than the geometrical losses 1- $(1/M^2)$ reported in the diagram, providing an active selection of the lowest order mode.

A general preference between the positive and negative branch resonators applies due to the high intensity confocal spot of the latter. Because of this usually for high power lasers the PBUR is preferred. The fraction of the light distribution which is not geometrically collected and reflected by M_2 remains outcoupled. The near-field has a typical annular shape, while the far-field distribution will exhibit a central lobe whose angular width is of the order $\lambda/2Md$, surrounded by a set of diffraction rings. The largest is M the most part of the energy is propagated within the central lobe. Coutts *et al.*[9] have measured the evolution of the beam quality for such PBUR along the pulse duration, finding an improvement of a factor of 100 from the initial 14 mrad of the superradiance to the 40 μrad of the last (eight) pass.

This feature is severely limiting the average quality of the CVL radiation, which can be unsatisfying for many application where the focusing properties are important.

For this reason other resonators can be considered which can be intrinsically diffraction limited for most of the time of actual oscillation.

3.3. SELF FILTERING UNSTABLE RESONATOR

One of these resonators is known as self filtering unstable resonator, SFUR [10]. In this solution a suitable aperture is set on the confocal plane of two mirrors M_1 and M_2. As shown in Fig. 7 it is a negative branch unstable resonator, in which the radius of the aperture *a* is matched to the first zero of the Airy pattern ρ_A which transforms with Fourier the plane wave which constitutes the outcoupled beam. The design conditions become

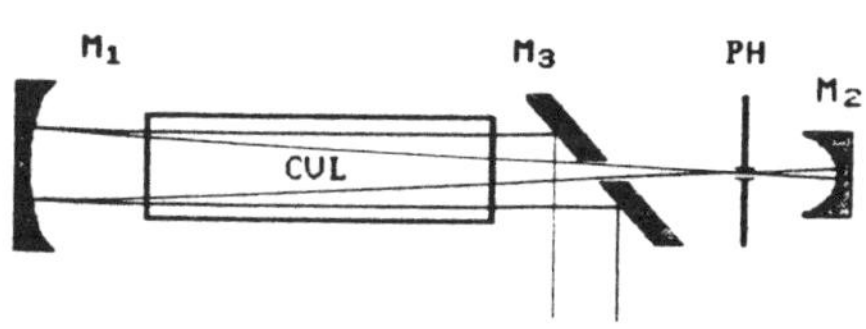

Figure7.

$$\rho_A = 1.22\,\lambda\, f_1/2a = a \tag{5}$$

$$1.5\ (f_2/f_1)\ 2a = 2d \tag{6}$$

These relations come out from the numerical analysis of the empty SFUR cavity starting from the Huygens integral equation.

The first one takes account of the intrinsic filtering of the lowest order single transverse mode, and the second describes the actual magnified size of the outcoupled beam which has to match the diameter of the amplifier 2d, for maximum energy extraction. The factor 1.5 which rises the effective value of the magnification f_2/f_1, takes in account the larger diffraction of a plane wave in respect of a gaussian one.

Being a negative branch resonator, nevertheless the SFUR limits the intensity at the aperture because of the diffraction, while the energy extraction is provided by a scraper mirror which deviated the first pass of the magnified beam. The output beam is collimated by M_2 and has very specific properties which implement the typical performance of other unstable resonators.

As reported in [11], the nominal dimensions were approximated with $2a$=200 μm, M=60, 2d=12 mm. The beam distribution in the far field showed a top hat shape regularly modulated by diffraction rings due to the 1.2 mm diameter scraper mirror hole, and the 12 mm pupils of the tube electrodes. Being 110 and 130 μrad at the two lines, the divergence evaluation was close to the diffraction limit $\theta = 2.44\ \lambda/d$, to be considered for a finite dimension plane wave, as the observed top hat distribution can be likely approximated.

Particular interest should be given to the temporal characteristics of the high quality output and to the relative power extraction in comparison to stable or other unstable resonators. In fact the increased losses of the URs rise the threshold for oscillation reducing the time duration of the emission pulse. During the off time for high quality oscillation some energy left is partially emitted as ASE. An eventual spatial filter can produce a better clean up of the beam. About the pulse shapes, they are dependent on the excitation circuitry in terms of discharge current dynamics, on tube dimensions because of several effects which determine different gain dynamics at the wall in respect of the centre. So generally speaking suitable designs for the oscillator should consider a long lasting gain with maximum axial distribution as an optimum to establish a diffraction limited, low noise emission.

3.4. GAIN SPATIAL EVOLUTION

Several studies on the gain evolution have been reported for different size lasers [12]. In effects for high power applications lasers size scale in amplifiers diameters, justifying this need. It has been found that for large size lasers the radial gain delay, the beam intensity and the pulse temporal profiles develop differently. Measurements of the lasing onset on the axis and at the wall have underlined that this delay for small bore lasers applies to the wall gain in respect to the bore centre. This condition is perfectly matched to the need of a fast rise of the gain on the symmetrically centred resonator axis. Differently, for large bore lasers the electric field penetration has been found to be limited by the plasma skin effect. The diffusion coefficient D is given by

$$D = 1/(\mu_o \sigma) = m_e v_m /(\mu_o e^2 n_e) \tag{7}$$

where μ_o is the permeability of vacuum, σ is the plasma conductivity, m_e and n_e are the mass and density of electrons, and v_m is the electron-heavy particle collision frequency for momentum transfer.

The characteristic time τ required to reach on the axis 1/e the value at the wall is

$$t \propto R_2/D \tag{8}$$

where R is the bore radius. The calculated values of τ can reach 40 ns delay for the onset of the gain on the axis. The consequence of this effect on the emission parameters is evident, providing annular intensity distribution and a peculiar choice of the resonator.

Kushner and Warner [12] investigated the large bore related issues, clarifying the heating of the centre of the discharge as a limit because of the thermal population of the metastable lower laser level, could be partially compensated by plasma skin depth effect, the radial distribution of the gain causing an elongation in time of the emitted pulse.

Blau [13] has modelled the radial dependence of the plasma conductivity, according to the measured electrons and neutral atoms densities and electron temperature, giving consistent description of the electric field penetration, indicating the problem to speed up the too slow recombination process in the inner region, which determine after the interpulse time a higher conductivity in the centre.

Chang [14] has studied how this radial delay varies with rising buffer gas pressures, finding a significant reduction of the radial delay from 45 ns at 30 Torr of Ne to only 8 ns if the pressure rises up to 110 Torr. Better radial beam profile turn out also in higher output power and efficiency.

The importance to control the inhomogeneities of conductivity have suggested also additions of a small pressure of H_2 in the discharge. Piper and coworkers [15] succeeded in the practical demonstration of implementation of the high quality component of the output, because of the earlier filling of the gain at the centre of the discharge.

For the same reason the recently enhanced performance of the copper hybrid lasers has been found to provide besides other relevant features, also a near-Gaussian transverse profile, as reported by Little [16].

Trying to take advantage of the time delay in the radial distribution of the gain Kearsley *et al.* [17] employed an off-axis resonator coupled with a matched round trip time, to achieve short pulse high power emission, particularly suitable for ultrashort pulse amplification.

In conclusion nowadays the use of large bore CVL can find efficient power extraction in both oscillator or oscillator-amplifier configurations with centred on-axis geometry.

3.5. OSCILLATOR-AMPLIFIER CONFIGURATIONS

Between the other investigations on the overall performance of a combination between a high quality emission master oscillator and a high power extraction amplifier, it has exemplary results the work by Brown and Coutts [18]. The high gain of CVL reduces the saturation fluence to about 60 $\mu J/cm^2$, so that a miniature master oscillator can be devised, capable to extract 90% of the energy in a single pass amplification with a 20 W diffraction limited output. The comparison between the performances of the same amplifier as a single oscillator with a high M PBUR or in a MOPA configuration, attributes a significant advantage to the miniature MOPA, essentially because of the intrinsic diffraction limited emission of the SFUR oscillator. The test of the spatial transverse coherence and the measurements on the intrinsic divergence confirm the full coverage of the pulse duration with optimum values of these parameters. This example demonstrates the possibility to use small size, master oscillators characterised by high quality and long duration emission for a single-pass amplification. Another choice is represented by injection-locked high power oscillators (ILO). An exemplary application has been studied by Chang and Solarski [19]. In the scheme presented in Fig. 8, the injection beam can lock the oscillating modes of the ILO if the injected wavefront is suitably matched to the adjoint button mirror reflection which self-images the amplified injected beam. The results appear implemented in respect of the simple phase matched injection. In this case the time duration of the master oscillator is not crucial for the locked oscillator control.

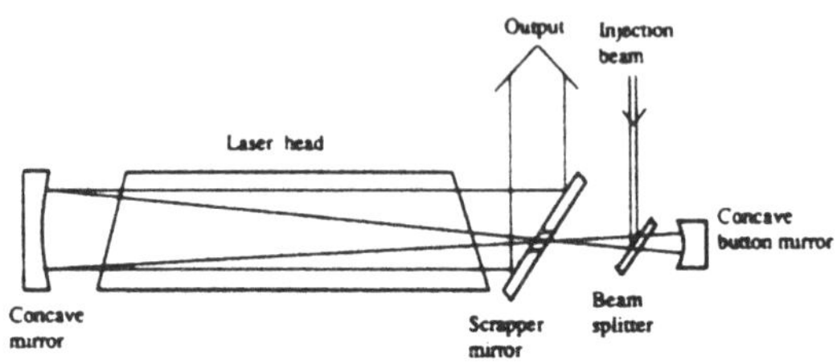

Figure 8.

Concluding this review the beam quality required for CVL application is a crucial factor to achieve the peculiar characteristics of precision shown in drilling and cutting tests. The specific CVL parameters provide viable solutions for the technology of the laser by the oscillator-amplifier scheme. Diffraction limited beams with hundreds of watts have been developed to supply the power for industrial scale utilisation of these features.

4. References

1. Bergmann H.W., Hartmann M. (1994) Drilling of metals with copper vapour lasers,in J.Mazunder, K.Mukherjes and B.L.Mordike (eds.), *Laser material processing IV*, The Minerals, Metals & Materials Society, pp. 33-44.
2. Pini R. Salimbeni R. Toci. G, Vannini M. (1992) Copper vapor laser high aspect ratio drilling process on transparent materials, in C.P. Wang (ed.), *Lasers 92*, STS Press, Mc Lean, pp. 932-936.
3. Kautz D.D. *et al.* (1992) Cutting and drilling studies using high power visible lasers, in C.P. Wang (ed.), *Lasers 92*, STS Press, Mc Lean, pp. 943-950.
4. Boyd G.D., Kleinman D. (1968) Parametric interaction of focused Gaussian light beams, J. Appl. Phys. 39, 3597-3639.
5. Srinivasan R., Leigh W.J. (1982) Ablative photodecomposition: action of far-ultraviolet (193 nm) laser radiation on poly(ethylene terephtalate) films, J. am. Chem. Soc., 104, 6784.
6. Pini R., Salimbeni R., Toci G., Vannini M. (1995) High quality drilling with copper vapor lasers, in press in OQE.
7. Abrams R.L. (1972), IEEE J. Quant. Electron., QE-8, 838.
8. Siegman A.E. (1974) Unstable optical resonators, Appl. Optics, 13, 2, 353-367.
9. Coutts D.W., Brown D.J. Piper J.A. (1993) Measurements of the divergence evolution of a copper-vapor laser output by using a cylindrical imaging technique, Appl. Optics 32, 12, 2058-2061.
10. Gobbi P.G., Reali G.C. (1984) A novel unstable resonator configuration with a self filtering aperture, Opt. Comm., 52, 195-198.
11. Pini R., Salimbeni R., Toci G., Vannini M. (1991) High efficiency diffraction limited operation of a copper vapor laser, Opt. Comm., 81, 138-144.
12. Kushner M.J., Warner B.E. (1983) Large-bore copper vapor lasers: kinetics and scaling issues, J.Appl. Phys., 54, 29700-2982.
13. Blau P. (1994) Power limitation of as large bore copper vapor laser due to inadequate penetration of the electric field, CLEO/EUROPE '94, paper CTuK97.
14. Chang J.J. (1993) Pressure dependence of copper laser output characteristics, Appl. Optics, 32, 5230-5235.
15. Withford M.J., Brown D.J.W., Coutts D.W., Piper J.J. (1994) Increased efficiency of high-beam quality extraction from a Cu-vapor laser with H_2/Ne admixtures, in CLEO, Vol.8 of OSA 1994 Techn. Digest series, paper CFD3.
16. Jones D.R., Akerboom F., Maitland A., Little C. (1993) Copper hybrid laser producing 149 W at 2.4 % efficiency and 112 W at 3.1 % efficiency, in CLEO, Vol. 11 of OSA 1993 Techn. Digest series, paper CThN8.
17. Kearsley A.J., Naylor G.A., Lewis R.R. (1987) Optimization of resonator design for large volume short-pulse copper vapor laser, in CLEO, Vol. 14 of OSA 1987 Techn. Digest series, Paper MG1.

18. Brown J.W., Coutts D.W. (1994) Efficient high-beam-quality Cu-vapor laser MOPA system with a miniature master oscillator, in CLEO, Vol. 8 of OSA 1994 Techn. Digest series, paper CFD2.
19. Chang J.J., Solarski M.E. (1993) Self-imaging injection locked copper laser oscillator, in CLEO, Vol. 11 of OSA 1993 Techn. Digest series, paper CThN5.

BEAM QUALITY ISSUES IN COPPER VAPOUR LASERS

D.J.W. BROWN and D.W. COUTTS*

Centre for Lasers and Applications
Macquarie University, NSW 2109
Australia.

1. Introduction

There has been significant recent interest in techniques for generating high beam quality (low divergence) output from copper vapour laser (CVL) systems, driven principally by the requirements of applications such as micromachining and nonlinear frequency conversion.

In this paper, we review our long-running studies of high beam quality generation from CVL devices and systems. In the following section, we discuss experiments elucidating the mechanism of output formation in CVL devices. In section 3, detailed measurements and modelling studies of the temporal evolution of beam quality from medium scale CVLs utilising unstable resonators are summarised. The inherent beam quality advantages of small-scale CVL devices are discussed in section 4, and the performance of a miniature CVL developed for high beam quality applications is presented. In the final section, fundamental limitations in high beam quality generation from single CVL active media are highlighted, and an efficient technique for scaling high beam quality power which involves using the miniature CVL in an oscillator-amplifier arrangement is discussed in detail.

2. Formation of Output from CVLs

One of the key issues in understanding beam quality issues in CVLs is a detailed knowledge of how output is formed in CVL devices. While it is generally agreed that the output from a CVL is initiated by an amplified spontaneous emission (ASE) seed or nucleus (see for example [1],[2]), there have been few investigations of the detailed interaction of this seed with the gain medium, or the spatial and temporal characteristics of the seed.

We have recently reported time-resolved measurements of the gain evolution in a CVL configured either as an amplifier or operating with an unstable resonator [3]. These gain measurements provide detailed information on both seed characteristics and the interaction of the seed with the gain

* now at: University of Oxford, Department of Atomic and Laser Physics, Clarendon Laboratory, Parks Road, Oxford OX1 3PU, UK.

C. E. Little and N. V. Sabotinov (eds.), Pulsed Metal Vapour Lasers, 241–254.

medium. A probe technique was utilised for these measurements, with one CVL configured to produce a short (5ns) pulse with high beam quality and contrast ratio, which was used to probe the gain of another CVL configured as an amplifier, or as an oscillator (with unstable resonator) using a bent-cavity arrangement. Two identical CVLs with 25mm diameter by 1m active regions were utilised for these experiments, operating at either 4kHz or 7kHz prf. Full details are given in [3]; a brief selection of the results are given below.

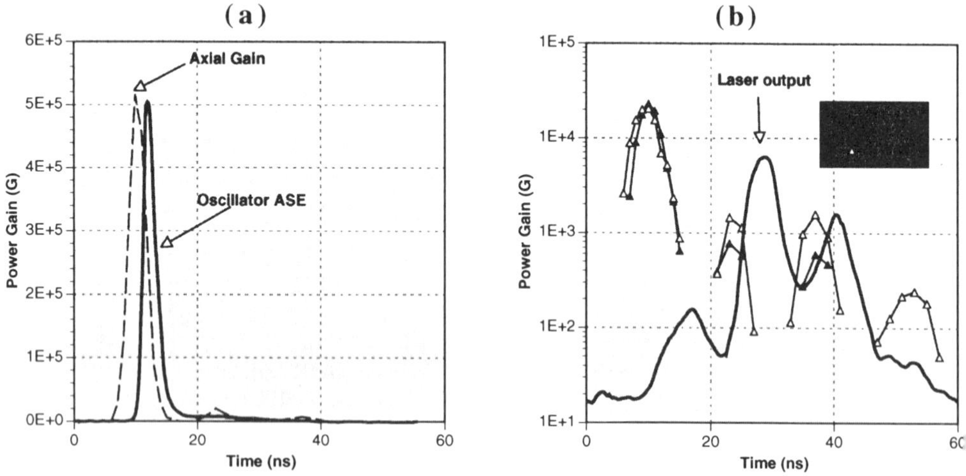

Figure 1: *(**a**) Temporal evolution of the single-pass power gain in the CVL oscillator operating at 4kHz prf, and the paraxial ASE output arising from that gain behaviour.*
*(**b**) Single-pass gain behaviour (log scale) overlayed with the measured laser output of the probed oscillator.*

Figure 1a shows time-resolved measurements of the single-pass power gain (on-axis) and the *paraxial* single-pass ASE output (ie ASE parallel to the tube axis) from a CVL that was operating with an M=62 unstable resonator. The temporal gain behaviour is dominated by short period (~5ns FWHM) of high gain, characteristic of *gain-switched* behaviour. The high optical gain in the CVL at this time causes a large photon flux to build up rapidly through amplification of spontaneous emission occurring during this period, as evident in this figure. The increasing photon flux (from both paraxial and non-axial or skew ASE) causes the gain to saturate so that the ASE output is also rapidly reduced in intensity, with the ASE output thus being restricted to a short (~3ns FWHM) spike. Gain saturation therefore occurs in less than one transit through the gain medium. There is little single-pass ASE output after this initial ASE spike. Figure 1b shows the single-pass power gain (log scale) and measured oscillator output for this configuration. After the initial gain spike, the gain never rises above ~5% of the peak gain due to the continued presence of a saturating optical flux (from the oscillator field or skew-ray ASE) at all later times. It is interesting to note that the minor peaks in gain are anti-correlated to the oscillator output, indicating that the gain recovers because of the drop in the optical field intensity in the gain medium at these times.

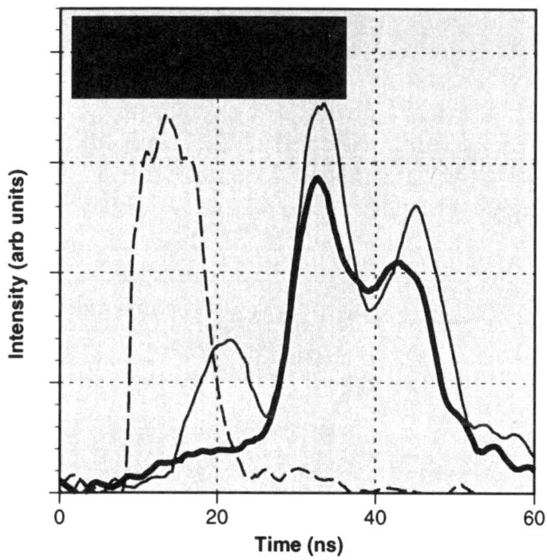

Figure 2: *Perturbation of the oscillator output by injection of a saturating optical pulse during the initial gain spike.*

Figure 2 shows the results of a perturbation experiment that demonstrates the significance of the initial ASE burst on the development of oscillator output. By heavily saturating the initial gain spike using an externally-applied short optical pulse (derived from the probing laser) the output of the probed oscillator is significantly modified. In particular, the initial peak in oscillator output is almost completely quenched when the initial gain spike is perturbed, and the subsequent output is changed in both peak intensity and depth of modulation. Perturbation of the gain at any later time throughout the CVL's gain duration does not produce any change in the oscillator output.

Based on these (and other) experiments, we confirmed that the output from a CVL derives from the initial burst of single-pass ASE (travelling parallel to the tube axis) that forms during the first few nanoseconds of gain. The observable output from the CVL is produced by the propagation and amplification of this *seed* on repeated round-trips within the resonator. Hence it is more correct to view the CVL as a multi-pass or travelling-wave amplifier of the seed radiation, rather than a true laser oscillator. The round-trip intensity modulation commonly observed in CVL output (particularly for the green line) is therefore a direct result of the propagation of the seed within the resonator, rather than due to any gain-recovery effect like relaxation oscillations. This picture of the development of CVL output is supported by the results of Carman [4], where sophisticated computer modelling was used to investigate the spatio-temporal evolution of gain and intracavity fields in a copper laser oscillator.

This mechanism for output formation is quite different to that in a "classical" (ie low gain-length product) laser, where many transits of the initial spontaneous emission noise is required before threshold is reached and gain saturation occurs. During this build-up to oscillation in "classical" lasers, the intracavity field is required to conform to round-trip phase boundary conditions, and hence resonator modes develop. In a CVL, gain saturation occurs before the resonator structure can impose any real boundary conditions, and so no laser modes develop. The sole function of the resonator in the CVL is to provide geometric constraints on the propagation and amplification of the seed.

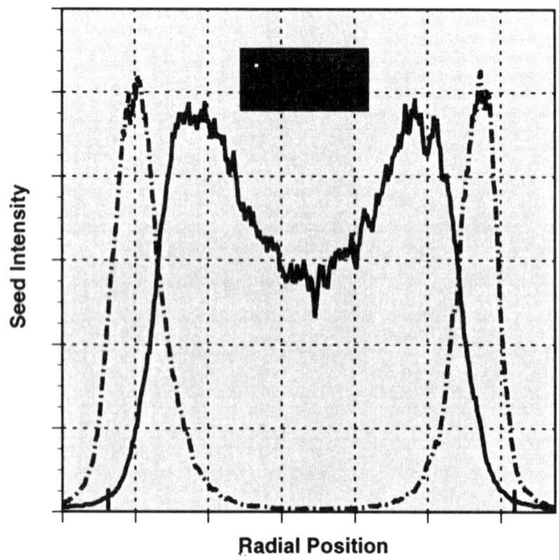

Figure 3: *Spatially-resolved seed intensity profiles for the CVL operating at 4kHz and 7kHz prf.*

Using the technique described in [3], we measured the spatial intensity profile of the seed (ie single-pass paraxial ASE) radiation, which is reproduced in figure 3. At 4kHz, the profile slightly annular, while it is highly annular at 7kHz. The differences in the seed profile are important when utilising high beam quality output, as the seed profile is important in determining the power extraction and spatial characteristics of subsequent output, as described in the next section.

3. Beam Quality from Single, Medium-Scale CVL Oscillator

It has long been appreciated ([1],[2]) that generating high beam quality output from the CVL necessitates the use of unstable resonators, because of the high gain (up to 0.25cm^{-1} [5]), short gain duration, and large Fresnel number of CVL active elements. Further, it is generally understood that the output beam quality from a CVL with unstable resonator evolves from highly divergent

ASE at the start of the laser pulse to diffraction-limited output by the end of the pulse (provided that the gain duration is sufficiently long and the resonator magnification is sufficiently high). However, until recently, the precise form of the beam-quality evolution had not been rigorously investigated, and many other aspects of high beam quality generation concerned with the detailed interaction of unstable resonator with the gain medium were not fully understood.

The first step to a more complete understanding of CVL beam quality came with the report of time-resolved measurements of the transverse coherence of CVL output with unstable resonator [6],[7]. These measurements demonstrated that the output beam quality from an unstable-resonator CVL evolves in a *step-wise* fashion towards full transverse coherence (ie diffraction-limited divergence), with the inter-step period corresponding to the cavity round-trip time. Later time-resolved measurements of divergence using a cylindrical imaging technique [8] confirmed similar step-wise evolution of the beam quality. This latter report showed that the stepwise evolution of beam quality is linked to the mechanism of output formation in the CVL. In particular, it was shown that propagation of the seed through one round-trip within the resonator (producing one output component) is equivalent to propagation through an expanding telescope of magnification M. Each sequential circulation of the seed produces another output component with divergence reduced by a factor M over the previous component. Note that the rapid transition time between one divergence component and the next is related to the time-scale of the ASE burst or seed.

An example of this divergence behaviour is shown in figure 4, which shows time-resolved far-field intensity profiles obtained using the cylindrical imaging technique [8],[9]. These far-field intensity profiles were recorded using a CVL (active region 25mm diameter by 1m long) operating with 2 different positive-branch unstable resonators (M=50 and 100, both on-axis aligned) [9]. Figure 5 shows the corresponding output pulse shapes. The CVL was operated at low prf (4kHz) for these measurements, and hence the spatial seed intensity profile (see figure 3) was reasonably uniform. There are 4 discrete temporal components in the output from both resonators. The initial ASE or two-pass output (ie arising from two passes of the seed through the gain medium) is highly divergent (>100 times diffraction-limited), and is not shown for brevity. The subsequent output component, which we call 4-pass output, arises from seed radiation having undergone 4 transits through the gain medium. Note that the angular distribution of the 4-pass output from the two resonators is different, with the M=100 output having half the angular width of the M=50 4-pass output. Based on a top-hat beam diffraction-limit criterion, the M=50 4-pass output was ~10× diffraction-limited (DL), and the M=100 4-pass output was ~5× DL. The 6-pass and 8-pass output components for both resonators have essentially diffraction-limited divergence. The non-ASE portion of the CVL output for these resonators therefore consists of 2 divergence components that appear sequentially in the output.

It is important to note the power vs beam quality trade-off when operating with these two resonators. The non-ASE green output is 8.2W when using the M=50 resonator, but this falls to 6.9W when the M=100 resonator is substituted. This fall in output is due the decrease in the effective reflectivity of

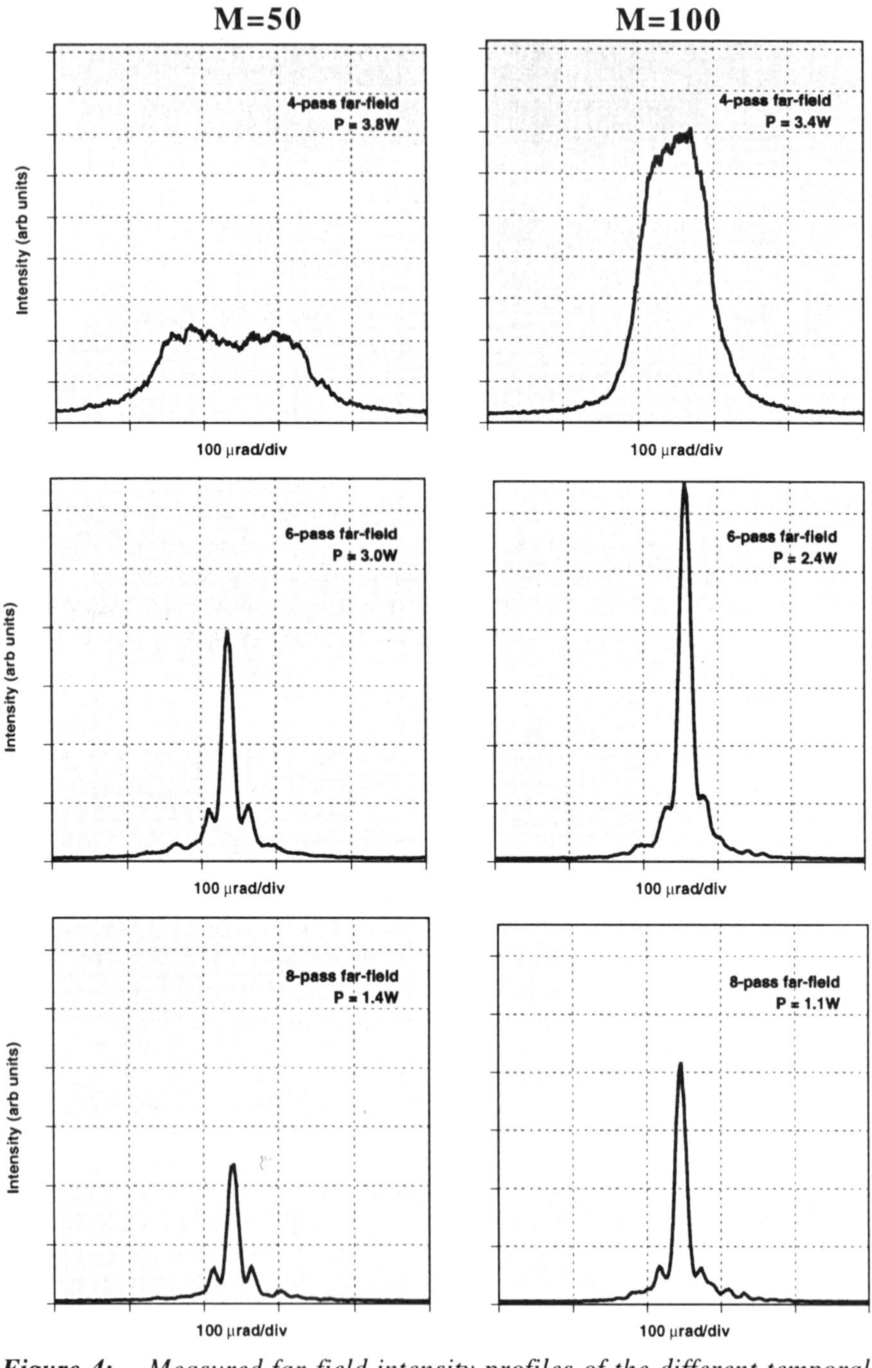

Figure 4: *Measured far-field intensity profiles of the different temporal components of the CVL operating at 4kHz prf, for M=50 (left column) and M=100 (right column) unstable resonators*

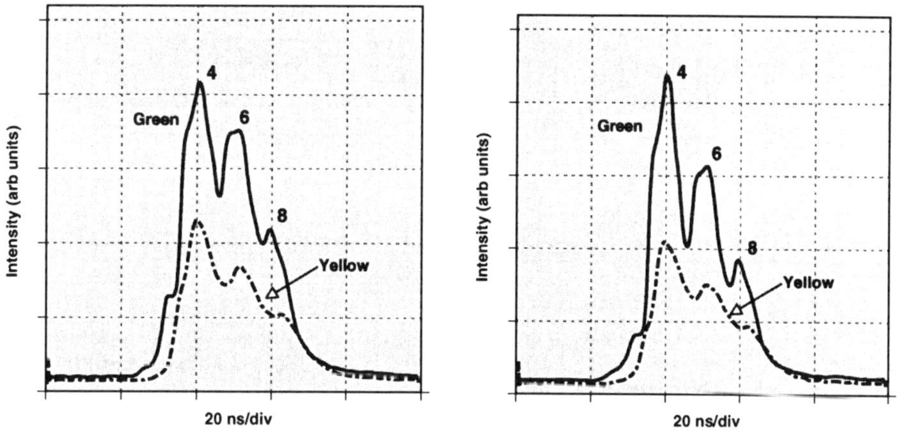

Figure 5: *Output pulse shapes for the CVL with M=50 (left) and M=100 (right) unstable resonator*

the output coupler with higher magnification ($R \propto 1/M^2$), which reduces the effectiveness of power extraction from the CVL. However, the pulse-average beam quality is higher for the M=100 resonator, because of the significant difference in the divergence of the 4-pass component which contains a large fraction (~45-50%) of the overall output for both resonator configurations. This trade-off highlights that care must be taken in selecting an unstable resonator, as the intended application needs to be considered. For example, for beam relay or solid-state laser pumping applications, a moderate magnification unstable resonator is appropriate, giving good power extraction and controlled divergence. For applications such as nonlinear frequency conversion which are critically dependent on beam quality, the highest magnification unstable resonator (for best beam quality) which gives reasonable power extraction is the most appropriate choice.

While the generation of high beam quality output from medium-scale CVLs at these low repetition rates is reasonably efficient, it becomes more difficult to effectively generate output with good beam quality as the repetition rate is elevated. This is due to the gain duration becoming shorter (which limits the number of round-trips within the resonator), as well as the seed profile becoming increasingly annular, as the repetition rate is increased. For instance, at sufficiently high repetition rates (15-20kHz) in this CVL, the gain duration shortens so that only 2-pass (ASE) and 4-pass output is generated. Further, the far-field profile of the 4-pass output, is annular (corresponding to the annular seed profile), which is undesirable for many applications.

There are two approaches to combating the difficulties inherent in high repetition rate operation. Firstly, operation with an off-axis aligned unstable resonator (ie resonator with axis translated to near the tube wall) may be preferred [10], as this alignment makes more effective use of the annular seed profile to provide better high beam quality power extraction. As shown in [9] however, the off-axis alignment means that the 4-pass output beam divergence is highly anisotropic, and it is important to be aware of this anisotropy when integrating the CVL into the application. Another subtlety of off-axis resonator operation is the sensitivity of the beam quality and power performance to the alignment, which is discussed in detail in [11]. Note that off-axis alignments are also commonly utilised in larger-scale CVLs, where the seed profile is also usually annular. The second approach to improving high beam quality operation at elevated prf is to utilise a small admixture of hydrogen in the neon buffer gas of the CVL. As shown in [12], the improvement in seed profile (ie reduced annularity) and gain characteristics when using a H_2/Ne mixture increased the yield of high beam quality power by 3.9× in a medium-scale CVL operating at 20kHz prf.

Modelling of the beam quality behaviour has also been undertaken in parallel with our experimental characterisation of the beam quality behaviour of medium-scale CVLs. Coutts [9],[11] has developed an extensive ray-tracing model of the development of beam quality in single-oscillator CVLs, which is based on the unfolded lensguide approach of Eggleston [13]. The model propagates rays (derived from the incoherent seed source) through the unfolded lensguide to the output aperture, taking into account all intervening apertures that can provide geometric constraints for propagation of these rays. Diffraction effects are only taken into account at the output aperture. This model has been shown to accurately predict the spatial characteristics of CVL output (ie far-field intensity profiles) for a wide variety of resonator topologies (including on and off-axis alignments) and seed intensity profiles (corresponding to operation at different prfs, or with CVLs of different tube diameters).

4. Beam Quality of a Miniature CVL Oscillator

The study of unstable resonator operation on medium-scale CVLs highlighted how beam quality performance is linked to the physical size of the CVL device. For instance, the round-trip time determines how quickly the stepwise improvements in beam quality occur, while the aperture of the CVL dictates the required resonator magnification to obtain a reasonable yield of diffraction-limited output from the CVL. Consideration of these "beam quality scaling" issues shows that scaling down the physical size of the CVL device offers some advantages for high beam quality generation. In particular, the round-trip time is reduced, which speeds up the beam quality evolution, while a smaller tube aperture means the 4-pass beam quality is higher for unstable resonators of the same magnification. Further, small tube apertures also allow the effective use of resonators with intracavity diffraction filtering such as Self-Filtering Unstable Resonators (or SFURs) [14], which are capable of producing diffraction-limited performance throughout the gain duration (apart from initial ASE). Such resonators are impractical for moderate-scale CVL devices due to their low feedback [15].

On the basis of these observations, we developed a "miniature" CVL that was purpose-designed as diffraction-limited source [16]. This CVL had an active volume of only $5cm^3$ (5.5mm tube diameter, 200mm long), with the entire plasma tube assembly being only 50mm diameter by 400mm long. This device was operated with a SFUR, with the cavity beam size carefully matched to the plasma tube aperture. The use of this resonator, coupled with the small plasma tube dimensions, ensured that the entire output pulse (apart from 3ns initial ASE) had diffraction-limited divergence. The output performance of the miniature CVL as a function of repetition rate (over the 4-24kHz range) is shown in figure 6a, and the green output pulses produced at 4kHz and 24kHz are shown in figure 6b.

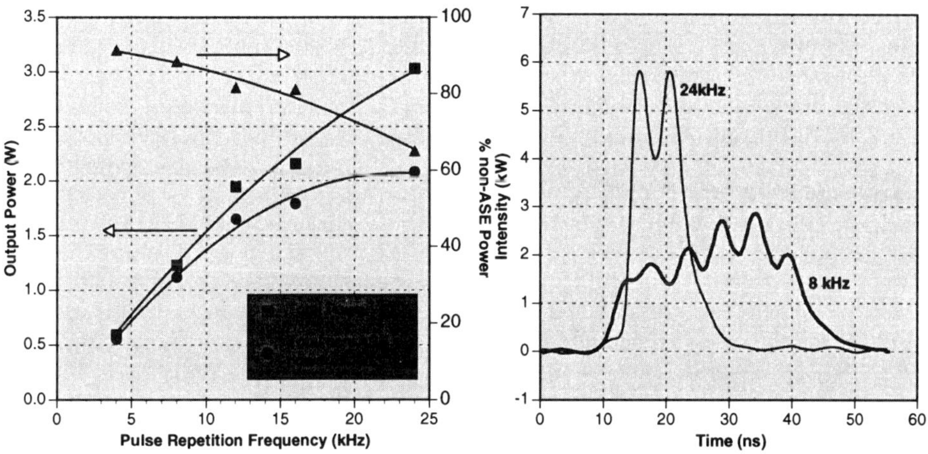

Figure 6: ***(a)*** *Miniature CVL power performance with repetition rate.*
(b) *non-ASE green output pulses produced by the miniature CVL at 4kHz and 24kHz.*

The total output of the miniature CVL increases almost linearly over the 4-24kHz prf range. However, the fraction of diffraction-limited output (ie 4 or greater pass) decreases from >90% at 4kHz to ~65% at 24kHz due to the shortening of the gain duration (figure 6b). As a result, the diffraction-limited output from the miniature CVL increases monotonically with increasing repetition rate, but begins to saturate at the highest prf's. The maximum diffraction-limited output power of 2W was obtained at 24kHz prf, which is comparable to the diffraction-limited power obtained from an 18mm diameter CVL operated with M=16 unstable resonator over its a range of repetition rates (4.3kHz to 19.7kHz) [12]. Note that at low prf's, the miniature CVL produces as much power with diffraction-limited beam quality (ie using the SFUR cavity) as when it is configured with a plane-plane resonator. We have successfully utilised the miniature CVL as a Master Oscillator for a high beam quality oscillator-amplifier scheme (described in the next section), and also as an effective stand-alone source for frequency doubling [17] and pumping of a Ti:sapphire laser [16].

5. Power Scaling of High Beam Quality Output from CVLs

One of the distinguishing features of CVLs is the ease of power scaling the output of these devices by increasing the active volume. Our previous investigations, however, have shown there are limits to the size of single CVL oscillator devices that can effectively utilise unstable resonators to produce high beam quality output. In particular, increasing the length of the active volume and hence the resonator length increases the time between step-wise improvements in beam quality, reducing the yield of high beam quality output. Further, increasing the tube aperture requires concomitant increase in resonator magnification M to provide the same rate of improvement in beam quality as in a smaller tube. The increasing magnification M of course rapidly causes difficulty with power extraction, as noted in section 3. Chang [18] has also discussed the problems associated with utilising single CVL oscillators with large Fresnel numbers for high beam quality applications. Hence there are practical limitations in the high beam quality output power that can be generated from a single CVL oscillator.

To scale high beam quality power from CVL systems, it is necessary to utilise multi-CVL arrangements and separate the requirement for high beam quality from the output power requirements. There have been two approaches of this type developed for scaling of high beam quality power from CVLs. In the Master Oscillator - Power Amplifier (MOPA) approach, high-beam-quality output generated by a Master Oscillator (MO) is amplified (in a single or double-pass) using a second larger CVL as a Power Amplifier. In the injection-seeded oscillator (ISO) approach, the high beam quality output of a Master Oscillator is injected to seed the output of a second CVL configured as a Power Oscillator. As these systems are generally assembled from pre-existing CVL active elements, there have been few prior attempts to optimise individual CVL elements as part of the system for maximum efficiency and simplicity.

We have recently reported an efficient MOPA system [19] utilising the miniature CVL described above as the MO. For maximum overall MOPA system efficiency, the CVL MO need only produce sufficient diffraction-limited injection power to saturate the amplifier throughout its gain duration. Measurements of power extraction as a function of injected signal power, shown in figure 7, show that the power requirements of the master oscillator are quite modest indeed, as >90% extraction from moderately-sized amplifiers only requires peak irradiance of ~200 W/cm^2 (~0.2W average power at 4kHz). The miniature CVL has ideal characteristics to act as an MO for moderately-sized amplifiers, as it produces diffraction-limited output of reasonable duration over a wide range of repetition rates, while producing about 2-3× the minimum power required to ensure full amplifier power extraction.

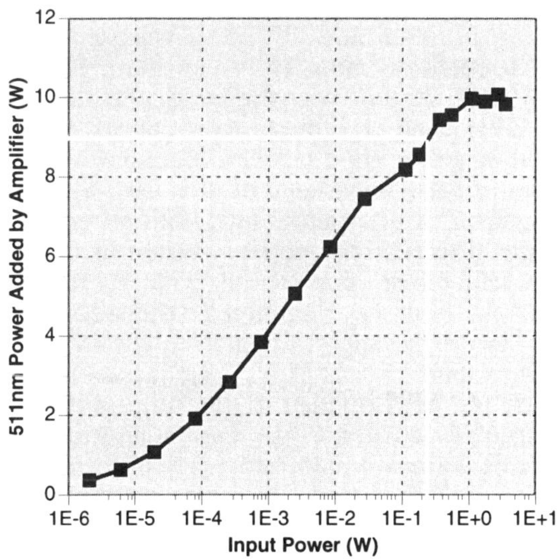

Figure 7: *Amplifier extraction as a function of input signal level for a medium-scale CVL amplifier operating at 4kHz.*

The performance of the mini-MOPA system using a nominally 20W amplifier (25 mm diameter by 1 m long active region) was investigated over a range of repetition rates (4-20.8kHz), with the output power performance being summarised in table 1. The output power available by operating the amplifier CVL with a standard plane-plane cavity is also presented for comparison. Note that the beam quality of the MO was preserved during the amplification process, so that the beam quality of the MOPA output was essentially diffraction-limited.

PRF	MOPA Power	Flat-Flat Power
4kHz	**17.1W**	16.7W
7kHz	**21.8W**	20.2W
20.8kHz	**15.2W**	16.5W
20.8kHz, H_2 additive to Amp.	**22.6W**	21.1W

Table 1: *Mini-MOPA system performance as a function of repetition rate, compared with the extractable power from the amplifier CVL (ie when operated with flat-flat resonator).*

The mini-MOPA produced in excess of 15W of diffraction-limited output at all investigated repetition rates for total electrical input power of ~4kW (wallplug efficiency > 0.4%). Perhaps the most significant result is the MOPA performance at 20.8kHz, when hydrogen was added to the amplifier CVL to improve its gain characteristics [12]. In this case, over 22W of diffraction-limited output was obtained at a wallplug efficiency approaching 0.6%. Such high prf operation with high beam quality output has previously been obviated due to the difficulty in obtaining diffraction-limited output from power oscillators at elevated repetition rates. Note that the high-beam-quality output power of the mini-MOPA system was similar to the flat-flat power available from the amplifier CVL at all repetition rates, confirming that the miniature CVL was effective in ensuring good power extraction from the amplifier CVL.

Table 2 shows a critical comparison of the mini-MOPA system with two other high-beam-quality CVL configurations, with all configurations operating at 4kHz repetition rate. The first comparison system is a nominally 20W CVL (25mm diameter by 1m long) configured with M=100 unstable resonator, whose performance was discussed in section 3. The second comparison system used the CVL with M=100 resonator as an MO for the amplifier CVL used with the mini-MOPA system.

Not unexpectedly, the performance of the mini-MOPA system is superior to that of the single CVL oscillator. In particular, the mini-MOPA system produced 17W of diffraction-limited output, while the medium-scale CVL oscillator produced only 11W of high beam quality power, of which only half was diffraction-limited (ie 6 or greater pass output). Further, the mini-MOPA configuration produced as much diffraction-limited output power as the larger MOPA, but with significantly higher wallplug efficiency. Note also that the mini-MOPA is more efficient in utilising the available CVL power than either of the other two approaches.

	20W CVL M=100	**2×20W CVLs MOPA**	**Mini-MOPA 20W Amp**
Flat-Flat Power Capability			
Green:	10 W		
Yellow:	6W		
Total:	**16W**	**35W**	**17W**
Power <5× Diffraction-limit			
Green:	7W	18W	11.5W
Yellow:	4W	9W	5.5W
Total:	**11W**	**27W**	**17W**
Diffraction-limited Power			
Green:	3.5W	12W	11.5W
Yellow:	2W	6W	5.5W
Total:	**5.5W**	**18W**	**17W**
Wallplug Efficiency for DL Output	**0.17%**	**0.3%**	**0.43%**
Fraction of Raw Power within DL	**34%**	**51%**	**>95%**

Table 2: *Comparison of the mini-MOPA with two alternative configurations for high beam quality generation.*

An alternative approach for generating high power diffraction-limited output from a multi-CVL system is to utilise an injection-seeded oscillator approach [20],[21], which has advantages if the high beam quality output duration from the MO CVL is significantly shorter than amplifier gain duration. The most recent advance in CVL ISO configurations was reported by Chang [22]. In his arrangement, the use of adjoint coupling of the injection beam allows this beam to be self-filtered by the scraper mirror to improve its beam quality before the power oscillator CVL switches into gain. Using this scheme with a 2cm diameter CVL with M=100 unstable resonator as MO and a 4cm diameter CVL as power oscillator, he obtained 30W of diffraction-limited output.

6. Conclusions

In this paper, we summarised the key results of our studies into beam quality in CVL devices and systems. Investigations of the temporal evolution of gain in CVL amplifiers and oscillators clarified that CVL output originates from an ASE seed formed during the first few nanoseconds of gain. The subsequent propagation and amplification of the seed gives rise to the observable output. Time-resolved far-field profile measurements showed that the output beam quality evolves in a step-wise fashion to the diffraction-limit when using an unstable resonator, as a consequence of the tight geometric constraints on the propagation and amplification of the seed within the resonator. The inherent beam quality advantages of small-scale CVLs were demonstrated with the development of a miniature CVL capable of producing diffraction-limited output with high efficiency over a wide range of repetition rates. Finally, efficient scaling of high beam quality power was demonstrated using the miniature CVL as Master Oscillator in an oscillator amplifier configuration.

7. Acknowledgments

Th authors would like to thank Professor J.A. Piper for many helpful discussions. DJWB would like to acknowledge financial support from the Australian Research Council in the form of an Australian Research Fellowship. DWC would like to acknowledge financial support from Macquarie University in the form of a Macquarie University Research Fellowship.

8. References

1. Anan'ev Yu.A. and Anikichev S.G. (1983) Kinetics of lasers with unstable cavities and short inversion times *Sov. Phys. Tech. Phys.* **28** 1205-1208
2. Isaev A.A., Kazaryan M.A., Petrash G.G., Rautian S.G., and Shalagin A.M. (1977) Shaping of the output beam in a pulsed gas laser with an unstable resonator *Sov. J. Quantum Electron.* **7** 746-752
3. Coutts D. W. and Brown D. J. W. (1995) Formation of output in copper vapor lasers *Appl. Opt.* **34** 1502-1512

4. Carman R.J. (1995) Modelling of the intracavity optical fields in a copper vapour laser *Opt. Commun.* **119** 415-423
5. Hargrove R.S., Grove R., and Kan T. (1979) Copper Vapor Laser Unstable Resonator Oscillator and Oscillator-Amplifier Characteristics *IEEE J. Quantum Electron.* **QE-15** 1228-1233
6. Coutts D.W., Ainsworth M.D. and Piper J.A. (1992) Observation of the temporal evolution of transverse coherence in copper vapour lasers *Opt. Commun.* **87** 245-248
7. Omatsu T., Kuroda K., and Takase T. (1992) Time-resolved measurement of spatial coherence of a copper vapour laser beam using a reversal shear interferometer *Opt. Commun.* **87** 278-286
8. Coutts D.W., Brown, D. J. W., and Piper J.A. (1993) Measurements of the divergence evolution of copper-vapor laser output by using a cylindrical imaging technique *Appl. Opt.* **32** 2058-2061
9. Coutts D.W. (1995) Time-resolved beam divergence from a copper vapor laser with unstable resonator *IEEE J. Quantum Electron.* **QE-31** 330-342
10. Kearsley A.J., Naylor G.A., and Lewis R.R. (1987) Optimisation of resonator design for large-volume, short-pulse copper vapor lasers *Tech. Dig. Conf. Lasers Electro-Opt.* Paper MG1
11. Coutts D.W. (1994) Influence of off-axis unstable resonator alignment on copper vapour laser output divergence *Opt. Commun.* **13** 283-291
12. Withford M.J., Brown, D.J.W., Coutts D.W., and Piper J.A. (1995) Increased Efficiency of High-Beam Quality Extraction from a Copper Vapor Laser with H_2-Ne admixtures *IEEE J. Quantum Electron.* **QE-31** 898–902
13. Eggleston J. (1988) Theory of Output Beam Divergence in Pulsed Unstable Resonators *IEEE J. Quantum Electron.* **QE-24** 1302-1311
14. Pini R., Salimbeni R., Toci G., and Vannini M. (1991) High efficiency diffraction-limited operation of a copper vapor laser *Opt. Commun.* **81** 138-144
15. Siegman A.E. (1992) Performance limitations of the self-filtering unstable resonator *Opt. Commun.* **88** 295-297
16. Knowles D.S. and Brown D.J.W. (1995) Compact 24kHz Copper Laser Pumped Ti:S Laser *Opt. Lett.* **20** 569-571
17. Coutts D.W. and Brown D.J.W. (in press) Production of high average power uv by second harmonic and sum-frequency generation from copper vapour lasers, *IEEE J. Sel. Topic. Quantum Electron.*
18. Chang J.J. (1994) Time-resolved beam-quality characterization of copper-vapor lasers with unstable resonators *Appl. Opt.* **33** 2255-2265
19. Brown D.J.W. and Coutts D.W. (1994) Efficient high beam quality copper vapor laser MOPA system with miniature master oscillator *Tech. Dig. Conf. Lasers Electro-Opt. 1994* Paper CFD2
20. Warner B.E. (1986) Injection locking of a copper laser oscillator *Tech. Dig. Conf. Lasers Electro-Opt. 1986* Paper WE1
21. Naylor G.A., Lewis R.R. and Kearsley A.J. (1988) Performance of high power copper vapour lasers in an injection-controlled oscillator-amplifier configuration *Proc. SPIE* Vol. 894 - Gas Laser Technology 110-117
22. Chang J.J. (1995) Copper laser oscillator with adjoint coupled self-filtering injection *Opt. Lett.* **20** 575-577

ADVANCES IN UNSTABLE RESONATOR TECHNOLOGY FOR PULSED METAL VAPOR LASERS

R. BHATNAGAR, S. K. DIXIT, P. K. SHUKLA, B. SINGH AND J. K. MITTAL

Laser Programme
Centre for Advanced Technology
Indore-452013, INDIA

Metal vapor lasers are characterized by high gain and short inversion times. The gain is non-uniform across the aperture of the active medium. It also builds up earlier around the periphery than along the axis [1]. The choice of resonators for obtaining diffraction limited output with good mode volume from such lasers is very complicated due to short inversion times. In this paper, results of studies on self-filtering unstable resonators of positive and negative branch type [PB & NBSFUR] and diffraction filtered resonators [DFR] with copper vapor laser [CVL] are presented. A new approach based on repeated imaging of output aperture in succcessive round-trips by the resonator optics is proposed to explain the output characteristics.

Conventionally, three types of resonators, namely plane-plane, concave-plane stable and confocal positive and negative branch unstable resonators have been used with CVL [2-8]. The divergence is typically in the range of few hundred times diffraction limit (DL) for plane plane resonators to few times DL for positive branch unstable resonators.

Hargrove et al. [2] established, by recording time resolved far-field intensity distribution, that the output divergence varies during the laser pulse. In a pulse of 22 ns, the divergence reduced from 10 to 2 times DL, giving time averaged divergence of 6 DL. The temporal dependence of beam characteristics was also confirmed by various authors using saturable absorber, reversal shear and Mach-Zender interferometers [9-11].

C. E. Little and N. V. Sabotinov (eds.), Pulsed Metal Vapour Lasers, 255–262.

The divergence with a plane-plane resonator also reduced from 220 DL to 120 DL over a pulse of 35 ns [12].

First significant departure from the use of conventional resonators was the application of self-filtering unstable resonator to metal vapor lasers [13,14]. In this resonator, the radiation diffracted by an intra-cavity aperture placed at the confocal plane in a NBUR, is spatially filtered by the aperture itself (Fig. 1). A plane wave, of wavelength λ, incident on the aperture, from the direction of M_1 is diffracted by it, reflected and Fourier transformed by M_2 into the far - field Airy's pattern as it comes back to the aperture. If aperture radius, a, is chosen such that it equals the radius of Airy's central

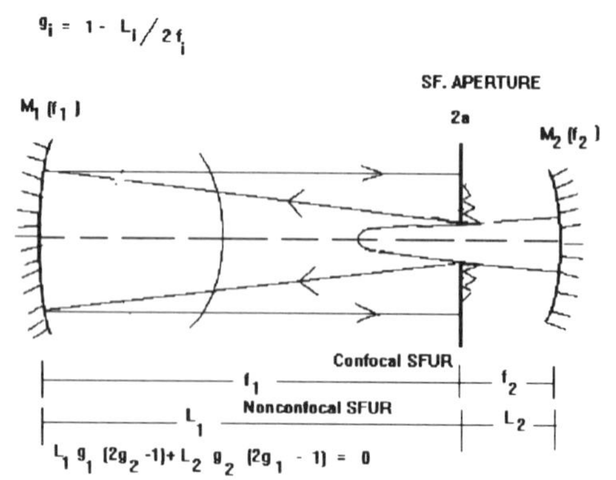

Figure 1.

lobe (1.22 λ f_2/ 2a), it allows only the central lobe to pass through. Radius a, is then given by, a = $(0.61 \lambda f_2)^{1/2}$. The output beam has a diameter, D = 3.M.a. Self-filtering can also be achieved in confocal PBUR by suitably changing the size of the aperture and replacing the concave mirror, M_2 , by a convex mirror. The radius of the aperture and the output beam diameter then is given by [14],

$A = (0.61 \lambda f_2)^{1/2} [1 + L_2 (1.5M - 1)/(f_1 - f_2)]$, and $D = 3. M. (0.61 \lambda f_2)^{1/2}$.

In confocal SFUR the beam size is restricted to (3.M.a). An attempt to increase the mode size by increasing either M or a, is largely offset by decrease in the other parameter. It can only be increased by increasing the resonator length but it leads to higher transit time. In order to increase the mode size, non-confocal NBSFUR were proposed [15]. In these the concave mirror, M_1 , is replaced by a convex or plane mirror. If L_1 and L_2 are the distances of convex and concave mirrors respectively from the aperture, these generalized self-filtering unstable resonators [GSFUR] are governed by the conditions,

$L_1 g_1 [2g_2 - 1] + L_2 g_2 [2g_1 - 1] = 0$, $a = (1.22 \lambda L_2 . g_2)^{1/2}$ and D = 3.M.a .

Since the role of mirror,M_2 is to transform the diffracted beam from the aperture in to its far-field, the objective can also be met by allowing the diffracted beam to travel beyond the Rayleigh range of the aperture and use another aperture to cut-off the wings of the Airy's distribution (Fig. 2) [16]. These resonators known as diffraction filtered resonators [DFR] retain the filtering properties of SFUR at the same time fill the cross-section of the gain medium as the size of the beam can be controlled by choosing appropriate size of the aperture. The output beam diameter $D = 2.44\, f_1 \lambda /2a$. is still controlled by the aperture size. Thus, in large cross-section lasers, attempt to fill the whole cross-section by reducing the aperture will lead to decreased feedback and hence poor efficiencies.

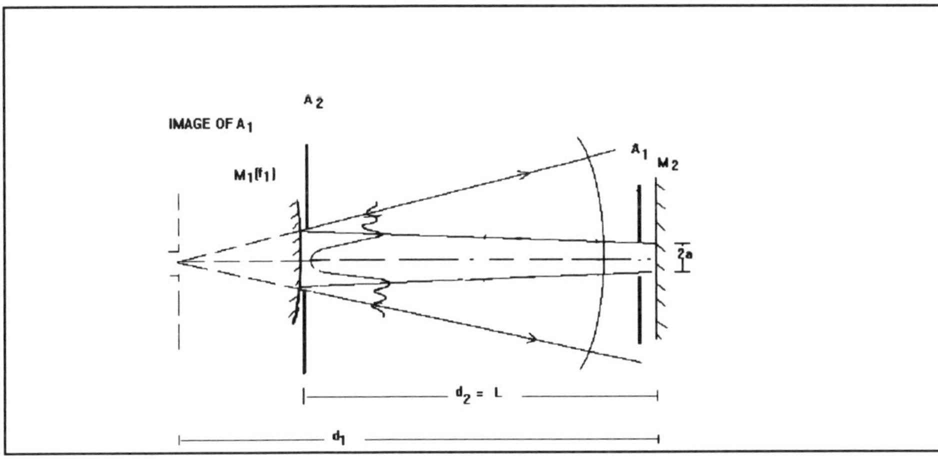

Figure 2.

Dixit et al. [17] proposed a generalized DFR [GDFR] scheme in which beam size was increased by replacing the concave mirror of DFR, by a convex or plane mirror, M_1 without changing other parameters (Fig 2). The condition now demands that
$d_1 = L + f_1.L/(f_1 + L) >> a^2 / \lambda$,

where d_1 is the distance between the virtual point source and the aperture at the plane mirror M_2 and $d_2 = L >> a^2/\lambda$. The diameter D_1 of the central lobe of the Airy's pattern at the the plane mirror, M_2 for the GDFR can be written approximately as $D_1 = (2 + L / f_1).\ 2.44\ \lambda\ L / 2a$. Thus with GDFR, depending on the focal length of the convex mirror, energy can be extracted from the gain medium of any cross-section without decreasing the aperture size or increasing the cavity length.

Table (1) lists significant results for SFUR and DFR with CVL along with results for other resonators. The divergence was less than 2 DL in all the SFUR/DFR configurations. In terms of output energy, the performance of GDFR was far superior to SFUR / DFR while the divergence was similar. Two types of GDFR (a) convex-plane (CP) $f_1 = 0.2$ m, $L = 2.3$ m, $2a = 0.4$ mm and (b) plane-plane (PP) $L = 2.3$ m, $2a = 0.4$ mm were used. About 2 watt average power in diffraction limited beam was obtained from convex-plane GDFR (CP) as compared to about 0.5 W from PBSFUR. The temporal characteristics for PBSFUR and equivalent PBUR were studied by recording the pulse shapes of the full beam (Fig.3a) and the part passing through anaperture corresponding to the diffraction limited beam size in the focal plane of a lens (Fig.3b). For PBSFUR CVL the duration of the full beam and the diffraction limited parts are almost identical. On the contrary, in PBUR-CVL, the shape of the

Table 1 Typical results on various resonators used with CVL by the authors. (2a, D, 2θ and P, are aperture diameter, output beam diameter, output diveregnce and output power respectively)

Resonator	2a (mm)	D (mm)	2θ (DL)	P (Watts)
SR		25	200	11.5
PP		25	40	10
PBUR Case 1		25	8	7
PBUR Case 2		25	3	3
NBSFUR	0.4	6.5	2	0.1
GSFUR	0.4	13	2	0.5
PBSFUR	0.4	8	1	0.5
DFR	0.4	6.5	1.3	0.4
DFR	0.15	18	2	0.7
GDFR (PP)	0.4	13	1.3	0.9
GDFR (CP)	0.4	25	2	2.0

diffraction limited part (Fig.3c,d) is significantly modified. The first two round-trips are completely cut-off and overall pulse duration reduced from 48 to 20 ns. This confirms that in SFUR CVL the diffraction limit of divergence is reached very early and then stays constant during the remaining period. Similar conclusions were arrived at with other filtering resonators used with CVL [13,14,17].

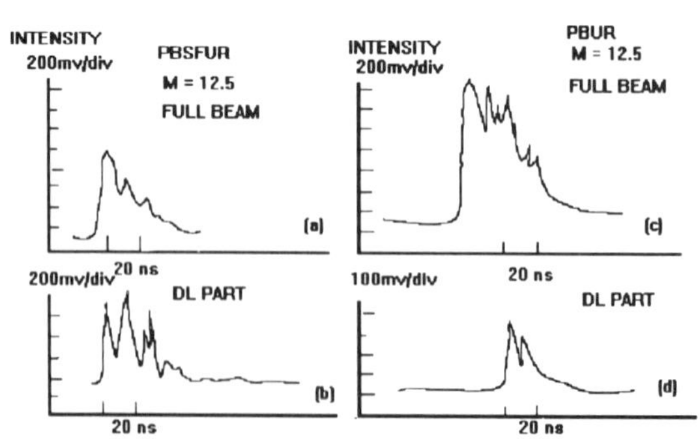

Figure 3.

It is clear that use of fast excitation pulses in short inversion time lasers leads to high extraction by the first transit as compated to later transits. Since the inversion times are comparable to transit time of radiation, conventional resonator approach fails to predict the behaviour of these lasers. Dixit et al. [18] proposed a new approach in which repeated imaging of an initial spontaneous emission source, equal in size to the cross-section of the medium, by the resonator optics in successive round-trips is used to predict the output characteristics. The distance and effective size of the image of the source in each round-trip, as seen from output plane, determine the divergence. Since the radiation in the first round-trip extracts heavily from the gain medium and its characteristics govern the average characteristics of the pulse, it is important to study the evolution of divergence in the first round-trip. In this transit, the medium encounters the radiation reflected back from one of the resonator mirrors. The situation is similar to a non-cavity laser with optics only at one end of the gain medium.

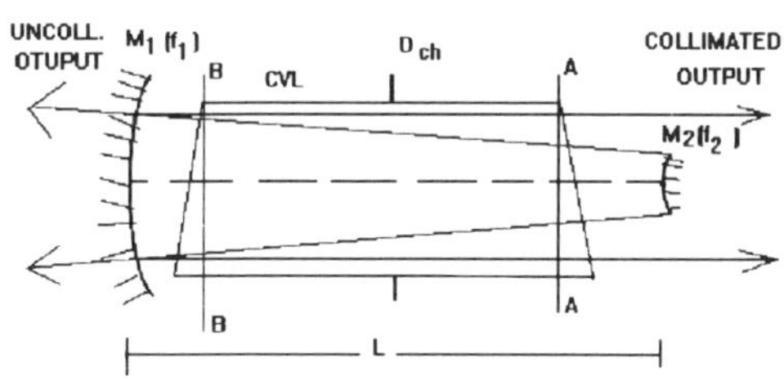

Figure 4.

Unstable resonator of figure (4) can now be examined afresh as each mirror sees the radiation coming only from the opposite ends of the gain medium. It is essential for the ray matrix analysis to define the location of the reference source plane. It was experimentally shown using non-cavity CVL that the source is located near the exit end of the discharge tube [19] .Thus planes AA and BB become location of sources for mirrors M_1 and M_2 respectively for a non-cavity first transit of radiation. The ABCD matrix for complete round trip, for the rays starting from this reference plane can be written [19]. The effective distance of the image of the source plane from the output aperture, L_e, and its effective size, D_e, is given by $L_e = B / D$, $D_e = D_{ch} / D$. The divergence is then given by , $2\theta = D_{ch} / (2 . L + 2 L^2 / R)$. where D_{ch} is the diameter of gain medium.

The matrix elements B and D will be different for mirrors of different radii of curvatures. Experiments were performed with convex, plane and concave mirrors at one end of a CVL discharge tube [19]. The measured and calculated divergences are given in table 2. Close agreement between experimental and calculated divergence indicates that the proposed geometrical approach can be successfully used to analyze the output from CVL. The divergence is about 14 times higher for the concave mirror than for the convex mirror. In unstable resonators with mirrors of different radii (Fig.4), the divergence of the first transit will therefore be different, introducing asymmetry in the characteristics of the out-put from the two ends [20].

Table (2). Calculated and measured divergences--noncavity CVL

Configuration	$2\vartheta_{exp}$ (mrad)	$2\vartheta_{calc}$ (mrad)
Convex mirror (R = 30cm, L = 210cm)	0.70	0.78
Concave mirror (R = -250 cm, L = 210cm)	10	10.2
Plane mirror (L = 210 cm)	7.35	7

The divergence of radiation from the two ends namely convex side [collimated, Case1] and from concave side [uncollimated, Case2] is therefore expected to be different. Using ray mattrix Dixit et al. has worked out the expression for the divergence of the output for the two cases [18]. The divergence as function of number of round trip is shown in Figure (5).

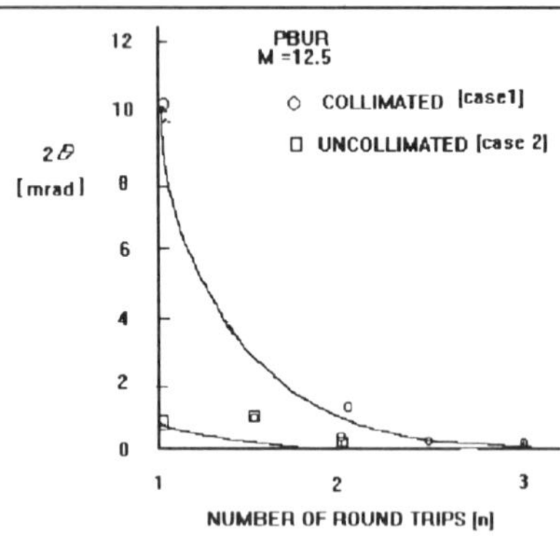

Figure 5.

The first round-trip divergence in case1 is much higher than that for case2. After first round-trip, the divergence decreases in successive round-trips by a factor of magnification M in both the cases. Since the average divergence is decided by the weighted average of the divergence of all the transits, and the first round-trip extracts heavily from high gain medium the average divergence is expected to be higher for case1 than for case2. In order to study the evolution of divergence in successive round-trips for case1 and case2, the temporal shape of the beam transmitted through apertures corresponding to various divergence angles between 3 mrad to 0.2 mrad were recorded for confocal PBUR (fig 6). For case1, the pulse width reduces rapidly from 48 ns to 14 ns for divergence angle from 1.6 to 0.2 mrad. On the other hand, for case 2, the pulse width changes from 48 ns to 46 ns only for divergence angles from 1.6 mrad to 0.17 mrad [20]. This could only be possible if the diffraction limit is established early during the pulse. This indeed is expected from earlier discussion where less number of transits are required for the radiation in case2 to reach diffraction limit as compared to that in case1. Similar behavior was observed for other unstable resonator geometries also [18].

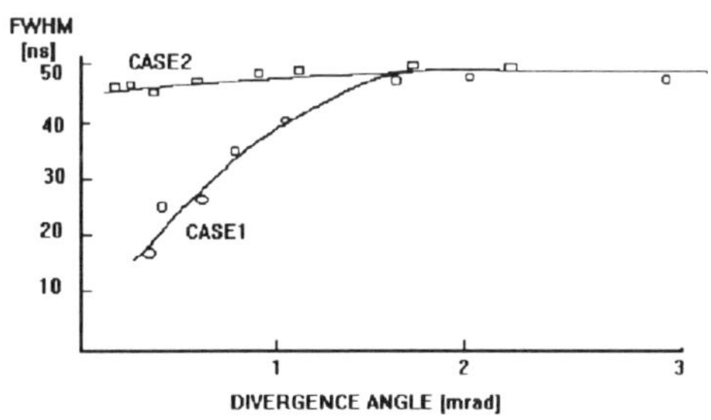

Figure 6.

The analysis presented above can also be used to make first hand predictions of the output divergence from a resonator containing a very small aperture close to one of the mirrors as is the case in SFUR and DFR. In these resonators the aperture rather than the cross-section of the gain medium is the limiting aperture. D_{ch} therefore needs to be replaced with the diameter of the aperture, 2a. It is easy to see that the divergence of radiation in an any round-trip is a factor 2a / D_{ch} less than that without intra-resonator aperture.

To summarize, the resonators for self-terminating lasers like CVL have come a long way from stable and unstable resonators to spatial-filtering resonators of various types. It is possible to obtain diffraction limited beams even from these super-radiant lasers. The divergence reduced from about 200DL of early resonators to < 2DL for recent resonators. A new approach to characterize the output from short duration high gain, self-terminating lasers is presented. It is shown that the output characteristics depend heavily on the development of radiation during the first transit. The radiation emanating from one end of the gain medium, behaves as if it is not bounded by the resonator mirrors during this transit, extracts heavily due to high gain available and controls the development of the beam through successive transits. The divergence depends on the sizes and the distances of the images of the aperture, offered either by the gain medium or by intra-resonator aperture. It is shown that the difference in the radius of curvatures of the two mirrors in unstable resonators leads to an asymmetry in the divergence of the counter-propagating beams in the resonator.

Acknowledgments- Authors are indebted to all the colleagues in the group for their excellent support in this work and gratefully acknowledge their unreserved cooperation.

References

1. Coutts, D.W. and Brown, D. J. (1995) Formation of output in copper vapor lasers, *Applied Optics* **34**, 1502-1512.
2. Hargrove, R. S., Grove, R., and Kan, T. (1979) Copper vapor laser unstable resonator oscillator and amplifier characteristics, *IEEE J. Quant. Elect.* **15**, 1228-1233.
3. Kuroda, K., Omatsu, T., Shimura, T., Chihara, M and Ogura, I. (1990) Second harmonic generation of a copper vapor laser in barium borate, *Opt. Commun.* **75**, 42- 46.
4. Belyaev, V.P., Zubov, V.V., Isaev, A.A., Lyabin, N. A., Sobolev, Yu. F., and Chursin. A.D. (1985) Spatial, temporal and energy characteristics of copper vapor laser radiation, *Sov. J. Quant. Elect.* **15**, 40-44.
5. Isaev, A. A., Kazrayan, M. A., Petrash, G.G., Rautian, S. G., and Shalgin, A. M. (1977) Shaping of the output beam in a pulsed gas laser with an unstable resonator,*Sov. J. Quant. Elect.* **7**, 746-751.
6. Coutts, D.W., Ainsworth, M. D., and Piper, J. A., (1990) Enhanced efficiency of uv second harmonic and sum frequency generation from copper vapor laser,*IEEE J. Quant. Elect.* **26**, 1555-1558.
7. Amit, M., Lavi, S., Erez, G., and Miron, E. (1987) Temporal and spatial properties of an oscillator-amplifier copper vapor laser, *Opt. Commun.* **62**, 110-14.
8. Dixit, S. K., Singh, B., Nakhe, S.V., Mittal, J.K., and Bhatnagar, R. (1990) Negative-branch unstable resonator copper vapor laser, *Optics Letters* **15**, 428-430.
9. Omatsu, T., Kuroda, K., Shimura, T, Chihara, M., Itoh, M., and Ogura, I. (1991) Time resolved measurment of beam diveregnce of a copper vapor laser using a saturable absorber,*Opt. Commun.* **85**, 343-349.
10. Omatsu, T., Kuroda, K., and Takase, T. (1992) Time resolved measurment of spatial coherence of a vapor laser using a recersal shear interferomter,*Opt. Commun.* **87**, 278-286.
11. Coutts, D.W., Ainsworth, M.D., and Piper, J. A. (1992) Observation of the temporal evolution of transverse coherence in copper vapor lasers, *Opt. Commun.* **87**, 245-248.
12. Buzhinskii, O. I., Kuznetsova, S.A., Slivitskaya, I.A., and Slivitskii, A.A., (1980) Investigation of the evolution of the divergence of radiation during an output pulse of a transverse discharge Cu laser, *Sov J. Quant. Elect.* **10,** 1548-1549.
13. Bhatnagar, R., Dixit, S.K., Singh, B., and Nakhe, S.V., (1989) Performance of a copper vapor laser with a self-filtered unstable resonator, *Opt. Commun.* **74**, 93-95.
14. Bhatnagar, R., Dixit, S.K., Mittal, J.K., Singh, B., Nakhe, S.V., and Sharangpani, K. K.(1991) A copper vapor laser with a positive branch self-filtering unstable resonator, *Opt. Commun.* **82**, 557-562.
15. Lazzaro, P. Di., Hersman, T. and Zheng, T. E. (1988) A generalization of self-filtering unstable resonator,*IEEE. J. Quant. Elect.* **24** 1543-1547
16. Pax, P. and Weston, J. (1991) Novel large mode volume resonator,*IEEE. J. Quant. Elect.* **27**, 1242-1246.
17. Dixit, S. K., Mittal, J. K., Singh, B, Saxena, P., and Bhatnagar, R. (1993) A generalized diffraction filtered resonator with a copper vapor laser,*Opt. Commun.* **98**, 91-94
18. Dixit, S. K., Singh, B., Mittal., J.K., Choube, R., and Bhatnagar, R. (1994) Analysis of the temporal and spatial characteristics of the output from short-inversion time self-terminating lasers with various resonators,*Opt. Engineering*, **33**, 1908-1920.
19. Dixit, S.K., Nakhe, S.V., Singh, B., Mittal, J. K., and Bhatnagar, R. (1991) Analysis of a Noncavity laser with a magnifying optics placed in front of the high reflectivity mirror, *IEEE. J. Quant. Elect.* **27,** 1143-1145.
20. Dixit, S.K., Mittal, J. K., Singh, B., Saxena, P., Vora, H., and Bhatnagar, R. (1992) Asymmetry in the characteristics of collimated and uncollimated output from a copper vapor laser with a confocal positive branch unstable resonator,*Opt. Commun.* **88**, 397-402.

BEAM QUALITY ISSUES FOR THE SECOND HARMONIC GENERATION OF COPPER VAPOUR RADIATION

TAKASHIGE OMATSU
Department of Image Science, Chiba University
1-33 Yayoi-cho, Inage-ku, Chiba, 263 Japan

KAZUO KURODA
Institute of Industrial Science, University of Tokyo
7-22-1, Roppongi, Minatoku, Tokyo, 106 Japan

Abstract

We investigate the evolution of spatial coherence of a copper vapour laser (CVL) during laser oscillation. Temporal rise of the spatial coherence is observed during the pulse, and the rise rate depends on lasing conditions. Moreover, in the beam quality of the CVL, we compared an off-axis resonator with an on-axis one, while the spatial coherence is the same in both resonators. In SHG, the off-axis resonator is more effective than the on-axis one.

1. Introduction

The CVL is a high average power and high repetition rate (~4 kHz) pulsed laser at 510.6 nm and 578.2 nm [1,2]. The second harmonic (SH) of the CVL at 255.3 nm will be a useful ultraviolet light in several fields such as semiconductor lithography and photo-chemical processing. Several reports have demonstrated SHG of the CVL using a barium borate for the nonlinear crystal and a high conversion efficiency over 20% was achieved [3–5].

In general, high power pulsed lasers such as the CVL, have poor beam quality, i.e. beam divergence or spatial coherence. Further, their beam quality changes temporally during the pulse oscillation. In SHG, the beam divergence or spatial coherence of a light source limits the maximum conversion efficiency. Therefore, measuring the beam quality of the CVL is important for estimating the maximum efficiency.

In this paper, we report the time-resolved measurement of the beam divergence and spatial coherence. Moreover, in the beam quality of the CVL, we compared an off-axis resonator with on-axis one. The beam divergence with the off-axis resonator is narrower than that with the on-axis one, while the spatial coherence is the same in both resonators. In the SHG, the off-axis resonator is more effective than the on-axis one.

C. E. Little and N. V. Sabotinov (eds.), Pulsed Metal Vapour Lasers, 263–268.

2. Second Harmonic Generation

The SHG of CVL was demonstrated. The CVL is equipped with an on- or off-axis type unstable resonator, for which the magnification factor is 60. The beam diameter, the resonator length, the maximum average laser power at 510.6 nm and the repetition rate are 20 mm, 3 m, 2 W and 4 kHz, respectively. We used β-BaB_2O_4 with 6.65 mm length as a doubling crystal. The focusing lens was a spherical lens with the focal length of 450 mm. In the off-axis resonator, the conversion efficiency reached 10.2% at the laser power of 1.9 W. However, the conversion efficiency was only 8% using the on-axis resonator. In order to clear up the cause of the difference of the efficiency between the off- and on-axis resonators in SHG, we carried out the time-resolved measurement of the beam quality in the laser pulse.

3. Beam Quality

3.1. SPATIAL COHERENCE

In order to compare the off-axis resonator with the on-axis one, we investigated the spatial coherence of CVL equipped with the on- or off-axis resonator. In general, the spatial coherence can be measured using a Young's interferometer. However, it is not suitable for a CVL beam, whose coherence changes during the pulse duration (~20 ns) [6], because we must measure the visibility of the fringes by varying the spacing of the two pinholes. We developed the time-resolved measurement method of spatial coherence using a reversal interferometer and a streak camera [7,8]. Using this method, the spatial coherence function of the CVL can be evaluated from a single shot measurement. The experimental setup is shown in Figure 1. The laser beam propagating in the optical path Σ_1 is reflected twice at mirrors M_1 and M_2. On the other hand, the laser beam propagating in the path Σ_2 is reflected three times at mirrors M_3, M_4 and M_5. Therefore the wavefront from Σ_2 is folded around the center axis of the wavefront from Σ_1. These two wavefronts interfere on the screen after passing the beam splitter BS_2. The shear is at right angles to the line of folding, and, at any point on the interferogram, is equal to twice the distance of that point from the line of folding. When the visibility of the fringe is measured from the folding axis to the edge at right angles, we can obtain the spatial coherence function. This fringe pattern is imaged on the slit of the streak camera.

The streaked image of the interference fringes is shown in Figure 2. The number of fringes observed along a spatial axis corresponds to the degree of coherence. We read the streaked image by a CCD camera, sampled this image every one nanosecond and analysed evolution of the coherençe. The experimental results in the on-axis and off-axis resonators are shown in Figure 3(a) and (b), respectively. We defined the zero point of the time scale at the first peak of the laser pulse. There was no difference between on-axis and off-axis resonators in the spatial coherence. The spatial coherence width began to change like a step function at 15 ns. Within the next 10 ns, the laser beam reached a perfect coherent beam. This build-up time of the coherence depends on the lasing conditions [9]. Each step of the coherence is related to the laser part that travels a different number of the round trips in the resonator.

3.2. BEAM DIVERGENCE

Second, we compared the off-axis type unstable resonator with the on-axis type unstable resonator in the view of beam divergence. The laser beam is focused on to the slit of the streak camera using a lens of focal length 3000 mm. We estimate the temporal change of the beam divergence of CVL using M^2 factor [10] from the streaked far-field pattern. The experimental results are illustrated in Figure 4. At the beginning of the laser pulse, there is no difference between the on- and off-axis resonators, and the M^2 is estimated as 2.5. In the off-axis resonator, the M^2 factor starts to decrease at 10 ns and reaches 1.2 within the next 10 ns. On the other hand, in the on-axis one, the M^2 factor doesn't reach 1.5 even at the end of the pulse. This difference in the narrowing of the beam divergence is caused by the difference of the near-field pattern between the two resonators. The near-field pattern of the on-axis resonator is annular. Therefore, the amplitude of the far-field pattern is given by subtracting the diffraction pattern of the small central hole from that of the whole beam [11]. Because the diameter of the diffraction pattern is inversely proportional to that of the near-field pattern, a centric side-lobe appears in the far-field pattern. On the other hand, in the off-axis, the small hole shifts to the edge of the near-field pattern. The shift of the hole adds the linear phase change to the diffraction pattern. As a result, after subtracting the diffraction pattern of the small hole from that of the whole beam, the diffraction pattern doesn't have a side-lobe. Figure 5 shows a theoretical diffraction pattern of the CVL beam with the off- or on-axis resonators. The far-field pattern of the off-axis resonator is like a Gaussian profile and its peak intensity is high. On the other hand, in the on-axis resonator, the far-field pattern has a side-lobe, and its peak intensity is lower compared with that in the off-axis one. Consequently, the M^2 factor in the off-axis resonator becomes smaller compared with that in the on-axis resonator. In the SHG, the conversion efficiency is proportional to the incident laser intensity as far as the depletion of the fundamental beam can be neglected. The off-axis resonator is more effective than the on-axis one. The beam divergence of the laser beam in the off-axis type was narrower than that in the on-axis type. We found that the difference of the efficiency in SHG resulted from the difference of the diffraction pattern of the laser beam between the two types of the unstable resonator.

4. Discussion

We identified the temporal behaviors of the beam quality of CVL in the laser oscillation. They suggest that the conversion efficiency of SHG changes during the pulse duration. In order to analyse the temporal change of the efficiency, the pulse shapes of the CVL, equipped with the off- or on-axis resonator, and its SH were observed. The pulse shapes of CVL P_{FW} (power in fundamental) and its

SH P_{SH} (power in second harmonic) at 1.5 W are shown in Figure 6(a). From these data, the temporal behavior of the efficiency η and the efficiency η_N normalized to the CVL power can be derived by P_{SH}/P_{FW} and η/P_{FW}, respectively (Figure 6(b)). The normalized efficiency η_N is inversely proportional to the M^2 factor, the square of the beam divergence, in neglecting the depletion of the CVL power. In the off-axis resonator, since the normalized efficiency at the end of the pulse reached twice that at the start of the pulse, the M^2 factor at the end became twice as small as that at the start. The increase of the normalized efficiency η_N corresponds with the intra-pulse decrease of the M^2 factor discussed in Figure 4. In view of the normalized efficiency, we compared the off-axis resonator with the on-axis one. There is no difference between both resonators at the start of the pulse. However, at the end of the pulse, the normalized efficiency in the off-axis resonator becomes 1.2 times larger than that in the on-axis one. At the end of the pulse, the ratio of the normalized efficiency between off- and on-axis resonators agrees with that of the M^2 factor. As a result, the total amount of the conversion efficiency during the laser pulse in the off-axis resonator becomes larger than that in the on-axis one. It is proved that in the SHG the off-axis resonator is more effective than the on-axis one.

5. Conclusion

In the evolution of the beam quality of CVL during the laser oscillation, we compared the off-axis unstable resonator with the on-axis one. There is no difference between the two resonators in the view of the evolution of the spatial coherence. On the other hand, at the end of the pulse, the far-field pattern in the off-axis resonator is narrower than that in the on-axis one, since the near-field pattern in the off-axis resonator is different from that in the on-axis one. This proves that the off-axis unstable resonator is more effective than the on-axis one in the SHG of CVL, since the conversion efficiency of the SHG is proportional to the input laser intensity.

6. References

1. Walter, W.T., Solimene, N., Piltch, M., Gould, G. (1966) *IEEE J. Quantum Electron.* **QE-2,** 474.
2. Isaev, A.A., Kazaryan, M.A., Petrash, G.G. (1972) *JETP Lett.* **16,** 27.
3. Kuroda, K., Omatsu, T., Shimura, T., Chihara, M. and Ogura, I. (1990) *Opt. Commun.* **72,** 42.
4. Coutts, D.W., Ainsmorth, M.D. and Piper, J.A. (1989) *IEEE J. Quantum Electron.* **QE-25,** 1985.
5. Omatsu, T., Takase, T. and Kuroda, K. (1993) *Opt. Commun.* **97,** 65.
6. Hargrove, R.S., Grove, R. and Kan, T. (1979) *IEEE J. Quantum Electron.* **QE-15,** 1223.
7. Omatsu, T., Takase, T. and Kuroda, K. (1992) *Opt. Commun.* **87,** 278.
8. Coutts, D.W., Ainsworth, M.D. and Piper, J.A. (1992) *Opt. Commun.* **87,** 245.
9. Omatsu, T., Takase, T. and Kuroda, K. (1992) *Opt. Commun.* **92** 50.
10. Siegman, A.E. (1990) *SPIE* **1224,** 2.
11. Born, M. and Wolf, E. (1964) *Principles of Optics,* MacMillan, New York, 260.

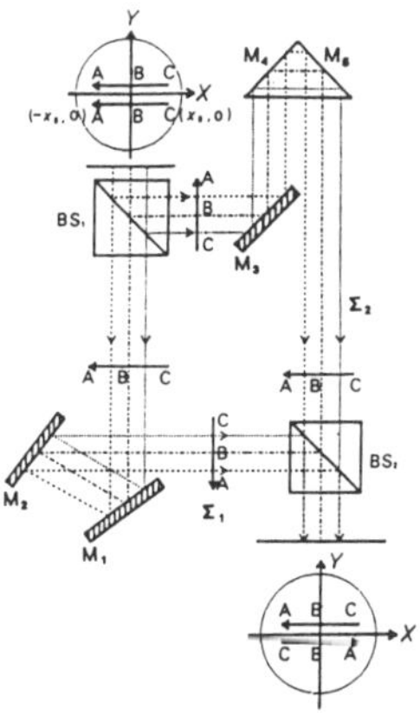

Fig.1 Reversal interferometer.

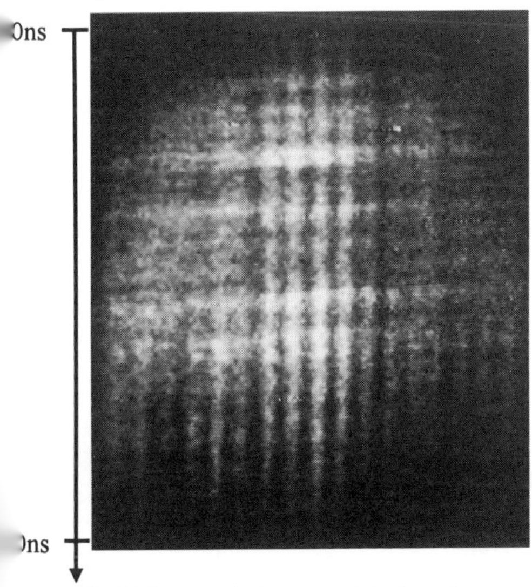

Fig.2 The streaked image of the interference fringe by the reversal interferometer.

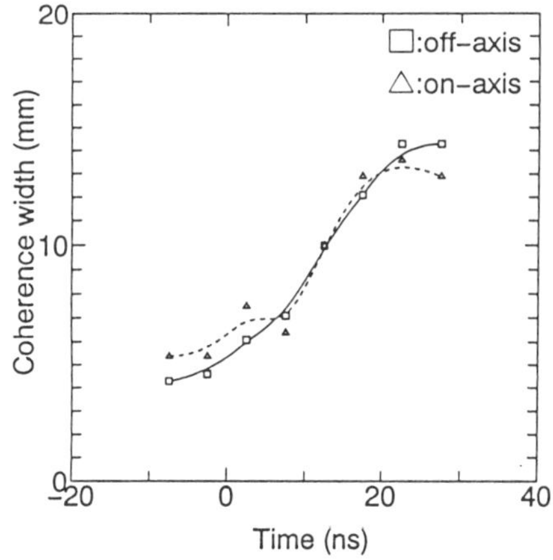

Fig.3 Temporal behaviors of spatial coherence. □ :the off-axis resonator. △: on-axis resonators.

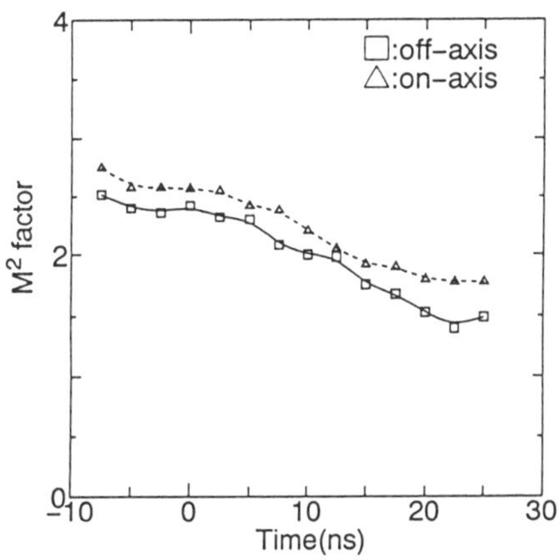

Fig.4 Temporal change of M^2 of the CVL. Solid and broken lines show results in off- and on-axis resonators, respectively.

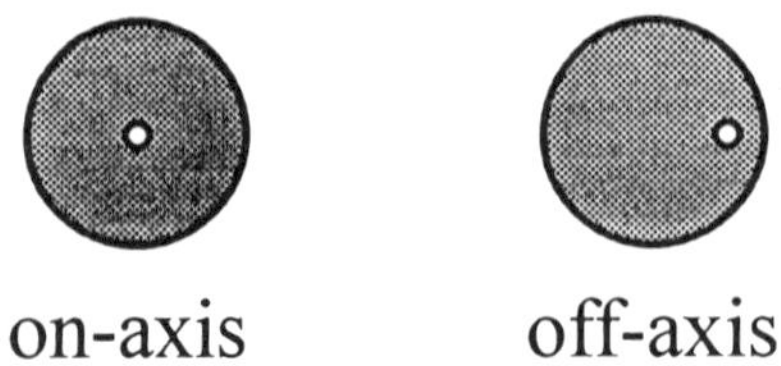

(a)

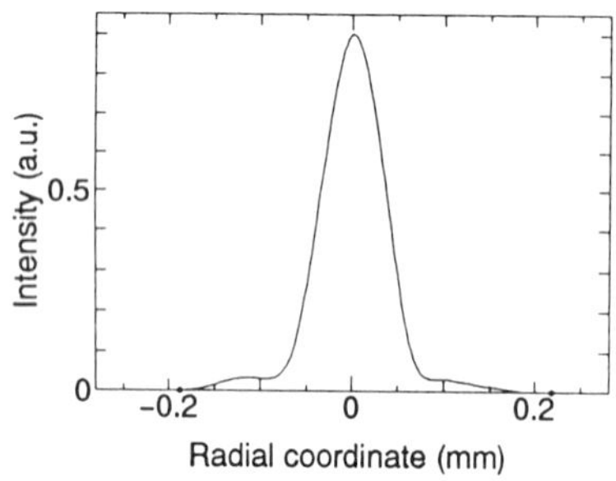

(b)

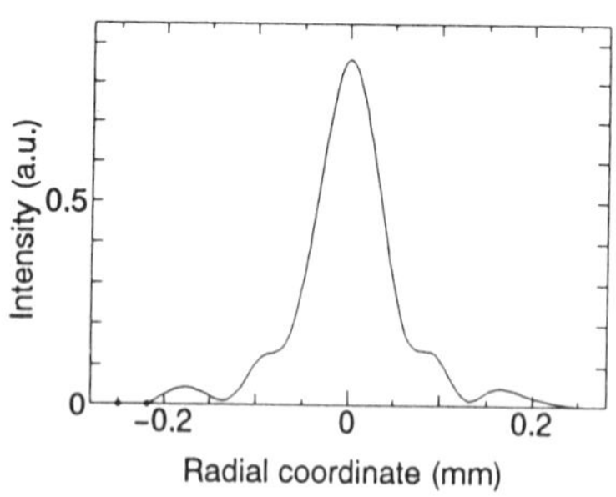

(c)

Fig.5 (a) Near-field patterns of the CVL with off- and on-axis resonators. (b) and (c) show the calculation results of the far-field patterns with off- and on-axis resonators, respectively. The coherence width, the beam diameter, and the hole diameter are 11mm, 20mm and 4mm, respectively.

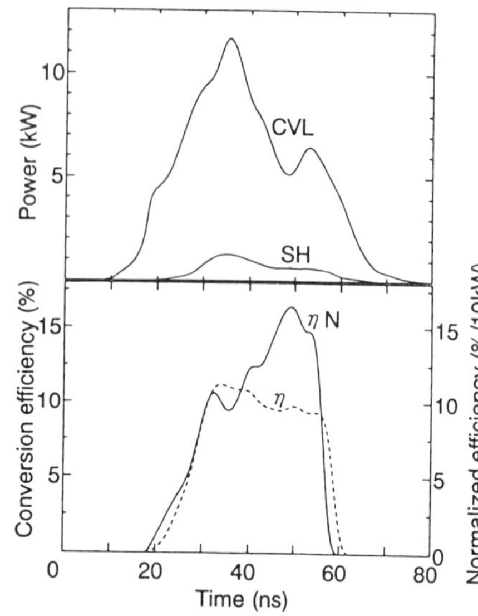

(a)

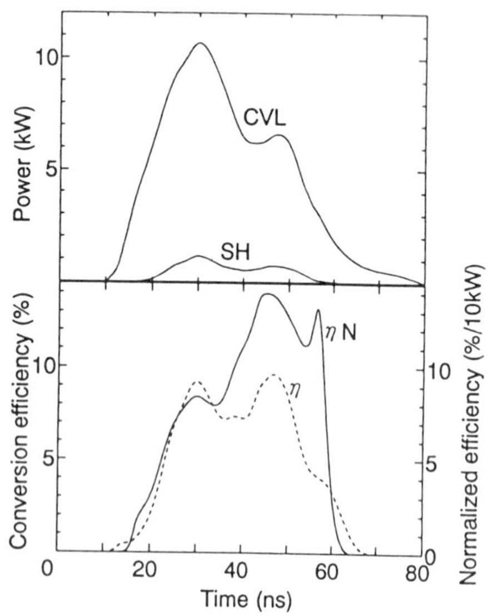

(b)

Fig.6 Temporal behaviors of conversion efficiency and normalized efficiency. (a) and (b) are with off- and on-axis resonators, respectively.

SPECTRAL CONTROL IN COPPER BROMIDE LASERS AND ITS APPLICATION IN HIGH SPEED HOLOGRAPHY

JIHONG GENG
Shanghai Institute of Optics and Fine Mechanics
Chinese Academy of Sciences, P. R. China

Abstract

Various spectral outputs in either oscillator or amplifier of the CuBr lasers can be obtained by controlling the oscillation of the longitudinal modes as well as injection seeding. The CuBr laser with narrow linewidth as a light source for holography can be used to record the hologram of high speed moving objects at a pulse rate from 8kHz to 25kHz.

1. Introduction

Copper vapor lasers (CVLs), being the most efficient source in visible region at high frequency, have currently been used as light sources for various applications such as optical imaging and measurement. Particularly, CVL with its inherent capability of high frequency and high average power is expected to be greatly promising for the application in high speed holocinematography [1]. However, these lasers usually have very complex spectral structures [2][3][4] because of hyperfine structure and isotope shift of the copper atoms and generally operate in multimode conditions. As a result, the coherence of CVL pulses is so low (usually only several millimeters) that their direct application to holography is severly limited [1], although several research groups have reported the experiments of holography using CVL [5][6][7]. W. Lauterborn et al [1] even transformed the coherence properties of the laser pulses by pumping a dye laser from a few millimeters to 7 cm for holography.

The copper halide laser operates at relatively reduced temperature of about 500°C. W. C. Kreye [8] indicated that the coherence of low-power CuCl lasers is much better than high-power CVLs. It should be reasonable that a small-scale copper halide laser is certainly more advantageous to the application of holography. In this paper we study experimentally how to control the spectral structure and the coherence properties of CuBr lasers including both an oscillator and an injection seeded amplifier. Then we report the preliminary holographic application by directly using a small-scale CuBr oscillator.

C. E. Little and N. V. Sabotinov (eds.), Pulsed Metal Vapour Lasers, 269–274.

2. Spectral Structure and Coherence Properties of CuBr Lasers

The CuBr lasers used here are homemade as previously discribed [9]. The spectral structure of the CuBr lasers is monitored by a scanning interferometer with a free-spectral-range of 2GHz (Spectra-Physics, Model 470). An usual Michelson interferometer is used to measure temporal coherence length of the laser beams. A reversal shear interferometer is used to measure spatial coherence of the beam [10]. We studied the spectra and coherence properties of both a small-scale CuBr laser oscillator and an injection-seeded amplifier systematically. The injection seeding configuration and alignment scheme have been discribed in Ref. [11].

2.1 COPPER BROMIDE LASER OSCILLATOR

CuBr laser is a high-gain pulsed laser. Generally, many high-order transverse modes can oscillate simultaneously in the free-running situation. At this time the spectral distribution of the laser line is continuous instead of the discrete longitudinal mode structure [9]. Fig.1 shows a typical lineshape of the laser. The linewidth (FWHM) is about 1GHz, which is close to the Doppler linewidth of copper atoms estimated in CuBr lasers [12]. This value is much smaller than that of high-power CVL (a few GHz) [2][3][4]. This is because the gain band of our CuBr lasers is formed by only one main Doppler-broadened component of the hyperfine and isotopic structures of the transition levels. The temporal coherence length of such a small-scale CuBr laser is measured to be more than 20cm for the green and 40cm for the yellow, which is long enough to produce a hologram.

Since high order transverse modes do not affect the total linewidth of the laser. When the oscillation of high-order transverse modes is suppressed by an aperture, the laser shows the discrete spectral structure with the same linewidth and its longitudinal modes can be recognized [9].

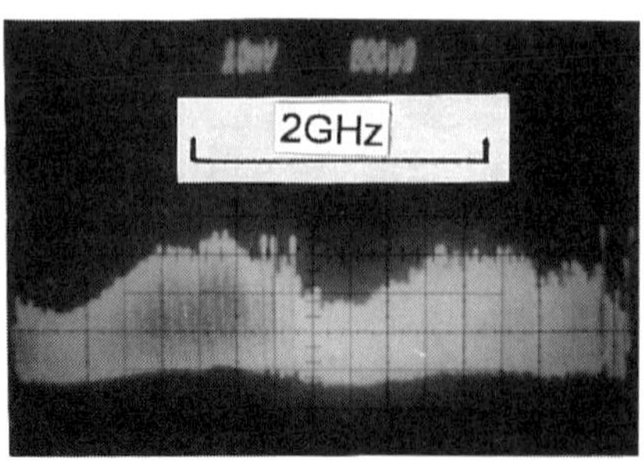

Figure 1. Typical spectral output of a free-running CuBr laser. The linewidth (FWHM) is about 1GHz.

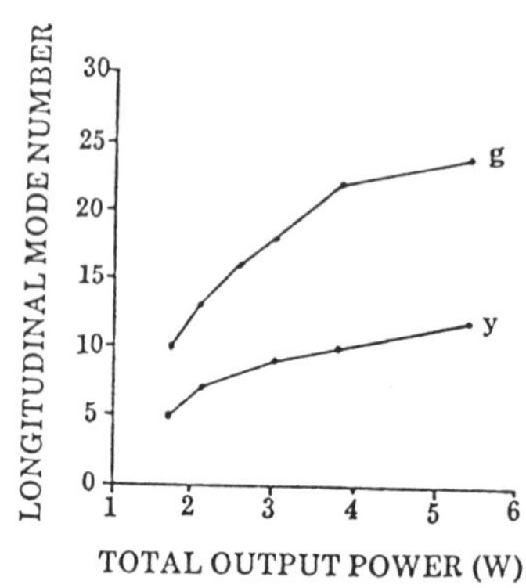

Figure 2. Experimental dependence of the mode numbers of the green laser (curve g) and the yellow (curve y) on total output power

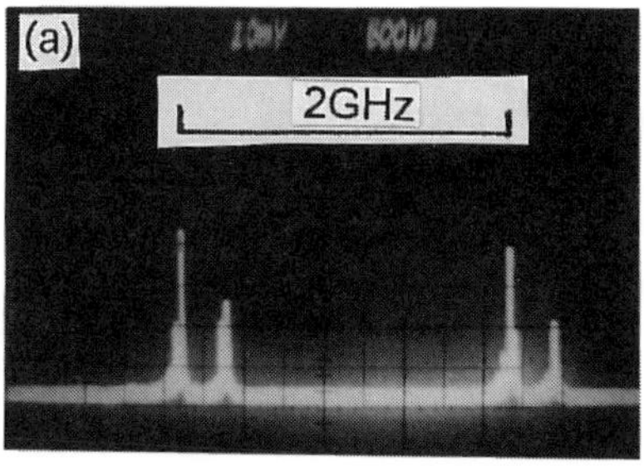

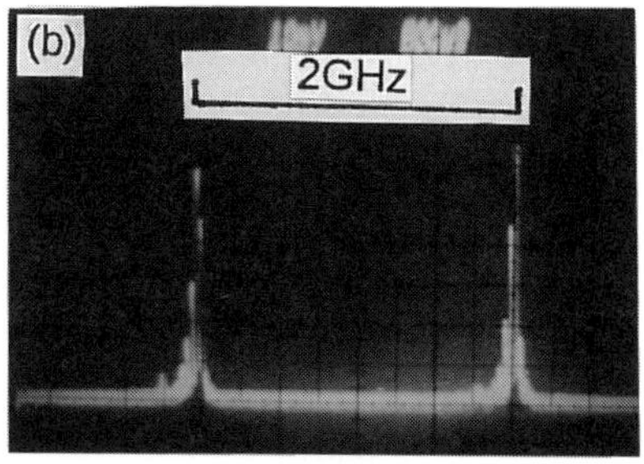

Figure 3. Spectra of (a) two longitudinal modes and (b) single longitudinal mode in the multimirror-cavity CuBr laser oscillator

While increasing the pumping excitation power of the laser, the gain level as well as the output power of the laser rapidly raises. Then the gain bands become wider and the numbers of oscillating modes obviously increase. Fig.2 presents the experimental dependence of the longitudinal mode numbers of the two laser lines on the total output power respectively. With increasing the output power the mode number is more (nonlinear added). The experiment indicates that the coherence length decreases with the output power but its variation trends to be less. These results are consistent with the theoretical calculation in our recent work [13].

It is found that the oscillating longitudinal mode number as well as the temporal coherence length is dependent on the reflectivity of the output mirror. When the reflectivity decreases the oscillating mode number becomes less and the coherence becomes better at the same gain level. However, a more efficient approach to improve the temporal coherence of the laser is the multimirror cavities. We have demonstrated that the oscillating longitudinal modes in CuBr lasers can be partly quenched and the mode number can be reduced greatly[12][14]. Fig.3(a) shows the oscillation of two longitudinal modes in an etalon three-mirror cavity CuBr laser. Since the oscillating bandwidth of the laser reduces, the temporal coherence of the multimirror cavity laser becomes better. Particularly, the single longitudinal mode laser can be realized in an appropriate multimirror cavity CuBr laser[12], as shown in Fig.3(b). In this case, the laser beam is temporal coherent during the whole pulse of the laser.

The spatial coherence of the beam from the stable resonator CuBr laser is very poor because of high gain and simultaneous oscillation of many transverse modes. Fig.4 shows the interference fringes through the reversal shear interferometer. The beam pattern of the photograph is annular because the beam undergoes a reflection by a flat mirror with a central hole[11] of a diameter of 4mm before entering the interferometer. The spatial coherence length is measured to be less than 1mm. However, it can be improved greatly by

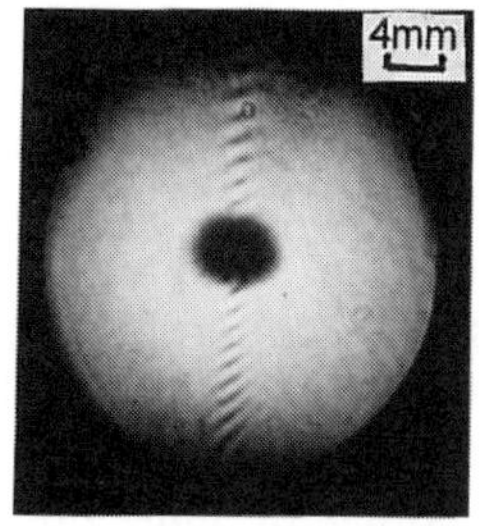

Figure 4. Spatial coherence of the beam from a stable laser

using a spatial filter or an unstable resonator.

2.2 AN INJECTION-SEEDED COPPER BROMIDE AMPLIFIER

As seen above we can obtain the spectral output with the controllable laser modes in the oscillator of CuBr laser. But the output power is extremely limited. Injection seeding is a simple, reliable and efficient way to obtain high power beams with controllable spectral properties from high gain lasers. The schematic of our injection seeding configuration has been discribed detailedly in Ref.[11]. Since the CuBr laser is a high-gain short-pulsed laser, compared with other kinds of lasers, the injection seeding behavior in CuBr lasers is quite different. It resembles more a multipass power amplifying process. The spectral properties of the injection-seeded CuBr laser depend on the seed. Therefore we can use all possible techniques mentioned above to control the oscillation of the seed laser modes transversely and longitudinally, and then amplify them in the injection-seeded amplifier.

Fig.5 gives the spectral properties of the amplifier by injection locking the seed laser beam from the oscillator as shown in Fig.3. In this case, narrow linewidth copper laser radiation with average power of a few hundred milliwatts and with bandwidth of less than 200MHz (shown in Fig.5(b)) can be obtained. The temporal coherence length of such narrow linewidth laser pulses is estimated to be more than 1.5 meters. This is to our knowledge the longest coherence length of copper laser radiation with so high average power.

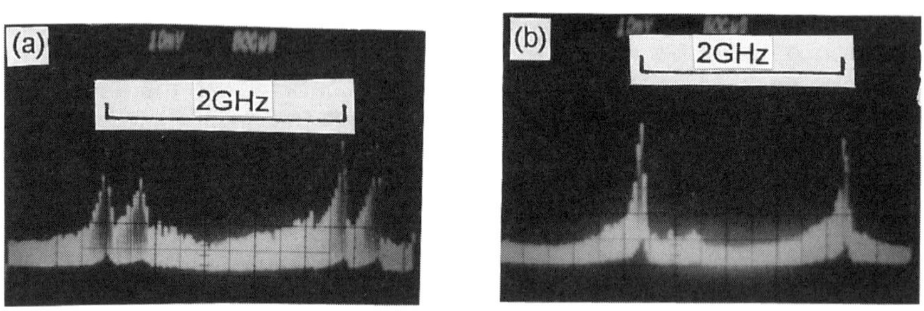

Figure 5. Spectral output of the amplifier by injection seeding the beam from the oscillator as shown in Fig.3 respectively.

Similarly, the spatial coherence of the laser can be also improved greatly by injection seeding [11]. Since the spatial coherence width of the oscillator in free running operation is about 1mm. Within this width the laser beam is transverse coherent. When we expand this part beam of the laser and lock a high-gain stable resonator CuBr laser, the spatial coherent radiation with high output power [11] from the amplifier can be obtained. Fig.6 shows the interference fringes of the injection-seeded beam

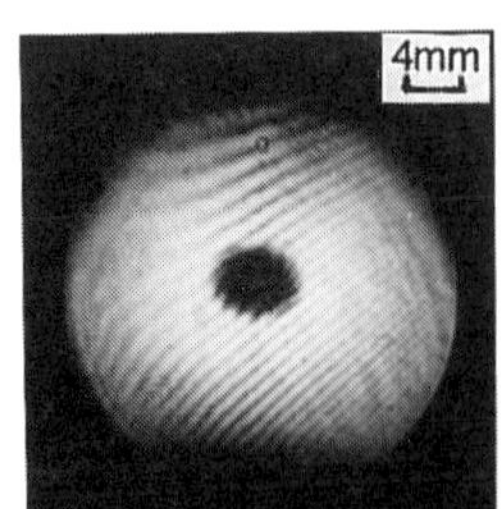

Figure 6. Spatial coherence of the injection-seeded beam

by the shear interferometer. It can be seen that the interference fringes are visible for the whole laser beam, indicating full spatial coherence (coherence diameter ~ 20mm). Such spatial coherent radiation particularly with narrow spectral linewidth can be expected to be much useful in high-speed holography.

3. Holography with a CuBr Laser

High-repetition-rate CVLs with high average power are expected to be a potential tool for high speed pulsed holographic recording. For CuBr lasers we have demonstrated above that the spectral linewidth of both an oscillator and an amplifier can be controlled from about 1GHz to less than 200MHz by using multimirror cavities and injection seeding technique. Their temporal coherence properties are good enough to produce a hologram. Here we report our preliminary experiments of directly using a small-scale CuBr laser oscillator with a total average power of about 1W as the light source in holography.

In order to obtain sufficiently high spatial coherence to record a hologram, we choice a parallel-plane resonator and an appropriate spatial filter outside the CuBr laser. After improving the beam quality and passing through an optical dispersion element, the power of the green laser to record holograms is about 300mW (the ratio of the green and yellow is about 2:1). An usual recording system for off-axis holography is used [1]. The object beam and the reference beam make an angle of about 15^0 and the path difference between them is about 10cm. The recording material is homemade (Tianjin III-type holographic plate)[15]. The exposure time is controlled by a mechanical shutter. The same CuBr laser is also used to reconstruct the images of the object.

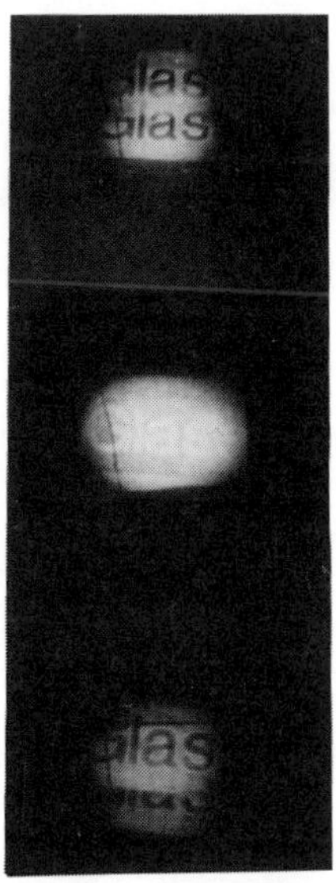

Figure 7. The reconstruction image of the off-axis hologram recorded by a CuBr laser

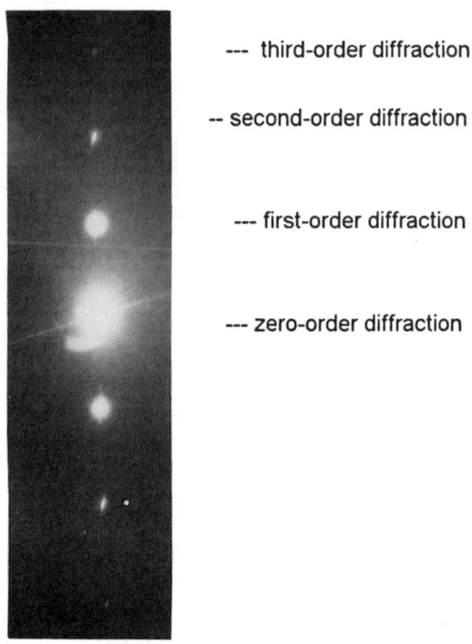

Figure 8. Fourier transform of the reconstruction image with multiple-order diffraction pattern

The holograms with high diffraction efficiency can be recorded by using this small-scale CuBr laser. Fig.7 shows the clearly reconstructed images of a transparent object with the word "Glass". The exposure time of this hologram is 2 milliseconds, so that the recording system is without the necessity for antivibration mounting. The diameter of the hologram is about 20mm. It has very high diffraction efficiency. For example, the efficiency of the first-order diffraction image is measured to be up to 7.5%. Fig.8 gives the multiple-order diffraction pattern of the above reconstructed image on the Fourier focus plane.

Since the holographic plate is fixed within the exposure time, the holograms recorded by tens of laser pulses overlap completely and exhibit a time-average hologram. Due to lack of a high speed motor to set the holographic plate into rotation, the holograms obtained in our experiment can not be coded. This will be our next step work in the future. We believe that high speed holography directly using a small-scale CuBr laser as well as injecton seeding technique is certainly superior to the system with a copper-vapor-pumped dye laser [1].

References:

[1] Lauterborn, W., Judt,A., and Schmitz,E. (1993) High-speed off-axis holographic cinematography with a copper-vapor-pumped dye laser, *Optics Letters* **18**, 4-6

[2] Tenenbaum,J., Smilanski,L., Gabay,S., Levin,L.A., Erez,G., and Lavi,S. (1980) Structure of 510.6 and 578.2 nm copper laser lines, *Optics Communications* **32**, 473-477

[3] Wang,Y., Shen,S., and Xie,T. (1988) Mutiple spectral structure of the 578.2nm copper laser line, *Applied Physics* **B 47**, 87-89

[4] Wen,C. and Tobin,R.C. (1990) Linewidth of Cu I (511nm) and Au I (628nm) laser transitions at room temperature, *Applied Physics Letters* **56**, 414-416

[5] Zemskov,K.I., Kazaryan,M.A., Mokerov,V.G., Petrash,G.G., and Petrova,A.G. (1978) Coherent properties of a copper vapor laser and dynamic holograms in vanadium dioxide films, *Soviet J. Quantum Electrom.* **8**, 245-247

[6] Harding,K.G., Gokay,M.C., and MacBain,J.C., (1983) High strobe-rate lasers for high-speed holographic testing, *Proceedings of Society of Photo-Optical Instrumentation and Engineering* **427**, 162-171

[7] Ruff,G.A., Bernal,L.P., and Faeth,G.M. (1990) High speed in-line holocinematography for dispersed-phase dynamics, *Applied Optics* **29**, 4544-4546

[8] Kreye,W.C. and Roesler, F.L. (1983) High-resolution line-shape analyses of the pulsed cuprous-chloride-laser oscillator and amplifier, *Applied Optics* **22**, 927-939

[9] Geng,J., Zhang,G., Song,X., Lin,F. (1992) Highly stable self-mode locking and longitudinal mode structure in CuBr laser, *Applied Physics Letters* **60**, 2969-2970

[10] Omatsu,T., Kuroda,K., Shimura,T., Chihara,K., Itoh,M., and Ogura,I. (1991) Measurement of spatial coherence of a copper vapor laser beam using a reversal shear interferometer, *Optical and Quantum Electronics* **23**, S477-S484

[11] Geng,J., Wu,Y., Zhang,G., Song,X., and Lin,F. (1994) Injection locking of a self-mode locking CuBr laser, *Applied Physics Letters* **64**, 692-694

[12] Geng,J., Zhang,G., Song,X., and Lin,F. (1993) Observations of the frequency limits of self-mode locking and the operation of single longitudinal mode in CuBr lasers, *Applied Pyhsics Letters* **62**, 3235-3237

[13] Geng,J., Zhang,G., Song,X., and Lin,F. (1994) Temporal coherence of a high-gain laser, *Acta Opt. Sinica* **14**, 480-485

[14] Geng,J., Wu,Y., Zhang,G., Song,X., and Lin,F. (1993) Mode quenching and locking in CuBr lasers, *Optics Communications* **98**, 105-110

[15] Wang,Q. and Da,X. (1993) Experimental research of Tianjin III type holographic plate, *Applied Lasers* **13**, 225-226 (in Chinese)

6. Frequency Extension

FREQUENCY EXTENSION OF METAL VAPOUR LASERS

J.A. PIPER
Centre for Lasers and Applications
Macquarie University NSW 2109
Australia

Abstract

Frequency extension of the fundamental wavelengths of cyclic pulsed metal vapour lasers to the tunable visible and infrared has been accomplished by pumping of liquid dye and crystalline solid state lasers, and to the ultraviolet by nonlinear frequency conversion in bulk materials. Progress for these techniques is reviewed and novel approaches to metal vapour laser frequency extension are discussed.

1. Introduction

Following the first report by Walter *et al* [1] of operation of cyclic pulsed metal vapour lasers (MVL) and the subsequent demonstration by Isaev *et al* [2] of efficient self-heated operation of a copper vapour laser (CVL) at high average powers, the science and technology of pulsed metal vapour lasers have advanced considerably. Small-scale (5-10W) sealed-off and medium-scale (10-100W) non-sealed CVLs are widely available commercially, while large CVL (MOPA) systems delivering kilowatts of power are installed in several laboratories worldwide. Although a number of applications employ the MVL output directly (eg for high speed photography and various medical procedures) many use the MVL output indirectly ie as a pump source in a frequency conversion scheme. Indeed the strongest motivation for CVL development historically has been Atomic Vapour Laser Isotope Separation (AVLIS) where copper laser pumped tunable dye lasers enable the crucial isotope-selective step in the process.

Dye laser pumping requires good pump beam quality for efficient beam transport and focusing characteristics, thus use of unstable resonators for MVLs dates from the earliest demonstrations of high average power operation. However it was not until the requirements of nonlinear frequency conversion of the CVL (and other applications requiring high beam quality) were confronted that comprehensive studies of beam quality issues were undertaken. As a result of these studies over recent years (see the article by D Brown in this volume) MVLs have been transformed from efficient high-power, but relatively poor beam-quality sources, to efficient, high-power, high-beam-quality (diffraction limited) sources.

C. E. Little and N. V. Sabotinov (eds.), Pulsed Metal Vapour Lasers, 277–287.

Efficient frequency-extension of MVLs by nonlinear processes has become a reality as a consequence of the high spatial power density now available from these sources. Thus, for example, instantaneous conversion efficiencies in second harmonic generation (SHG) at 255 nm from the 510 nm CVL line are now as high as 48% and average uv powers of several watts may be achieved at wall plug efficiencies around 0.1% (see section 4).

The high beam quality available from CVLs also makes them ideal sources for pumping solid-state laser materials (where issues of mode matching are critical). Significant progress has been made in the past five years in development of CVL-pumped tunable titanium sapphire (Ti:S) lasers and studies of CVL pumping of other solid-state laser materials are continuing.

This paper briefly reviews the techniques of frequency extension of high power metal vapour lasers (the CVL in particular) and reports on the state-of-the-art in each case. The techniques are covered in four following sections, the first relating to CVL pumped dye lasers and amplifiers, the second to CVL pumped solid-state lasers, and the third to nonlinear SHG and SFG (sum frequency generation) of CVLs. The final section relates to other nonlinear frequency conversion techniques which have been the subject of recent but incomplete study. These include Raman conversion in optical fibres and bulk solid state media and give considerable promise for future efficient extension of CVL and other MVL outputs to various regions of the spectrum.

2. CVL pumped dye lasers and amplifiers

Undoubtedly the most developed MVL frequency conversion technique is the CVL pumped dye laser/amplifier. Because this is a critical technology in the AVLIS programs of several countries the publicly accessible literature relating to CVL/dye lasers is comparatively sparse. However the reader is directed to an excellent recent review by Webb [3] focusing on smaller scale CVL dye laser systems for spectroscopic and medical applications.

The first report of a CVL pumped dye laser was by Decker *et al* in 1975 [4]: a low-power (250mW) CVL was used to obtain weak broadband laser emission from Rhodamine 6G using a simple colinear pumping scheme. Zemskov *et al* [5] reported 200 mW of broadband output for a 1.5W CVL pumping an R6G dye jet shortly thereafter. Broadband conversion efficiencies as high as 50% for CVL pumped R6G were reported by Morey [6], and high efficiencies for various other dyes by Masarnovskii *et al* [7] in 1979.

The milestone paper by Hargrove and Kan [8] subsequently established the basis for choice of either colinear or transverse pumping configurations for high power CVL dye oscillators or amplifiers. Colinear pumping was shown to be unsatisfactory if high single pass gains are to be maintained in the pumped dye medium owing to the effects of excited-state absorption. Thus transverse pumping configurations are used for high power CVL pumped amplifiers and, in most cases, in low-Q oscillator cavities.

Single-frequency narrowband spectral output is required from CVL-dye lasers used for isotope-selective optical pumping in the AVLIS scheme and most other spectroscopic applications. In 1977 Pease and Pearson [9] from Lawrence Livermore National Laboratories analysed the output of a narrowband CVL-dye laser oscillator employing a then-standard Hansch-type cavity (Littrow grating with intracavity beam-expanding telescope and etalon). Although 1.3 Ghz time-averaged bandwidth was achieved, the time resolved output contained several longitudinal modes with randomly fluctuating intensity. Stable single mode CVL-dye laser output with Fourier-limited bandwidth was later reported by Bernhardt and Rasmussen [10] for a compact cavity in which the beam-expanding telescope was replaced by a prismatic beam expander. Almost simultaneously Zherikin *et al* [11] reported similar single mode operation of a CVL-R6G dye laser oscillator incorporating a 6-prism expander and etalon.

In 1982 the present author with F. Duarte [12] reported the results of a comparison between Littrow and grazing-incidence (Littman) grating cavities incorporating prism expanders for CVL pumped dye laser oscillators. These and later experiments demonstrated superior performance of prism-expander grazing-incidence grating cavities in terms of narrow-bandwidth, single-mode operation, low ASE component and good conversion efficiencies without the necessity of intracavity etalons. The advantage of the prism-expander grazing-incidence cavity is that high spectral resolution is obtained at good grating efficiency by reducing losses at the grating using moderate expansion with low loss at the prisms.

Although little technical detail is available publicly, it is likely most high-power CVL pumped dye laser systems developed for AVLIS use compact prism-expander grazing-incidence grating cavities for the dye laser master oscillator. Certainly this is the case for the high-power CVL pumped MOPA dye laser system reported in detail recently by Singh *et al* [13]. The bandwidth obtained from the oscillator with a 4-prism expander was ~2Ghz at a conversion efficiency 19% for 10W pump power in R6G. Extraction efficiencies from R6G dye amplifiers transversely-pumped at 30W power levels were typically 30% in saturation.

Although the requirements of AVLIS have rather dominated the development of CVL pumped dye lasers, there has been significant progress also in development of simpler, lower-power devices for commercial applications. In some cases, for example in large-scale projection display or in medical applications such as vascular photocoagulation or photodynamic therapy, efficient frequency conversion from the CVL pump beam to a fixed wavelength within a given spectral band is all that is necessary. This may be achieved using simple tunable cavities, however injection-seeded dye oscillators and amplifiers offer an approach which is simple, efficient and overcomes problems of wavelength drift.

Perhaps the simplest example is that of "green-to-yellow" conversion in which the yellow CVL output is amplified directly in an R6G dye cell pumped by the green CVL output. Kravchenko *et al* [14] first reported this technique in 1979, achieving an optical conversion efficiency of a little over 25%. Subsequent optimisation of the dye-amplifier pumping and signal injection parameters by Coutts *et al* [15] resulted in green-to-yellow conversion efficiencies exceeding 50%. Thus some 70% of the maximum total output power available from the CVL can be obtained in the yellow.

A second fixed-wavelength device investigated by the Macquarie group is a CVL-pumped dye oscillator injection-seeded by low-power cw lasers (eg He-Ne lasers) [16]. The optical arrangement comprises three mirrors, two of which form the cavity of the injection source (they may indeed be the integral mirrors of a He-Ne laser) and a third mirror which, with the shared central mirror, makes up the cavity for the dye oscillator. Appropriate choices must be made of the spectral reflection characteristics of the third (dye laser coupling) mirror so that the free-running wavelength of the dye laser is close to the wavelength of the injection source, and of the spacings between the mirrors so that the longitudinal mode spacings for the injection source and the dye laser are approximately the same. In seeded operation at 632.8 nm using a 1-5mW He-Ne laser injection source, around 75% of the R640 dye laser power occurs on a few modes at the seeded wavelength for CVL pump power 2W. Conversion efficiency into the fixed wavelength component is about 12%. Injection-seeded CVL pumped R640 dye lasers have also been operated at 605 and 612nm based on a prism tunable He-Ne laser injection source.

3. CVL pumped solid state lasers

The availability of tunable crystalline solid-state laser materials of high optical quality over the past decade has resulted in displacement of dye lasers by solid-state lasers for broad spectral regions. The titanium sapphire (Ti:S) laser operates with good efficiency over the region from 700nm to a micron and its spectral absorption characteristics are well matched to pumping at the 510 nm CVL output wavelength. However the absorption coefficient is low (~1 cm^{-1} at 510nm) compared with dyes and the gain is also low. This presents problems in matching the pumped volume in the Ti:S crystal to the mode volume of the Ti:S resonator for low beam-quality pump sources and also restricts the use of high-loss cavity tuning elements (eg gratings).

Operation of a CVL-pumped Ti:S laser was first reported in 1986 by Kruglik *et al* [17] in the USSR; subsequently Bartoshevich *et al* [18] reported details of a Ti:S laser pumped by a 25W CVL MOPA system. Adequate pump beam quality was obtained by spatially filtering the output of a CVL oscillator fitted with a plane-plane resonator before injection into the CVL amplifier. Approximately 1.5W of broadband output was obtained from the Ti:S laser for 15W absorbed in the crystal. Later refinements included use of a Littrow grating tuner with intracavity prism expander and etalon to achieve narrowband tunable operation.

Concurrently studies of CVL-pumped Ti:S lasers were undertaken in the US, UK, Japan and Australia. Kangas *et al* [19] reported some 300mW of broadband output from a Ti:S laser pumped colinearly through a cavity mirror by the output of a large aperture (40mm) CVL fitted with a standard plane-plane resonator. However the poor beam quality of the CVL proved a serious limitation to achieving good mode-matching without attendant damage to the optics. These problems were overcome by Takehisa and Miki [20] who used a 45 degree centre-hole reflector to couple the strongly-focussed beam from a stable-resonator CVL into the Ti:S crystal. In this way Ti:S laser powers over 3W (broadband) were obtained at 20% slope efficiency for a simple near-concentric resonator structure. This latter method has its own limitations however, and the most effective way of achieving good mode-matching in the Ti:S laser is to use a high beam quality CVL, either an oscillator fitted with a high magnification unstable resonator or a MOPA system.

Knowles and Webb [21] have reported extensive studies of Ti:S lasers colinearly pumped by high beam quality CVL oscillators and MOPA systems. Concentric Z and ring cavities with the Ti:S Crystal placed at the intracavity beam waist are shown to give stable, efficient high power operation. Over 5.5W of broadband Ti:S laser output was obtained for 22W CVL pump power (which was split into two beams incident on different ends of the laser crystal). Over 1W of narrowband (<0.3nm) tunable output was obtained using a 3-plate birefringent tuner inserted in the Z-configuration cavity. Narrowband operation was also demonstrated using injection seeding from a diode laser.

Experiments undertaken by the present author with J. Dawes in 1990 used the dispersion of Ti:S to couple a low divergence CVL beam directly into a Brewster cut Ti:S crystal. Slope conversion efficiencies of 30% were observed for incident powers up to 4 Watt in broadband operation using a hemispherical linear Ti:S laser cavity. This form of coupling has subsequently been used by Knowles and Brown [22] to achieve highly efficient operation of a Ti:S laser pumped by a mini-scale CVL at high repetition rate. The "mini-CVL" fitted with a self-filtering unstable resonator generates 2W of diffraction limited output at a pulse repetition rate of 24 kHz. Threshold pump energies of only 20 μJ per pulse were obtained; maximum average power of the Ti:S laser in broadband operation was 320mW at a slope efficiency of 33%.

One significant disadvantage of Ti:S as a laser material is that it has little storage capability due to its rather short (3μs) upper laser level lifetime. Thus the Ti:S laser operates only in gain-switched mode. A number of other solid-state laser materials which might be pumped effectively by the CVL offer considerably better energy storage (> 100μs). This energy storage may be used to enhance the solid-state laser performance when pumped by a high repetition rate pulsed source such as the CVL. In a technique called 'pulse stacking', the energy of several consecutive pump pulses is accumulated in the upper laser level of the solid-state laser medium using an intracavity Q-switch to suppress lasing. When the Q-switch is opened the solid-state laser output has increased peak power and improved extraction efficiency over single pulse gain-switched operation because the laser is operating well above threshold.

Pulse stacking has been demonstrated by French *et al* [23] for a high prf mini-CVL pumping the solid-state laser material Cr, Nd:GSGG. The Cr^{3+} dopant ion absorbs strongly in the green and energy is transferred rapidly to the Nd^{3+} lasing ion, for which the upper laser level lifetime is 260 μs. For a pump laser pulse repetition rate of 24 kHz, this allows some 5-6 pump pulses to be stacked within the upper level storage time, with enhanced laser performance observed in those cases. Current experiments are directed towards demonstrating this technique in a variety of other materials with potential to achieve further wavelength extension of the CVL pump.

4. Nonlinear SHG and SFG of CVLs

Perhaps the most dramatic developments in frequency extension of MVLs in the past five years have been those based on nonlinear optical frequency conversion techniques. Although SHG and SFG are quite standard techniques for high power pulsed solid state lasers, MVLs present special difficulties. In particular, the comparatively low peak power and large beam diameters of MVLs require that the beam be focused into the nonlinear medium to achieve sufficient power densities for efficient frequency conversion. Thus developments in nonlinear conversion of MVLs have proceeded hand-in-hand with the improvements in MVL beam quality which have occurred over the same period, and also the availability of nonlinear materials (eg. BBO) of high optical quality. An excellent and comprehensive review by Coutts and Brown [24] of production of high average power uv by second harmonic and sum frequency generation from CVLs has been published recently, thus only a brief historical perspective will be given here.

The prospect of creating an efficient high average power uv source based on the CVL for applications such as uv photolithography stimulated quite early attempts at SHG and SFG of these lasers. The very first report of frequency doubling of CVL output in ADP by Isaev *et al* [25] in 1980 was followed in 1987 by a report by Polunin and Troitskii [26] of significant power (600mW) and optical conversion efficiency (5%) for SFG of the green and yellow at 271nm in KDP using a high beam quality CVL MOPA system. Zhang *et al* [27] had first demonstrated SHG of the 510nm CVL line in BBO in 1984 but it was not until experiments by Kuroda *et al* [28] that the potential for high conversion efficiency in SFG or SHG of the CVL became clear. Using a small-scale (2W) CVL fitted with a positive branch unstable resonator, Kuroda initially obtained 35mW at 255 nm by SHG of the 510nm line corresponding to 4% optical conversion efficiency. Shortly later 230mW at 255nm corresponding to over 8% optical conversion efficiency was obtained [29]. In both cases simple spherical focusing of the CVL beam into the nonlinear medium was employed.

Following initial studies [30] comparing KDP and BBO for rather low beam quality CVL oscillators the present author and colleagues in 1990 reported SHG of both the green and yellow at 255nm and 289nm and SFG at 271nm in BBO with good optical efficiencies (9.6%, 6.4% and 5.5% respectively) for a 20W CVL oscillator fitted with high magnification unstable resonators [31]. These latter experiments, which used spherical focusing into the nonlinear medium, demonstrated the need for alternative focal geometries at high incident powers (the narrow acceptance angle of BBO requires low numerical aperture leading to high focal power densities and consequent damage at the crystal surface in spherical focusing). We subsequently developed a line-focus geometry for which the incident CVL beam is collimated in the tuning plane but tightly focused in the other plane (which has higher angular tolerance). This geometry has the added advantages of reduced beam walk-off losses and reduced power densities at the crystal surface. Employing this focal geometry for BBO we obtained 1.3W at 255nm, 0.9W at 289nm and 1.0W at 271nm at optical conversion efficiencies 24%, 28% and 12% based on the same 20W CVL with M=51 unstable resonator [32]. Detailed optimisation of the line focussing geometry by Coutts [33] has resulted in further enhancements to uv powers and efficiencies: 1.75W (255nm), 1.2W (289nm) and 1.5W (271nm) with maximum instantaneous optical conversion efficiencies up to 48%. Average optical conversion efficiencies were up to 34% for SHG and wall plug efficiencies to the uv were 0.06%.

The latter results represent record conversion efficiencies for SHG and SFG of CVL oscillators. However the longer pulse duration and higher wall plug efficiencies of CuHyBrID lasers may yet see significant improvements in average power and conversion efficiencies as well as wall plug efficiencies for uv generation based on CuHyBrID oscillators. For small scale CVLs giving diffraction-limited output using self filtering unstable resonators, peak instantaneous optical conversion efficiencies of 30% and average optical conversion efficiencies of 20% have been reported by Pini *et al* [34].

Power scaling of uv SHG/SFG from CVLs beyond the 2-3W level requires use of CVL MOPA or injection-controlled oscillator (ICO) systems since aperture scaling of CVL oscillators becomes problematic due to increased difficulties with beam quality control and wavefront integrity. Naylor *et al* [35] reported 600mW at 255nm by SHG in BBO from a 100W CVL MOPA system in 1989 and subsequently 2W at improved optical conversion efficiencies (11%), limited by crystal damage. Using the line-focus geometry discussed above with two 20W CVLs operating in MOPA configuration, the present author and co-workers obtained 3.6W at 255nm, 2.0W at 289nm and 3.8W at 271nm corresponding to average optical conversion efficiencies in the BBO of 20%, 22% and 14% respectively and wall-plug efficiencies ~0.06% [36]. Substitution of a mini-scale CVL as the master oscillator is expected to improve wall-plug efficiencies to the uv to 0.1%.

Power scaling to the 10W level and beyond will require solution of a number of wavefront integrity and thermal problems which arise from volume heating of the

BBO due to small residual absorption in the ultraviolet. Indeed a record 9W at 255nm has recently been reported by Molander [37] for a very-large scale CVL MOPA system using a line focussing geometry and two BBO crystals in alternating-Z configuration, however average optical conversion efficiencies were only 8%, limited by thermal and various optical problems, and wall plug efficiencies were very low. Continued improvement in the optical quality of BBO may also be an important factor in further progress in high-power uv generation from CVLs.

5. Other nonlinear frequency extension techniques for MVLs

The evolution of the CVL from a low beam-quality to a high beam-quality device over the past several years has stimulated investigation of a number of nonlinear frequency extension techniques. Guasti *et al* [38] and the present author with H. Pask [39] simultaneously reported Stimulated Raman Scattering (SRS) of CVL and barium vapour laser (BVL) radiation in silica optical fibres. For a 2.5W CVL pumping a 200-micron core multimode fibre of 330m length Gausti observed 73% optical conversion efficiency from the 510nm line into 5 or more Stokes lines ranging up to 580nm, and 43% conversion efficiencies from the 578nm line into 4 Stokes lines up to 650nm. Conversion efficiencies into the first Stokes lines were 58% and 38% for the green and yellow respectively. Using 2.5-micron core single mode fibre 12.6km in length, Pask observed strong SRS emission on 3 Stokes lines at 1.61, 1.74 and 1.90μm for pumping by the strong 1.50μm line of the BVL. Average output powers were low due to difficulties of coupling significant power into the small core, but conversion efficiencies into the first Stokes line reached 35%.

Studies of SRS of CVL output in fibres are ongoing, however the prospects for high average power output with significant frequency shifts are limited by coupling and materials problems. Perhaps a more rewarding approach may be by way of stimulated Raman scattering in bulk solid-state materials. Indeed a highly promising result has been reported recently by Vitsinskii *et al* [40] for SRS of CVL output in bulk crystalline barium nitrate. An optical conversion efficiency of 32% corresponding to output power 600mW was achieved on the first Stokes line at 539.4nm for SRS of the green, and 18% (300mW) on the first Stokes line at 615.4nm for SRS of the yellow CVL line. These results demonstrate both significant conversion efficiencies and wavelength shift, and perhaps point the way to a new range of efficient sources based on frequency extension of metal vapour lasers.

Acknowledgments

The author wishes to gratefully acknowledge the contributions of Dr M Ainsworth, Dr D Brown, Dr R Carman, Dr D Coutts, Ms S French, Ms A Glover, Dr D Knowles, Dr H Pask and Dr M Withford to work in frequency extension of metal vapour lasers within the Centre for Lasers and Applications, Macquarie University. Financial support from the Australian Research Council is also gratefully acknowledged.

References

1. Walter, W.T., Piltch, M., Solimene, N. and Gould, G. (1966) Efficient pulsed gas discharge lasers, *IEEE J. Quantum Electron.* **QE-2**, 474-479.
2. Isaev, A.A., Kazaryan, M.A. and Petrash, G.G. (1972) Effective pulsed copper-vapor laser with high average generation power, *JETP Lett.* **16**, 40-42.
3. Webb, C.E. (1991) High-power dye lasers pumped by copper vapor lasers, in F.J. Duarte (ed.), *High-power dye lasers*, Springer-Verlag, Berlin, 143-182.
4. Decker, C.D., Fahlen, T.S. and Falk, J. (1975) Amplification and laser action on a laser dye pumped by a copper vapor laser *J. Appl. Phys.* **46**, 2308-2309.
5. Zemskov, K.I., Isaev, A.A., Kazaryan, M.A., Petrash, G.G., Adamushko, A.V., Belokon, M.V., Rubinov, A.N. and Evtukhovich, P.G. (1976) Jet dye laser pumped by copper vapor laser, *Sov. J. Quantum Electron.* **6**, 727-728.
6. Morey, W.W. (1980) Copper vapor laser pumped dye laser, in V.J. Corcoran (ed.) *Proc. Int. Conf. Lasers '79*, STS Press, McLean, VA, 365-373.
7. Masarnovskii, L.V., Soldatov, A.N. and Sukhanov, V.B. (1979) Excitation of dye solutions and their mixtures by copper vapor laser radiation, *Sov. J. Quantum Electron.* **9**, 900-902.
8. Hargrove, R.S. and Kan, T. (1980) High power efficient dye amplifier pumped by copper vapor lasers, *IEEE J. Quantum Electron.* **QE-16**, 1108-1113.
9. Pease, A.A. and Pearson, W.N. (1977) Axial mode structure of a copper vapor pumped dye laser, *Appl. Opt.* **16**, 57-60.
10. Bernhardt, A.F. and Rasmussen, P. (1981) Design criteria and operating characteristics of a single mode pulsed dye laser, *Appl. Phys.* **B26**, 141-146.
11. Zherikin, A.N., Letokhov, U.S., Mishin, V.I., Belyaev, V.P., Evtyunin, A.N. and Lesnoi, M.A. (1981) High repetition rate tunable dye lasers pumped by copper vapor lasers, *Sov. J. Quantum Electron.* **11**, 806-808.
12. Duarte, F.J. and Piper, J.A. (1982) Comparison of prism-expander and grazing-incidence grating cavities for copper laser pumped dye lasers, *Appl. Opt.* **21**, 2782-2786.
13. Singh, S., Dasgupta, K., Kumar, S., Manohar, K.G., Nair, L.G. and Chatterjee, U.K. (1994) High-power high-repetition-rate copper vapor pumped dye laser, *Opt. Eng.* **33**, 1894-1904.
14. Kravchenko, V.I., Livinenko A.Y. and Smirnov, A.A. (1979) Dye solution conversion of the output from a copper vapor laser to 5782 A, *Sov. Tech. Phys. Lett.* **5**, 277-278.
15. Coutts, D.W., Ainsworth, M.D. and Piper, J.A. (1990) Efficient green/yellow conversion of copper laser output, *Opt. Commun.* **75**, 301-306.

16. Ainsworth, M.D., Glover, A.C.J. and Piper, J.A. (1995) Fixed-wavelength operation of a copper laser pumped dye laser injection-seeded by low-power He-Ne lasers, *Appl. Opt.* **34**, 6108-6113.
17. Kruglik, G.S., Skripko, G.A., Shkadarevich, A.P., Kondratyak, Y., Urbanovich, V.S. and Nasavenko, P.N. (1986) Generation of Ti^{3+}:Al_20_3 crystals in the continuous and quasi-continuous regime, *Zh. Prikl. Spektrosk.* **45**, 567-573.
18. Bartoshevich, S.G., Zuev, V.V., Mirza, S.Y., Nasarenko, P.N., Polunin, Y.P., Skripko, G.A. and Sukhanov, V.B. (1989) Wide-band conversion of copper laser radiation in an Al_20_3:Ti crystal, *Sov. J. Quantum Electron.* **19**, 138-141.
19. Kangas, K.W., Muller, C.H., Lowenthal, D.D., Picarello, S. and Liebeskind, J. (1989) Six-kilowatt copper vapour laser pumped Ti:sapphire oscillator, *Conf. on Lasers and Electro-Optics, Baltimore, USA,* paper WF3.
20. Takehisa, K. and Miki, A. (1992) Method for pumping a Ti:sapphire laser with a stable resonator copper vapor laser, *Appl. Opt.* **31**, 2734-2737.
21. Knowles, M.R.H. and Webb, C.E. (1992) Cavity configurations for copper vapour laser pumped titanium sapphire lasers, *Opt. Commun.* **89**, 493-506.
22. Knowles, D.S. and Brown D.J.W. (1995) Compact 24-kHz copper-laser-pumped Ti:sapphire laser, *Opt. Lett.* **20**, 569-571.
23. French, S.E., Knowles, D.S., Brown, D.J.W. and Dawes, J.M. (1995) Short-pulse high-repetition-rate operation of a copper vapour laser pumped Cr, Nd:GSGG laser, *Conference on Advanced Solid State Lasers, Memphis USA,* paper MF11.
24. Coutts, D.W. and Brown, D.J.W. (1995) Production of high average power uv by second harmonic and sum frequency generation from copper vapour lasers, *IEEE J. Sel. Topics Quantum Electron.* **3**, in press.
25. Isaev, A.A., Lemmerman, G.Y. and Malfeeva, G.L. (1980) Second harmonic generation from pulsed copper vapor radiation, *Sov. J. Quantum Electron.* **10**, 983-986.
26. Polunin, Y.P. and Troitskii, V.O. (1987) Nonlinear frequency conversion of copper vapor laser radiation, *Sov. J. Quantum Electron.* **17**, 1433-1434.
27. Zhang, G., Lin, F., Chen, C. and Wu, B., (1984) Second harmonic generation of Cu vapor laser by using a beta-BaB_20_4 crystal, *Acta Optica Sinica* **4**, 513-516.
28. Kuroda, K., Shimura, T. Omatsu, T., Chihara, M. and Ogura, I. (1988) Second harmonic generation of a copper vapor laser in β-BaB_20_4, *Conference on Lasers and Electro-Optics, Anaheim USA,* paper WV4.
29. Kuroda, K., Omatsu, T., Shimura, T., Chihara, M. and Ogura, I. (1990) Second harmonic generation of a copper vapor laser in barium borate, *Opt. Commun.* **75**, 42-46.
30. Coutts, D.W., Ainsworth, M.D. and Piper, J.A. (1989) Sum frequency mixing of copper vapour laser output in KDP and β–BBO, *IEEE J. Quantum Electron.* **QE-25**, 1985-1987.
31. Coutts, D.W., Ainsworth, M.D. and Piper, J.A. (1990) Enhanced efficiency of uv second harmonic and sum frequency generation from copper vapor lasers, *IEEE J. Quantum Electron.* **QE-26**, 1555-1558.
32. Coutts, D.W. and Piper, J.A. (1992) One watt average power by second harmonic and sum frequency generation from a single medium-scale copper vapor laser, *IEEE J. Quantum Electron.* **QE-28**, 1761-1764.

33. Coutts, D.W. (1995) Optimisation of line focusing geometry for efficient nonlinear frequency conversion from copper vapor lasers, *IEEE J. Quantum Electron.* **QE-31**, in press.
34. Pini, R., Salimbeni, R., Toci, G. and Vannini, M. (1992) Efficient second harmonic generation in β–barium borate by a diffraction-limited copper vapor laser, *Appl. Opt.* **15**, 2747-2751.
35. Naylor, G.A., Lewis, R.R. and Kearsley, A.J. (1989) Frequency doubling of a 100W copper laser for high power uv generation, *Conference on Lasers and Electro-Optics, Baltimore USA,* paper TuJ2.
36. Coutts, D.W., Brown, D.J.W. and Piper, J.A. (1993) Nonlinear frequency conversion of CVL fundamental and second-harmonic outputs, *Conference on Lasers and Electro-Optics, Baltimore USA,* paper CThNT.
37. Molander, W.A. (1994) High average power second-harmonic generation using a copper laser oscillator/amplifier source, *Conference on Lasers and Electro-Optics, Anaheim USA*, paper CThM2.
38. Guasti, A., Pini, R. and Salimbeni, R. (1991) Stimulated Raman scattering of copper vapour laser radiation in optical fibres, *Opt. and Quantum Electron.* **23**, S555-S562.
39. Pask, H.M. and Piper J.A. (1991) Stimulated Raman scattering of barium laser output in a silica fibre, *Opt. and Quantum Electron.* **23**, S563-S568.
40. Vitsinskii, S.A., Isakov, V.K. Karpukhin, S.N. and Lovchii, I.L. (1993) Stimulated Raman scattering of the beam from a copper vapour laser in a barium nitrate crystal, *Quantum Electron.* **23**, 1001-1004.

TRANSFORMATION OF COPPER AND COPPER BROMIDE LASER RADIATION IN NON-LINEAR PROCESSES

A. A. ISAEV
P. N. Lebedev Physical Institute, Russian Academy of Sciences,
53 Leninsky Prospect, 117924, Moskow, Russia.

1. Introduction

Copper vapor lasers (CVLs) are efficient visible pulsed gas discharge lasers. CVLs radiate both green (510.6 nm) and yellow (578.2 nm) light with pulse repetition rates of between 5 kHz and 20 kHz. Average laser output power is typically several tens of watts and may reach more than 100 W. CVLs are widely used in different fields of scientific and industrial applications.

Expansion of their applications is possible with the transformation of CVL radiation into another spectral region, for example into the ultraviolet (UV) and infrared (IR) spectral regions by means of optical non-linear processes. Many fundamental problems of the optical non-linear process have been successfully resolved by using solid-state lasers, which enable high power densities to be reached. The essential features of CVL radiation are low pulse energy (1-10 mJ) and high pulse repetition rates (PRFs). These properties are however a disadvantage in reaching high efficiency in non-linear conversion but conversely are an advantage in the field of high precision machining of materials[1]. The disadvantages are associated with the short intervals between successive laser pulses which are insufficient for residual phenomena removal, such as crystal heating [2].

Well-known non-linear processes are: Second Harmonic Generation (SHG); Stimulated Raman Scattering (SRS) and Four Wave Mixing Generation (FWMG). These processes are discussed here from the point of view of practical CVL applications. The paper consists of three parts: CVL output beam quality; the CVL SHG as a source for submicron photolithography and the transformation of CVL radiation frequency in short and long fibers by SRS and FWMG.

2. CVL Output Beam Quality

Non-linear optical processes are proportional to the square of the fundamental beam power density. Therefore, for low pulse power CVL output beams, beam quality is the key to high overall efficiency of the processes. For CVLs with stable resonators (SR),

C. E. Little and N. V. Sabotinov (eds.), Pulsed Metal Vapour Lasers, 289–301.

output beam divergence is large due to the very short laser active-medium gain time. For a typical 15-20 ns pulsewidth and resonator length of 100 cm the number of cavity round-trips is 2-3, this is insufficient for the formation of SR modes because of the low diffraction losses. The unstable resonator[3] (UR) has large optical-geometrical losses and the modes are formed after a few cavity round-trips.

The first CVL with diffraction-limited output beam (using an unstable resonator) was demonstrated in 1974-75 [4-6]. Since then many investigators have used various UR designs to efficiently extract low divergence beams from high-gain lasers, for example: the self-filtering unstable resonator (SFUR) and diffraction filtered resonator (DFR). [7-10]. Various theories have been put forward to elucidate how an output beam of low divergence is produced in a CVL when fitted with an unstable resonator. Here we consider a very simple idea using a formalism which involves a main UR mode derived from self-reproducing wavefronts.

From this point of view, a parallel beam propagating initially along the axis inside the resonator is first telescoped down by successive passes through the resonator. Thus, the beam diameter after reflection from both mirrors (one cavity round-trip) is decreased M times, where M is the magnification of the resonator. Such a decrease of the beam diameter is limited by diffraction. After compression to the diffraction limit, the beam is transformed into a magnifying beam and after further passes through the resonator it exits the resonator as a parallel beam. Consequently, the parallel beam with a diffraction limited divergence is the main mode of the UR.

The compression of the parallel beam by successive passes through the resonator is very important for the transient regime of the output beam. Such a demagnifying beam has no optical-geometrical losses that are inherent in a steady-state mode of the UR. Therefore, the power density (W/cm^2) of the converging beam is increased because of both the decrease in the beam diameter and the active medium gain. Since saturation of the active medium is defined as a power density value - the converging beam is saturated first - then all other beams (modes), initially propagating at different angles to the resonator axis and having higher losses, are saturated. Thus, we see that the output beam with a low diffraction-limited divergence is formed from seed spontaneous emission radiated along the resonator axis at the beginning of an excitation pulse.

The number of cavity round-trips n_0 required for the formation of a diffraction-limited beam from seed spontaneous emission within an UR is given by the following equations [4]

$$M^{n_0-1} = \frac{D^2}{2\lambda f_1}; \; n_0 = 1 + \frac{\ln M_0}{\ln M}; \; M_0 = \frac{D^2}{2\lambda f_1} \tag{1}$$

where D is the output beam diameter, f_1 is the focal length of the mirror M_1, λ is the wavelength and M_0 is a Fresnel number.

It is evident that the establishment of the output beam parameters in a two mirror resonator system are only possible when the seed beam passes through the active medium four or more times. On the other hand, eq.(1) shows that an UR with $M=M_0$ will form a diffraction limited output beam after the minimum number of cavity round-trips, i.e. $n_0=2$ (four transits through the active medium). This is of a great importance for CVLs, which have short laser pulse lengths and are high gain systems. In this case, the time of the main UR mode formation coinsides with that of the output beam for plane-plane resonator. For such an UR we have

$$M = M_0 = \frac{D^2}{2\lambda f_1}; \; f_2 = \frac{f_1}{M_0} = \frac{2\lambda f_1^2}{D^2} \tag{2}$$

where f_2 is the mirror M_2 focal length.
In this case, the second mirror focal length is equal to the diffraction length of the beam "waist" and the second mirror surface coinsides with the beam wavefront. Thus, the beam reflected by mirror M_2 repeats the incident beam.

It is clear, that the diffraction-limited output beam power is a maximum if the seed beam power reaches the saturation level at the end of a main UR mode formation. Therefore, a criterion for the diffraction limited divergence of the UR output beam is

$$\frac{2n_0 L}{c} \geq \tau_d \tag{3}$$

where L is the resonator length, $\tau_d = A/c\alpha_{eff}$ is the seed beam saturation time, $A \cong 15$ is a saturation parameter, c is the light velocity. The effective gain α_{eff} is an average gain of the cavity round-trip for diverging beam

$$\alpha_{eff} = (\alpha L_a - \frac{\ln M}{2}) / L \tag{4}$$

where L_a is the active medium length. Thus, the active medium gain α must not exceed the critical value α_{cr}

$$\alpha \leq \alpha_{cr} = \frac{\ln M}{2L_a}(\frac{A}{\ln(MM_0)} + 1) \tag{5}$$

If $\alpha > \alpha_{cr}$, the output beam divergence will exceed the diffraction limit. With an UR where $M=M_0$ it is possible to use an active medium with the maximum gain value, this is equal to

$$\max\langle\alpha_{cr}\rangle = \left(\frac{A}{2} + \ln M_0\right) / 2L_a \tag{6}$$

If the gain is higher than that of (6), the diffraction-limited divergence of the output beam can be reached by reducing the active medium length L_a. For CVL $L_a \approx 100$ cm, M=200 and

$$\max\langle\alpha_{cr}\rangle \approx 0.1 cm^{-1} \tag{7}$$

This is much less then the experimental value, which is 1 cm^{-1} [6]. Therefore, for CVLs it is very difficult to have a pure diffraction-limited output beam. Thus, URs with $M=M_0$ are characterised by minimum time of the diffraction-limited output beam formation and maximum active medium gain.

This paper is concerned primarily with results obtained using a CVL with an unstable resonator. The CVL used was capable of 10 W average output power at a

PRF of 10-20 kHz when operated with a plane-plane cavity. In all self-heating CVLs the average electrical input power governs the copper vapor density which is depended on the tube wall temperature. The active medium gain and output power depend on the excitation pulse power and the copper vapor density. These increase with increasing input power. There is however an optimum tube wall temperature, at which maximum output power is obtained.

The UR consisted of a concave mirror M_1 of focal length 150 cm and a set of convex mirrors of focal lengths 0.5-25 cm. The convex mirror diameters were 2-5 mm, the resonator lengths 125-150 cm. The output beam was focused at a distance of 14-40 m from the output mirror by altering the position of the mirror M_2 near the confocal position. The far-field intensity profile was recorded in the focal plane by a photocamera. The exposure time was of 1/500 s. The distribution exhibited an Airi pattern profile. The calculated output beam divergence from this distribution is (3-4)$_*10^{-5}$, this coincides with the plane-wave diffraction of 2 cm diameter.

Typical experimental results of the output beam divergence behaviour are summarised as follows. Diffraction distribution in the focal plane for an UR with M=6 was observed at an output power of less than 0.3 W. The output power with a SR was not in excess of 0.5 W. At higher output power the focal spots had the intermediate sizes in between the diffraction-limited spot and plane-plane resonator spot. At output powers of 2-3 W with the SR the diffraction-limited output beam was observed only with an UR with a magnification of 20-30. When the output power using a SR had the maximum value of 10-12 W, an UR with M=100 was required to achieve a diffraction-limited output beam. In all cases, the diffraction-limited output beam was observed with some high divergence background output power present. The output power of the background beam at fixed M increases with electrical input power. With increasing M the background beam power first decreases, then increases again. For each fixed input excitation power there is an optimum value M_{opt} for which the output beam divergence is almost diffraction limited and the output power is still high. For example, if the SR output power is 5 W, M_{opt}=60-100, then the UR output power is 3 W. The maximum output power in this experiment with the SR was 10 W. In this case M_{opt} was 250 and the output power was 5 W. It is worth noting that any defects present on the UR mirror surfaces will be reproduced by the wavefronts of the output beam and beam quality will be reduced.

Knowledge of the temporal behaviour of the UR output pulses is important in order to understand the operational characteristics of these systems. The temporal behaviour was recorded with a fast vacuum photodiode and an oscilloscope. The overall response time of the detection system was 0.5 ns. The pulse shapes of the output laser emission for a few cases are shown in Figure 1. The upper trace shows the Amplified Spontaneous Emission (ASE) pulse shape with a single 95% reflectivity rear mirror. It shows the seed beam power increasing after one or two transits through the active medium. These seed beams have 100% emission losses after one active medium transit. The next trace is a typical output pulse shape obtained with a SR composed of a plane 95% reflectivity rear mirror and an uncoated flat. It shows the seed beam power increasing after three or four transits through the active medium.

The seed beam emission losses are 92%. Comparison of the pulse shapes shows that the seed beams that pass through the active medium three or four times - exit the resonator earlier and have higher power than beams that only pass through the gain medium for one or two transitions. The time for four transits in our case was 16 ns. In other words the seed beams that formed the SR output pulse occured in the active medium 16 ns earlier. This position in time is taken as the zero point, i.e. this is the sweep start point of the oscilloscope which is triggered by the current pulse. Fig. 1d shows the output pulse of the SR composed of two 95% reflectivity mirrors. Since the output pulse is formed by seed beams which have no radiation losses, then the increasing intensity of the output pulse can coincide with that of the main mode of the UR which is also formed from seed beams which have no losses.

The UR output pulses with M=30 (Fig.1e) and M=6 (Fig.1g) occur 22 ns and 33 ns later than the sweep oscilloscope start respectively. The exit times coincide with the UR main modes formation times calculated by eq.(1). Fig.1a and Fig.1f (UR with M=6) shows that the main mode exit occurs at the same time as those seed beams that exhibit one or two transit through active the medium and also reach high power. Therefore, for high power excitation (P_s=2W) the diffraction-limited beam is lost against the background beams with higher divergence. For UR with M=6 the diffraction-limited beam is observed only at a very low excitation power levels when

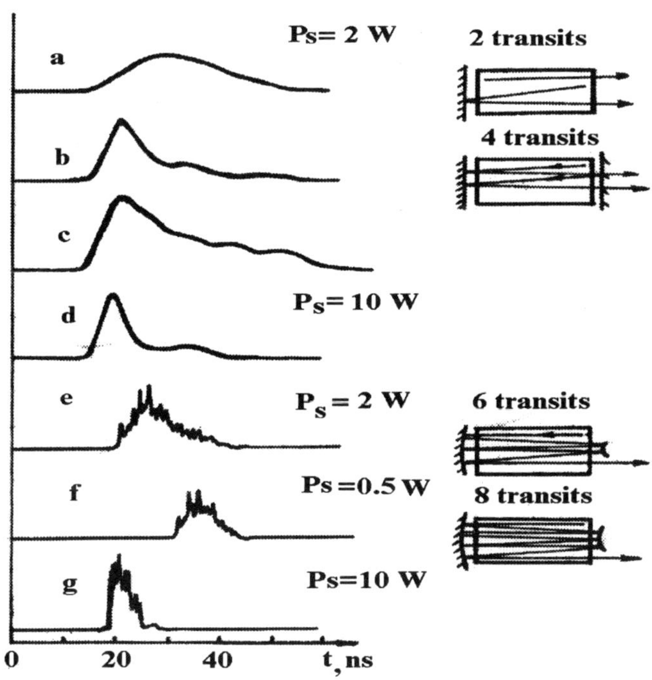

Figure 1. Output pulse profile for different resonators. P_s is the SR output power.

P_s=0.5 W and the seed beams after one or two transits through the medium are very weak. For maximum excitation power the gain time is shorter (see Fig.1d), therefore all URs with small M form an output beam in the time interval after the high gain has ceased, and we have to use URs with M=100 or greater. Figure 1g shows the UR laser output pulse with M=200 when the SR average output power is 10 W.

Thus, maximum output power of the diffraction limited beam is realised with URs when the magnification M and length L are matched to the temporal pulse gain behaviour. From this point of view, the increase in the efficiency of the UR with a copper-bromide active medium (in this case UR output power reached 89-90% of the SR output power) is explained by the better matching of the temporal gain shape with the seed transit through the active medium for the main UR mode.

Recently a self-filtering unstable resonator (SFUR) and diffraction filtered resonator (DFR) have been used as resonators for CVLs. These types of resonators use a self-filtering and diffraction-filtered intracavity aperture which transmits only seed beams forming the UR main mode. All other beams transit through the active medium only one or two times. Since, the SF aperture size is

$$a = \sqrt{1.22\lambda f_2 / 2} \tag{8}$$

where f_2 is the focal length of mirror M_2, the output beam diameter D onto mirror M_1 is

$$D = 1.22\frac{\lambda}{a} \tag{9}$$

The M_0 value for this D is

$$M_0 = \frac{D^2}{2\lambda f_1} = \frac{f_1}{f_2} = M \tag{10}$$

and the diffraction-limited SFUR output beam is formed by the same cavity round trip n_0=2 as for the UR. Thus, the minimal cavity round trip number of the UR is reached by increasing the resonator magnification up to M_0, and for the SFUR it is reached by reducing M_0 to M defined by cavity geometry.

3. CVL Second Harmonic Generation as an UV-Source for Submicron photolithography

Submicron photolithography with a minimum line-width of 0.3-0.5 μm requires Deep Ultra Violet (DUV) laser sources with wavelengths shorter than 300 nm. The 249 nm KrF excimer laser radiation is commonly used in UV-steppers [11]. Below 300 nm there is only one suitable optical material available for building the stepper lens - fused silica. Therefore, the spectral line-width has limitations because of the high chromatic dispersion of silica. The defocusing depth Δf of the silica lens due to chromatic aberration can be estimated as follows

$$\frac{\Delta f}{\Delta \lambda} = \frac{f}{(n-1)} \frac{dn}{d\lambda} \tag{11}$$

where f is the lens focal length, Δλ is the spectral bandwidth of the illumination source, n is the refractive index and dn/dλ is the chromatic dispersion of silica.

If Δf=0.5 μm, which is acceptable for a 0.3 μm feature size, focal length f=10 cm, n=1.5, dn/dλ=10^{-3} nm^{-1}, then Δλ according to (11) would be equal to 2.5 pm. The bandwidth (fwhm) of a free running KrF excimer laser is approximately 300 pm. As discussed above, an excimer stepper requires the output of the laser to be spectrally narrowed. Spectral narrowing for excimer lasers can be accomplished by introducing wavelength dispersive optical elements in the resonator - a fairly complex and difficult method to implement.

Atomic gas lasers emit narrow spectral lines. Of the atomic gas lasers available, CVLs are the most efficient (> 1%) and generate the highest output powers (> 100W). They are commercially available, inexpensive, uncomplicated to use and have comparitively low running costs. Previous attempts to generate frequency-doubled UV light at 255 nm wavelength by pumping non-linear crystals has been unsuccessful. This is due to the unsatisfactory properties of the crystals. Recent progress has been achieved with the development of new non-linear BBO crystals.

The results of CVL second harmonic characteristics are presented below from the point of view of UV photolithography requirements, i.e. line-width and output stability of the lines generated. The main advantage of a CVL SHG UV source against a KrF excimer laser is the CVL linewidth-stability which is ensured by atomic transition constrants and is not influenced by external circumstances.

A copper-bromide vapour laser (CBVL) with unstable resonator (M=60) was used as a pumping source for the SHG. A CBVL was chosen because of its high UR power efficiency (80-90% of SR output power compared with the CVL's 50% of SR output power). The PRF was 14-18 kHz. For effective SHG it is necessary for the fundamental beam to be polarized. CVL radiation is in fact almost completely unpolarized because of the high gain and a polarizer was therefore placed between the discharge tube and rear mirror. In this case output radiation polarization was nearly 100%, while power losses did not exceed 10%. To examine the spatial coherence of the beam a beam splitter removed some of the radiation to two fibers scanning along each wave front.

The spectral line profile of the SHG was examined by means of a Fabry-Perrot (FP) interferometer. The FP fringe pattern was examined through a 0.5 mm diameter pinhole by focussing the beam with a 300 mm focal length lens onto a photomultiplier placed at the lens focal plane. The SHG spectral line was analysed using a stroboscopic oscilloscope enabling the profile to be examined at any time during the laser pulse. Final spectral scanning was recorded by two co-ordinated plotters controled by the oscilloscope output analog pulse. Time resolution of the system was limited mainly by photomultiplier time resolution and was nearly 3 ns in this case.

The SHG line profile is governed by the fundamental spectral line profile, which in turn depends on the spontaneous emission line (510 nm) profile. The 510 nm line consists of 18 hyperfine and isotopic structure components. The resulting profile was calculated as a sum of all Doppler broadening components. The calculation was performed for 1.5 kK, 2.2 kK and 2.5 kK gas temperature. The results show that the FWHM as well as the peak position are independent of the gas temperature. This implies a weak dependence of the absolute line frequency and FWHM from excitation conditions, for example, the average imput power. As has been estimated a 1000 K change of gas temperature results in fundamental wavelength instability of ±0.15 pm and a FWHM of 6.5 pm. A fundamental linewidth of 2 pm was achieved experimentally, at the onset of lasing (that is, at the start of the reservoir temperature regime of laser operation) the SHG beam linewidth was 1 pm at an average fundamental power of 1.5-2 W. At optimal excitation conditions and with a green line average output power of 5-6 W the fundamental linewidth was 4 pm. The SHG (255 nm) linewidth was 2 pm at an average power of 0.5 W. The experimental line profiles are shown in Fig.2.

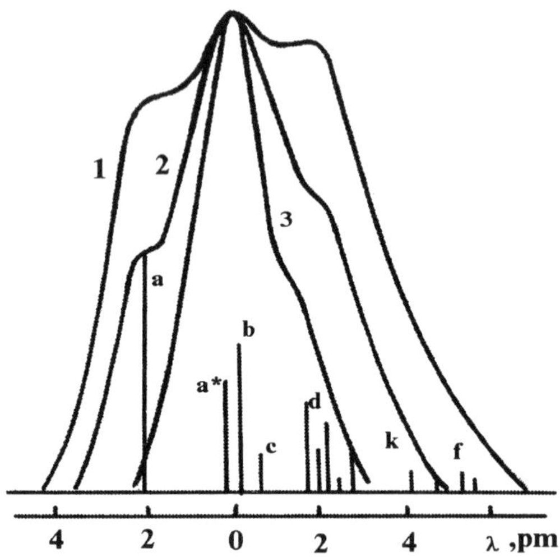

Figure 2. Experimental spectral line (510 nm) profiles: 1 is the spontaneous emission; 2 is the fundamental beam; 3 is the SHG

Spatial coherence measurements has shown nearly complete beam coherence in the horizontal direction, transverse SHG beam coherence and pumped beam spatial coherence with a coherence radius of 5 mm are the same. It is difficult to observe high quality optical images formed from high-coherence radiation beams due to the presence of speckle. To overcome this problem the coherence radius on the photomask plane must be smaller than the minimum size of the element transmitted. Submicron lithography with 0.4-0.5 μm resolved feature sizes and a projection system with a 5:1 or 10:1 reduction requires the coherence radius in the photomask plane to be less than 2-4 μm. To eliminate spatial coherence a quartz fiber was employed. Fibers form a round spot of transmitted light. A high degree of intensity homogenity within the spot is an additional advantage of fibers. A fibre undergoing temperature or pressure deformation changes its refraction index. As a result, transmitted fiber modes 'run' in the field of view and the speckle-structure is averaged.

In the experimental UV photolithography unit an emittance source (an end of the fiber) was imaged by a condenser lens onto the projection objective centre. A photomask placed in the condensor lens plane was projected onto a photoresist covered plate. As a projecting objective we used a mirror-lens micro-objective with quartz-fluoride optics OK-40 of NA=0.5. The aperture diaphragm was 5 mm. Therefore, the limit of its resolution was Δ_l=0.3 μm, DOF Δ_f=0.5 μm. The objective ensured a reduction factor of 40 for a working length of 165 mm. The reduction ratio was regulated by the working length. A standard photomask element was generated by evaporating Al film onto a quartz plate to produce a transparent square 30 x 30 μm^2. Only 10% of the available SHG power passed through the fiber (50 mW), due to the rough surface at the fiber end, but it was sufficient to iluminate the photoresist surface. Analysis of the square angle sharpness of the multiple images formed in the photoresist of 0.5 μm thickness demonstrated that the system achieved 0.5μm resolution. Thus the experiment demonstrates the photoresist process using a UV source based on a CVL SHG of 0.5 μm resolution.

4. Transformation of CVL Radiation by Optical fibers

The optical fibers represent a typical example of a guide geometry where the light waves interaction length is extremely long and consequently the threshold of the non-linear optical process is low. Fiber eigen modes which have no diffraction and, therefore, don't change their optical field distributions along the length are an important feature of optical fibers. The number of such modes is varied by varying the fibre parameters. Thus, the optical fibres extend the application sphere of relatively low power CVLs by non-linear optics. Recently, wideband frequency conversion of CVL radiation has been achieved by stimulated Raman scattering in multimode silica fibers. Up to nine Stokes' orders have been generated in the 511 to 649 nm spectral region [12]. Three Raman Stokes' lines 1.6, 1.74 and 1.89 μm have been observed from single-mode silica fiber pumped at 1.5 μm by the pulsed barium vapor laser [13].

This part of the paper examines CVL radiation frequency conversion by passing through both long and short low-order mode fibers. The PRF in all the experiments was 18.7 kHz. Maximum combined average output power (both lines) was 10-12 W. The laser output beam was coupled into a fiber with a single lens of focal length 12 cm. In order to separate pumping of the fiber either by green (510 nm), or by yellow (578 nm) lines we used a 45° prism placed in front of the focusing lens. The fiber used was a fused quartz core of 7 μm diameter dopped with germanium oxide (3 mol %). The fiber numerical aperture (NA) was 0.1 and the cut-off wavelength was 1.05 μm.

The lengths of the fibers were 5-10 m (short fiber) and 1300 m (long fiber). Measured optical losses for the green line were of the order of 25 dB/km.

Radial distribution of the laser output beam power for the green line and the fraction of power coupled into the fibre were measured with a help of short piece of fiber (about 0.5 m). In these experiments the input end of the fibre could be accurately moved along three axis near the focal spot. The average power of the fibre output was measured using a calorimeter. The focal spot waist was 20 μm. The measured distribution was slightly wider than the actual distribution because it was measured using a fiber with a 7 μm core. Measurements of the power fraction coupled into the fibre showed that at a total average input power of 2.8 W (2.3 W green and 0.45 W yellow) the average fibre output power was 0.28 W in the green and 0.05 W in the yellow line. This corresponds to a coupling efficiency of 12% for green and 11% for the yellow line. The efficiency estimates made from a measured focal spot power distribution at the fibre input gave nearly the same result. The pulse shapes with respect to time were measured by using a photomultiplier tube and an oscilloscope with an overall time response of 3 ns. The pulse duration was about 10 ns for both lines. Varying the output power from 1 W to 2.5 W by increasing the input electrical power did not change the pulse shapes or duration. In this condition the time delay between the green and yellow pulses was 10-13 ns, the yellow pulse delay was the longer. It should be mentioned that all experiments were made with average powers of less than 2-2.5 W at the fiber input face because at higher power levels breakdown of the fiber was frequently observed. Taking into account the pulse duration and pulse energy it is possible to evaluate the threshold of the fiber breakdown, which occurred at a peak power density of $3.7\ 10^9$ W/cm^2. This is close to the well known limit.

The spectrum measured at the output of the 10 m long fiber when pumped by the green line is shown in Fig.3. The spectrum was measured with a prism monochromator. In this case a few Stokes' and antiStokes' components are observed. Wavelengths and frequency shifts from the green line are listed in Table 1.

TABLE 1. Emission wavelengths and frequency shifts of output lines

Line	Wavelength (nm)	Shift (cm^{-1})	Line	Wavelength (nm)	Shift (cm^{-1})
A6	499.3	439	SR1	523.6	486
A5	501.7	343	S6	522.4	446
A4	503.7	267	S5	519.6	341
A3	505.3	201	S4	517.6	267
A2	506,8	143	S3	515.8	201
A1	508.9	63	S2	514.3	142
Green	510,6	-	S1	512.2	63

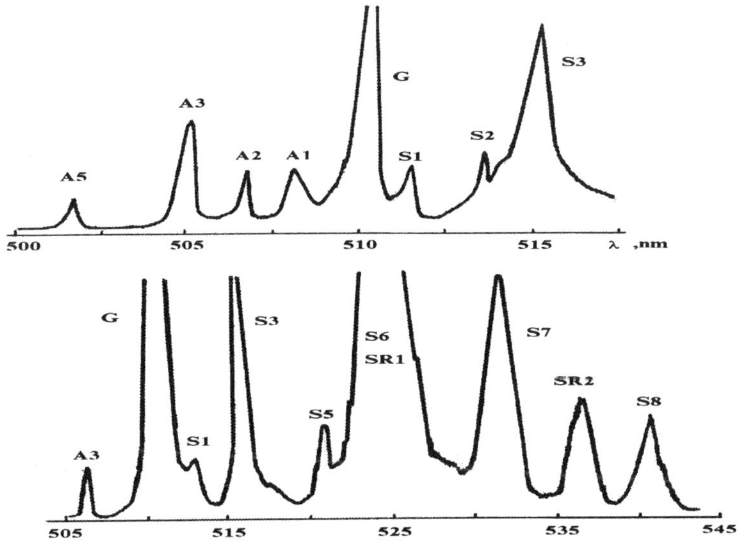

Figure 3. Spectrum at the output of the 10 m long fiber at different incident angles

The spectra consists of pairs of components which are equidistant from the green line. This results from stimulated four-photon processes which occur in multimode optical fibres where phase mode-matching takes place. Two fundamental photons are transformed into one photon with low energy (Stokes' $h\omega_s$) while the second photon has high energy (anti Stokes' $h\omega_a$). The phase mismatch is expressed by

$$h\omega_{f_1} + h\omega_{f_2} = h\omega_s + h\omega_a \qquad (12)$$

In a bulk media, phase matching occurs at definite angles. For optical fibres, mode phase matching is achieved by overlapping the optical mode fields through a long fibre and, consequently, the threshold power of such a process is low. The intensity and number of the spectrum components depend on the angle of the fundamental beam incidence into the fiber.

The measured spectrum at the output of a 1300 m long fiber, pumped by both green and yellow lines is shown in Fig.4. Pump-power to SRS conversion efficiency did not exceed 10%. This comparatively low efficiency was due to the high level of losses for both the pump lines and several of the first Stokes' components in the long fiber. The green and yellow output powers were negligible - with all the output power concentrated into the SRS components. As one can see from Fig.4 the output spectrum of the fibre covered a spectral region from the green line up to 0.9-1 μm. The maximum frequency shift for the SRS components from the green line was about 9000 cm^{-1}. This corresponds to 20-22 Stokes' components. Up to 18 Stokes' components have been observed from the yellow line. The output spectrum of the SRS of the fibre was strongly dependent on the coupling of laser light into the fiber and on input power. Reduced input power and poor focusing resulted in a decrease in the number of

SRS components excited and a considerable shift of the spectrum to the shorter wavelength region. By varying the CBVL input power it was possible to change considerably the spectrum at the output of the fiber

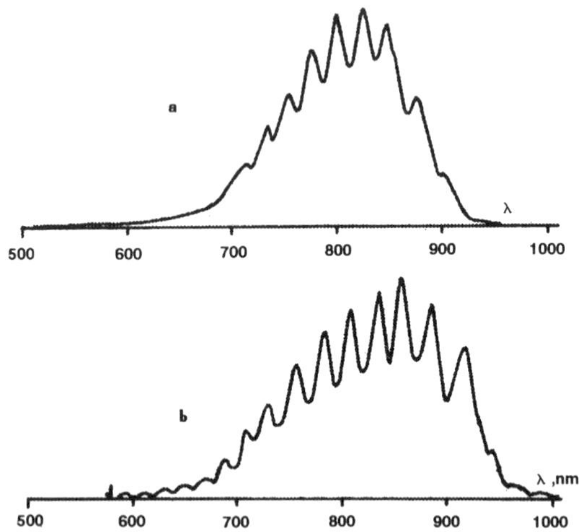

Figure 4. Spectrum at the output of a 1300 m long fiber: a is the green fundamental line; b is the yellow fundamental line.

5. Conclusions

Maximum UR output power is reached by matching the UR main mode formation time with the active medium saturation gain time. To realise this condition the resonator magnification M must be increased up to M_0. In the case of SFUR or DFUR the M_0 is decreased up to M by placing self-filtering or diffraction-filtering apertures inside the resonator cavity.

SHG (255 nm) of a CVL is an alternative UV source for submicron lithography. The advantage of the CVL against the KrF excimer laser is the narrow and stable atomic spectral lines.

Stimulated four-photon processes under the mode phase matching condition give narrowband spectral re-tuning in short fibers. Wide-band spectral re-tuning up to 9000 cm^{-1} is possible by SRS in long fibers.

6. References

1. Ramsay D.J., Glover A.C.J., Coutts D.W., Piper J.A. (1994) High-speed precision uv drilling with frequency-dobled CVLs, *SPIE* **2062**, 22-29.
2. Isaev A.A., Lemmerman G.Yu., Malafeeva G.L. (1980) Second harmonic generation from pulse copper vapor laser radiation, *Sov. J. Quantum Electron.* **10**, 983-985.
3. Siegman A.E. (1974) Unstable optical resonators, *Applied Optics* **13**, 353-367.

4. Zemskov K.I., Isaev A.A., Kazaryan M.A. et al. (1974) Use of unstable resonator in achieving the diffraction divergence of the radiation emitted from high-gain pulsed gas lasers, *Sov. J. Quantum Electron.* **4**, 474-477.
5. Isaev A.A., Kazaryan M.A., Petrash G.G. et al. (1974) Convergin beams in unstable telescopic resonators, *Sov. J. Quantum Electron.* **4**, 761-766/
6. Isaev A.A., Kazaryan M.A., Petrash G.G. et al. (1975) Evolution of gaussian beams and pulse stimulated emission from lasers with unstable resonator, *Sov. J. Quantum Electron.* **5**, 607-614.
7. Bhatnagar R., Dixit S.K., Singh B. et al. (1989) Performance of a copper vapor laser with self filtered unstable resonator, *Optics Communications* 74, 93-95).
8. Dixit S.K., Singh B., Nakhe S.V. et al. (1990) Negative-branch unstable resonator copper vapor laser (1990), *Optics Letters* **15**, 428-430.
9. Pini R., Salimbeni R., Toci G., et al. (1991) Highe efficiency diffraction limited operation of a copper vapor laser, *Optics Communications* **81**, 138-144.
10. Dixit S.K., Mittal J.K., Singh B. et al (1993) A generalized diffraction filtered resonator with a copper vapor laser, *Optics Communications* **98,** 91-94.
11. Sengupta U.K. (1993) Krypton fluoride excimer laser for advanced microlithography, *Optical Engineering* **32**, 2410-2420.
12. Guasti A., Pini R., Salimbeni R. (1991) Stimulated Raman scattering of copper vapour laser radiation in optical fibres, *Optical and Quantum Electronics* **23**, S555-S562.
13. Pask H.M., Piper J.A. (1991) Stimulated Raman scattering of barium laser output in a silica fibre, *Optical and Quantum Electronics* **23**, S563-S568

HIGH POWER DYE LASER PUMPING BY COPPER VAPOUR LASERS

DENIS DOIZI
CEA, CEN SACLAY
DPE/SPL/SCLO, Bât 467, 91191 Gif Sur Yvette Cedex, France

1. Introduction

France has developed a major electronuclear industry which nowadays supplies about 75% of the nation's electricity. It comprises 55 power reactors which use enriched Uranium produced by the enrichment plant Eurodif located in Pierrelatte, Drôme. This modern plant uses gaseous diffusion separation; it has a production capacity of about 10.8 MSWU/year, for an estimated world market of about 30 MSWU/year.

This market is today characterised by a large overcapacity but world needs will increase in the next 15 years. At that time, the gaseous diffusion plants will be replaced by other new competitive plants based on different separation processes such as ultracentrifugation or laser processes.

France has been studying different processes since the 80's and chose in 1985 to focus on the selective photoionisation process called AVLIS (Atomic Vapour Laser Isotope Separation) [1].

We shall here briefly recall the general principles of this separation process and we shall describe our pilot facility ASTER designed to demonstrate the technical and economic feasibility of the process.

2. Physical Features of the Laser Separation Process

The SILVA process is based on the selective photoionisation of U^{235}. It uses the tiny differences in the excitation frequencies between U^{235} and U^{238}. The ionisation of U^{235} needs 6.18 eV and is achieved with a three step scheme using four wavelengths delivered by dye lasers as indicated on the scheme below:

C. E. Little and N. V. Sabotinov (eds.), *Pulsed Metal Vapour Lasers*, 303–307.

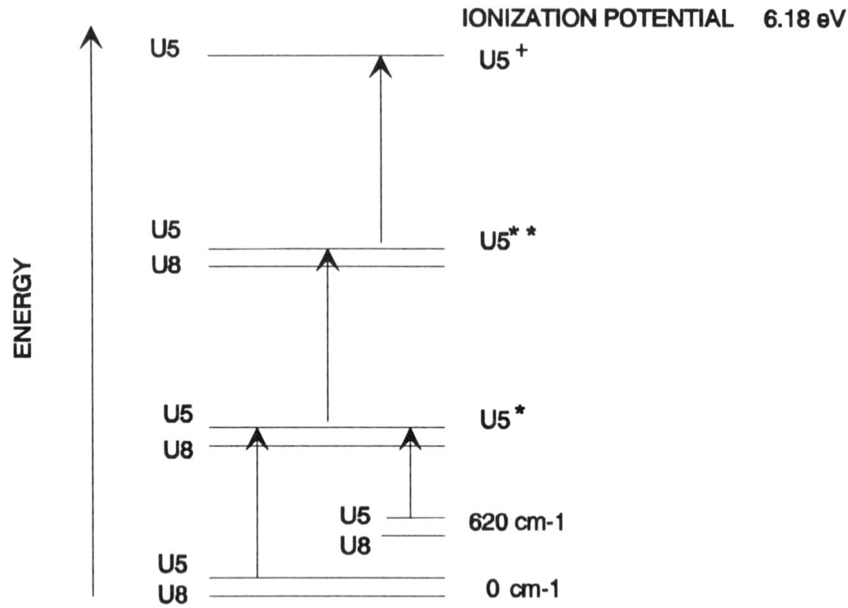

3. Schematic Diagram of a SILVA Plant

A SILVA plant is composed, as shown below, of two major subsystems:

- a separator subsystem

- a laser subsystem

The first one produces the atomic Uranium vapour. Natural Uranium is vaporized in a crucible with electron beam guns at a temperature of about 3000 °C. The Uranium vapour interacts with the laser light in irradiation zones where selective photoionisation occurs. U^{235} is ionized, extracted on charged plates and then collected in liquid phase.

The second subsystem delivers the four wavelengths necessary to realise the photoionisation. The general features required for the lasers are fixed by the technicoeconomy of the process and are, for example:

- tunability: the choice of a three step scheme requires wavelengths in the domain 550 – 700 nm.

- spectral width: the first two steps are selective and require a narrow spectral width but sufficient to overlap the hyperfine structure of U^{235}.

ATOMIC VAPOR LASER ISOTOPE SEPARATION

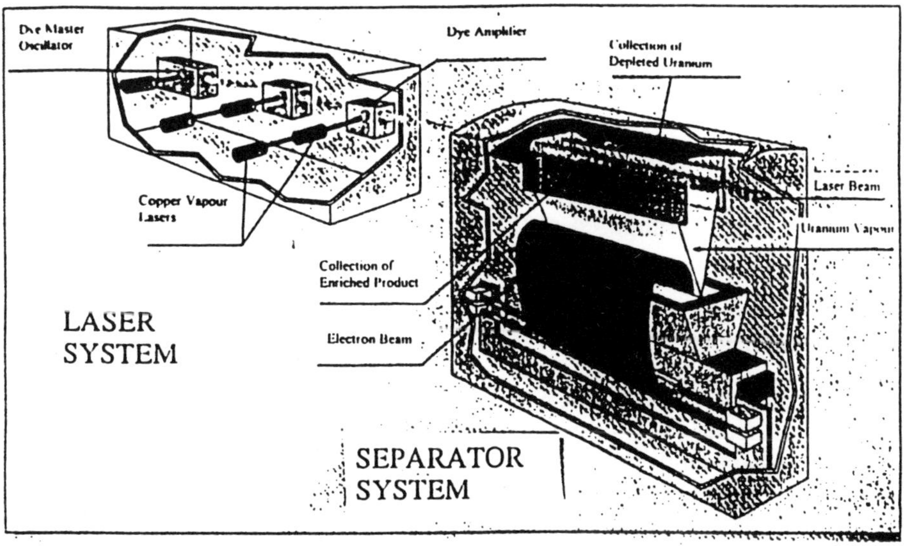

- repetition rate: the Uranium vapour rises at a velocity of about 1000 m/s. The photoionisation of all the U^{235} atoms in the irradiation zones requires repetition rates above 10 kHz.

- pulse duration: the laser pulse duration must be lower than the radiative lifetimes of the excited levels.

- pulse energy: the laser energy must be sufficient to saturate the atomic transitions.

The lasers which have been chosen and which satisfy many of the above criteria are dye lasers pumped by copper vapour lasers.

R&D on the SILVA process has been carried out in close cooperation with Cogema. Dye lasers were developed in CEA laboratories and copper vapour lasers in collaboration with Cilas Company. A general assessment will be made in 1997. This assessment will consist of demonstrations intended to prove technical performances and to give an evaluation of the economics of the process.

For example, enrichment performances, handling of liquid uranium, long term operation of dye laser chains, reliability of specific components will have to be demonstrated.

In order to realise these demonstrations, a pilot facility called ASTER is under construction. It comprises two subunits: the separator and the laser systems.

We shall now briefly describe the laser system which is being installed in the pilot facility.

4. Description of ASTER Laser System

The laser system is composed of two main parts: dye lasers and copper vapour lasers.

4.1. DYE LASER SUBUNIT

There are four dye laser chains which deliver the four wavelengths necessary to achieve the selective photoionisation. All of these are composed of a dye laser source and power amplifiers. The dye laser source comprises a master oscillator and one or two preamplifiers. The master oscillator is composed of a Hänsch oscillator which generates a single-mode tunable beam frequency stabilized with a 10^{-7} relative precision obtained by means of a specially developed Fizeau wedge wavemeter. The output spectral width of this device is about 50 MHz. It is then extended to a few GHz by means of an electro-optic modulator in order to cover the hyperfine structure of U^{235}. The master oscillator is fibre pumped by a specifically devoted pump laser. It delivers about 25 μJ at a repetition rate of 5 kHz. The energy is then boosted with preamplifiers working an unsaturated regime and power amplifiers working in a saturated regime. These power amplifiers were scaled and designed taking into account hydrodynamics (careful design of the dye flow cell, proper choice of the dye flow loop), nature of the active volume (modelling of the gain volume, synthesis of new laser dyes). Optical modelling has been developed to understand the propagation of the dye laser beam in the dye laser chain and to optimize the optical coupling between amplifiers in order to maximise the extraction efficiency.

All along the dye laser chain, sensors enable control of the pulse energy, the pulse wavefront and the pulse duration.

The dye power amplifiers are fibre pumped by copper vapour lasers. This enables safe beam handling and decreases the number of optical components.

The four colours coming out of the four dye laser chains are spectrally multiplexed before being sent to the separator module.

4.2. PUMP LASER SUBUNIT

The pump laser subunit is composed of copper vapour lasers installed in MOPA configuration. A pump chain is composed of an injection seeded oscillator and two or three power amplifiers.

This equipment has already been described elsewhere [2]. Briefly, the injection seeded oscillator, developed by Cilas, is a single unit delivering about

30 W with a divergence lower than 300 μrad. The pulse duration is about 40 ns (FWHM), the jitter is lower than 2 ns. The green to yellow ratio is about 1.4. Its dimensions are about 4 m length, 0.6 m height and 0.8 m width. The laser is driven by two CX1535 EEV thyratrons. The measured wallplug efficiency is higher than 0.4%. The laser MTBF is higher than 1000 hours.

The copper vapour laser amplifiers are 100 W Cilas amplifiers. They deliver more than 100 W in amplifier configuration with a power decrease of about 0.6 to 1.4 W per 24 hours. The beam diameter is 50 mm, the pulse duration is 50 ns (FWHM), and the jitter is lower than 4 ns. The external dimensions are 3.9 m length, 0.7 m width and 0.5 m height. This laser is driven by two CX1535 thyratrons. The wallplug efficiency is 0.7% and the MTBF is close to 1000 hours.

The next generation of copper vapour lasers delivering more than 400 W is under development. It will use solid-state power supplies developed by GEC Alsthom Company.

In the ASTER laser facility, 4 master oscillators and 16 power amplifiers have been installed in order to obtain 6 copper vapour laser chains devoted to the optical pumping of the dye power amplifiers via fibre optics.

The master oscillators are multiplexed energetically in order to increase the reliability of the installation. The copper laser beams can also be temporally multiplexed to increase the repetition rate of the facility

The total copper laser power installed is about 2 kW.

A fully integrated command control has been developed. It controls optical power, voltages, synchronization, pulse duration, green to yellow ratio, spatial profile, for each master oscillator and for the entire laser chain.

The pump laser subunit has now been working for nearly one year with good technical performance and reliability. At the present time, the pump power is being delivered to the dye power amplifiers.

The whole laser facility is expected to work in a few weeks ready to begin the first demonstrations of the general assessment.

5. References

1. Camarcat, N., Lafon, A., Pèrves, J.P., Rosengard, A., and Sauzay, G. (1993) *Laser Isotope Separation SPIE* **1859**, 14–23.

2. Bettinger, A., Neu, M., Maury, J., and Chatelet, J. (1993) *Laser Isotope Separation SPIE* **1859**, 108–113.

AMPLIFICATION OF THE TEMPORALLY STRETCHED YELLOW BEAM OF THE COPPER VAPOUR LASER

SUNITA SINGH, K. DASGUPTA, S. KUNDU, S. KUMAR, L.G. NAIR AND U.K. CHATTERJEE
Laser & Plasma Technology Division, B.A.R.C.
Bombay 400085, India

Abstract

Amplification of temporally stretched yellow beam of the copper vapour laser in a Rh6G dye amplifier is reported. Effect of high signal intensities on the amplifier extraction efficiency was investigated experimentally and theoretically. The efficiency does not saturate but shows a maximum and then decreases with increasing signal intensity. Tolerance on synchronization between pump and signal pulses improved from ~2 ns (unstretched) to ~8 ns (stretched).

1. Introduction

Pulsed dye laser oscillator-amplifier systems are useful and versatile sources of high power, narrow band, tunable radiation and have been reported extensively in the literature [1–4]. These systems are becoming increasingly important in research areas such as high resolution spectroscopy, isotope separation and coherent pulse spectroscopy.

CVL pumped pulsed dye laser oscillators have optical conversion efficiencies of 10%–20%. In general, higher efficiencies from the CVL pump to the dye laser output require the use of oscillator-amplifier arrangements. An important design consideration for such systems is that the signal intensity should be large enough to saturate the amplifier gain. The commonly used model of Hargrove and Kan [1] predicts that the amplifier extraction efficiency would also saturate. However, our results show that efficiency actually starts decreasing beyond an optimum signal intensity because of a nonlinear loss at the signal wavelength arising from excited state absorption (ESA) and ground state absorption (GSA).

In the work presented here, we have experimentally and theoretically investigated the significance of this nonlinear loss on a transversely pumped dye amplifier at high signal intensities. To obtain high signal intensities, we have utilized the yellow beam (λ = 578.2 nm) of the CVL as the input signal in a double-sided transversely pumped Rhodamine 6G dye laser amplifier, whereas the green (λ = 510.6 nm) beam from the same CVL is utilized as the pump beam. With this arrangement up to 4 watts average power of the yellow beam (as signal) could be obtained.

C. E. Little and N. V. Sabotinov (eds.), Pulsed Metal Vapour Lasers, 309–314.

2. Experimental Set Up

A schematic of the experimental arrangement used is shown in Figure 1. The output beam from a copper vapour laser was temporally stretched and smoothed. This was achieved by multiple passing of the beam through a ring resonator arrangement comprising a beam splitter BS1 and high reflectivity mirrors M1, M2, M3. The stretched CVL beam was split into its yellow and green components using a dichroic beam splitter DBS1. The green component formed the pump beam and the yellow component formed the signal beam.

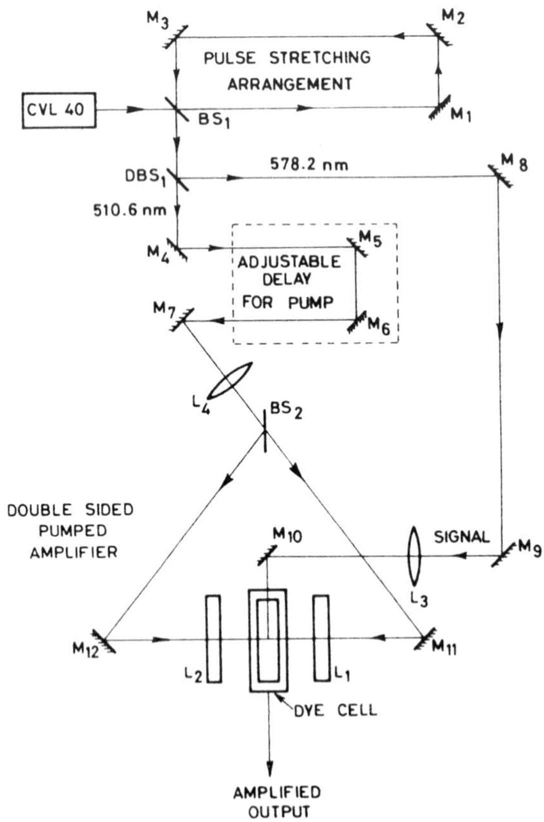

Figure 1. Schematic of the transversely pumped dye laser amplifier.

The signal beam ($\lambda = 578.2$ nm) was directed by the mirrors M8 and M9 via a focusing lens L3 and a mirror M10 into the flowing dye cell. The dye used was Rh6G in ethanol at a concentration of 0.9 mM. The average pump powers used for the experiment were 4.0 watts to 7.0 watts. The input signal power was varied from 100 milliwatts to 4.0 watts. The stretched pump and signal pulse durations (FWHM) were 36 ns and 33 ns respectively. Temporal stretching also leads to smoothing of the pulse shapes. The inherent delay of the yellow component with respect to the green component is 7 ns. The position of the mirror combination M5 and M6 was varied to change the delay between the two components. For a certain fixed delay between the two pulses both the signal power and pump power were varied and the corresponding amplifier extraction efficiencies were measured. The amplifier extraction efficiency was also measured as a function of delay between the pump and signal pulses.

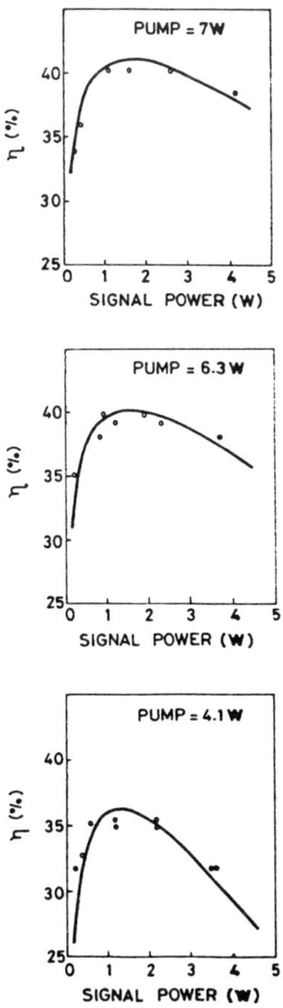

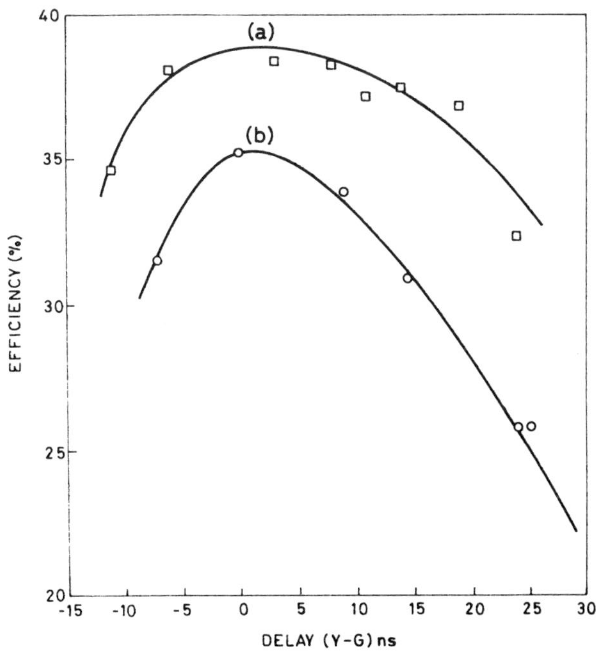

Figure 2 (left). Extraction efficiency (η) as a function of the input signal power for pump powers of 7 W, 3 W and 4.1 W. The solid lines are theoretical curves.

Figure 3 (right). Amplifier extraction efficiency (η) as a function of delay between the yellow and green beams for 3.5 W (578.2 nm) and 6.2 W (510.6 nm).

3. Results and Discussion

The extraction efficiency is shown as a function of input signal power in Figure 2 for three different pump powers. The solid lines are the theoretical curves matched to the experimental data as discussed below. A maximum extraction efficiency of 40.3% was achieved at a pump power of 7 watts and an input of 1.6 watts. It is clear that an optimum value of the signal intensity exists which depends on the pump power and the signal wavelength. The decrease in amplifier extraction efficiency at high input powers is attributed to ground and excited state absorption at the signal wavelength [5], as discussed below.

We have studied the extraction efficiency as a function of the time delay between the pump and signal pulses for temporally stretched and unstretched pulses. Figure 3 shows the amplifier extraction efficiency as a function of delay between the yellow and green beams when they are temporally stretched. The delay was measured as the difference between the 10% of the peak values of the two pulses. The extraction efficiency curve shows a plateau over a large variation of the delay between the yellow and green components. This indicates a large tolerance (~8 ns) on synchronization between the pump and signal beams when they are temporally stretched.

To explain the experimental results, theoretical modelling of the transversely pumped dye laser amplifier was done using rate equations [5]. The following rate equations can be written down for the transversely pumped dye laser amplifier with ground state absorption (GSA) and excited state absorption (ESA) at signal wavelength:

$$dN_1 / dt = \sigma_{oP} N_o I_P + \sigma_{oL} N_o I_L - \sigma_e N_1 I_L - N_1 / \tau \,, \tag{1}$$

$$dI_1 / dt = \sigma_{oe} N_1 I_L - \sigma_{oL} N_o I_L - \sigma_{12} N_1 I_L \,, \tag{2}$$

$$N = N_o + N_1 \,. \tag{3}$$

In the above equations σ_{oP}, σ_{oL}, σ_{12}, σ_e are the cross-sections respectively for GSA at the pump wavelength and GSA at the laser wavelength, ESA and stimulated emission at the signal wavelength; τ is the spontaneous decay lifetime of the first excited singlet state and I_P and I_L are respectively the pump and signal intensities expressed in photons/(cm^2.s); N, N_o, N_1 are total, ground and excited state molecular densities expressed in molecules/cm^3. Triplet state effects are neglected because of the short pulse duration. Also the presence of amplified spontaneous emission (ASE) has been neglected. Throughout the above treatment, a transversely pumped gain region of length L with uniform pump intensity was assumed. Under steady state conditions, the saturated gain $G = (I_{out}/I_{in})$ of the amplifier is given [5] as

$$G = g_o / SI_{in} - \left[(g_o / SI_{in}) - 1 \right] \exp\left[(\ln G - g_o L) / (1 + g_o / SI_s) \right] \tag{4}$$

where g_o, L, I_s and I_{in} are the small signal gain, length of the amplifying medium, saturation intensity and input signal intensity. The factor S in equation (4) represents the nonlinear absorption loss due to GSA and ESA at the signal wavelength and is given by

$$S = N\sigma_{12}\sigma_{oL}\tau / (1 + \sigma_{oP} I_P \tau)$$

The amplifier extraction efficiency is given by

$$\eta = (P_{out} - P_{in}) / P_{pump} ,$$

where P_{out}, P_{in}, P_{pump} are average amplifier output power, input signal power and pump power respectively. Equation (4) was solved numerically. The parameters used were [6]:

$\lambda_L = 578.2$ nm, $\lambda_P = 510.6$ nm, Conc. $= 0.9$ mM, $L =$ 1 cm,

$\sigma_{oP} = 1.72 \times 10^{-16}$ cm^2, $\sigma_{oL} = 7.5 \times 10^{-19}$ cm^2, $\sigma_{12} = 8.2 \times 10^{-17}$ cm^2,

$\sigma_e = 1.52 \times 10^{-16}$ cm^2, $\tau = 4.8$ ns [1].

The pump pulse duration was 80 ns (full width) and the signal pulse duration was 72 ns (full width). The pump beam was focused to an area of 0.05 cm^2. In calculating the pump intensity, the averaged pump and the signal intensities were multiplied by two fitting parameters FP and FIN respectively, for a best fit with the experimental results. We have found that for FP = 1.95 and FIN = 0.889, the theoretical curves (solid lines) closely match the experimental values of the extraction efficiency over three different pump powers, as shown in Figure 2.

4. Conclusion

The effect of high signal intensities on the amplifier extraction efficiency has been studied both experimentally and theoretically. It has been shown that nonlinear absorption at the signal wavelength induced by ground state absorption (GSA) and subsequent excited state absorption (ESA) results in a reduction in the amplifier extraction efficiency at high signal intensities. Hence an optimum value of the input intensity exists for which the extraction efficiency is maximum.

5. References

1. Hargrove, R.S. and Kan, T. (1980) High power efficient dye amplifier pumped by copper vapor lasers, *IEEE J. Quantum Electron.* **QE-16**, 1108-1113.
2. Nair, L.G. (1982) Dye lasers, in *Progress in Quantum Electronics*, **Vol. 7**, Pergamon Press, England, 153.

3. Lavi, S., Amit, M., Bialolankar, G., Miron, E. and Levin, L.A. (1985) High-repetition-rate high-power variable-bandwidth dye laser, *Appl. Optics* **24**, 1905-1909.
4. Sunita, S., Dasgupta, K., Kumar, S., Manohar, K.G., Nair, L.G. and Chatterjee, U.K. (1994) High power high repetition rate CVL pumped dye oscillator amplifier, *Optical Eng.* **33**, 1894.
5. Dasgupta, K. and Nair, L.G. (1990) Effect of excited-state absorption at signal wavelength in pulsed-dye-laser amplifiers, *IEEE J. Quantum Electron.* **QE-26**, 189-192.
6. Sahar, E. and Treves, D. (1977) Excited singlet-state absorption in dyes and their effect on dye lasers, *IEEE J. Quantum Electron.* **QE-13**, 962-967.

7. Applications in Industry

PRECISION MACHINING WITH COPPER VAPOUR LASERS

H.W. BERGMANN[2], C. KÖRNER[1], M. HARTMANN[1], R. MAYERHOFER[1]
[1] *Universität Erlangen, Lehrstuhl Werkstoffkunde und Technologie der Metalle, Martensstraße 5, 91058 Erlangen, Germany*
[2] *Applikations- und Technikzentrum für Energieverfahrens-, Umwelt- und Strömungstechnik ATZ-EVUS, Rinostraße 1, 92249 Vilseck, Germany*

1. Abstract

In the present paper criteria are presented for precision machining with laser beams. It is outlined that a copper vapour laser system (CVL) consisting of an unstable resonator, subsequent amplifiers, a suitable synchronizing unit, an optical system for beam forming and guiding and an CNC-handling system fulfills most of these requirements. Machining speeds are given for different materials. Finally, examples are shown for possible applications of CVL-processing.

2. Introduction

Today laser machines have changed manufacturing even in medium sized and small companies. More recently, one can observe a transition from simple processing tasks which do not require a precise beam towards more complex applications where beam quality is essential (caustic, intensity distribution as function of position in space and time). In precision machining the laser beam has to compete with several conventional techniques. For spark erosion, mechanical machining and electron beam ablation the minimal size of structure elements is currently limited to about 100μm. Below that only chemical etching and laser techniques can be used. Numerous processes that either use lasers or laser supported techniques are known. They either build up or remove material via local and well defined energy input. A selection of typical techniques used is given in Table 1. In this paper only ablation processes without chemical support will be described. Many of the presented criteria are also applicable for other precision machining processes.

C. E. Little and N. V. Sabotinov (eds.), Pulsed Metal Vapour Lasers, 317–330.

TABLE 1: Different processes of precision machining with lasers

geometrically different processes	ablation
	modification (and past processing)
	deposition
optically different techniques	focussing
	projecting
	interference
number of stages	single stage processes
	multiple stage processes
	- two-pulse technique
	- chemical activation before illumination
	- chemical removal after irradiation
laser activated processes	chemical etching or deposition
removal strategies	single pulse drilling
	multiple pulse drilling
	drill hole cutting

3. Criteria for precision machining

Precise manufacturing of a structure demands a handling system with a high degree of accuracy, a well defined tool and a controlled process. For precision machining, a stationary process, selectivity and a minimum degree of damage by secondary effects is essential. Typical **structure sizes** (S_G) in micro-machining require the removal of small masses of the material and hence pulsed processing. In order to be also economic an adequate output power is necessary (**scalability**). For the selection of the laser light it is of importance whether or not the process can be sufficiently stationary, which wavelength is favourable and how the energy can be transfered to the material as a function of time. Finally, investment and running costs have to be considered.

3.1. STATIONARY PROCESSING

The energy E to remove a unit volume V is approximately constant:

$$\frac{E}{V} = \frac{P_{Av}}{v_f \cdot b \cdot d} = const\ . \tag{1}$$

P_{Av} = mean power = $P_L \cdot \tau \cdot f_P$, P_L = pulse power, τ = pulse duration, f_P = repetition rate, v_f = feed rate, d = cut curve depth, b = cut curve width.

The relative geometric deviation for cutting a sheet of given thickness d is given by:

$$\left|\frac{\Delta b}{b}\right| = \left|\frac{\Delta v_f}{v_f}\right| + \left|\frac{\Delta P_{Av}}{P_{Av}}\right| \tag{2}$$

Stationary processing means that $\Delta b = 0$.
Stationarity is certainly fulfilled for constant laser parameters, surrounding conditions and homogeneity of the material. Apart from the fact that the material is never homogeneous on a small scale, the detection and control of the laser parameters in real time is very difficult. A more realistic alternative is to work in a repetition mode and to level statistical deviations by the number of pulses. For short pulsed lasers with high repetition rates like the CVL it is possible to overcome acceleration problems by an analog coupling of feedrate and repetition rate. Moreover inaccuracies of the handling system can be avoided by a flash on the flight technique. For a MOPA-chain [4] both is possible by a suitable synchronization unit (master timing system).
Stationarity of the drilling process is given if the drilling speed v_{dr} during a single laser pulse is constant over the whole pulse length. This is fulfilled by processing with copper lasers with high intensity, see Fig. 1.
It should be mentioned that the feed rate v_f, the drilling speed v_{dr} (the actual velocity at which the solid /vapour-interface moves) and the ablation speed v_{abl} (mean velocity of the ablated material) have to be distinguished.

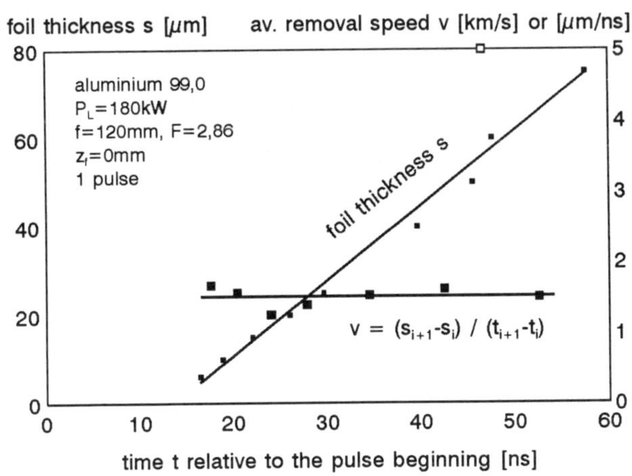

Figure 1: Measured drilling speed for drilling aluminum foils of different sheet thickness by a single CVL laser pulse.

3.2. SELECTIVITY AND ASPECT RATIO

The structure size S_G for a simple meander is given by the sum of web width and the width of the cut curve. A minimal width of the cut curve needs small focal radii and a high selectivity. For material processing with a CVL, cut curve width and drill hole diameters of less than 8μm and web width of about 10 μm are achievable, Figure.2. To achieve ablation a material dependant threshold value ϕ_{th} of the fluence ϕ is necessary. Precision therefore requires also a high gradient of the fluence across the beam at the threshold ϕ_{th} (where $\phi_{th} = \phi(r_{th})$). The more the intensity distribution on the target is top-hat like, the higher the **selectivity**:

$$\textit{selectivity} \quad S: \qquad S := \left. \frac{\partial \phi}{\partial r} \right|_{r_{th}} \tag{3}$$

To fulfill this condition throughout the depth of the work piece a suitable **aspect ratio** must be given:

$$\textit{aspect ratio} \quad A_V: \qquad A_V = \frac{d}{b} \propto \frac{z_R}{r_f} \tag{4}$$

z_R = Rayleigh-length = $0{,}3721 \cdot \lambda \cdot f^2 / R^2$, f = focal length, $2 \cdot R$ = beam diameter at the lens, d = cut curve depth/drill hole depth, b = cut curve width/drill hole diameter, r_f: focal radius

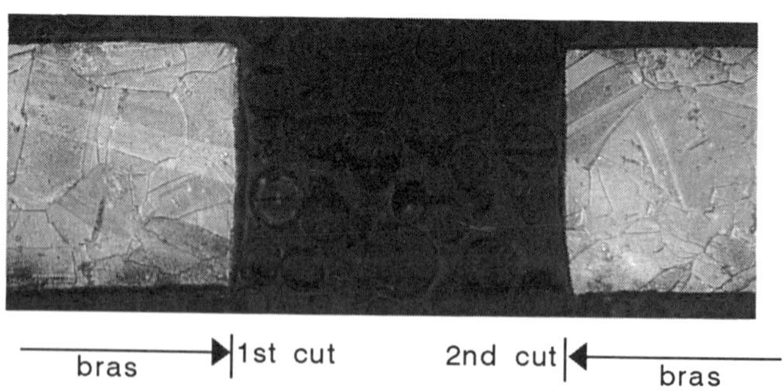

Figure 2: Two cuts with the CVL in a 200μm thick sheet bras.

Therefore, small focal radii and a long Rayleigh length are requested. Together with the desired structure size this means that diffraction limited beams have to be used together with lenses of long focal length. For different divergencies of CVL beams Figure 3 shows radii and Rayleigh lengths as function ot the focal number F. Figure 4 gives drill holes with a small diameter and a high aspect ratio, respectively. More generally it is necessary for a high selectivity to avoid processing with a point source since three dimensional heat conduction will always lead to semi-spherical temperature distributions. A plain source (1-d-heat conduction) with a diameter of the structural size allows the deduction of the maximum pulse duration τ for which one dimensional heat conduction is given: $d_{focus} >> \sqrt{k\,\tau} = \delta_{th}$ (k: thermal diffusivity, δ_{th}: thermal penetration depth). For a top hat distribution in the focused beam, a focus diameter of 10μm and typical numbers for the thermal properties of the material (copper) it turns out, that pulse durations on the nanosecond scale and smaller are suitable while microsecond pulses do not fulfill the above condition.

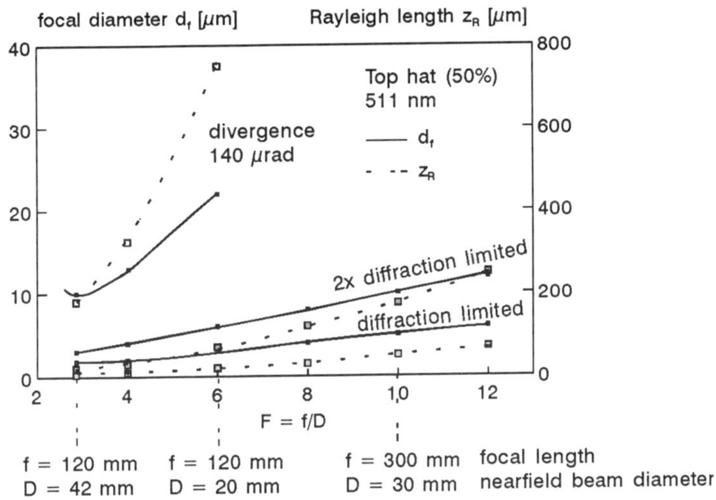

Figure 3: *Focal diameter as a function of focussing number and Rayleigh length.*

Figure 4: *Drill holes with a) small diameter b) large aspect ratio*

3.3. QUALITY OF LASER MACHINING

Talking about quality in laser machining one first of all considers micro-contour deviations and thermal damage of the target in the **heat affected zone (HAZ)**. During a stationary drilling process the heat affected zone is continuously removed resulting in an effective time independent value of k/v_{dr}. In principle, there are two ways to minimize the HAZ. One can either try to achieve a stationary drilling process or to reduce the pulse length to a degree where the calculated room temperature value for the thermal penetration depth is smaller than that of the optical penetration depth δ_{th} [1].

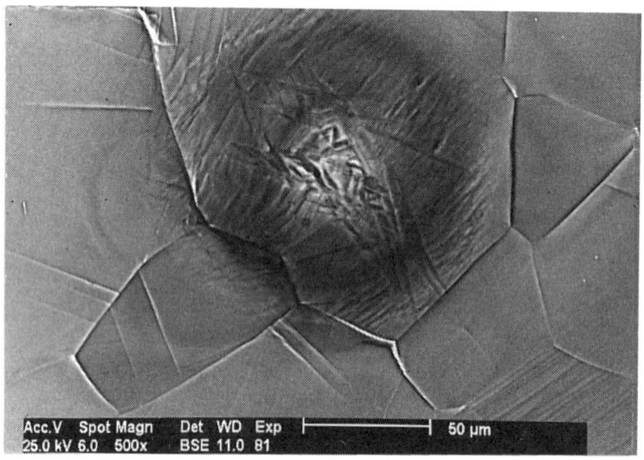

Figure 5: Deformation affected zone (DAZ) on the rear side of a titanium sheet drilled with the CVL at high intensity.

However, since for short pulses about 100fs the ablation occurs much later than the end of the pulse, but is still of thermal nature [3], pulse shortening is not a real alternative. Stationary drilling requires high power flux densities to achieve a sufficiently high velocity of removal from the target. The limiting factor is the drilling speed into the target which has to be considerably smaller than the speed of sound (1-10km/s) otherwise shock phenomena will occur, see Figure 5. If this is the case, the mech anical damage will limit the obtainable precision as a **deformation affected zone (DAZ)** results. One can calculate that $5 \cdot 10^9 W/cm^2$ of absorbed intensity is sufficient to generate a DAZ larger than the HAZ. Considerable shock effects will form for pulse durations below 100ps if the same ablation rate is achieved as for ns-pulses ($v_{dr} >$ 10km/s). One could argue that a single pulse with small pulse energy leads to uncritical deformations. However, in order to remove a certain mass of material either higher energies per pulse or a higher number of pulses are required. As the mechanical influence per pulse is nearly additive, both alternatives will lead to a similar DAZ.

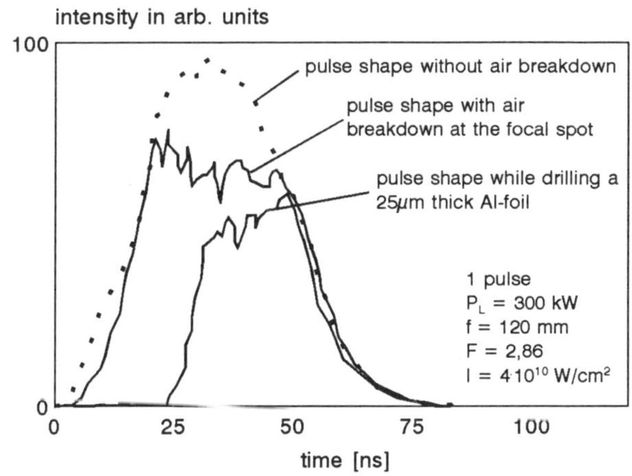

Figure 6: *Plasma transparency during optical air breakdown . Shown are the incident and shielded pulse.*

Technical processes are mainly carried out in air. High intensities could than lead to undesired plasma effects like shielding (loss of efficiency), scattering (loss of precision) or secondary effects damaging the target (plasma heating, recoiling, debris deposition). One can minimize transparency losses by using shorter and shorter wavelengths. However, it was found that best plasma transparency occurs in the visible region [2] of the spectrum, see Figure 6. Much smaller wavelengths cause direct ionization due to the higher photon energy while for a longer wavelength (e.g. IR) absorption via inverse Bremsstrahlung dominates. Both can cause the formation and propagation of dense air plasmas which lead to shielding. Two physical phenomena could be used to overcome plasma effects at high intensities. One is pulse-shortening (see above) the other is to reduce the focal volume (irradiated area ·Rayleigh length) so that an air plasma can not build up due to losses of electrons diffusing out of the focal volume [2]. As precise machining needs small focal diameters anyhow, it is obvious to choose that way. Due to the variety and complexity of machined parts numerous quality criteria depending on the actual geometry exists. As a more general number to characterize the **quality Q** it is suggested to use the ratio of ablated volume V_{abl} to damaged volume V_{dam}:

$$\textit{quality number } Q : \qquad Q := \frac{V_{abl}}{V_{dam}} \qquad \mathbf{(5)}$$

Here the damaged volume means the sum of the largest micro-contour deviation generated and the maximum of either HAZ or DAZ.

3.4. GENERAL CRITERIA FOR PRECISION MACHINING

For a given task it is always possible to describe precision by a geometry dependent characteristic number. As a more general criterion it is suggested to use two different numbers. One of them is the above mentioned **quality number Q** the other is the **degree of miniaturization M** (Figure **7)** which can be derived from a meanderlike structure as the product of aspect ratio and inverse structure size S_G. Which of the both numbers is more essential for a given process depends on the application.

$$\textit{degree of miniaturization} \quad M: \qquad M = A_V \cdot \frac{1}{S_G} \tag{6}$$

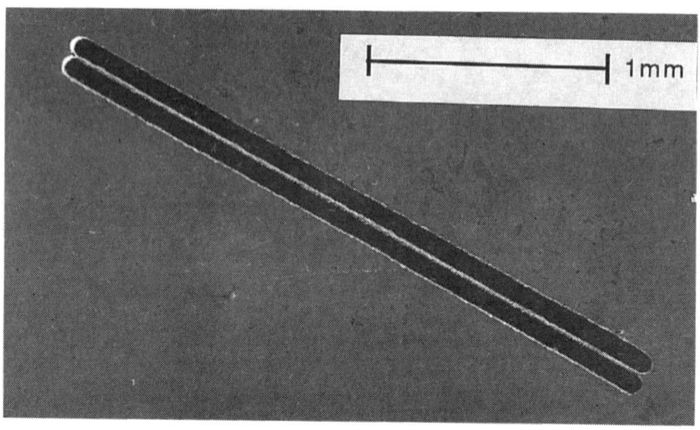

Figure 7: Cuts with the CVL demonstrating the degree of miniaturization.

3.5. MATERIAL ASPECTS AND EFFICIENCY OF REMOVAL

The amount of material that can be removed during a laser pulse depends on the material processed and how efficiently the energy of the laser pulse is converted into removal of material. So far, we found for all materials investigated that with increasing intensity and pulse duration the drilling velocity can be described schematically as shown in Figure 8. At a given intensity four different stages of drilling can be distinguished. These are: heating without drilling, non-stationary drilling above a material dependent threshold value (time-dependent drilling speed), stationary drilling (time independent drilling speed) and plasma limited drilling (reduction of the drilling speed due to shielding effects).

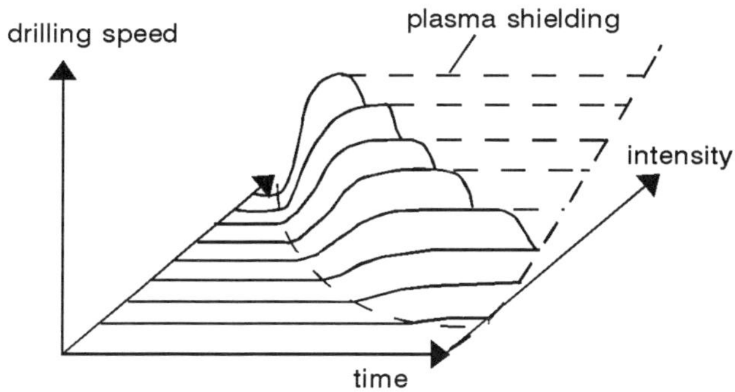

Figure 8: Schematic drawing of the drilling speed as function of time and intensity.

Increasing the intensity it is only possible to increase the drilling speed to the degree of heat transport into the target. For conductors heat transport occurs via electrons moving with Fermi velocity (10^6m/s) much larger than the drilling speed, hence no limitation occurs. Most dielectric materials have a high starting transparency until with increasing intensity the dielectric breakdown occurs. From then onward heat conduction limits the drilling speed in non-conductors where heat transport occurs via phonons whose maximum speed is limited by the speed of sound. As the irradiated energy can no longer be used to increase the drilling speed the energy is lost for the drilling process and converted into the ablation speed, multiple ionization and build up of an air plasma.
Staying below critical drilling speeds, which is for metals certainly the case for intensities up to 10^{11}W/cm^2, one empirically finds that materials are drilled more easily the lower is its hardness, shear modulus G (Figure 9) and binding energy per volume E_{spec}. Each of these properties is linked with the strength of the lattice. With the empirical rule that the specific energy per volume E_{spec} required for removal is proportional to the shear modulus. It is possible to define the **efficiency** η of converting the laser energy into the removal of material. For copper vapour laser processing with the above intensities we found numbers of about 10% for single pulses and all metals investigated, equation (7).

$$V_{abl} \propto \frac{1}{G} \quad \wedge \quad E_{spec} \propto G \qquad \Rightarrow \qquad \eta \approx const. \tag{7}$$

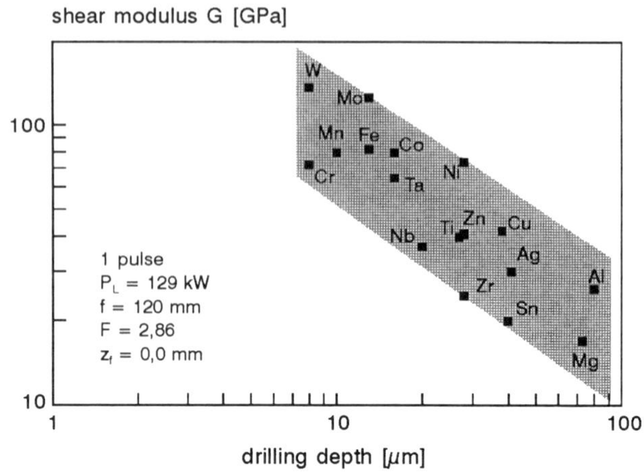

Figure 9: *Single pulse ablation depth as function of the shear modulus for different metals.*

3.6. FURTHER CRITERIA OF INDUSTRIAL RELEVANCE

Technical applications require a high productivity. With the above defined material independent efficiency of the process, the **productivity γ** is given by the product of the above defined efficiency η and the mean output power P_{Av}:

$$\textit{productivity} \quad \gamma : \qquad \gamma = \eta \cdot P_{Av} \tag{8}$$

Since structural size and sheet thickness limits the scalability of the energy/pulse, the average power can only be increased by the repetition rate. It is than a question whether high frequencies cause increased shielding due to permanent plasma formation. Processing with copper vapour lasers, the opposite behaviour was found in the range of 100-6500Hz, see Figure 10.

A further technical important criterium is **flexibility λ**:

$$\textit{flexibility} \quad \lambda : \qquad \lambda = \frac{P_{max} - P_{min}}{\Delta t_{control}} \tag{9}$$

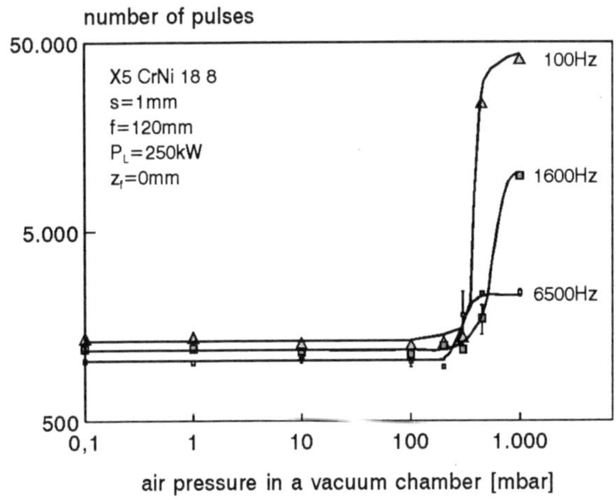

Figure 10: *Number of pulses required for drilling through a given sheet thickness as function of atmospheric pressure and repetition rate.*

Process flexibility is typical for laser machining. Machine flexibility is characterized by the amplitude of the intensity that can be varied without changing the beam properties (focal diameter, focal position) and the shortest time in which the amplitude can be changed. In general, it is much easier to alter the mean power and to keep the beam constant, if gas lasers are used instead of solid state lasers (deviation from the focal point position due to thermal distortion). For copper vapour laser processing the average power can be altered over five orders of magnitude from pulse to pulse by changing the repetition rate without changing the focal diameter and position (fast changing of the pulse duration is only possible, however with slight changes of the beam properties).
Lasers in manufacturing have to be integrated in a production line. One possible measure of the **integratability** ι is the mean power of the low divergent beam devided by the overall size (volume) of the laser :

$$\textit{integrability} \quad \iota : \qquad \iota = \frac{P_{Av}}{V_{laser}} \tag{10}$$

For a copper vapour laser MOPA-chain and a low divergent beam of 140µrad currently 100W/1.5m^{3}is typical.
Good integratability numbers are believed to be one of the challenges of direct pumped Q-switched solid state lasers. However, scaling the output power to a reasonable size the space for power supply and external cooling systems will restrict scaling with the same number of Watt/volume.

A last but important criterium is the **up-time** ξ of the laser source. It is given by the ratio of access time t_{access} to the working time t_{work} in a company:

$$up\text{-}time\ \ \xi: \qquad \xi = \frac{t_{access}}{t_{work}} \tag{11}$$

Today only limited information on the up-time of CVL's is published. Our own experience are 75% for one shift and 95% for three shifts in a company. The difference can be explained by the heating-up time of the laser.

4. Desired properties of a laser source for precision machining

A profile of a desired laser source for precision machining can be derived. It should be a visible, pulsed, diffraction limited beam source with pulse energies of 5-200mJ and repetition rates of about 10kHz giving a mean output power of about 100W. To achieve small HAZ and DAZ values the pulse length should be between 100ps and 200ns.

TABLE 2: Desired properties of a laser source for precision machining

property desired	property of CVLs	comment
visible wavelength**	511 and 578 nm	fulfilled
repetition rate 1-15kHz	6,5-13 kHz	fulfilled
pulse length 100 ps - 200 ns	6-150 ns	fulfilled
pulse energy 5-200mJ	1-25 mJ	fulfilled
mean power > 100 W	0-250 W	fulfilled
diffraction limited beam	divergency < 140 μrad	fulfilled
fast shutter	master timing system	fulfilled
delay-free triggering	4 ns	fulfilled
precise and flexible beam delivery	wave guide (glass fiber)	fulfilled
integrability	100W /1.5m^3 diffraction limited	improvement req.
low running costs	100 W diffr. lim. DM 18/h	current value
long inspection intervals	$4 \cdot 10^{10}$ pulses /thyratron	improvement
	copper refill all 500h	required
instant access	2 hours heat-up time	developement
flexiblity	fast power control	fulfilled
	via rep. rate P_{max} in 10^{-5} sec.	constant divergence
	via pulse duration in 10^{-5} sec.	changing divergence

** It should be mentioned that for some materials or specific media a specific wavelength is required.

For coupling the source with a handling system a fast delay-free shutter and a precise and flexible beam guiding system (glass fiber) is essential. Table 2 summerizes the desired properties showing where the CVL fulfills the desirements and where further developement is necessary.

5. Possible applications of CVL-machining

In spite of the short period in which CVL´s have been used for precision machining a number of promising requests from the manufacturing industry have been registered. Table 3 summarizes different areas of industry and tasks to be carried out with CVL´s. In Fig. 11-13 typical examples are shown.

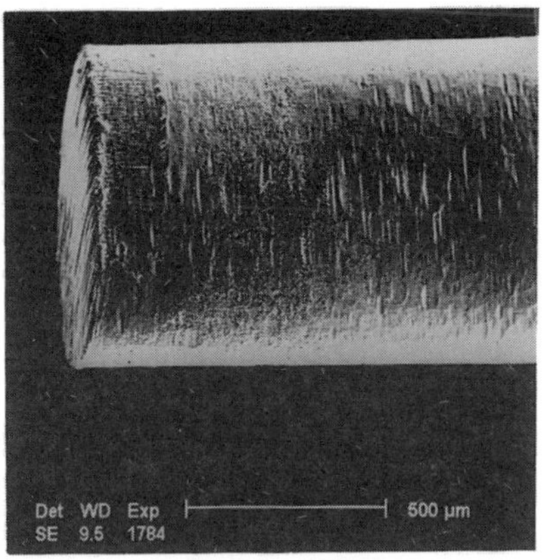

Figure 11: Cutting wires into sections

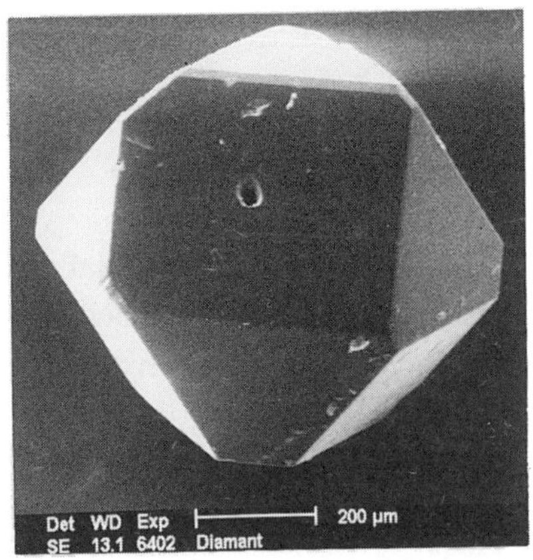

Figure 12: Drilling of diamonds.

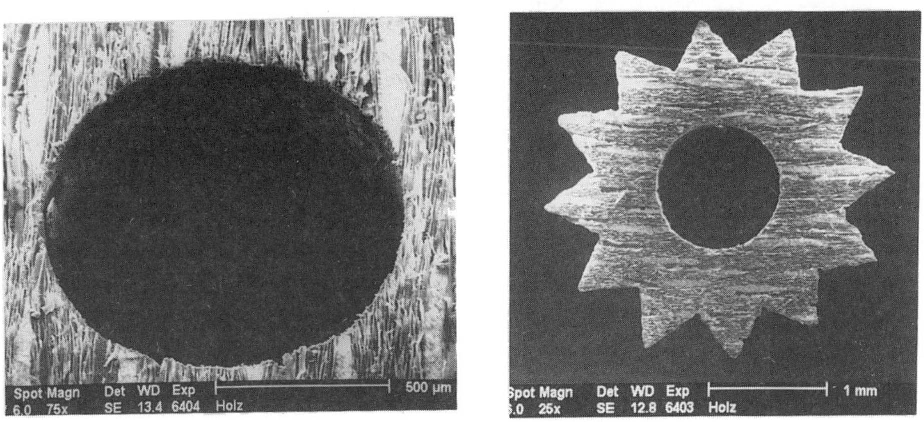

Figure 13: Cutting and drilling of wood

TABLE 3: Possible applications of CVL precision machining

branch of industry	typical task
medical technique	scratching or cutting of ceramics
automotive industry	scratching of connecting rods
	drilling of injection nozzles
(micro-)electronics	cutting SMD templates
measuring instruments	drilling of sample holders for balances
wood processing	cutting of decorative inlays
spray technique	drilling of nozzles
precision mechanics	cutting wires into sections
jewellery	drilling and cutting gem stones and noble metals
microsystem technique	cutting of micro parts like gears
packaging	cutting of foils without damaging goods
metal forming	drilling of drawing dies

6. Acknowledgements

The authors acknowledge financial support from the following organisations: DFG for fundamental research regarding beam solid interaction, BMBF for fundamental research concerning precision machining, the bavarian government for application related research and development.

7. Literature

[1] Preuss, S., Demchuk, A., Stuke, M., *Sub-picosecond UV laser ablation of metals*, Appl. Phys. A 61, 33-37 (1995)

[2] DeMichaelis, *Laser Induced Gas Breakdown: A Bibliographical Review*, IEEE Journal of Quantum Electronics, 188-202 (1969)

[3] Luft, A., private communication

[4] Bergmann, H.W., Hartmann, M., Mayerhofer, R., Körner, C., *Kupferdampflaser für die Mikromaterialbearbeitung*, Proc. of the 12th International Congress (LASER'95), Meisenbach, ed. Geiger, M. (1995)

INDUSTRIAL APPLICATIONS OF HIGH-POWER COPPER VAPOR LASERS

BRUCE E. WARNER, CHARLES D. BOLEY, JIM J. CHANG, ERNEST P. DRAGON, MARK A. HAVSTAD, MARK MARTINEZ AND WILLIAM MCLEAN II
Lawrence Livermore National Laboratory
P.O. Box 5508, MS L-467, Livermore, CA 94551, USA

Abstract

A growing appreciation has developed in the last several years for the copper vapor laser because of its utility in ablating difficult materials at high rates. Laser ablation at high rates shows promise for numerous industrial applications such as thin film deposition, precision hole drilling, and machining of ceramics and other refractories.

1. Introduction

Since its invention in the late 1960s [1], the copper vapor laser has been developed predominantly for laser isotope separation [2]. In this application the laser is valued because of its wavelength, 50 ns pulses, multi-kilohertz repetition rate, and 1 - 2% electrical to optical efficiency. As this laser system has matured into an industrial tool, new material processing applications for the laser have developed. When applied to materials processing, the visible wavelength is found to couple well to most materials, the short pulse length causes ablative material removal, and the high repetition rate increases the material removal rate to a moderate level. A further advantage of the laser is near diffraction-limited operation at hundreds of watts average power, allowing precise spatial control.

This paper is segmented into three major sections; a review of modeling and measurements that help illuminate the physics of copper laser beam - material interaction and a discussion of two applications of high-power copper lasers, precision hole drilling and pulsed laser deposition. In each of these applications the unique suitability of the copper laser will be discussed along with results of tests performed at LLNL.

2. Physics of copper laser beam - material interaction

Over the past couple of years LLNL has investigated both a one-dimensional model and experimental measurements of copper laser beam/material interactions. This effort has helped aid our understanding of the influence of

C. E. Little and N. V. Sabotinov (eds.), Pulsed Metal Vapour Lasers, 331–346.

processing parameters (irradiance and material type) and on material processing (material removal rate). First in this section is a discussion of the model with results and subsequently, a discussion of measurements of a plume expansion off an aluminum target.

The model involves a one-dimensional description of heat transport below a surface, hydrodynamic expansion of the vapor and compressed air, and light propagation through the vapor to the vaporization surface [3,4]. Figure 1 details the geometry of the model. On the left lies the solid material. Laser light strikes the surface (from the right in the figure), causing a melt layer to develop and propagate inward into the solid material. When the vaporization temperature is reached, the vapor begins to blow off, preceded by a plug of compressed air. This is bounded by a shock front. When the temperature becomes sufficiently high, both the incident beam and the reflected beam (to the extent that it is appreciable) can be absorbed in the vapor by photoionization and inverse bremsstrahlung.

The pressure and energy of the vapor require an equation of state. Since substantial ionization may be present in the bulk of the vapor, the electronic contributions to the pressure and energy density must be considered. Simple estimates show that, for time scales longer than a few nanoseconds, the assumption of local thermodynamic equilibrium is fairly well justified. Hence we have developed and utilized a Saha equation of state package, assuming hydrogen-like levels between ionization limits. It agrees with more sophisticated codes in the regime of interest. The total pressure is $p = (n + n_e)kT$.

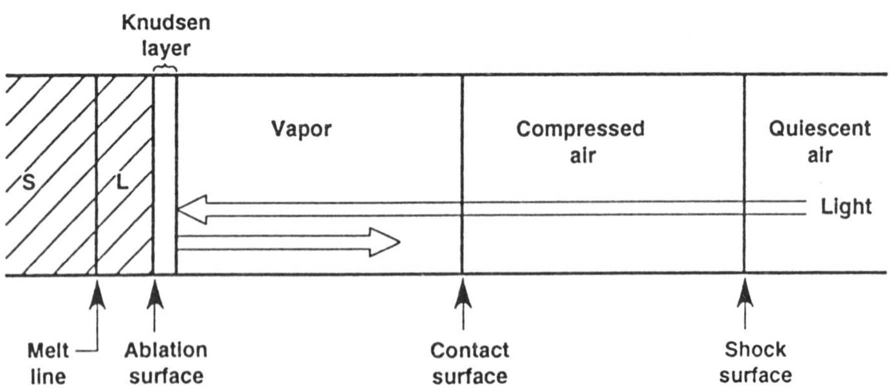

Figure 1. Geometry of the laser ablation model (not to scale).

For the calculations reported here, a 510.6 nm beam is incident on stainless steel at an average irradiance of 10^{10} W/cm^2 (i.e. 45 W at 4.5 kHz with a spot size of 50 μm). The temporal light pulse rises rapidly to a maximum in 8 ns, and falls off to 10% of maximum at 70 ns. The ambient atmosphere is assumed to be air, though no chemical reactions are modeled. Figure 2 displays the material temperature spatial profiles for three times after the light is first incident on the surface. Note that z = 0 is the original surface position. At 30 ns, the ablating surface temperature is 6500 C, having already peaked at 12,000 C after 5 ns of expose to the beam. At 30 ns, 0.5 μm of the steel has been removed by vaporization. As displayed by the 30 ns curve, the temperature gradient into the steel is quite steep; the material 2 μm from the original surface is only starting to be heated!

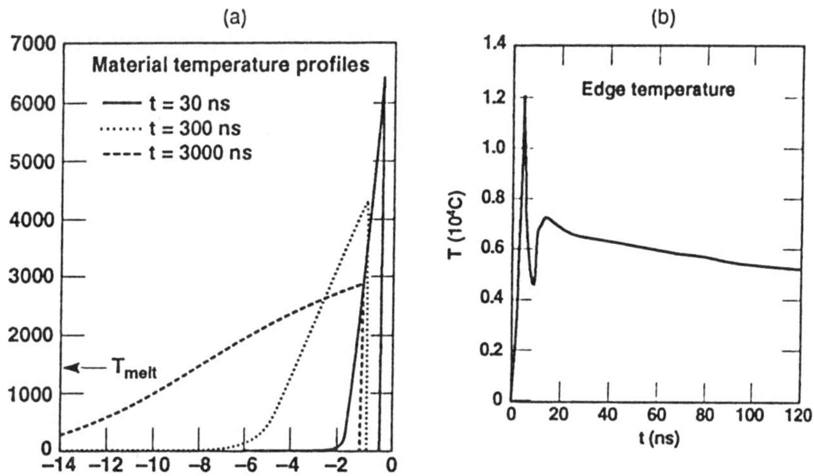

Figure 2. Temperature profiles of the stainless steel material. Note z is the dimension into the surface.

At later times the light pulse is over and consequently the surface temperature has dropped. In addition, our model indicates that at the longest time, 1.1 microns of steel has been ablated with 8 microns of material that have been raised above the melting temperature but not vaporized.

The response of the material during the laser pulse is dominated by two phases at 10^{10} W/cm^2. Early in the pulse the surface rapidly heats by direct laser illumination to 12,000 C, quickly propagating a vapor front with air being pushed by it. Our code indicates that under these conditions, a second phase occurs. At 8 ns, the laser beam is then nearly completely absorbed by this propagating vapor plume because of photoionization of the vapor. From 8 ns until 12 ns the surface receives negligible direct laser heating. As the vapor continues to expand, a small fraction of the laser beam (about 5% of the incident laser energy) transmits through the vapor to the ablating surface. Common terminology for this phenomena is laser detonation, see reference [5]. Our calculations indicate that 15% of the laser pulse energy directly heated the vaporization surface, and 85% of the energy was conducted to it by electron thermal conduction. The thermal conduction phase starts shortly after the laser detonation wave (12 ns) and lasts several microseconds. During this phase the surface temperature slowly decays from 6000 C to 3000 C.

Our model indicates that laser detonation occurs on steel at irradiance above 3×10^8 W/cm^2. Calculations for aluminum show similar thresholds. In the high intensity regime most of the energy transported to the surface is by electron thermal conduction from the vapor plume. Our calculations (and measurements) indicate that at 50 ns after initial illumination, the vapor front can be a large fraction of a mm in lateral dimension. This is many times the diameter of the laser spot, implying a loss of spatial control after laser detonation occurs.

We have measured important properties of copper laser beam/material interaction. From these measurements, estimates of vapor density, vapor temperature, surface temperature and pressure can be made through hydrodynamic relationships. In addition, onset of plasma shielding (laser detonation) are easily observed.

Standard shadow and Schlieren techniques were used in our experiments. A Q-switched and frequency-doubled YAG laser was used as a light source with 20 ns duration. In addition, a short pulse flash lamp was also used in combination with a gated (20 ns) camera. In each case the vapor plume was imaged onto the cameras by an achromatic lens with sufficient magnification to yield ~50 micron resolution. Fresh target material was constantly supplied after each laser pulse via a rotating target. The vapor plume was produced under ambient air without any assist gas.

Figure 3 illustrates images of vapor expansion when laser light was focused on an aluminum target. At a laser intensity of 0.93×10^9 W/cm^2, the vapor leaves the surface in an expected spherical expansion with a well defined vapor front. As the laser intensity is increased, an additional cone-shaped plume starts to appear. At 7.3×10^9 W/cm^2, the cone shaped expansion is well defined (figure 3). This feature is not correlated with the position of the surface, but with the direction of the incoming laser beam. This strongly suggests that the cone-shaped plume was initiated through a strong plasma absorption of laser light that generated a laser supported detonation wave. Our measurements at various laser intensities indicate that the laser threshold for this plasma ignition is approximately 1.5×10^9 W/cm^2 for carbon steel and 2×10^9 W/cm^2 for aluminum. These measured thresholds are within a factor of 5 of those predicted by our one-dimensional computer code.

We have measured vapor expansion dynamics and electron temperature and density by line emission spectroscopy. By iterative application of hydrodynamic relationships for adiabatic shock expansion, vapor density and temperature can be inferred. This results in a vapor density of the primary plume of ~ 3×10^{20} cm^{-3} at the vapor front and 10^{21} cm^{-3} near the surface. The electron temperature and density measurements yield 3.5 eV (peak) and 4.2 x 10^{18} cm^{-3} at a laser intensity of 10^{10} W/cm^2. With these values of vapor density and electron temperature and density, photoionization is predicted to dominate the laser absorption over inverse bremsstrahlung, consistent with our one-dimensional model. At 10^{10} W/cm^2 and just after the laser pulse bas terminated, we infer from our measurements an absorption depth of only 20 microns versus a vapor plume dimension of greater than 500 microns.

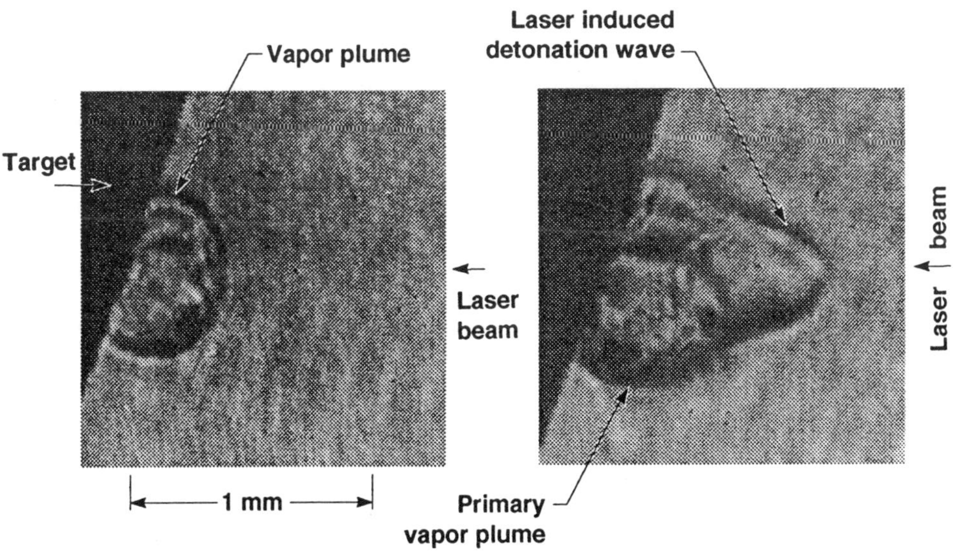

Figure 3. Shadow image of a copper laser produced vapor plume from an aluminum target. Both images produced at t = 50 ns (a). Image produced with a laser intensity of 0.93 x 10^9 W/cm^2 (b). Image at a laser intensity of 7.3 x 10^9 W/cm^2.

3. Precision hole drilling with a copper vapor laser

Laser drilling has been widely used in industry because of its high production rate, ability to rapidly vary hole size, ability to drill holes at shallow angles, and ability to drill traditionally hard to work materials such as ceramics and composite materials. We have extended this utility to sub-millimeter dimensions by use of a near diffraction-limited copper laser and precision wave front tilting technology.

The ability to focus a laser beam is one of the most important characteristics of a laser micro-machining system. In order to produce a small, circular laser focus spot with a reasonable focal depth (20—30 times the spot size), the laser beam needs to be near diffraction limited with wavelength in the visible or UV range. To minimize the heat affected zone, the duration of laser pulse needs to be less than a few microseconds such that heat conduction to the material during drilling is negligible. Multi-kilohertz operation is preferred to remove the material in a more controllable fashion, such that material removed during each pulse is small while the processing speed is maintained because of high pulse rate. A copper laser, with its high repetition frequency (4—20 kHz) and short output pulse (30—100 ns) in the visible spectrum, has been considered an ideal tool for laser micro-machining [6,7,8]. However, the difficulties in control of CVL beam quality and in optimization of processing parameters has prevented it from generating repeatable high precision results. Our recent development of high-beam quality CVLs and precision wave front tilting technology has led to success in laser precision micro-fabrication.

The schematic of our copper laser based micro machining system [9] is illustrated in Figure 4. An injection controlled copper laser oscillator [10] with near-diffraction-limited beam quality is used in this system. A power-in-the-bucket measurement indicates ~73% of the laser output is within diffraction limited beam divergence. The laser is designed to be operated between 4–9 kHz for various material processing requirements. This laser generates about 30 W output at wavelengths of 511 nm (60%) and 578 nm (40%). The self-terminated CVL used in our system typically has a pulse duration of 50–70 ns. The laser beam is focused by a diffraction limited achromat on the working piece that typically generates a spot size of approximately 20 microns when an F/10 system is used (i.e., a diffraction limited spot size of about 15 microns). As shown in Figure 4, a precision two-axis scanning mirror is used to tilt the laser wave front before passing through the achromat with an angular resolution better than 5 microradians. This scanning mirror is equipped with a feedback loop to compensate for hysteresis effects. The beam scanning frequency on the work piece can be varied from 1 Hz to about 100 Hz depending on application requirement. A camera is used to observe the progress of material processing on the working piece based on reflected laser light.

In copper laser percussion drilling, a stationary laser beam (scanning mirror turned off) is used to drill. Straight parallel holes with aspect ratio better than 40:1 have been repeatedly produced in a variety of steels; 20–25 micron diameter holes percussion drilled in 1 mm stainless steel are typical. Because the material removal is mostly through laser ablation, these holes show no measurable heat affected zone. Smaller hole sizes can also be achieved using a lower laser power or faster focusing optics, but this also reduces material penetration with the possibility of generating tapered holes. Since the coupling between the side wall of a drilled hole and the laser beam is very poor due to high incidence angle, our experience indicates that the laser peak power on the hole entrance must be more than a few GW/cm^2 to avoid hole taper. This intensity is just at the laser detonation threshold.

Assist gas was not found to be essential in CVL micro drilling, but there is evidence that oxygen increases drilling speed due to exothermic reactions. Although straight holes free of recast can be produced by a simple CVL percussion drilling at a fairly fast speed, the hole dimension control and hole repeatability are not satisfactory for high precision applications. Typically a tolerance of hole size and roundness of approximately 10% of its diameter is expected. Higher precision micro drilling must be accomplished by laser trepanning.

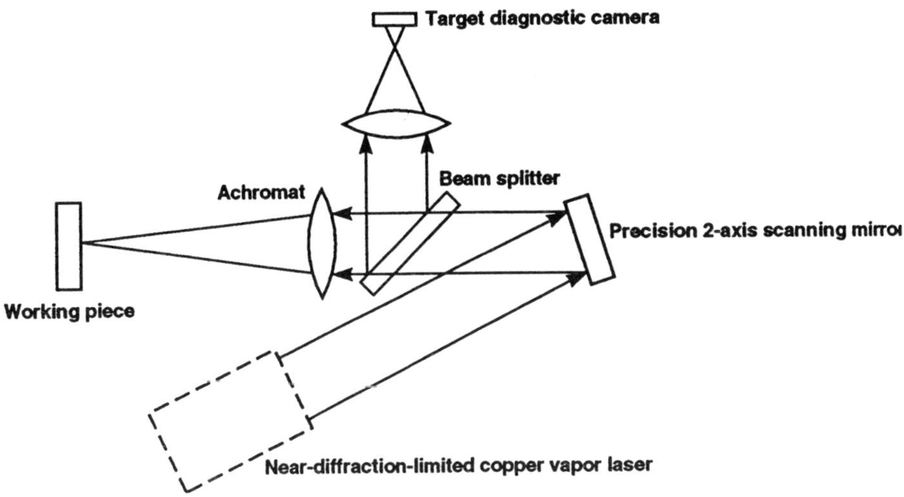

Figure 4. The schematic of a CVL micro-machining system.

Laser trepanning has long been applied in industry to either improve the hole accuracy or to generate large holes, mostly for hole aspect ratio less than 1:1. Precision micro trepanning for higher aspect ratio holes has rarely been accomplished because of difficulties in material penetration and removal. In our investigation, laser trepanning is achieved by tilting the X-Y scanning mirror sinusoidally such that the laser spot generates a circular pattern on the work piece. Trepanned holes down to 50 microns in diameter have been drilled this way.

Figure 5 illustrates three CVL trepanned holes with 185 micron diameter in 1 mm stainless steel. The sample was ultrasonically cleaned after drilling without additional post processing. Minor edge erosion with depth less than 10 microns was found on the laser entrance side as shown in Figure 5a. However, examination of sectioned holes indicated the material underneath of this area was not affected. The holes on the laser exit side, shown in Figure 5b, demonstrate extremely well defined circular hole pattern without erosion. Measurements indicate that both the entrance and exit hole have a roundness error of about 5 microns caused by various mechanisms.

It is obvious that the geometry of the entrance hole is predominately determined by the laser scanning pattern and the shape of a laser spot. Because of the existence of some residue hysteresis of our scanning mirror system, the laser scanning pattern actually has a ~2% roundness error that explains the dimension error of the entrance holes shown in Figure 5a. On the other hand, the geometry of the exit holes, although primarily controlled by the laser scanning pattern, is also strongly affected by the polarization state of the laser beam. Because of the difference in surface absorption between S and P polarized beams, especially at high incident angles, the laser beam material removal is more effective along the beam polarization because of its more effective coupling with the side wall. On the material entrance surface, the laser-material coupling is the same for both S and P polarization because of the normal incidence angle. The geometry of an entrance hole is thus not affected by the state of beam polarization.

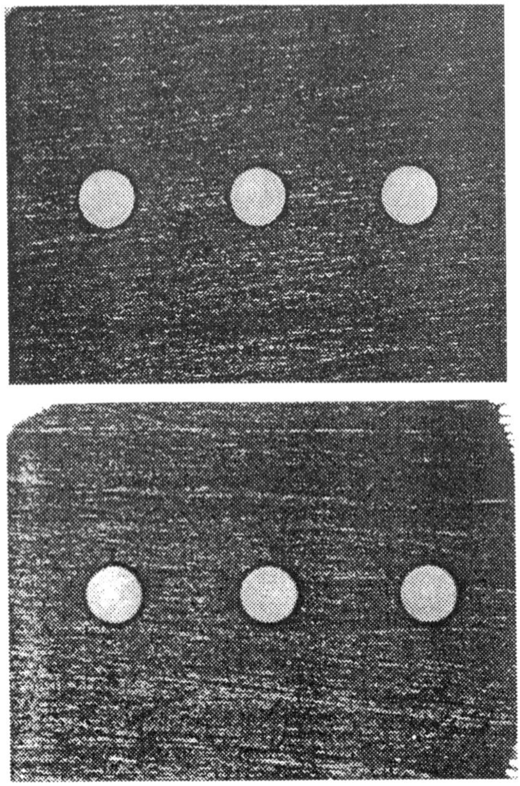

Figure 5. Precision laser trepanned holes with hole diameter of ~200 microns; (top) laser entrance side, (bottom) laser exit side.

Repeatability of hole dimension has been significantly improved with CVL trepanning as demonstrated in Figure 5. The variations of hole size and shape were almost impossible to measure based on our microscope measuring system. This striking improvement on hole repeatability is believed to be mainly due to the fact that laser trepanning not only performs material removal during the initial drilling phase, but also engages in material trimming and side wall polishing during the later drilling phase.

A section of the trepanned holes indicate a fairly straight side wall with surface roughness measured at 1-2 microns (i.e., peak to valley) which is comparable to the low end of a grinding finish. A minor curvature on the side wall may be caused by the caustic surface of the focused laser beam and can be corrected with a longer focus lens. Detailed examination of the hole section reveals that the recast layer and heat affected zone are minimized to a non-measurable level (i.e., sub-micron). SEM pictures of a CVL trepanned hole clearly show an unaltered grain structure on the side wall as a consequence of effective laser ablation process during CVL trepanning.

In addition to circular holes, we have demonstrated non-circular holes with various geometries on a 1 mm thick stainless steel, as illustrated in Figure 6. These holes were cut by modifying the scanning pattern of the trepanning system to draw the desired shape on the target. Hole aspect ratio higher than 10:1 with no taper has been achieved with these non-circular trepanned holes. Corner rounding effects become more severe as hole size is reduced to smaller than 100 microns. This demonstration of non-circular micro holes with high aspect ratio reveals the great potential in laser trepanning applications and may have significant impact on engineering design that is traditionally limited to circular holes.

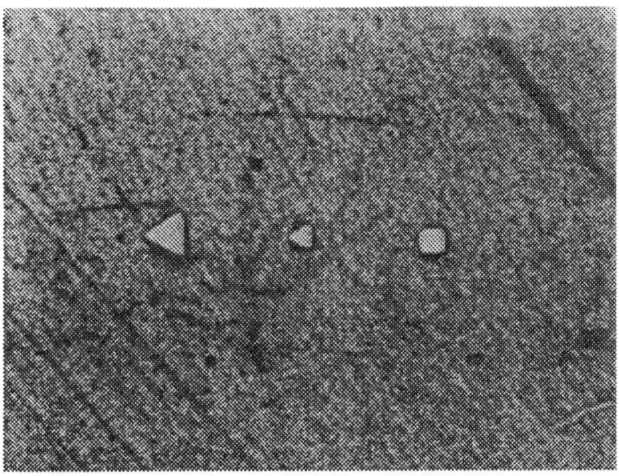

Figure 6. CVL trepanned holes on a 1 mm thick stainless steel that show a variety of hole geometries can be drilled with aspect ratio larger than 10:1. Hole dimensions from left to right are 156 μm, 84 μm, and 90 μm.

Copper laser trepanning represents a great potential in laser precision micro-drilling. Non-circular holes have been drilled by simply changing the laser scanning pattern. The hole quality drilled by this laser micro-machining system is found to be comparable or better than holes drilled by electric discharge machining (EDM). With the flexibility in hole geometry and size, and scalability to higher speed, this micro-machining system offers a superior technique to EDM machines which currently are widely used in industry.

We have found that there are four groups of industries that are interested in this precision drilling technology. They are the automobile and truck engine manufacturers for fuel injectors, the jet engine manufacturers for turbine blade cooling holes, industries that require precision orifices (ink jet, biochemical, etc.), and the electronics industry for "vias" in multi chip rnodules. Within these industrial segments, demand is for micron precision and drilling rates measured in holes per minute or higher.

4. Pulsed laser deposition utilizing a high-power copper laser

We like to think of pulsed laser deposition (PLD) with a copper laser as good utilization of the vapor ejected during precision hole drilling. This is not strictly true, but the same physical attributes that create ablative material removal also cause the transfer of high quality materials in a deposition chamber. As discussed in the first section of this paper, illumination of a target with a copper laser beam quickly heats the surface of targets to above 10,000 C. At this temperature most materials have very similar vaporization rates. In PLD the vaporized material is allowed to expand in a vacuum chamber before striking a substrate to be coated. Source material is typically replenished by moving the target relative to the laser beam.

Many high-value coatings have been produced by PLD. Examples are: high Tc superconductors ($YBa_2Cu_3O_7$) [11], ferroelectric films for nonvolatile memories ($PbZr_xTi_{1-x}O_3$) [12], and low work function field emitters (DLC) [13]. The majority of studies utilize short wavelength (193 nm [ArF] to 308 nm [XeCI]) light arriving in 20–50 ns pulses with peak irradiance of 10^8–10^{10} W/cm^2 at rates of 5–30 Hz. However, some authors report positive results with green (532 nm frequency doubled Nd:YAG) radiation sources operated under similar conditions. Typical film formation rates with these laser systems are on the order of 10 μm-cm^2/hr.

We have utilized a single copper vapor laser (CVL) oscillator and one high power amplifier for our PLD studies. In this paper we will only review our results on depositing diamond-like-carbon (DLC), ablated from a graphite target. Nominal operating parameters are summarized in Table 1. The optical delivery system was comprised of a discrete optics delivery system and a single 500 mm focusing lens located adjacent to the vacuum deposition tank. Relatively low irradiance (10^8 W/cm^2, below the laser detonation threshold) on the target was necessary to achieve high deposition rates and good materials properties. This was accommodated by defocusing the beam on the target.

TABLE 1. Operating Parameters.

Spot size on target:	50–300 μm
Irradiance	$2 \times 10^{10} - 4 \times 10^8$ W/cm^2
Pulse width:	50 ns (FWHM)
Target speed:	8 cm/s
Target - Substrate Distance:	7.6 cm
DLC deposition rate:	5.6×10^{-2} Å/pulse at 4×10^8 W/cm^2
	1.0×10^{-3} Å pulse at 2×10^{10} W/cm^2

A schematic drawing of the deposition chamber is shown in Figure 7. The laser light is introduced through a fused silica window. The graphite target is a rod of 1-inch diameter that is mechanically rotated at approximately 1 RPS and can be manually moved axially to keep a relatively fresh surface. The laser is aligned to strike the rod off center so that the laser plume is at 45 or 60 degrees to the incident beam. In general, the ablation plumes expand normal to the surface. The substrate to be coated is mounted within a sample port that is easily removed for deposition on multiple substrates. The graphite target to substrate distance is 7.6 cm. Viewports orthogonal to the laser and plume and in the scattered laser direction were used to help diagnose laser alignment, plume generation, and qualitatively assess macroparticle generation. A turbomolecular pump was used to achieve a base pressure of 5 x 10^{-8} Torr. For each set of deposition runs the vacuum chamber was heated up and outgassed. Typical vacuum levels during deposition episodes were in the 10^{-6} Torr range. To date, no assist or buffer gases have been used. Several different substrates have been coated with DLC. However, consistent with results of others, adherence of the films are substrate dependent.

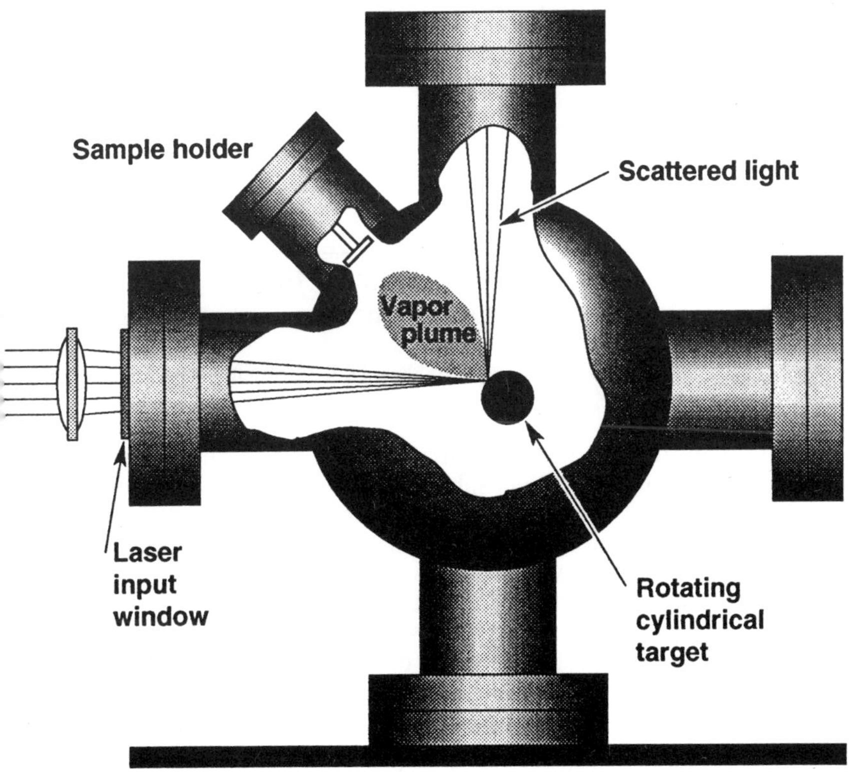

Figure 7. Developmental PLD chamber schematic.

Our early experiments investigated the deposition of DLC at laser irradiance of 10^8 W/cm^2 to 10^{10} W/cm^2. We have utilized both electron energy loss spectrum (EELS) and Raman scattering off of the DLC films to help quantify its bond structure. Figure 8 is a representative Raman spectrum of our DLC film grown with the above parameters. The EELS spectra and Raman spectra verify the DLC nature of the film. In addition, it is very hard. We have used a micro indentor to measure the hardness of these films to be 60% of crystalline diamond, ~58 GPa.

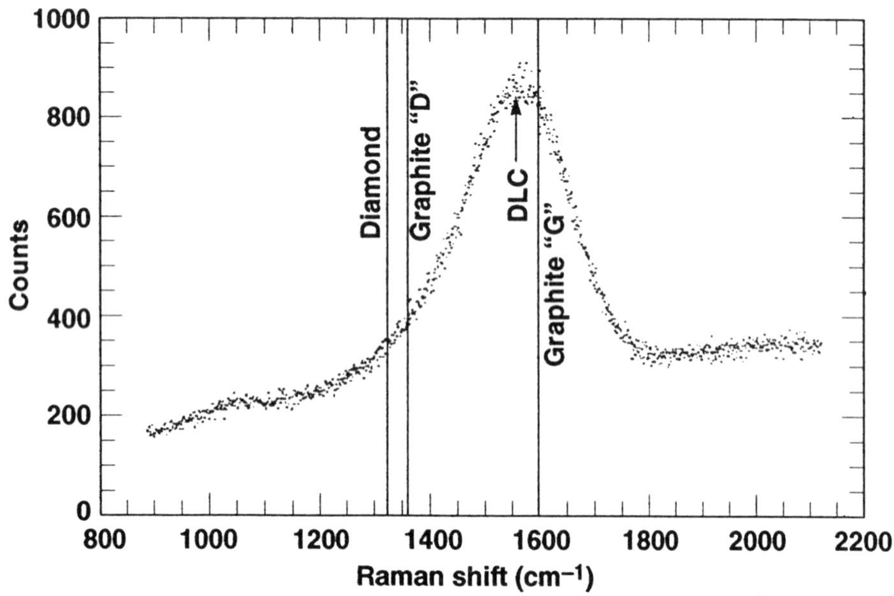

Figure 8. Raman spectra of copper laser produced DLC spectra indicates DLC.

We typically control both the target irradiance and number of laser pulses (in combination with a masked silicon substrate) to quantify the volume of DLC grown on the substrates. Its spatial distribution and its growth rate were determined by measuring film thickness and distribution with a diamond stylus profilometer at the step-up provided by the masked area Figure 9 displays a measured DLC deposition thickness across the substrate. The deposition has a very peaked character typical of PLD. It is this peaked film deposition that makes uniform film growth over large areas difficult. The deposition profile in Figure 9 has been fit with a plume distribution proportional to $\cos^{10.5}\phi$ (ϕ is the angle measured from the target surface normal). After 10^5 laser pulses at 4 x 10^8 W/cm^2 (190 W average power), we achieved 410 nm thickness at the center with half thickness at a radius of 4 cm. Since our copper lasers operate at 4.4 kHz, 10^5 pulses represents 23 seconds of elapsed time. Integrating the full volume of material and dividing by the elapsed time yields an average deposition rate of 2 x 10^3 μm-cm^2/hr. We have subsequently operated at 100 W average power at slightly lower target irradiance and have achieved 2.6 x 10^3 μm-cm^2/hr. At this rate we produce DLC coatings in minutes that would normally take hours by either low repetition rate lasers or RF plasma deposition (note that materials can have significantly different properties with RF plasma deposition). We have recently produced a 2.5 μm thick DLC film on a silicon substrate, with good surface characteristics. This film was grown with 58 seconds of laser on time.

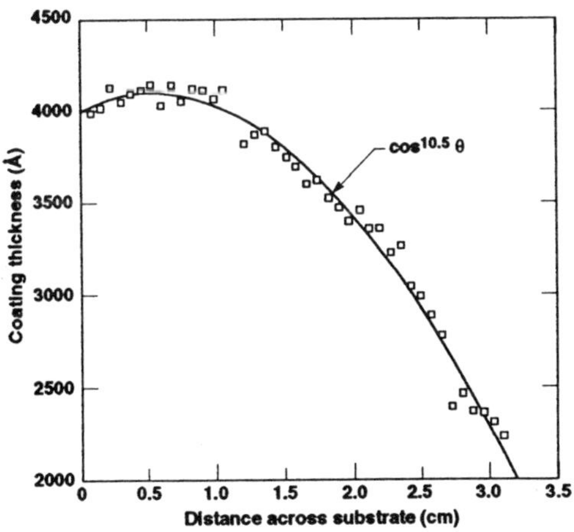

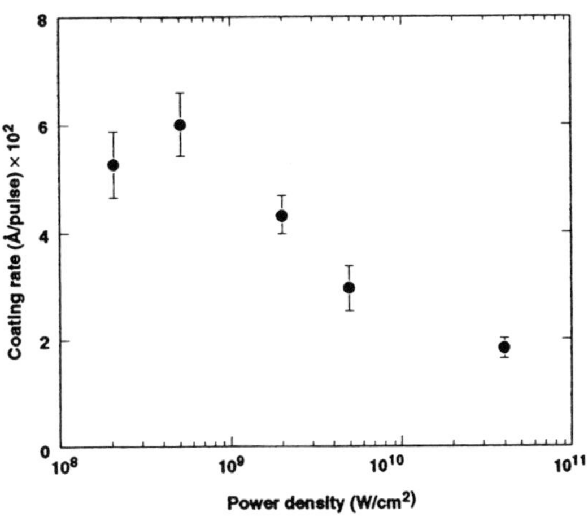

Figure 9. (top) DLC spatial distribution after 10^5 shots. (bottom) Coating rate at 100 W average copper laser power.

Figure 9(b) displays the measured coating rate for DLC versus laser irradiance on the target. Variations of irradiance were accomplished by defocusing the laser, but total energy on the target was kept the same. Notice the peak in coating rate occurs at just below the measured laser detonation thresholds for aluminum and steel. We expect to measure the threshold for carbon in the near future, but it appears that detonation slows down the vaporization rate of the graphite. We did not measure significantly different DLC properties at these intensities. Atomic Force Microscopy (AFM) was used to measure surface roughness. A 400 nm thick DLC film has a measured surface roughness (as measured by the AFM) of ≤10 nm. We have produced some films with RMS surface roughness as low as a few nanometers.

A promising pulsed laser deposition technology utilizing copper lasers has been demonstrated. Trial runs producing diamond-like-carbon have been encouraging. This technology appears to have significant advantages over other methods in deposition rate and surface morphology. More work is required to understand the physical characteristics of these coatings to verify their utility. Several proposed uses of DLC, or crystalline diamond films are: low work function emitters for flat panel displays, hard low-friction coatings for machine tool bits or infrared optics, and corrosion resistant coatings for chemical processing systems.

TABLE 2. Materials with difficult stoichiometries that have been successfully coated by PLD.

Material	Laser	Use
$BaFe_{16}0_{19}$	KrF	High density hard disk coatings
$BaTi0_3$	Nd:YAG	Optical coatings
Cubic BN	KrF	High temperature, radiation resistant semiconductors Cutting tool coatings
Amorphic C	Nd:YAG XeCl	Tribological coatings Heat sinks Field emitters for FPD's
$Cd_xZ_{1-x}Te$	KrF	Ir detectors
$CoSi_2$	KrF	Low resistivity interconnects for fast switching Schottky rectifiers
MoS_2	XeCl	High resistivity lubricating coating
$NbSe_2$	KrF	Low resistivity lubricating coating
$Pb(Zr_{0.54}Ti_{0.46})0_3$	ArF Nd:YAG	Piezoelectric devices Non volatile memory
SiC	XeF	High temperature, wide band gap semiconductors
$Ti0_2$	KrF	Antireflection coatings
$YBa_2Cu_30_7$	ArF, KrF	High Tc superconductors
Zn0	KrF Nd:YAG(2x)	Piezoelectric devices Phosphors

One of the distinct advantages of the PLD process is the near stoichiometric transfer from the ablation target to the desired substrate. We believe that the use of copper lasers allows high deposition rates and consequently a relatively inexpensive coating technique for a large number of high-value thin films. Table 2 serves as an indication of potential thin-film systems and their applications.

5. Conclusion

The copper laser has been in existence for nearly thirty years now. With the significant technology development that has gone on in those years, the laser is poised to become a significant source for several emerging industrial applications in addition to laser isotope separation. In this paper we have reviewed the physics of copper laser beams/materials interaction. The laser is capable of high intensity, causing ablative material removal. At sufficiently high intensities, the vapor blowoff causes laser absorption by photoionization and prevents the laser beam from being fully transmitted directly to the part. In this case, electron thermal conduction over longer periods heats the part to lower temperatures. Ablation above the laser detonation threshold does not appear to be as efficient as below it.

The copper laser, because of its ability to produce short pulses and near diffraction-limited beams, is a good tool for micro-machining. In particular, it has the ability to drill holes in difficult materials with very high aspect ratios with virtually no heat affected zone. When the copper laser is utilized in combination with fast tip-tilt mirrors, one can precision trepan holes with 1% diameter precision and almost arbitrarily shaped cross section.

The same virtues that make the copper laser a good drilling source also make it a good source for pulsed laser deposition. LLNL has utilized 100–200 W copper lasers to vaporize carbon from a graphite target, producing diamond-like-carbon thin films. Since the pulse repetition rate of the laser is several kilohertz, we can deposit films at 10–50 times other reported PLD rates. The properties of these films indicate high value for electronics, chemical resistance, and tribological applications. In addition to carbon, there are many high-value thin film materials that may provide a significant market for copper laser technology.

6. Acknowledgements

The authors would like to thank E. Fehring, G. Huete, and C. Cochran for excellent technical support and M. Balooch for his outstanding AFM measurements. This work was performed under the auspices of the U.S. Department of Energy by Lawrence Livermore National Laboratory under contract no. W-7405-Eng-48.

7. References

1. Walter, W., Solimene, N., Piltch, M., and Gould, G. (1966) Efficient pulsed gas discharge lasers, *IEEE J. Quantum Electron.* **QE-2**, 474-479.
2. Warner, B. (1987) Overview of copper-laser development for isotope separation, *Proc. SPIE* **737**, 2-6.
3. Boley, C. and Early, J. (1994) Computational model of drilling with high radiance pulsed lasers, paper presented at the *International Congress on Application of Lasers & Electro-Optics (ICALEO)*.
4. Kar, A., and Mazumbder, J. (1994) A mathematical model for nanoscale particles formed during laser ablation, *Phys. Rev. E* **49**, 410-419.
5. Poprawe, R. (1984) *Disintegration of Material and Plasma Formation in the Radiation Field of UV-Lasers*, Ph.D. Thesis, Darmstadt Technical University.
6. Kupfer, R., Bergmann, H. and Lingenauer, M. (1991) Material influence on cutting and drilling of metals using copper vapor lasers, *Proc. SPIE* **1598**, 46-60.
7. Bergmann, H.W. and Hartmann, M. (1993) Drilling of metals with copper vapour lasers, *Proc. TMS Annual Meeting on Laser Processing*, Denver, Colorado.
8. Pini, R., Salimbeni, R., Vannini, M. and Toci, G. (1992) Copper vapor laser high aspect ratio drilling process on transparent materials, *Proceedings Int. Conf. LASERS'92, Houston*, 932-936.
9. Chang, J., Martinez, M., Warner, B., Dragon, E., Huete, G. and Solarski, M. (1994) Precision micro drilling with copper vapor lasers, paper presented at the *International Congress on Applications of Lasers & Electro-Optics (lCALEO)*.
10. Chang, J.J. and Solarski, M.E. (1993) Self-imaging injection-locked copper laser oscillator, *Conference on Lasers and Electro-Optics (CLEO) Digest*, 458.
11. Paine, D.C. and Brarman, J.C. (eds.), (1990) *Materials Research Society Symposium Proceedings* **191**, 3.
12. Paine, D.C. and Brarrnan, J.C. (eds.), (1990) *Materials Research Society Symposium Proceedings* **191**, 25.
13. Collins, C.B., Davanloo, F., Juengerman, E.M., Osborn, W.R. and Jander, D.R. (1989) Laser plasma source of amorphic diamond, *Appl. Phys. Lett.* **54**, 216-218.

EXPLORING HIGH POWER MATERIAL PROCESSING APPLICATIONS WITH COPPER VAPOUR LASERS

G.J. DICKINSON AND I. MADDOCKS
Urenco (Capenhurst) Ltd
Capenhurst Works, Chester, CH1 6ER, UK

Abstract

Urenco have a large scale Copper Vapour Laser Facility, designed to investigate the potential for a commercial laser isotope separation process. Recently, materials processing applications have been explored using CVL light from a MOPA chain. This paper presents results on the drilling and cutting of Aluminium, Copper and Stainless Steel.

1. Introduction

The Copper Vapour Laser (CVL) Facility has been developed at Urenco (Capenhurst) primarily to demonstrate the engineering and economics of a light source for Uranium Isotope Separation. The facility comprises 4 Master Oscillator Power Amplifier (MOPA) chains of CVLs housed within a cleanroom and with supporting maintenance facilities. The output from the CVLs may be used to pump dye lasers, for the isotope separation process, or may be used for materials processing applications. The engineering of the facility has been described elsewhere [1].

CVLs are capable of being used to process a wide range of materials including metals, ceramics, diamond and fibre reinforced composites. The potential advantages that the CVL has over alternative laser systems include high levels of precision with limited associated heat affected zones and greater absorption than in the IR for many materials. The relatively high operating costs of CVLs however

C. E. Little and N. V. Sabotinov (eds.), Pulsed Metal Vapour Lasers, 347–352.

requires that commercial applications have a high added value associated with the processing operation. Many applications can be serviced with relatively modest power requirements of a few 10's of watts. The Urenco facility does however allow the use of more than 100 W of light. This paper discusses some of the issues associated with using CVLs at such high power and presents some results for cutting and drilling relatively thick metals.

2. Materials Processing Workstation

A materials processing workstation has been sited within an enclosure built adjacent to the end of one of the CVL MOPA chains, including an extract fan with a hood that can be positioned over the workpiece. Figure 1 shows a schematic

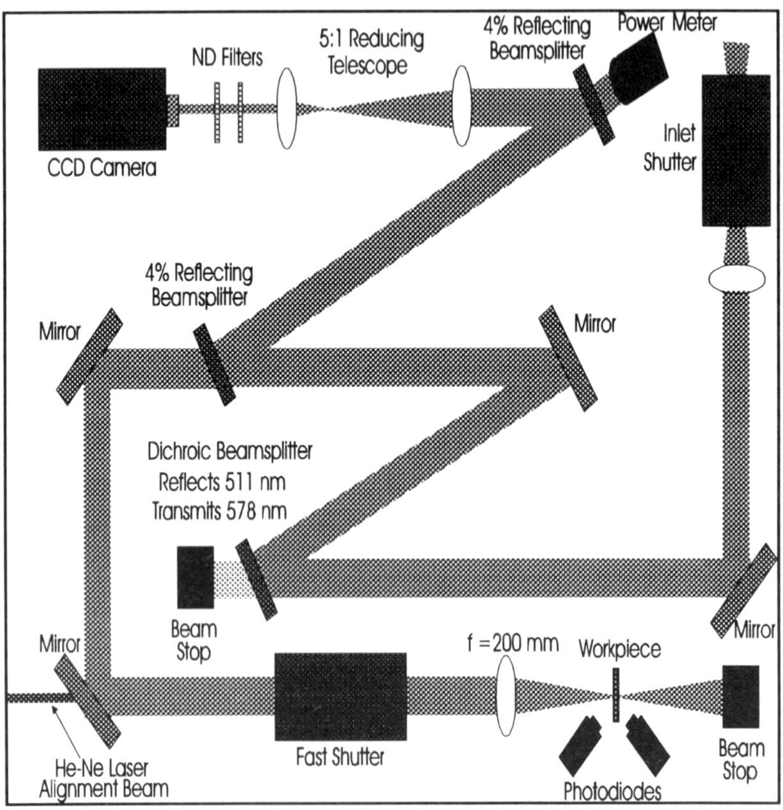

Figure 1. Schematic of the materials processing workstation.

of the optical layout of the workstation.

The CVL beam is reduced to about 20 mm diameter at the end of one chain. The inlet shutter is water cooled and controlled, either manually or by computer, from within the workstation enclosure. As achromatic optics are not yet deployed in the workstation the yellow light is separated from the beam and dumped. The facility has on-line power monitoring and capability to monitor the beam quality using a CCD Camera. Initial alignment to the workpiece is carried out using a He-Ne beam. A fast shutter, capable of switching in 15ms is used to switch the beam on to the workpiece.

3. Experimental Measurements

The following describes the measurements required to characterise the drilling and cutting of copper, aluminium and stainless steel with thickness in the range 0.9 to 3.2 mm. Firstly the focal length of the delivery lens is determined by timing the penetration of the beam as a function of focal length. All the samples gave a minimum drill time with a 200mm focal length lens. It should be noted that the choice of this lens is also dependent on the quality of the beam. The effective depth of field is also dependent on material thickness. For 0.9 mm thick material under these conditions drill through time was found to be relatively insensitive within +/-1 mm of the focus of the beam. For 3.2 mm thick material positioning of the sample needed to be within +/-200 μm of the focus.

For each of the three materials drill through times (s) and effective cutting feedrates(μm s^{-1}) were tabulated as a function of material thickness and beam power. (Beam power was varied using a commercial beam attenuator.) The results are shown in tables 1 to 3. To give an indication of potential results at higher powers, attainable using achromatic optics, cutting feedrate has been extrapolated to 200W.

Drill times are the mean of 10 measurements made by timing the application of the beam to the sample and penetration of the sample by fast photodiodes. The optimum effective cutting feedrate has been found to result from trepanning the beam along a cut. In the reported measurements, the sample was moved through the beam at a constant rate of 1000 μm s^{-1}. The number of trepans to cut through the beam can be calculated by dividing the absolute rate of movement by the quoted cutting rate. Thus for copper 1.6 mm thick, cut using 20 W of light 100 trepans were required to cut the sample.

TABLE 1. Drill Time (s)/ Cutting Feedrate (μm s^{-1}) as a function of Power and Material Thickness for Copper

Thickness (mm)		Power (Watts)					
		20	40	60	80	100	200 (extrapolated)
0.9	DT	2.5	1.2	0.7	0.4	0.3	
	CF	29	196	455	588	880	2500
1.0	DT	3.1	1.8	0.9	0.5	0.2	
	CF	-	-	-	-	-	
1.6	DT	15.2	6.1	3.1	2.4	1.6	
	CF	10	42	62	82	154	270
2.0	DT	-	10.7	6.1	4.5	3.9	
	CF	-	-	-	-	-	
3.2	DT	-	34.7	15.5	10.5	9.3	
	CF	-	9.3	16	22	31	85

TABLE 2. Drill Time (s)/ Cutting Feedrate (μm s^{-1}) as a function of Power and Material Thickness for Aluminium

Thickness (mm)		Power (Watts)					
		20	40	60	80	100	200 (extrapolated)
0.9	DT	1.12	0.78	0.31	0.13	0.02	
	CF	50	270	400	833	1800	7200
1.7	DT	-	5.99	3.6	1.8	0.94	
	CF	-	20	91	118	154	350
2.0	DT	-	27.6	6.8	6.4	5.5	
	CF	-	-	-	-	-	
2.9	DT	-	-	14.2	10.9	7.4	
	CF	-	-	27	44	57	150

TABLE 3. Drill Time (s)/ Cutting Feedrate ($\mu m\ s^{-1}$) as a function of Power and Material Thickness for Stainless Steel

Thickness (mm)		Power (Watts)					
		20	40	60	80	100	200 (extrapolated)
0.9	DT	0.23	0.11	0.11	0.1	<0.1	
	CF	30	370	400	526	1053	3370
1.7	DT	4.74	0.83	0.36	0.34	0.31	
	CF	-	23	45	143	286	1500
2.0	DT	25.1	2.8	1.25	0.91	0.59	
	CF	-	-	-	-	-	
3.3	DT	-	11.6	5.16	2.57	1.38	
	CF	-	-	-	11	45	320

All the results presented in Tables 1-3 have been obtained using a 3 amplifier MOPA chain and with no gas assist. Preliminary analysis of the use of gas assists with CVL cutting of copper and aluminium, in the parameter range under review, has suggested that gas assists may influence the quality of the cut edge, but do not significantly modify the cutting rate.

The available beam quality has a major impact on the processing capability of the CVL. This is illustrated (Table 4) by comparing the drill through time for copper, 0.9mm thick, when using an oscillator alone and when using a 3 amplifier chain.

TABLE 4. Comparison of Drill Time for 0.9 mm Copper a) using oscillator only, b) using a 3 amplifier MOPA chain.

	Oscillator Only			3 Amplifier MOPA Chain				
Power (W)	4	6	9	20	40	60	80	100
Drill Time (s)	3.3	1.7	0.7	2.5	1.2	0.7	0.4	0.3

It can be clearly seen that the drill time using the 3 amplifier chain can be considerably longer even at higher power. This is not surprising as the oscillator beam quality is significantly degraded by passing through successive hot amplifiers.

Experience at Urenco has shown that the following factors affect the beam quality from a CVL chain.

a) Alignment of the Injection Controlled Oscillator Cavity and drift in the relative timing of the Oscillator laser tubes.
b) The laser window temperatures, which can be affected by input power or cooling arrangements.
c) Air flows around the beam.
d) The number of amplifiers in the chain.
e) The stability of the electrical discharge in the lasers within the chain.
f) The distribution of copper within the lasers.

For any given application involving the cutting of relatively thick materials with copper vapour lasers, the user must trade off the conflicting requirements of 'high power' and good beam quality in optimising the process.

4. Concluding Remarks

An experimental arrangement and methodology has been identified to investigate the cutting and drilling of relatively thick metals using multi-amplifier MOPA chains. Cutting of 2 - 3 mm thick metals has been found to be relatively slow, therefore any commercial application must demand the high quality of processing possible using CVLs. Potential users of CVL MOPA chains must give due consideration to beam quality which, if not optimised, can swamp the benefit of high available beam power.

5. References

1. Dickinson G.J. (1992) Development Laser Light Facility for Uranium Isotope Separation, *The Nuclear Engineer* **33**,6,162-166

COPPER LASER MACHINING OF CERAMICS

A.J. KEARSLEY, M. KNOWLES AND R. FOSTER-TURNER
Oxford Lasers Ltd, Abingdon Science Park, Barton Lane, Abingdon, Oxon. OX14 3YR, UK

1. Abstract

The applicability of copper lasers to ceramic machining has been investigated. Very clean, crack free machining has been demonstrated in a range of ceramics. Cut quality was excellent and the density of machined features far above that attainable with traditional lasers. Process speed was far higher than with an excimer. The copper laser has demonstrated that it can fill the gap between the high power but low precision CO_2 and Nd:YAG lasers and the ultra-high precision but low speed excimer laser.

2. Project Aims and Background

2.1 AIM

The outstanding physical and chemical properties of ceramics has led to increasing use of these materials in industry. However, their hardness and brittleness make them difficult to machine using conventional methods. The project aim was to investigate the efficacy of ceramic machining with a copper laser and verify that it did indeed have the potential to be developed into a viable industrial process.

2.2 LASER MACHINING OF CERAMICS

2.2.1 *CO_2 Laser Machining*

The CO_2 laser is used most frequently. for ceramic machining In the electronics industry CO_2 lasers are often used to drill via holes in alumina which are later plated to form interconnections. However, the glassified layer and cracking around the hole wall can lead to adhesion problems during plating. The cracking and HAZ also restricts the density of holes which in turn reduces the interconnection density that is possible. Despite the cracking, glassification and HAZ, processing of ceramics with CO_2 is relatively widespread but is limited in its ability to address many applications.

2.2.2 *Nd:YAG Laser Machining*

The Nd:YAG laser can be focused to a smaller spot size than the CO_2 laser.

C. E. Little and N. V. Sabotinov (eds.), Pulsed Metal Vapour Lasers, 353–358.

Investigations into the effect of pulse length have shown that recast thickness, crack length and crack number are reduced as the pulse length shortens. Q-switched Nd:YAG lasers can produce pulses of 100 - 1000 ns length at pulse repetition frequencies of 10 kHz. The resulting samples have relatively little cracking or recast layer. Nd:YAG lasers are used when the component must have reduced HAZ and cracking and the value of the component can tolerate the slower processing and hence increased cost.

2.2.3 *Excimer Laser Machining*

Excimer lasers are pulsed ultra-violet lasers. The short pulse and UV light ablate a thin layer of material in a very controlled manner with much better edge quality than with CO_2 and Nd:YAG lasers. However, the low material removal rate, typically 100 $\mu m^3/s$, makes the process very expensive and not suitable for mass production.

2.3.4 *The Copper Laser*

In recent years there has been a growing interest in using medium and high power (20 - 1,000 W) copper lasers for high precision materials processing. The interest stems from the ability of the copper laser to drill and cut very fine features with little or no thermal damage to the surrounding material. The very good beam quality obtainable from copper lasers coupled with the short wavelength enables a very small focal spot to be achieved and very small features to be machined. Holes as small as 1.5 μm have been drilled and slots down to 4 μm width have been machined. In addition the stand-off distance between the focusing lens and workpiece can be very much larger than with conventional lasers. This large stand off distance gives very parallel sided holes and eliminates optics contamination problems.

The low energy short pulses give an exceptionally small HAZ because the heat from a single pulse is unable to diffuse into the surrounding material during the pulse. In general the HAZ is less than 1 μm in thickness in most metals.

2.3.5 *State-of-the-Art for Machining of Ceramics Prior to this Study*

Prior to this work there were no publications in this field. Work at Oxford Lasers had consisted of a few preliminary trials. The results of these trials indicated the potential for the use of copper laser in the machining of ceramics.

It had been found that percussion drilling could be used to drill holes in alumina sheets of up to 1 mm thickness in less than 1 second. However, due to the poor spatial profile of the focused beam used at the time, the holes were very elliptical and showed significant taper (approximately 10° full angle). However, there was no visible sign of micro-cracking, glassification or heat affected zone. In order to form larger holes, a trepanning technique was used. The ceramic sample was moved in a circular motion by a CNC XY table. The non-circular focused spot led to slightly oval holes, especially at smaller trepan diameters. When drilling an array of holes, the minimum hole pitch was 3-4 hole diameters before macro-cracking of the sample occurred. Cracking was least likely in thinner substrates when the diameters were greater than the substrate thickness.

Crack-free cutting was possible when high table speeds (50 - 100 mm s^{-1}) were used with many overcuts in order to achieve penetration. These high table speeds essentially limit the cutting to long straight cuts where the high table speeds may be used. The quality of the cut edge was very good with no micro-cracking or glassification.

3. EXPERIMENTAL RESULTS

A number of ceramics have been studied. These include alumina, super conducting ceramics, tungsten carbide, natural diamond, synthetic diamond film, sapphire and ceramic composites. Most of the results presented below are for alumina.

3.1 *Laser Beam Quality and Profile*

A new MOPA system was constructed (fig 3.1). The output of a low power oscillator was spatially filtered before amplification to nominally 50W. The beam quality was about 5 times the diffraction limit and produced a near perfect round focused spot.

It was noted that the new system produced rounder holes with shorter drill times. Using the improved system the drill time in 1 mm thick alumina was more than halved to 0.35s. It was also noted that the cut speed was now independent of the cut direction.

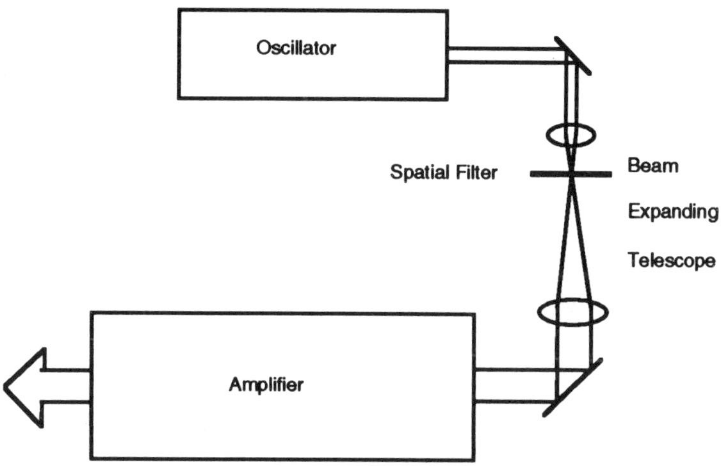

Figure 3.1. Master Oscillator Power Amplifier Configuration

3..2 *Lens Focal Length*

To a first approximation, the focal length of the lens determines the size of the focused spot. With standard lenses when the f-number is less than about 5, aberrations start to become significant. The maximum laser intensity is produced with f-numbers in the range of 4 - 8 depending upon the lens quality and laser beam quality.

It is desirable to set the confocal parameter of the focused beam to be greater than the material thickness. This ensures that the beam diameter does not vary greatly through the thickness of the material which in turn reduces the process time. For thick materials the optimum processing usually occurs with a f-number that is a compromise between achieving the maximum intensity and the longest confocal parameter.

The effect of the lens f-number was investigated in alumina as a function of the sheet thickness by measuring the drill time. The results are shown in figure 3.2. The laser power used was 44 W and the beam diameter was 40 mm. The figure shows that for sheet thicknesses of less than 0.4 mm the shortest drill time is achieved with the f-5 lens. As the sheet thickness increases then the longer lens (f-10) shows increasing benefit up to the maximum thickness tested (1 mm).

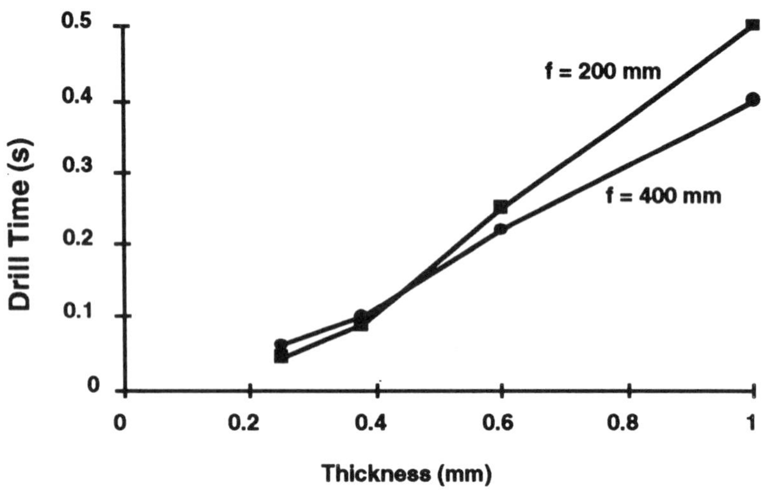

Figure 3.2. Drill time in alumina sheet for f=200 mm (f5) and f=400 mm (f10) lenses.

3.3 *The Effect of Traverse Speed and Number of Overcuts*

In general there exists a maximum speed at which the sample is just cut through with a single cut. At higher traverse speeds, complete penetration of the sheet is not achieved by a single cut but usually can be achieved if multiple passes are used (*overcutting*). When cutting ceramic with a copper laser, it was found be advantageous to use a high traverse speed with a large number of overcuts. This produced two benefits. First, double the maximum effective cut speed was achieved under these conditions. Second the thermal macro-cracking of the ceramic was eliminated.

3.4 *The Effect of Laser Power and Material Thickness*

Figures 3.3 and 3.4 show the effect of laser power upon the drill time and cutting speed in alumina. These show that process speed is not a linear function of the laser power.

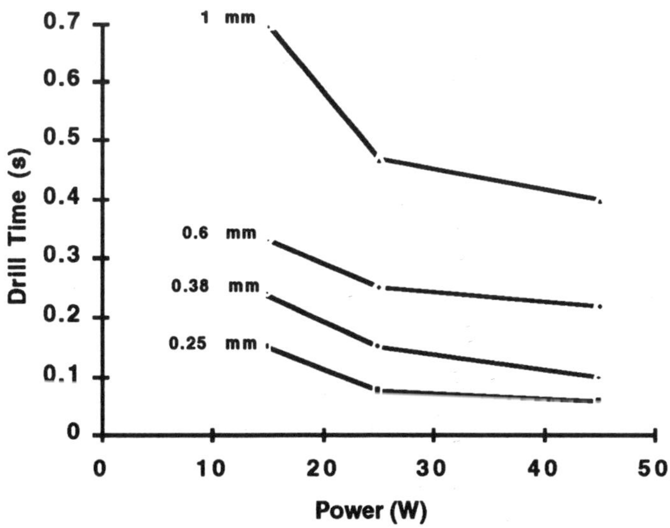

Figure 3.3 Drill time in alumina as a function of laser power for 0.25, 0.38, 0.6 and 1 mm thickness sheet.

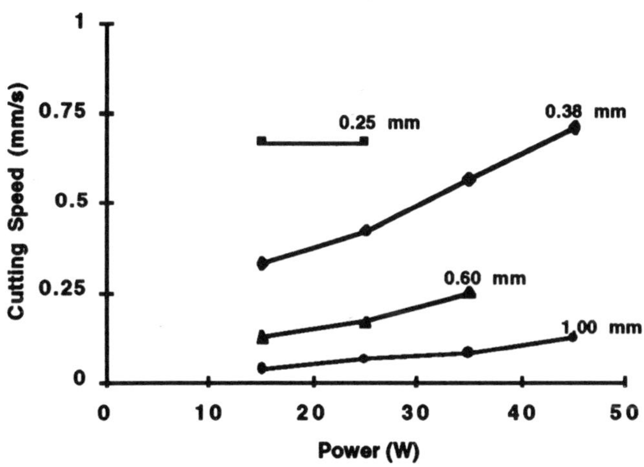

Figure 3.4 Cutting speed as a function of laser power for 0.25, 0.38, 0.6 and 1 mm thickness alumina.

Figure 3.5 shows the cutting speed as a function of material thickness at laser powers up to 45 W. For a given thickness the results indicate that above some power level the process speed becomes a linear function of the power. In the 10 - 50 W power range it appears that the thickness for which the process speed would be a linear function of laser power would be somewhat less than 250 μm. Figure 3.6 shows the drill time as a function of material thickness. To date the thickest ceramic cut with our 50 W system is 3.7 mm tungsten carbide with an approximate cut speed of 0.2 mm/min.

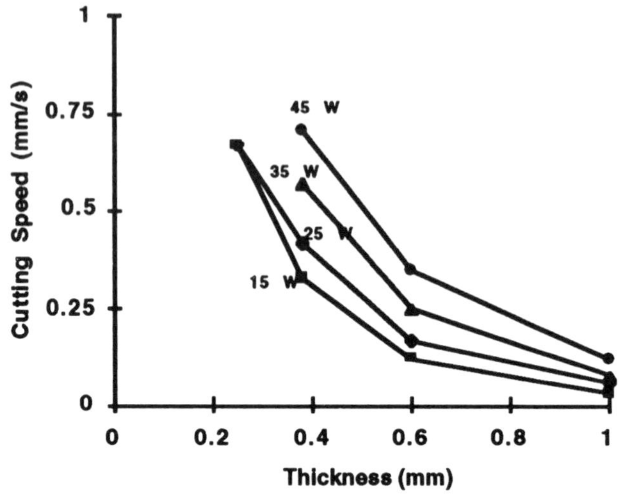

Figure 3.5 Cutting speed as a function of material thicknesses for laser powers of 15, 25, 35 and 45 W.

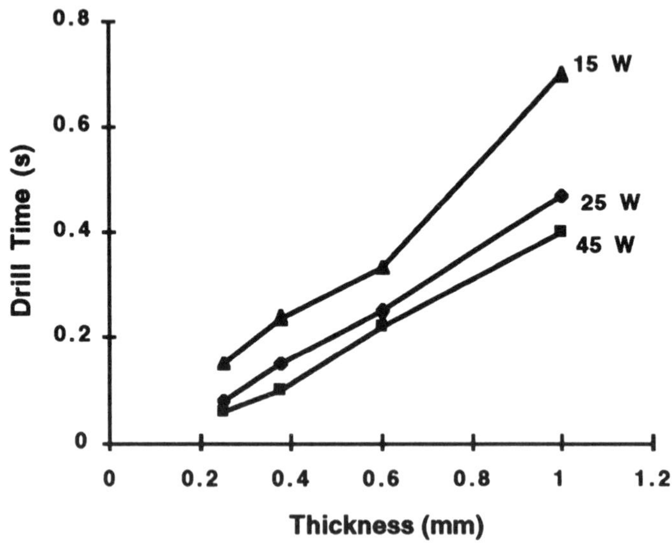

Figure 3.6 Drill time in alumina as a function of sheet thickness.

4. CONCLUSION

The copper laser has been shown to be able to process ceramics with a quality and speed which are of commercial interest. Specific application studies are now required.

Oxford Lasers would like to thank the European Commission for funding this study.

DRILLING AND CUTTING TRANSPARENT SUBSTRATES WITH COPPER VAPOUR LASERS

R. PINI
Istituto di Elettronica Quantistica del CNR
Via Panciatichi 56/30, 50127 Firenze, Italy

1. Introduction

The possibility to employ Copper Vapour Lasers (CVLs) lasers for micro machining and micro drilling of transparent optical substrates has been recently investigated by our group [1] for its intrinsic relevance in industrial applications, as well as to provide new insights on the mechanism of material processing, that can be useful to model similar processing in other materials, as metals, alumina, silicon, etc. In this respect, a particular concern has been devoted to the comprehension of the physical phases of pulsed laser interaction that lead to material drilling and cutting. This was done by using imaging and spectroscopic analysis, taking advantage of direct visualisation and control into the bulk permitted by transparent materials.

2. The drilling process

The first observation of CVL drilling in fused silica occurred by chance in 1988 during some trials of tight coupling of laser radiation into a 200 μm optical fiber. Instead of the typical catastrophic damage on the input face due to optical breakdown induced by high power pulsed lasers as excimers and Nd:YAG, we observed the formation of a straight hole of about 1 mm in length and 20 μm in diameter with smooth and glazed inner walls. Since the early CVL drilling trials that followed this first observation some relevant issues appeared evident, as the possibility to obtain holes with high aspect ratios (due to the guiding effect enhancing the hole length) and with walls of good optical quality. Furthermore we realised that this drilling process occurring in transparent materials was in close relationship with the combination of the emission parameters (as pulse peak power, average power, repetition frequency) that are characteristic of CVL devices.

The drilling process is initiated by local absorption of laser radiation into the sample surface. In substrates that are transparent to the CVL emission wavelengths, this can be produced only by nonlinear processes as colour centre formation and surface optical breakdown. Considering that these processes take place at high power densities (5-10 GW/cm^2 for fused silica) the requirements of a consistently high pulse peak

C. E. Little and N. V. Sabotinov (eds.), Pulsed Metal Vapour Lasers, 359–364.

power and a good beam quality are mandatory. Usually, a sequence of 10-50 surface breakdown events is necessary to raise the local temperature up to the material melting point. The high repetition frequency of the laser is thus necessary to increase the initial temperature and to maintain the melted core at the hole bottom pulse after pulse during channel formation. Laser pulse duration (40 ns) is much shorter than the typical heat diffusion time, so heating effects are limited to a very thin layer of irradiated material (less than 2 μm in SiO_2). Sudden temperature increase and subsequent fast vaporisation of melted material at the hole bottom induce the formation of a micro plasma at each laser pulse. Plasma expansion induces a supersonic pressure wave associated to vapor ejection that escapes from the channel in a few hundreds of nanosecond. This behaviour has been put in evidence with time-gated spectroscopy measurements (150 ns sampling window) that revealed the vapor fluorescence composed of strong Si and Si^+ emission lines over a wide band background (see Figure 1), suggesting the occurrence of optical breakdown processes not only at the drilling start but also during the hole deepening. Beam deflection measurements and time-resolved imaging of the ejected vapours (with a pump-and-probe arrangement [2]) confirmed the formation of a supersonic shock wave (see Figure 2) with high pressure peaks (50-100 bar at 200-300 ns after laser pulse)[3].

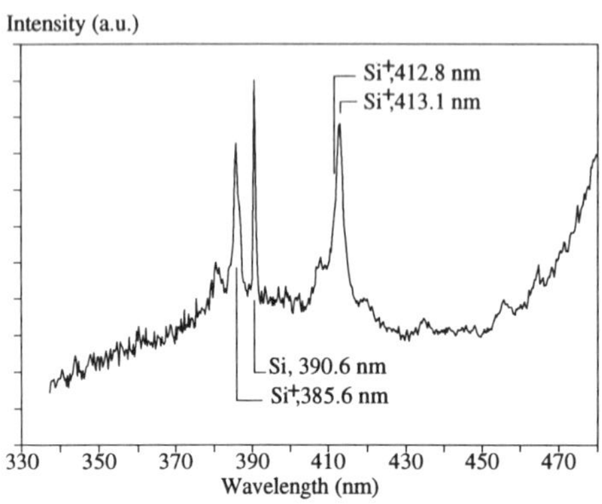

Figure 1. Fluorescence spectrum of the ejected vapor during CVL drilling in a fused silica sample.

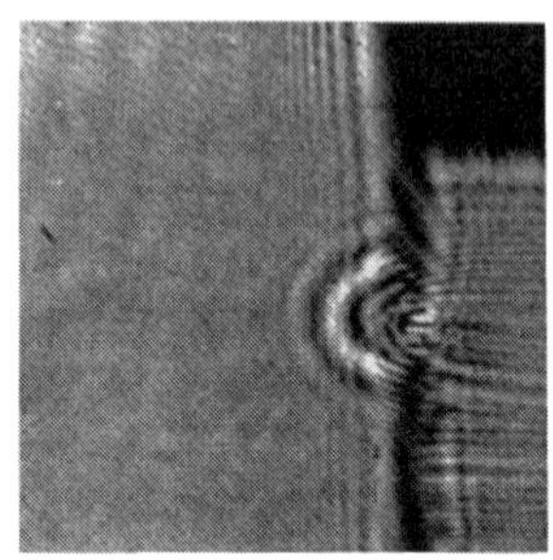

Figure 2. Time-resolved image of the shock wave front emerging from the hole during CVL drilling of fused silica (the probe beam is a XeCl-pumped dye laser of 10 ns duration at a delay of 0.6 μs).

3. Drilling and cutting tests

Typically the CVL drilling technique can produce micro holes with diameters of 10-30 μm, lengths up to 3 mm and aspect ratios even greater than 100, with a good reproducibility . These results have been obtained with a small scale home-made CVL (90 cm discharge length, 2 cm bore diameter) operated at repetition frequencies in the range 1-4 kHz and average power of about 1-5 Watts. The laser optical cavity was

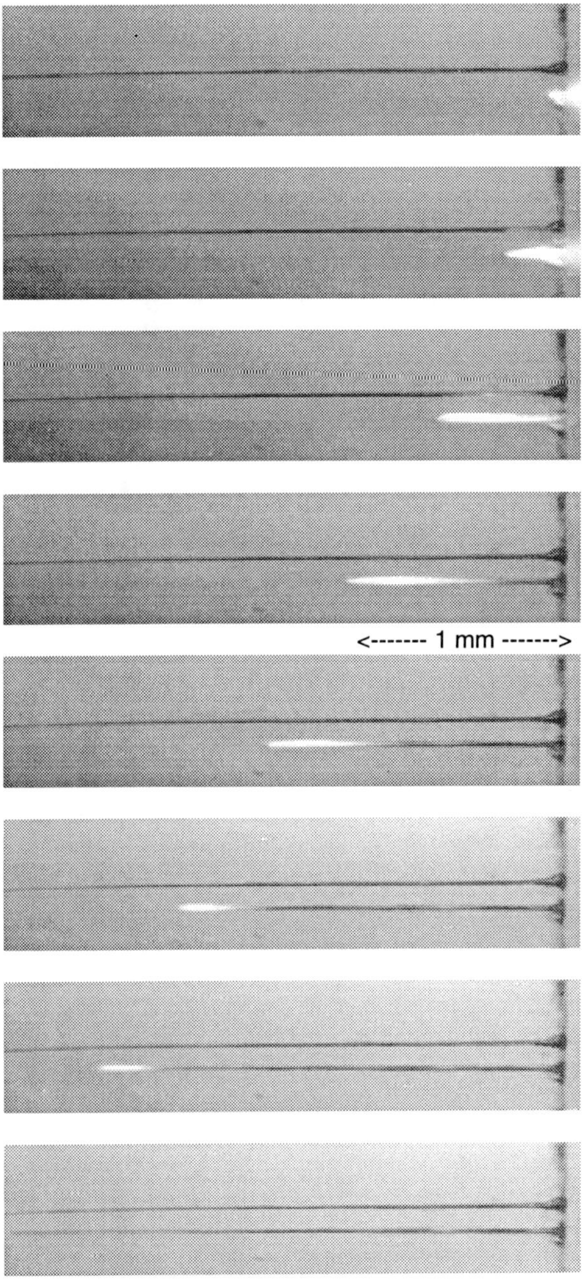

Figure 4. Drilling sequence in fused silica.

arranged with a self-filtering unstable resonator that provided almost diffraction limited emission (110 μrad divergence, 1.1 x D.L.) [4]. During the drilling process, the laser beam was focused perpendicularly onto the sample surface with short focal length achromatic lenses (f = 5 and 10 cm), resulting in peak power densities as high as 10-15 GW/cm^2 at the focus.

Materials under test were crystals as white sapphire (corundum) and ruby, and amorphous materials as optical grade fused silica and BK7 optical glass. Samples were prepared as cubes of 5 mm side with optically polished facets. This was done to allow side vision directly during hole drilling. To this aim we arranged a set up (Figure 3) that provided microscopic imaging of the drilling process, employing a gateable CCD camera with minimum exposure time of 125 μs to record the effect of a single laser pulse. This live video monitoring allowed precise targeting of the sample surface with respect to the beam focus and direct measurements of both drilling speed and volume removal rate. Figure 4 shows a sequence of 8 frames representing drilling in quartz with a laser power of 1.5 W and 2.4 kHz PRF.

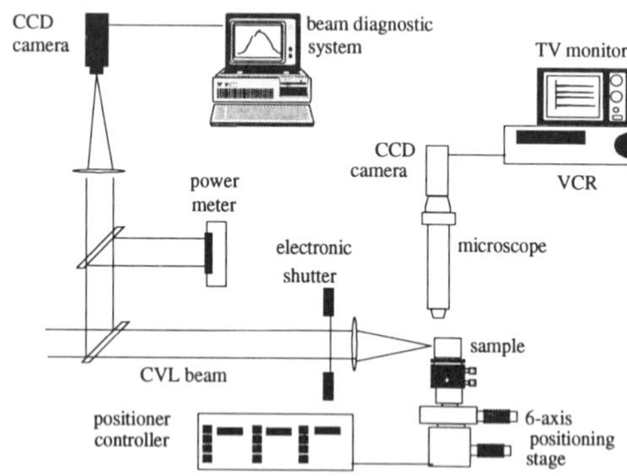

Figure 3. Set up for microscopic imaging of drilling.

Macroscopically , the drilling start is witnessed by a bright light scattering and a light buzzing. The first breakdown events sometimes produce also a limited surface damage (50-100 μm in diameter) that can be reduced by moving the beam focus a little apart from the sample surface. When the melted core is formed, the hole deepening proceeds at almost constant speed (180-250 nm/pulse in fused silica), that does not seemingly depend on the PRF in the range 1-4 kHz. The role of the waveguide effect enhancing the hole length is clearly recognisable, by considering that the final hole is usually more than three times longer than the confocal parameter of the focused beam. This effect is mainly due to the good optical quality of the hole walls that allow laser beam propagation by grazing reflection as in a hollow waveguide. In this concern, the choice of a longer focal length can optimise the coupling into the waveguide and enhance the maximum achievable hole length, provided that the consequently larger focal spot does not decrease the focal intensity below the breakdown threshold.

Our observation indicated also that even if the drilling speed is almost independent of PRF, high repetition rates can represent a limiting factor to the drilling of long holes. The maximum achievable hole length was actually observed to decrease of about 30% when the CVL pulse repetition frequency was increased from 1 to 4 kHz. Trying to explain this unexpected behaviour, we hypothesised and verified by beam deflection measurements [5] that this was caused by air refractive index modulations inside and nearby the hole, as produced by the non equilibrium air density distribution following shock wave formation, perturbing coupling and transmission of the subsequent laser pulse into the waveguide. In other words, at increasing repetition rates it seems that the time necessary to restore air equilibrium after the shock wave expansion and the following turbulent re-collapsing becomes comparable with the time separation between two consecutive pulses.

We also tested CVL cutting of fused silica and optical glasses sheets. Linear cutting as well as trepanning were obtained with good reproducibility by using the same irradiation parameters of laser drilling. The partial lack of guided propagation in comparison with hole drilling limited the maximum thickness (≤ 1 mm in fused silica) to obtain continuous and high quality cuts. In our experimental conditions we recorded typical linear cutting speeds of 8 mm/min.

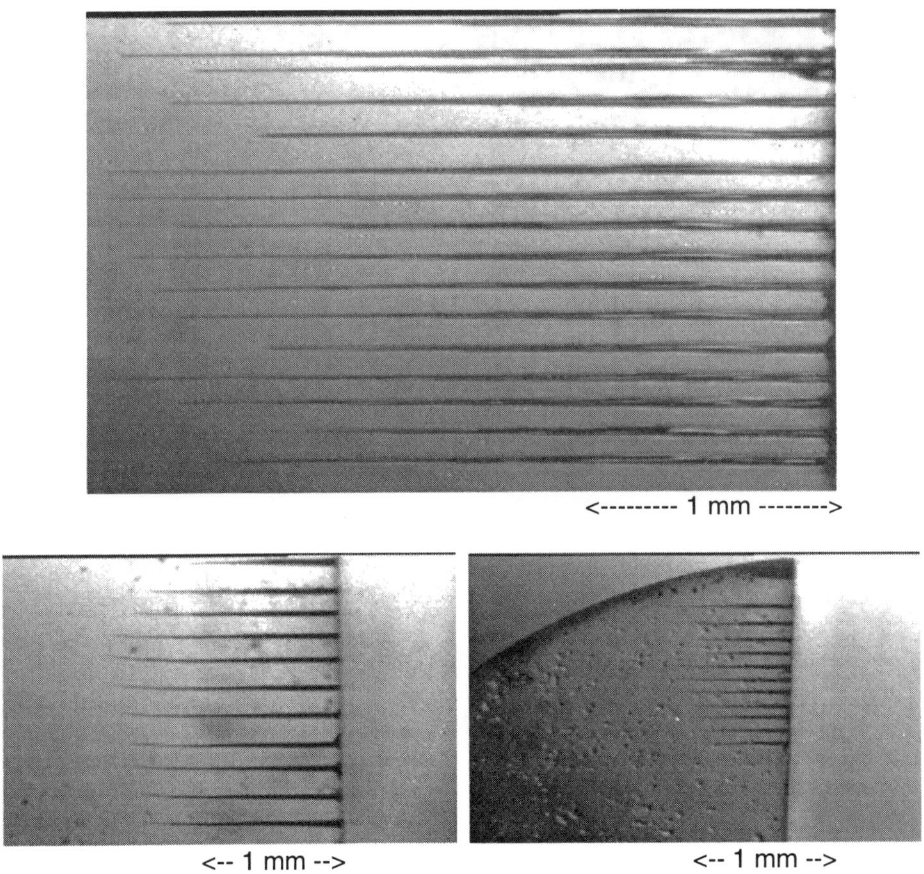

Figure 5. Holes drilled in fused silica (upper), white sapphire (left) and synthetic ruby (right). Average power: 1-1.5 W, PRF: 2.4 kHz, 10 cm focal length focusing lens.

4. References

1. Pini, R., Salimbeni, R., Vannini, M. and Toci, G. (1992) Copper vapor laser high aspect ratio drilling process on transparent materials, *Proceedings of Int. Conf. LASERS 92*, pp. 932-936.
2. Siano, S., Pini, R., Salimbeni, R.and Vannini, M. (1994) Refractive fringe diagnostics during excimer laser photoablation , *Proceedings of Int. Symposium on High Power Lasers and Laser Applications V*, SPIE Vol. **2207**, pp. 585-589.
3. Pini, R., Salimbeni, R., Siano, S., Vannini, M. and Toci, G. (1994) A general survey on hole drilling in transparent substrates by copper vapor lasers, *Proceedings of Int. Conf. LASERS 94*, in press.
4. Pini, R., Salimbeni, R., Toci, G. and Vannini, M. (1991) High efficiency diffraction limited operations of a copper vapor laser, *Opt. Commun.* **81**, 138-144.
5. Pini, R., Salimbeni, R., Vannini, M. and Toci, G. (1995) Probe beam deflection diagnostics of shock waves generated during laser drilling, *Appl.Phys. B*, in press.

UV MICROMACHINING USING COPPER VAPOUR LASERS

D. W. COUTTS*, A. C. J. GLOVER, E. K. ILLY,
D. J. W. BROWN and J. A. PIPER

Centre for Lasers and Applications
Macquarie University, NSW 2109
Australia.

Abstract

UV output powers in excess of 1W at 255nm can readily be generated by SHG from single, medium-scale copper vapour lasers. The high PRF and high beam quality makes the frequency-doubled CVL an ideal source for many UV micromachining applications such as drilling micro-orifices in polymers at high production rates.

1. Introduction

High power lasers are a widely accepted tool for use in industrial manufacturing applications such as machining, welding and surface processing where typically CO_2 and Nd:YAG systems operating in the infrared are used. However, the laser-materials interactions in many machining cases are more favourable when using lasers operating at shorter wavelengths. For example, micromachining of many metals is best performed using visible wavelengths. With its high average power, high pulse repetition frequency and high peak power, the copper vapour laser has been shown to excel in visible laser machining of metals, and other materials ranging from ceramics to diamond [1].

High precision micromachining requires a uv laser in order to achieve the necessary resolution. Furthermore laser machining of organic polymers is best performed with a uv source where the photon energy is sufficient to produce photochemical ablation rather than melting and subsequent vaporisation. Thus

* now at: University of Oxford, Department of Atomic and Laser Physics, Clarendon Laboratory, Parks Road, Oxford OX1 3PU, UK.

C. E. Little and N. V. Sabotinov (eds.), Pulsed Metal Vapour Lasers, 365–370.

uv lasers can be used to produce clean precision cutting and drilling without producing a large heat affected zone.

At present excimer lasers are the main industrial laser source used for micromachining, however the high pulse energy and low PRF of these lasers means that they are not well suited to machining applications requiring high production rate serial processing. Recently, waveguide excimer lasers which offer high PRF (2kHz) but low pulse energies (up to 50μJ) have been used for direct write machining applications, however total ablation/machining rates from these lasers are limited by their low average power.

Frequency doubled copper vapour lasers offer an attractive alternative for uv micromachining applications[2,3]. Recent advances in nonlinear uv generation from CVLs has enabled uv powers of over 1W to be readily obtained at all three uv wavelengths (255 nm = SHG green, 289 nm = SHG yellow, and 271 nm = sum-frequency) from a single medium-scale CVL with average conversion efficiencies of up to 35% [4] [5]. Yet higher average uv powers (up to 9W [6]) has been obtained from frequency doubled CVL MOPA systems. For micromachining applications, a single frequency-doubled CVL has sufficient peak power and pulse energy for efficient ablation of target areas of up to hundreds of microns square, while at the same time having very high PRF capability (4-20kHz) resulting in very high machining rates. In this paper the current status of micromachining with frequency doubled copper vapour lasers is discussed, in particular characterisation of the source, ablation rates for various polymers, and results of pre-production trials of precision hole drilling in polymer tape will be presented.

2. The Frequency Doubled Copper Vapour Laser / Machining System.

A ~1W uv source based on a single frequency-doubled CVL has been developed in the Centre for Lasers at Macquarie University as a prototype micromachining facility for high speed production hole-drilling of polymer films. The system comprised a single CVL, frequency doubling system and machining platform as shown in figure 1. The CVL incorporated an in house built plasma tube (25mm id x 1m long active region) fitted to a commercial power supply/subsystem manufactured by Visiray Pty Ltd. The laser was typically operated at a PRF of 4.25kHz with a high-magnification (M=100) on-axis unstable resonator and 0.8% hydrogen additive to the neon buffer gas in order produce maximum high beam quality output. Under these conditions the laser typically gave 6.1W low divergence green output, about half of which had divergence ~ 5 times the diffraction-limit, and the remainder essentially diffraction-limited beam divergence.

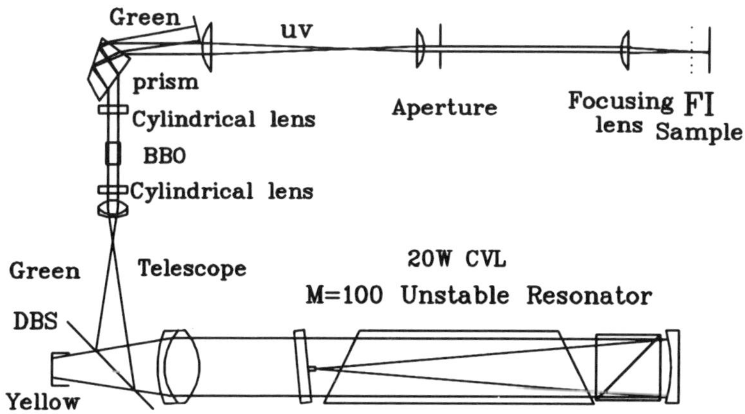

Figure 1. Optical layout for complete CVL-UV and machining system

Efficient frequency doubling was achieved using a line focusing arrangement as described in detail in [4]. An achromatic beam compression telescope was used to reduce the CVL beam diameter to ~5 mm followed by a cylindrical lens (f = 100mm) which produced a line focus in a 6 mm long BBO crystal. After recollimation with a uv grade cylindrical lens, the uv was separated from the residual green using a Pellin-Broca prism. Using this arrangement uv powers at 255nm of ~ 1W were readily achieved at an average conversion efficiency of ~18%. Note that higher conversion efficiencies have been obtained, however the system was optimised for long term reliability and uv power stability (±5% variation over 8 hours).

The micromachining platform incorporated uv beam shaping optics consisting of a telescope to compress the uv beam to ~1.5mm diameter, followed by a hard aperture and an imaging/focusing lens (typically f = 100mm). Machining could be performed at either the focus of this lens, or in the image plane of the aperture as is indicated in figure 1. A fast reed shutter was linked to the laser pulse driving electronics such that uv pulse trains from single shot up to any desired number could be produced for ablation studies.

3. Polymer Ablation Characteristics

Ablation studies were conducted for both PETG and brown pigmented Kapton. Surface profiles for two-shot uv ablation patterns were measured using an atomic force microscope. An example of such a profile for PETG irradiated at $1.9 J.cm^{-2}$ (per pulse) with the sample placed at the image plane of the circular aperture is shown in figure 2 together with a SEM photograph of the ablated sample. Diffraction effects from the hard aperture can be seen as ripples in the ablation profile. Average ablation depths per pulse were measured for different numbers of incident pulses at different intensities in order to characterise the average ablation rate per pulse. For the first ~20 shots in a train the ablation depth per shot is constant, however after ~50 shots the ablation depth per pulse was observed to fall to about one tenth of the initial rate principally due to loss of power by scattering off uncleared debris from the earlier shots.

Initial ablation rates for both PETG and Kapton for the CVL operating at PRFs of 4.25, 10 and 20 kHz are plotted as a function of fluence in figure 3, where the ablation threshold is ~0.8 J/cm^2. While the ablation depth per pulse is significantly lower at higher PRFs due to the lower average uv power, the greater PRF leads to faster average machining rates.

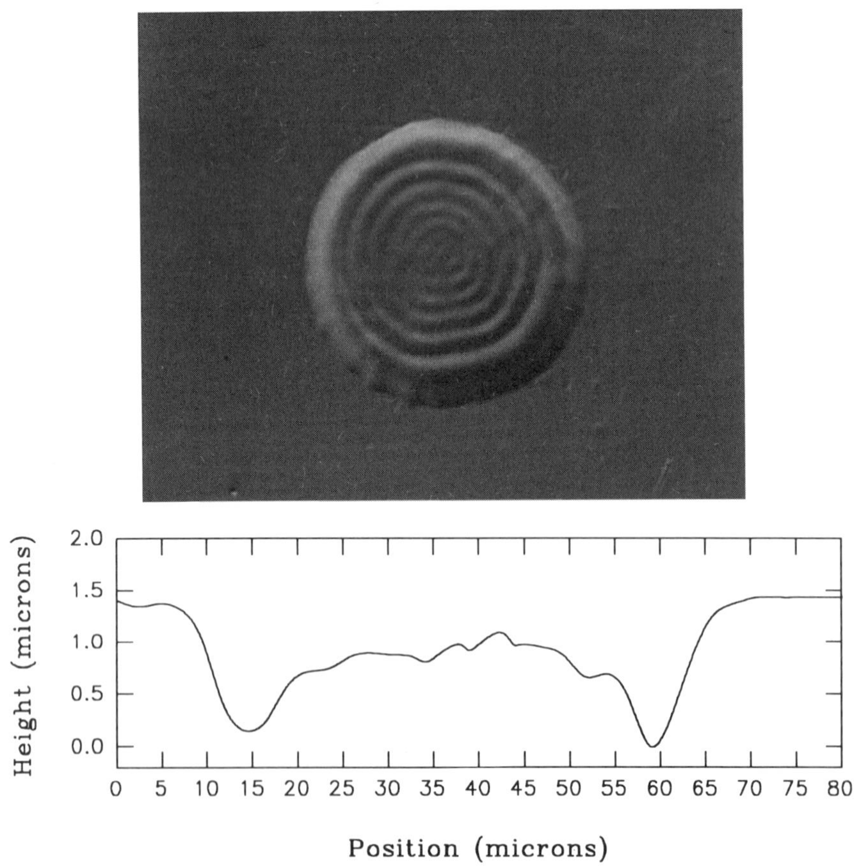

Figure 2. SEM micrograph and Atomic force microscope profile for a PETG sample located at the image plane of a circular aperture. Uv energy density = 1.9 $J.cm^{-2}$

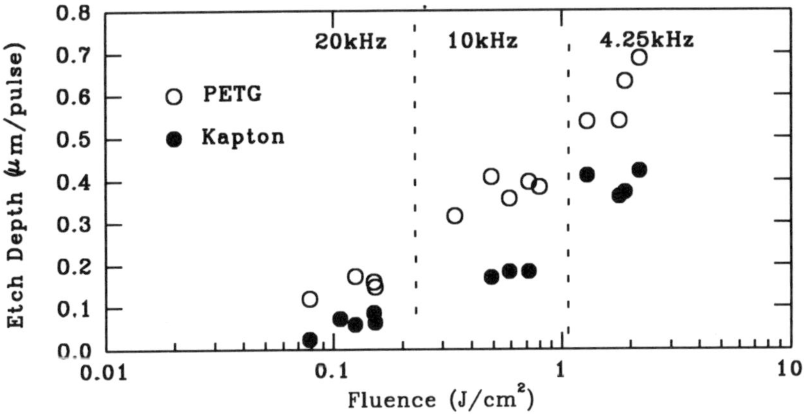

Figure 3 Variation of ablation rate with fluence (and PRF).

4. Hole Drilling in Polymers

Precision holes ranging in size from 5-300µm have been drilled in PETG and Kapton films ~125 µm thick. Figure 4 shows electron microscope photographs of 100µm holes drilled in PETG. The holes have very sharp edges with a minimal heat affected zone indicating clean photolytic ablation.

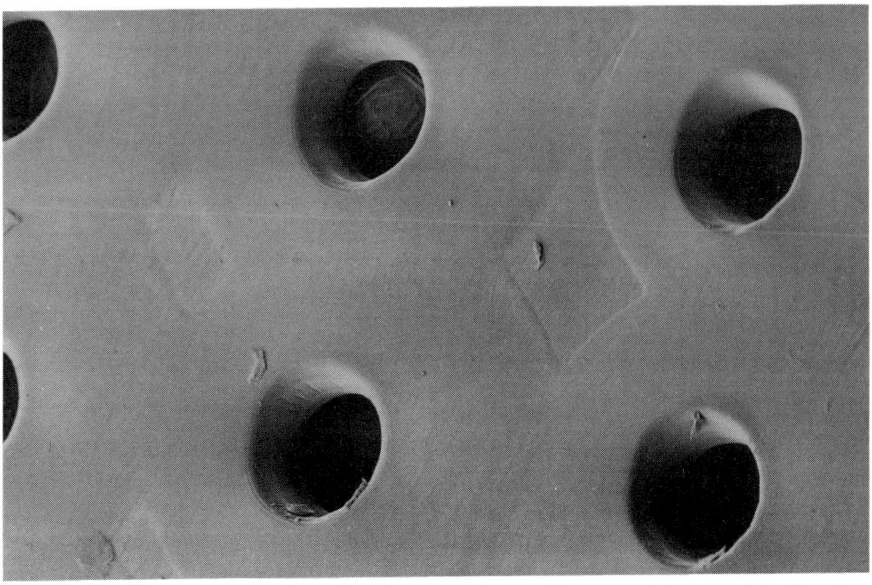

Figure 4. SEM photographs of 100 µm holes drilled in 125µm thick PETG.

Typical drill times are ~50 ms corresponding to ~200 shots, thus production rates are determined principally by substrate repositioning times. However drill speeds can be increased if necessary by operating the CVL at higher PRF where the pulse energies are lower but remain well above ablation thresholds for machining ~100 μm structures. Note that uv powers ~5 times higher than used for the initial ablation studies have been achieved in our laboratory from a CVL of the same specifications operating at a PRF of 20 kHz [7].

5. Conclusions

Frequency doubled CVLs have been shown to be ideal sources for uv micromachining of polymers. Typical ablation rates for PETG are 0.1-0.7μm per pulse with PRFs up to 20kHz. High precision reproducible holes 10-300 μm diameter have been drilled in PETG tape with drill times typically 40ms corresponding to production rates limited only by tape repositioning times (up to 40 Hz at present). Drilling and scribing has been performed on materials such as silicon, glass, PMMA, silicone rubber, santoprene and mylar.

Acknowledgments

The authors would like to acknowledge the support of M. Phillips of the Microstructure Analysis Unit at the University of Technology, Sydney, and the staff of the Microscopy Unit of Macquarie University School of Biological Sciences.

References

1. Knowles, M. R, H,. Evans, I. J., Foster-Turner, R., and Kearsley, A. J. (1995) Multi-wavelength copper vapour lasers for novel materials processing application, *Proc. 6th Int. Symp. Advanced Nuclear Energy Research*, JAERI, Japan, 236-245.
2. Glover, A. C. J., Coutts, D. W., Ramsay, D. J., and Piper, J. A. (1994) Progress in high-speed uv micromachining with high repetition rate frequency-doubled copper vapour lasers, *Proc., ICLAEO,* Laser Inst. Amer. (Orlando, FL), paper 505.
3. Glover, A. C. J., Illy, E. K., and Piper, J. A., (in press), High-speed uv micro-machining of polymers with frequency-doubled copper vapour lasers *IEEE. Sel. Topic. Quant. Electron.*
4. Coutts, D. W. (In press) Optimisation of line focusing geometry for efficient nonlinear frequency conversion from copper vapour lasers, *IEEE J. Quantum Electron.*
5. Coutts, D. W., and Brown, D. J. W., Production of high average power uv by second harmonic and sum-frequency generation from copper vapour lasers, (in press), *IEEE J. Sel. Topic. Quantum Electron.*
6. Molander, W. A. (1994) High-average-power second-harmonic generation using copper laser oscillator/amplifier source, *Tech. Dig. Conf. Lasers Electro-Opt.* Paper CThM2.
7. Withford, M. J. and Brown, D. J. W., Improved ultra-violet second harmonic generation at elevated repetition rates from a medium-scale copper vapour laser, (in press) *IEEE Sel Topic Quantum electron.*

UNDERWATER DIRECT METAL PROCESSING BY HIGH-POWER COPPER VAPOUR LASER

C. KONAGAI, Y. SANO, N. AOKI
Nuclear Engineering Laboratory, TOSHIBA Corporation
8 Shinsugita, Isogo-ku, Yokohama 235, Japan

Copper Vapour Lasers (CVLs) have been applied to underwater direct metal cutting and surface stress improvement. CVL's features of high peak power, high repetition rate and visible wavelength enable fast, clean and thermal-effect free processing. The CVL based processing is expected as an alternative repairing or maintaining method in nuclear facilities.

1. Introduction

CVL has features of visible wavelength, high repetition rate, high average power and high conversion efficiency from electricity to light. From these features, CVL is considered as the most effective light source to pump Dye Laser, which will be used for Atomic Vapour Laser Isotope Separation (AVLIS). Single unit CVL power of 500 W, the target power in Japanese AVLIS program [1], has been achieved through our development [2].

One of the most promising CVL applications other than AVLIS would be material processing, especially surface stress investment utilizing high-pressure plasma induced by ablative interaction of laser light with metal. Thermal effect on processed material could be reduced drastically through the short pulse duration of CVL. In addition, CVL's higher transmission through water and higher absorption on metal surface than such infrared lasers as CO_2 or Nd:YAG lead to underwater direct material processing, that is especially important for repair work in nuclear facilities to reduce outage time and radiation dose.

In this paper, the development of high-power CVL and the application to underwater direct processing are described.

2. Development of High-Power Copper Vapour Laser

For extracting as much power as 500 W, following developments have been carried out.

(1) Increase of laser gain volume,

C. E. Little and N. V. Sabotinov (eds.), Pulsed Metal Vapour Lasers, 371–376.

(2) Uniform temperature distribution along the laser tube,

(3) Discharge circuit with high voltage and current in short duration.

The above three items must be achieved simultaneously, in order to obtain the projected laser power.

2.1. APPROACH TO HIGH–POWER CVL

An optimization of the thermal insulator is necessary with the extension of laser tube length. The combination of usual thermal insulator and high–ability one has been adopted in our CVL, and the gain volume has been increased effectively while keeping the temperature distribution within 1500 ± 50 degree centigrade [2].

Improvements on discharge circuit have been performed to obtain higher laser output power with the fixed gain volume. As commonly known now[3], the extracted power at the central region of tube cross–section decreases in large bore CVL due to the decrease of both excitation and deexcitation at the center.

It is effective to increase the voltage across the CVL head with constant discharge power into the tube. To meet this requirement, magnetic pulse compression units or pulse–transformers utilizing amorphous cores have been adopted in the discharge circuit. The applied voltage up to 60 kV has been achieved by introducing the new discharge circuit with these components.

2.2. ACHIEVED PERFORMANCE

Figure 1 shows the experimental and calculated laser output powers with laser tube length. Each line indicates analytical result obtained by the simulation codes [2] when its electric field in the discharge plasma is assumed as described in this figure. From the results on 8 cm bore CVL, it is clear that the extracted output power per tube length can be increased effectively by increasing the electric field.

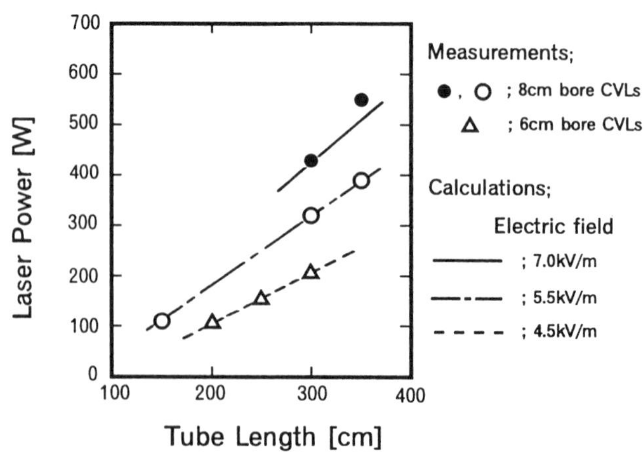

Figure 1. Laser power dependence on laser tube size and electric field.

TABLE 1 shows the summary of experimental results in two types of CVL with different laser tube length and different discharge circuit. The maximum oscillator output power of 550 W could be obtained with a laser tube of 350 cm length. However, in the CVL with such a short current duration and a long tube, it is difficult to extract optical energy effectively as an oscillator due to its short gain duration and long round–trip time.

TABLE 1. Experimental results in 8 cm bore CVLs

Item	Laser tube length	
	300 cm	350 cm
Voltage across a CVL head	36 kV	69 kV
Discharge peak current	2700 A	2800 A
Current rise time (10%–90%)	60 ns	40 ns
Laser pulse duration (10% of peak)	80 ns	65 ns
Oscillator output power	430 W	550 W
Amplifier output power	449 W	615 W

When the incident laser beam of 85 W was put into the same device, the maximum laser power of 615 W could be extracted as an amplifier. The maximum extractable power (MEP) is estimated to be about 750 W.

3. Stress Investment by Laser Irradiation

Residual compressive stress on metal surface extends fatigue life of mechanical components and reduces susceptibility of stress corrosion cracking (SCC) [5]. The output light of CVL is penetrable in water and is transmissible through optical fiber. By taking advantage of these features, water–immersed components could be remotely processed without drainage work.

3.1. PRINCIPLE OF STRESS INVESTMENT

When material is exposed to intense laser light, the material surface absorbs the laser energy and a plasma forms by ablative interaction of the laser light with the material through a multi–photon process. The effect of water confinement on the plasma is to intensify plasma pressure significantly, namely ten to one hundred magnifications compared to irradiation in air [6]. The plasma pressure reaches 1 to 10 GPa for a typical underwater laser irradiation. Since yield stress of metals is in order of several hundreds MPa, the impulse of such high–pressure plasma generates a plastic wave, which loses energy as it propagates to create dynamic strain in the material, as shown in Fig.2. In consequence, a residual compressive stress in the surface layer of irradiated metal is built uniformly by scanning the laser beam to cover the whole surface to be processed.

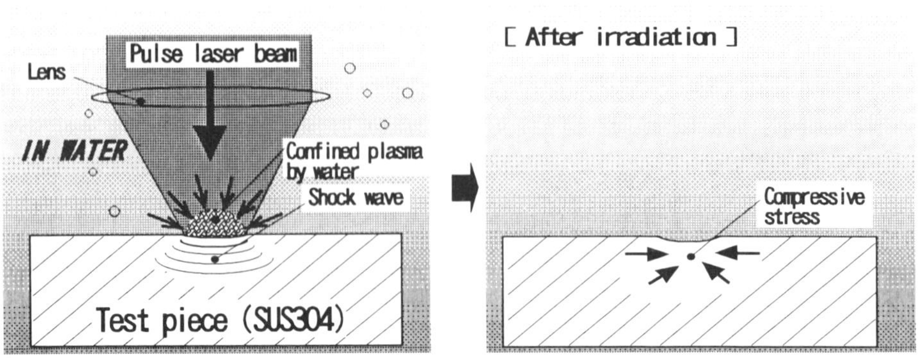

Figure 2. Principle of stress investment by laser irradiation.

3.2. RESULTS OF UNDERWATER LASER IRRADIATION

A residual compressive stress was successfully built in the surface layer of austenitic stainless steel (SUS304) by underwater irradiation of a CVL and a second harmonic of Nd:YAG (SH–YAG) laser. The irradiation conditions are summarized in TABLE 2. Most parameters are similar between the two lasers except for the pulse duration and the pulse repetition rate.

The residual stress generated on the metal surface was measured by the X–ray diffraction method. The in–depth profile of the residual stress was obtained while repeating the stress measurement and surface layer removal by electrolytic polishing.

TABLE 2. Experimental conditions of underwater laser material processing

Wavelength	532 nm (SH–YAG) 511 nm (CVL)
Laser pulse energy	15 ~ 166 mJ
Pulse duration	5 ~ 80 ns
Focal spot diameter	0.2 ~ 1.1 mm
Energy density	75 ~ 375 kJ/m^2
Peak power intensity	15 ~ 75 TW/m^2
Coverage factor	500 ~ 8000 %

3.3.1. SH–YAG laser Irradiation

Prior to CVL irradiation, a SUS304 test piece was irradiated in water by SH–YAG laser with a condition of 130 mJ of pulse energy, 5 ns of pulse duration, 0.75 mm of focal spot diameter, and 10 Hz of pulse repetition rate. Figure 3 shows the in–depth profiles of the residual stress compared with that of a non–irradiated test piece, which was pre–conditioned by grinding work. A residual compressive stress exceeding 400 MPa was developed over 100 μm in depth by the irradiation in water. Thermal effects on the material are minimized because of the ablative interaction of such short–pulse laser with the material, together with water cooling. On the other hand, for the irradiation in air, a large tensile stress remained on the surface layer of the test piece, which might be caused by the thermal effect of laser irradiation.

3.3.2. CVL Irradiation

A series of underwater irradiation was carried out using a SUS304 test piece. The typical result is shown in Fig.4, where the CVL irradiation condition is 18 mJ of pulse energy, 60 ns of pulse duration, 0.5 mm of focal spot diameter, and 4,000 Hz of pulse repetition rate. A residual compressive stress with the maximum of 400 MPa was built except for the surface layer of 10 μm. The tensile stress at the surface might be caused by the thermal effect due to the long pulse duration and high repetition rate of CVL compared to SH–YAG laser.

By using the CVL with the feature of high repetition, an area of 1 m^2 can be processed within one hour. The CVL should be used for practical applications to reduce the time required for the processing.

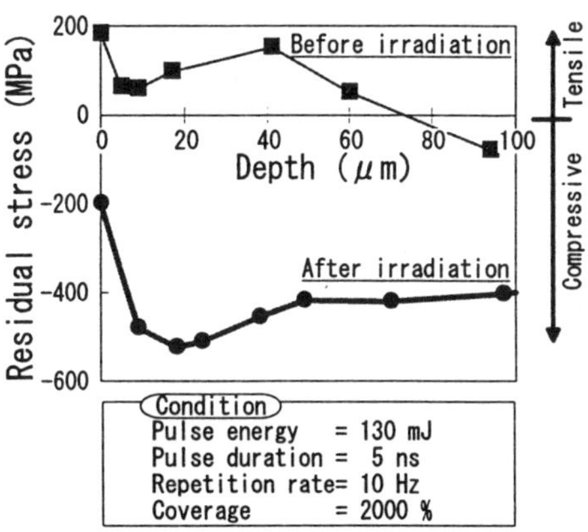

Figure 3. Residual stress after SH–YAG laser irradiation

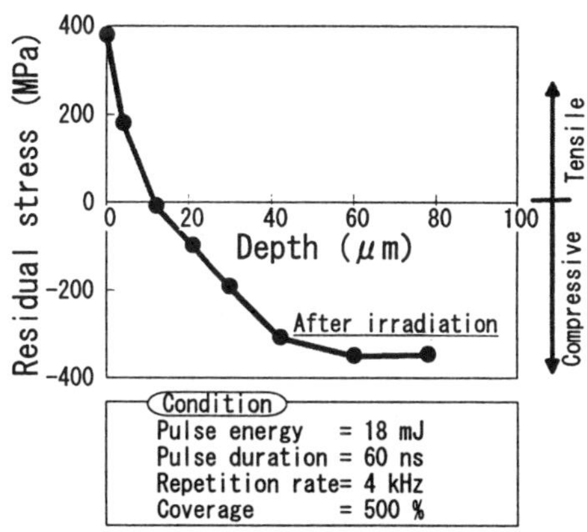

Figure 4. Residual stress after CVL irradiation

4. Direct Metal Cutting by Underwater Laser Irradiation

Laser irradiation abrades material due to ablative interaction. By measuring the mass reduction of a SUS304 test piece, the removal rate was estimated to be approximately 10 μg/J. This effect is applicable for cutting or eliminating a micro–crack which would be a cause of SCC. It was confirmed that SUS304 test pieces up to 20 mm thickness could be cut in water using a CVL with output of 110 W. Time required for cutting a 5 mm wide SUS304 test piece with thickness of 20 mm was about one hour. The thermal effect on material could not be observed in the cross section of the test piece, which is similar to the result of the stress investment.

5. Conclusion

The CVLs have been developed as the light source for the AVLIS system. The stable output power of 550 W for the oscillator, and 615 W for the amplifier was attained successfully.

The feasibility of underwater direct metal processing with a CVL was investigated, taking the advantage that the output light of the CVL was penetrable in water and transmissible through optical fiber. A residual compressive stress developed in the surface layer of SUS304 with significant magnitude and depth when a laser pulse of CVL or SH–YAG laser was incident. The feasibility of direct metal cutting was demonstrated using a CVL and a water–immersed SUS304 test piece with 20 mm thickness. Since laser irradiation induces no reactive force, water–immersed components could be easily accessed and processed.

Acknowledgement

The authors wish to thank Mr. N. Kobayashi, Mr. N. Mukai and Mr. K. Okazaki of Toshiba Corporation for valuable discussion.

References

1. Morioka, N. (1993) New Japanese AVLIS program, *Proc. of the International Society for Optical Engineering*, Los Angeles, **1859**, 2–13.
2. Konagai, C., Kimura, H., Aoki, N., Kobayashi, N., Iizuka, S. and Baba, Y. (1995) Development of High–power Copper Vapor Laser Amplifier, *Proc. of the 15th Annual Meeting of Laser Society of Japan*, Osaka, **15**, 112.
3. Aoki, N., Kimura, H., Konagai, C., Shirayama, S., Miyazawa, T. and Takahashi, T. (1991) High–power Copper Vapor Laser Development, *Proc. of International Society for Optical Engineering*, Los Angels, **1412**, 2–11.
4. Chang, J. J. (1993) Pressure dependence of copper laser output characteristics, *Applied Optics*, **32**, 5230–5235.
5. Obata, M. and Sudo, A. (1993) Effect of shot peening on residual stress and stress corrosion cracking for cold worked austenitic stainless steel, *Proc. of the 5th International Conference on Shot Peening*, Oxford, 256–263.
6. Mukai, N., Aoki, N., Obata, M., Ito, A., Sano, Y. and Konagai, C. (1995) Laser processing for underwater maintenance in nuclear plants, *Proc. of the 3rd JSME/ASME Joint International Conference on Nuclear Engineering*, Kyoto, **3**, 1489–1494.

APPLICATIONS OF CVLs IN THE ELECTRONICS INDUSTRY

M.A.LESNOI, V.V.ZUBOV, A.D.CHURSIN, S.S.GAPONOV
RPC "Istok"
Fryasino, Moscow Region, Russia

L.S.GLIKIN
RPC "Mekhatron"
Zelenograd, Moscow, Russia

Abstract

The requirements for copper vapour laser (CVL) used in electronic industry are determined. Main characteristics of the lasers and laser discharge tubes used in electronic industry in Russia are described. High efficiency of CVL application for processing of different materials and structures was demonstrated. Several original methods of laser processing and control of electronic industry elements and products have been suggested and realised in industrial equipment.

1. Introduction

Copper vapour laser (CVL) due to favourable combination of its parameters: pulsed mode of operation, high efficiency, high average and peak power have now rather wide applications in different fields of science and technology. Electronic industry where precision of processing is very important gives an example of successful application of CVL. High peak power and possibility of fine focusing allow to treat with CVL actually any material.

Realisation of the great potential of CVL applications in electronic industry is determined, to a great extent, by existence of reliable and not expensive laser tubes and efficient methods of processing. Russia has many important results in development of reliable, long life-time CVLs and different equipment utilising CVLs. Sealed-off long life-time copper vapour laser tubes (RPC "Istok" [1,2]), laser projection microscope with brightness amplifier, projection systems for laser processing of materials (Lebedev Phys. Institute [3]) were first developed in Russia.

C. E. Little and N. V. Sabotinov (eds.), Pulsed Metal Vapour Lasers, 377–382.

2. Characteristics of CVLs used in electronic industry

As our experience showed for successful applications of CVL in electronic industry it is desirable to have laser tubes with large enough aspect ratio and without drops of liquid copper on the tube wall. The total length of the tube should be as close as possible to the length of discharge tube. For industrial applications the laser tube should be sealed-off and should have small size and low weight. Stability of laser parameters, its reliability and time of service, and all this is very important for applications in industry, are also strongly dependent on the discharge tube used.

Investigations carried out at RPC "Istok" resulted in development of several copper vapour discharge tubes having average output power from 1W up to several tens of W [4]. For processing of almost all materials used in electronic industry it is enough to have average output power in the range 1 - 10W. A set of laser tubes "Kulon" and also the tube UL-102 were specially developed for applications in electronic industry. Main characteristics of these tubes and of lasers utilising these tubes are presented below. The widest application has the laser tube "Kulon-10", therefore the majority of the characteristics are for this tube.

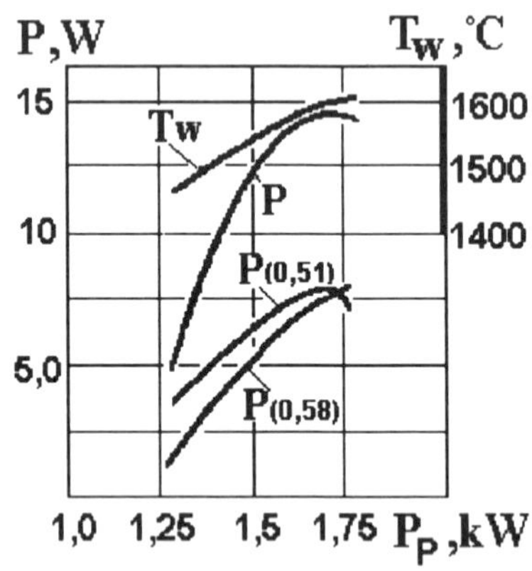

Figure 1. Dependencies of average output power (P), power green (P(0,51)) and yellow (P(0,58)) wavelength, tube wall tempe-rature (Tw) on power from power supply (Pp).

Fig.1 shows the dependencies of average output power P and tube wall temperature T_W on the power consumed from the DC power supply P_P. Fig. 2 illustrates typical dependencies of output power P, tube wall temperature T_W and consumed power P_P on the pressure of buffer gas neon. In practice the choice of buffer gas pressure is determined, as a rule, by a compromise between output power and life time of discharge tube because life time is increased with the buffer pressure rise but output power is decreased.

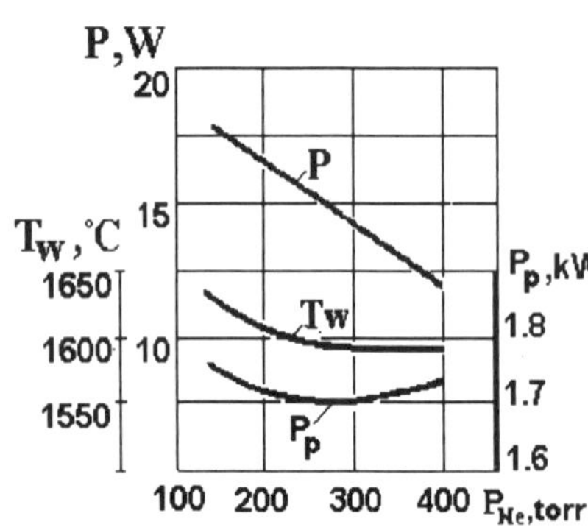

Figure 2. Typical dependencies of output power (P), tube wall temperature (Tw) and consumed power (Pp) on the pressure of buffer gas neon (p_{Ne}torr).

Fig. 3 shows the results of life time tests of some laser tubes produced by "Istok". Curves 1 and 2 correspond to two UL-102 laser tubes operating as brightness amplifiers (with one mirror). The majority of the tests are carried out in this regime since it is the most efficient in material processing applications. Curve 3 is for "Kulon-3" laser tube, also operating as an amplifier, and curve 4 is for “Kulon-10” laser tube in laser regime. The different character of power decrease for the last two curves is connected, first of all, to the fact that “Kulon-3” laser tube was tested as amplifier. The decrease of transparency of the tube windows due to deposit from discharge, especially from the mirror side, has less influence on the output power because of high gain of the active medium. There was no obscuring of laser tubes discharge channels by copper droplets even after long time of operation.

Main characteristics of the lasers utilising “Kulon” and Ul-102 laser tubes are listed in the table. The lasers are produced by RPC “Mekhatron”. The characteristic features of these lasers are: air cooling, small size and low weight.

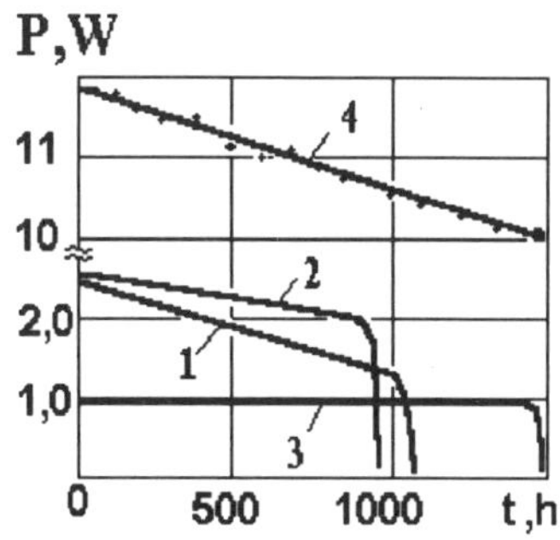

Figure 3. Results of life time tests of: 1 and 2 - two laser tubes UL-102 operating as brightness amplifiers (with one mirror), 3 - laser tube "Kulon-3" as ampifier, 4 - laser tube "Kulon-10 " in laser regime.

N	Laser / Parameter	CVL-1	CVL-3	CVL-5	CVL-10
1	Laser tube	Kulon-1	Kulon-3	UL-102	Kulon-10
2	Wavelength, mm	0.5106 and 0.5782			
3	Beam diameter, mm	6.0	12.0	20.0	12.0
4	Beam divergence, mrad	from 0.2 to 5			
5	Output power, W: At the beginning After 1000 h	 1.0 - - 0.8	 4.0 - - 3.0	 10 - - 5.0	 14 - - 10
6	Pulse repetition rate, kHz	16 - 18	14 - 16	8 - 12	14 - 18
7	Laser pulse length, ns	15±5			
8	Power consumption, kW	0.7	1.2	2.2	2.2
9	Warm up time,min.: normal forced	 25 20	 45 30	 60 40	 50 35
10	Size, mm	470x410 x210	750x490 x260	950x490 x260	950x490 x260
11	Mass, kg	15	35	40	40

3. Applications of CVL for material processing

Very high gain (up to 40dB) of CVL active medium allows to use it effectively in active projection systems to realise different processes of laser treatment of many materials and structures. On this basis a multipurpose automated laser complex (MALC) has been developed for processing and control of materials and products of electronic industry. The diagram of this complex is shown in Fig. 4. An image of the object to be treated 3 is formed by the lens 1 in the plane of the spherical mirror 2. Aperture diafragm 4 is placed before the lens. The laser tube 5 is located between the lens 1 and mirror 2. The mask 6 is placed in front of the mirror 2. In this system a bright light spot repeating the form of the mask 6 is formed on the object. With the help of beam splitter 8 and spherical mirror 9 an image of the treated object is formed on the screen 7 for visual control of the treatment process. Thus the object is observed as in a usual laser projection microscope.

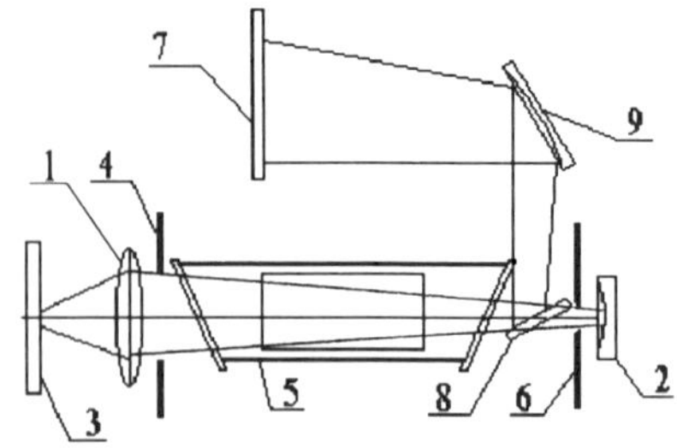

Figure 4. Optical diagram of laser technological equipment based on active projection system with "conjugate" resonator.

Unlike traditional projection systems of laser processing, where there are usually great losses of light on mask, in this system only useful beam carrying the image of the mask is amplified. This means that the losses on mask are eliminated and active medium is used much more efficiently.

Active medium in this system under typical conditions operates at rather high level of saturation which enable to use almost all inversion stored in the volume of the active medium occupied by the processing beam. In this case the power of processing beam weakly depends on the open area of the mask and power density on the object to be treated is inversely proportional to the area.

Low losses, high efficiency and possibility to concentrate the processing beam on an area of micron size on the object allows to achieve average power density on the object as high as 10^7 W/cm^2 and peak power density up to 10^{11} W/cm^2 with a laser tube providing in a laser 5-10 W of average power. It allows to realise such operations as cutting, hole drilling, evaporation of metal layers, thermotreatment and so on.

Depending on the field of applications the laser complex (MALC) is equipped by one of the following laser tubes: “Kulon-3”, “Kulon-10” or UL-102. The complex consists of separate blocks which allows easy rearrangement of the system for different processing operations. The most frequently used of them are listed below:

Laser lithography with resolution up to 2 microns. Thin layers of Cr, Al, Cu, Au, iron dioxide, polysilicon can be easily removed from substrates. The operation is usually controlled visually on laser projector screen. It is used for trimming electronic components, precision drilling, etc.

Marking of silicon plates. A liquid crystal panel is used as a mask in this case. Characters of about 1 mm size are formed on the surface by evaporation of surface layer. When marking in SEMI standard the productivity of the equipment is 400 plates per hour.

Pulsed treatment of semiconductor structures for modification of their properties. The most typical operations are: annealing of semiconductor layers after ion implantation; treatment of “silicon on sapphire structures” in order to improve

their crystallographic properties; preepitaxial treatment of silicon plates in order to decrease concentration of microdefects; laser gettering of semiconductor plates.
Besides the operations listed above it is possible to perform many other operations which is difficult to do using traditional equipment.
CVL can be also effectively used in testing and measuring equipment used in electronic industry. Laser projection microscope is for example used for visual control on a screen, including large screen, of different electronic elements, masks etc. The magnification of the microscope with different objectives is usually from 100 to 1000. The advantage of the microscope using laser amplifier is that at the same brightness of the screen illumination of an object can be orders of magnitude lower than in traditional microscope. This allows to perform non-destructive testing of practically any objects including biological ones.
One more example is Laser thermowave microscope FM-5. It is developed on the basis of "Kulon-1" laser tube and is used also for non-destructive control of electronic elements and devices. It comprises three channels: optical, acoustic and thermowave which allows to measure simultaneously optical, acoustic and thermophysical characteristics of the sample under control. The diagram of the microscope is shown in Fig. 5.

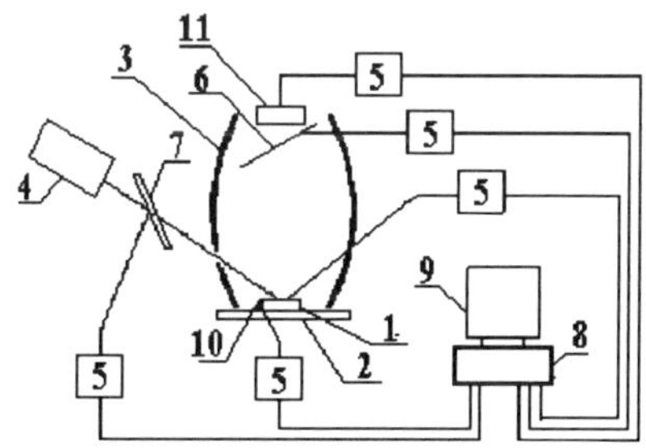

Figure 5. The diagram of the laser thermowave microscope.
1 - object under control, 2 - coordinate table, 3 - ellipsoid, 4 - laser, 5,8 - receiving-treating devices, 6 - mirror, 7 beam splitter, 9 - monitor, 10 - acoustic-receiving devices, 11 - two-spectral IR receiver.

In the optical channel radiation reflected and scattered from the sample is registered. Acoustical channel allows to control the structure of the sample at a depth of up to 10 mm with resolution about 100 microns. If the thickness of the layer to be controlled is 1 mm the resolution may be about 1 micron. The most informative channel is the thermowave one. It registers spectra of thermal radiation of the sample surface after each laser pulse.
An analysis of information from all three channels allows to control the results of soldering, welding, quality of adhesion of thin films on substrates, nonuniformity of thin film thickness and results of some other operations in many products of electronic industry.

4. Conclusion

Experience of CVL applications in equipment for electronic industry in Russia shows that the applications are rather efficient. New applications and new equipment are now under development.

The authors thank Professor G.G.Petrash for useful discussions and help in preparation of this paper.

5. References

1. Burmakin V.A., Lesnoi M.A., Yevtiunin A.N., Bylkin V.I., (1978), Sealed-off copper vapour laser with long life time, *Kvantovaja Elektronika (Quantum Electronics)*, **5**, 1000-1004. (In Russian)
2. Belyaev V.P., Zubov V.V., Lesnoi M.A., Lyabin N.A., Chursin A.D., Vorobyov O.D., (1981), Application of active elements of copper vapour lasers in technological equipment for electronic industry products control, *Elektronnaya Promyshlennost' (Electronic Industry)*, No 5-6, 82-83. (In Russian)
3. Blinov I.G., Valiev K.A., Petrash G.G., Rebrov S.I., Savin V.V., (1976), Laser projection microscope, *Elektronnaya Promyshlennost' (Electronic Industry), No 3, 28.*
4. Zubov V.V., Lesnoi M.A., Lyabin N. A., Ugol'nikov S.A., Chursin A.D., (1994), Copper vapour lasers with sealed-off active elements, in G.G.Petrash (ed), Proc. SPIE, **v. 2110**, *Metal Vapor Lasers and Their Applications,* 78-89.

LASER INDUCED CHEMICAL ETCHING OF SILICON WITH SF_6 USING A COPPER BROMIDE VAPOUR LASER

B. IVANOV, D. PHILIPOV, V. SHANOV AND G. PEEV
Higher Institute of Chemical Technology, Dept of Semiconductors,
8 Kliment Ohridski Street, Sofia 1756, BULGARIA

1. Introduction

Over the last 15 years, a wide range of laser induced chemical etching systems have been investigated [1]. Typically these studies have been concerned with etching that is initiated with either a tightly focused, cw scanning laser beam, or with a large area pulsed laser beam. In general, laser induced gaseous etching may be either photolytic - the main mechanism is the photodissociation of precursor or adsorbed reactant molecules, or pyrolytic - the basic mechanism is thermal. We present here our initial results on direct write dry pyrolytic etching of silicon with SF_6 using Copper Bromide Vapour Laser (CBVL). To the best of our knowledge this the first results of dry etching of any materials using Copper Vapour Laser (CVL) or CBVL. The CVL and the CBVL have a group of characteristics - wavelength, repetition rate and pulsewith, which make it unique to application in laser induced surface chemical processes. Our previous experience in the field of Laser Chemical Vapour Deposition of Al and Si using CVL and CBVL shows quite different effects with comparison with cw Ar+ laser [2 - 10]. In the literature there is information only about etching of Si with SF_6 molecules vibrationaly excited by CO_2 laser via multiple photon absorbtion [11].

2. Experimental

The experimental set-up for of laser induced chemical etching of silicon with SF_6 has been described in earlier works [3,5]. The beam from Copper Bromide Vapour Laser with repetition rate 20 kHz and pulse length 60 ns was focused up to 20 μm on the silicon (100) monocristaline wafer at room temperature. The scanning speed was varied from 40 to 400 μm/s, the laser power between 0.72 and 2.0 W and the pressure of the SF_6 from 10 mbar to 1000 mbar. A Quadrex - 200 (Leybold - Inficon) quadrupole mass spectrometer equipped with a Faraday cup was used. The experimental arrangement for

C. E. Little and N. V. Sabotinov (eds.), Pulsed Metal Vapour Lasers, 383–388.

sampling the gaseous products during laser etching has been presented in previous paper [4].

2. Results and Discussion

Mass spectrometric study of laser induced chemical etching of Si with SF_6 using CBVL is presented at Fig. 1. As one can see the main conclusion from this figure are: 1) consummation of SF_6 corresponding to 108 SF_4^+ and 127 - SF_5^+ AMU peaks decreasing, 2) formation of SF_4, SF_2, S and Si corresponding to 54 - SF_4^{2+}, 35 - SF_2^{2+}, 32 - S^+ and 28 - Si^+ increasing, 3) absence of 85 AMU peak corresponding to the SiF_4 formation. A possible explanation of this unusual mass spectrometric result is formation of solid compound, containing Si, S and F due to unconventional way of etching chemical process caused by pulsed laser irradiation and secondary reaction of products. An SiF_xO_y layer on elemental silicon was found to be formed as cited in [12 -13]. Our opinion is that in this case is possible production of similar SiF_xS_y layer. As a result of this there is not SiF_4 generation in the gaseous phase. The scanning electron micrographs Fig. 2 - Fig. 4 show the influence of different process parameters on the morphological structure of etched lines. The decreasing of SF_6 pressure presented on Fig. 2 a) and Fig. 2 b) show the resultant morphological changes. The decreasing of laser power presented on Fig. 3 a) and Fig. 3 b) also exhibit deterioration of etch line quality. At high scanning speed as we can see from Fig. 4 a) and b) the formed solid compound disturbs the etching process and does is not allow production of defined groove. The best quality of etched lines is at low scanning speed - 40 or 80 μm/s, high SF_6 pressure - 1000 mbar and high laser power - 2.0 W - Fig. 2 a) and Fig. 3 a). The reason for this behaviour is probably the necessity for desorption of obtained SiF_xS_y solid compound from Si surface. The additional evidence for above conclusion are the scanning electron micrographs of all lines indicating formation of solid residue within and along etched lines. Fig. 5 and Fig. 6 display etched depth behaviour as a function of scanning speed and laser power for different SF_6 pressure. An unusual etching regime is observed where the etching depth anomalously increases with increasing of scanning speed and shown in the papers of Tang and Herman [14 - 15]. We can define three regimes in the process of etching. In the first regime with increasing of scanning speed we increase the rate of solid compound formation. Finally the rate of solid compound formation exceeds the rate of its desorption from Si surface that cause decreasing of the etch depth. In the second regime the increasing of scanning speed facilitate the reaching to fresh Si surface where is not necessary desorption of already formed solid compound. This effect gives rise etching depth. In the third regime the increasing of scanning speed cause decreasing of dwell time of laser irradiation. This lead to reducing of the rate of etching chemical process. The highest value of etching rate was about 100 μm/s , obtained at 10 mbar SF_6, 2.0 W laser power and 120 μm/s scanning speed.

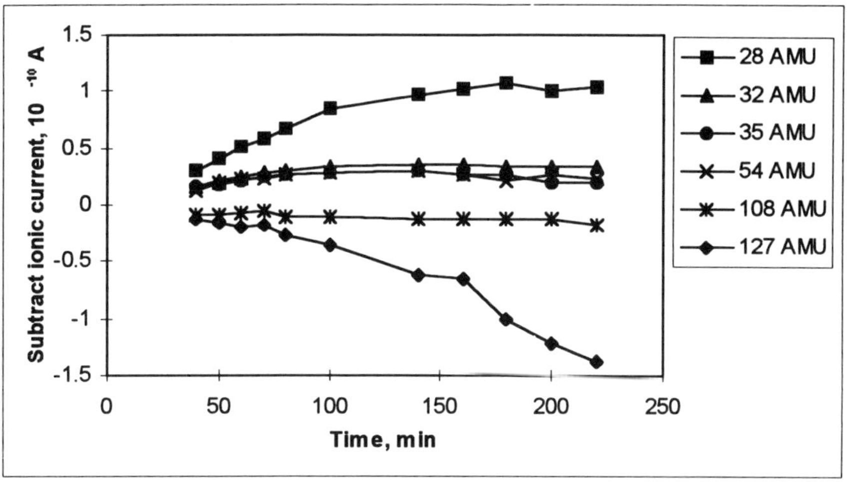

Fig. 1 Time evolution of ionic current for different Atom Mass Units (AMU) 28 - Si^+, 32 - S^+, 35 - SF_2^{2+}, 54 - SF_4^{2+}, 108 - SF_4^+, 127 - SF_5^+ at 2.3 W laser power and 40 μm/s scanning speed.

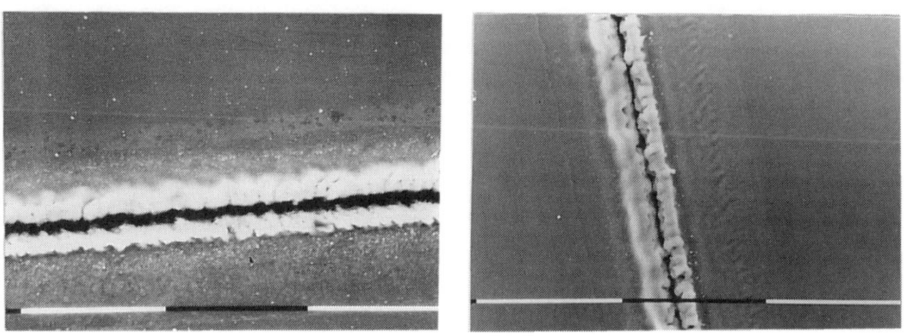

a) The mark ―――――― is 100 μm b)

Fig. 2 Scanning electron micrographs of silicon etched lines

a) SF_6 pressure 1000 mbar - flow system, scanning speed 40 μm/s, laser power 2.0 W.

b) SF_6 pressure 10 mbar - flow system, scanning speed 40 μm/s, laser power 2.0 W.

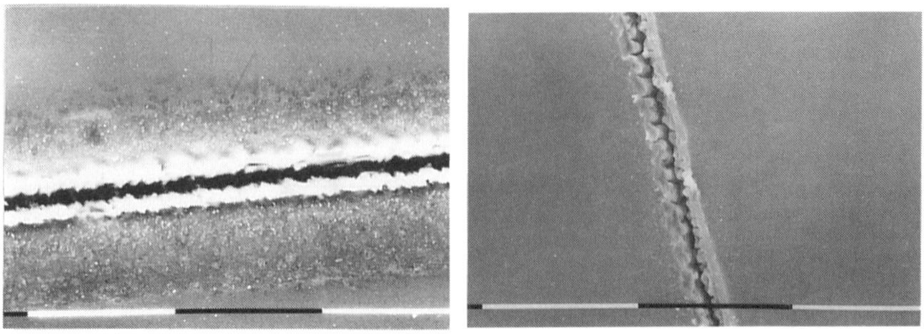

a) The mark ———— is 100 μm b)

Fig. 3 Scanning electron micrographs of silicon etched lines
a) SF_6 pressure 1000 mbar - closed system, scanning speed 80 μm/s, laser power 2.0 W.
b) SF_6 pressure 1000 mbar - closed system, scanning speed 80 μm/s, laser power 1.33 W

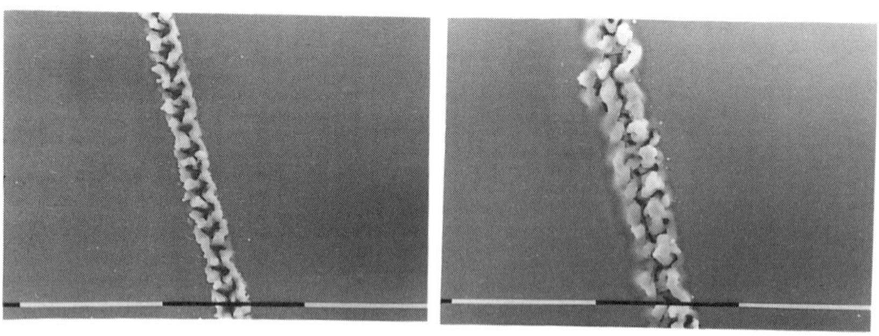

a) The mark ———— is 100 μm b)

Fig. 4 Scanning electron micrographs of silicon etched lines
a) SF_6 pressure 1000 mbar - closed system, scanning speed 400 μm/s, laser power 1.33 W.
b) SF_6 pressure 10 mbar - closed system, scanning speed 400 μm/s, laser power 2.0 W.

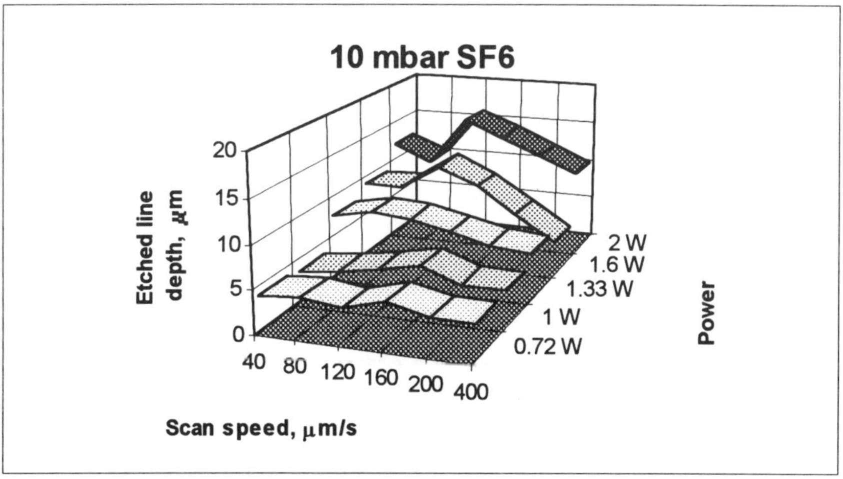

Fig. 5 Etched line depth as a function of scanning speed and laser power at 10 mbar SF_6.

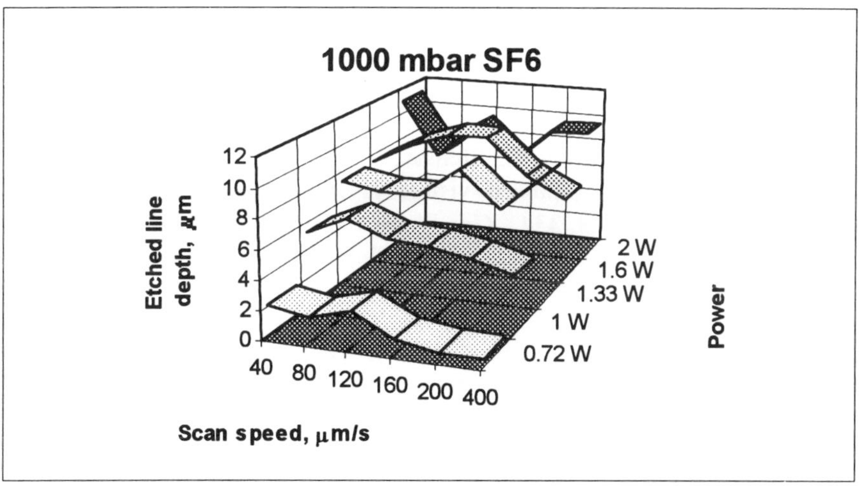

Fig. 6 Etched line depth as a function of scanning speed and laser power at 1000 mbar SF_6.

3. References

1. Ritsko, J.J., (1989), Laser Etçhing, in Ehrlich, D.J. and Tsao, J.Y. - (eds.), *Laser Microfabrication - Thin Film Processes and Lithography*, Academic Press Inc., Boston, pp. 333 - 383.

2. Shanov, V., Ivanov B. and Popov, C. (1991) Laser Induced Direct Writing of Aluminium, *Journal de Physique II* **1**/7, 373 - 380.

3. Shanov, V., Ivanov B. and Popov, C (1992) Laser Chemical Vapour Deposition of Thin Aluminium Coatings , *Thin Solid Films* **207**, 71 - 74.

4. Ivanov, B., Popov, C. and Shanov, V., (1992) Mass Spectrometric Study of Laser Induced Pyrolytic Decomposition of TMA, *Journal of Advanced Materials for Optics and Electronics* **1**, 287 - 292.

5. Shanov, V., Ivanov B. and Popov, C (1993) Growth Rate Behavior of LCVD Aluminium, *Processing of Advanced Materials* **3**, 41 - 44.

6. Shanov, V., Popov, C, Ivanov, Souleva, A. and Peev, G. (1993) An experimental approach for growth rate determination of LCVD written aluminium stripes, *Journal of Materials Science: Materials in Electronics* **4**, 55 -58.

7. Shanov, V., Popov, C and Ivanov B. (1993) LCVD of aluminium stripes obtained by pyrolysis of TMAA and TMA, *Journal de Physique IV,* **3** 255 - 260.

8. Popov, C., Ivanov, B. and Shanov, V., (1993) Mass Spectrometric Study of Laser Induced Pyrolytic Decomposition of TIBA and TMAA, *Journal de Physique IV* **3**, 107 - 112.

9. Popov, C., Ivanov, B. and Shanov, V., (1994) Laser-induced chemical vapor deposition of aluminium from TMAA, *Journal of Applied Physics* **75**, 3687 - 3689.

10. Ivanov, B., Popov, C., Shanov, V. and Filipov, D., (1993) Pyrolitic LCVD of silicon using a pulsed visible laser - experiment and modelling, Mat.Res.Soc.Symp. W Vol. **334** - *Gas Phase and Surface Chemistry in Electronic Materials Processing* - November 29 - December 4, 1993, Boston, USA.

11. Chuang, T.J., (1982) Laser-enhanced gas-surface chemistry: Basic proces and applications, *J. Vac. Sci. Technol.*, **21** , 798 - 806.

12. Legtenberg, R., Jansen, H., de Boer, M. and Elwenspoek, M., (1995) Anisotropic Reactive Ion Etching of Silicon Using SF_6 /O_2 /CHF_3 Gas Mixtures, *J. Electrochem.Soc.*, **142**, 2020 - 2028.

13. Oehrlein, G., Robey, S. and Lindstrom J. (1988) Surface processes in CF_4 /O_2 reactive etching of silicon, *Appl. Phys. Lett.* **52**, 1170 - 1172.

14. Tang, H. and Herman, I. (1990) Laser-induced room temperature etching of copper films by chlorine with analysis by Raman spectroscopy, *J. Vac. Sci. Technol.*, **A 8** , 1608 - 1617.

15. Tang, H. and Herman, I. (1992) Anomalous local laser etching of copper by chlorine, *Appl. Phys. Lett.* **60** , 2164 - 2166.

8. Applications in Medicine

TREATMENT OF PORT-WINE STAINS WITH THE 578 NM LIGHT FROM A COPPER VAPOUR LASER AND AN AUTOMATED SCANNING DEVICE

PHILIP H. BUTLER†, DEREK J. SMITHIES AND
NATHAN W. MEHRTENS
Invited Speaker†
Physics and Astronomy Department,
University of Canterbury,
Christchurch,
New Zealand.

Abstract. Like the CO_2 and Nd:YAG lasers, the copper vapour laser may be used to ablate tissue. However, when used with a computerised scanning system, it is ideal for the treatment of port-wine stains.

The yellow light produced can be used to damage just the enlarged capillaries in the skin. These capillaries are necrosed, and no longer contain blood. The volume of blood in the skin is reduced, and the skin takes on a more normal appearance.

Theoretical calculations that we have carried out have shown that the illumination time needs to be in the 2-8 ms range, which is consistent with the clinical results of many workers. We describe how these illumination times can be achieved with a computer controlled treatment system and the response of the skin to such a system.

1. Introduction

Since their inception in the 1960's, the use of lasers in medicine has steadily increased. Provided that sufficient power is available, they are ideal for cutting and ablation. Several lasers, such as the CO_2 and Nd:YAG provide sufficient power to cut the tissue in an acceptable time span. The copper vapour laser may be used for this, but the time required may be too long as the available power is usually in the order of watts. However, it is in

C. E. Little and N. V. Sabotinov (eds.), Pulsed Metal Vapour Lasers, 391–402.

the treatment of port-wine stains that the copper vapour laser has found a niche.

Port-wine stains are vascular malformations in the skin present in approximately 0.3% of the population [6, 7]. The key difference between normal skin and port-wine stain skin is the size of the capillaries. Normal capillaries have approximately 6 μm diameters. Port-wine stains capillaries can be up to 300 μm in diameter [12]. As the capillaries are much larger, there is more blood in the capillaries, and the skin takes on the colour of blood.

Treatment with laser light has been the most effective means of removing port-wine stains [20]. The effect of the laser light is to heat the blood in the capillaries and the capillary wall is heated by thermal conduction. If there is sufficient heating of the capillary wall, the capillary will die, and no longer contain blood. This lowers the amount of blood in the skin, and the red colour of the port-wine stain will fade.

In section 2 we summarise the calculations in the literature for the ideal treatment parameters of wavelength, fluence, and illumination time. These results show that yellow light (577-585 nm) should be used. Light in this region of the visible spectrum is strongly absorbed by the blood in the enlarged capillaries, but not strongly absorbed in the other skin structures.

During treatment, the laser beam is incident on each point of the port-wine stain for a period of time referred to as the illumination time. For a given fluence, if the illumination time is too short, then there will be insufficient damage to the capillary wall. Conversely, too long an illumination time will result in unwanted damage to the dermal tissue which surrounds the capillaries. The calculations described in section 2 show that the illumination time should be in the range of 2-8 ms, with a resulting fluence of approximately 10 J/cm^2.

To achieve illumination times in the 2 to 8 ms range, a mechanical device is necessary to direct the light onto the tissue. Several devices have been reported in the literature (for example [15, 1, 19]). We describe how the automated treatment device reported by Smithies *et al.* [19] is used to scan the light generated by a copper vapour laser over a port-wine stain in section 3.

The clinical changes as a result of treatment with this automated treatment device are given in section 4. In section 5 we present some photographs which show that the automated treatment system can be used successfully to remove port-wine stains.

2. The ideal treatment parameters

In this section, we describe the calculations for the ideal parameters for the treatment of port-wine stains.

The Monte Carlo method was used by Smithies and Butler [17] to describe the passage of light through a model of skin. The model of skin used was as complete as the available optical parameters for the skin structures in the model allowed. In these calculations, the light is represented by neutral particles. The wave-like nature of light is ignored, so effects resulting from polarisation and diffraction are not described. The paths of the individual particles are generated from probability distributions, absorption points of the particles are recorded, and thus a map of absorption in the skin is generated.

These results show that when the wavelength is 577 nm, absorption in the blood in the capillaries is maximised, relative to the absorption in the epidermis. This wavelength corresponds to an absorption peak of oxyhaemoglobin, the oxygen rich pigment haemoglobin which is contained in red blood cells and is responsible for the red colour of blood [3].

These theoretical calculations show that changing the wavelength to 585 nm reduces the absorption in the blood. The absorption within the capillary is more uniform, and the difference between the absorption at the top and bottom of the capillary is less than when using 577 μm. Absorption in the very deep capillaries can be a little higher when the wavelength is 585 nm. Less light is absorbed in the shallower capillaries, and so more reaches the very deep capillaries.

Clinical work suggests that 585 nm is the ideal wavelength to use with pulsed dye lasers (illumination time 0.45 ms), in part because 0.45 ms is too short for thermal averaging to heat a capillary uniformly, as is necessary with 577 nm light. However, absorption in the very deep capillaries is only a tiny fraction of the absorption in the shallow capillaries. If sufficient energy is deposited in the very deep capillaries so that they reach an average temperature of 70 °C, then the average temperature of the blood in the shallow capillaries will be several hundreds of degrees (ignoring phase changes). The surplus of energy in the shallow capillaries will be conducted to, and thermally damage, the surrounding non-vascular tissue.

To carry out optimal treatment, there should be no damage to the non-vascular tissue. Sufficient energy, at either wavelength, should therefore be placed in the skin so that only the uppermost capillaries are damaged. Some energy will be absorbed by the deeper capillaries but it will be insufficient to cause thermal damage.

Setting the wavelength to 577 (or 585) nm does not ensure that optimal treatment is carried out. The illumination time must also be set so that there is minimal damage to non-vascular tissue but the vascular tissue is thermally damaged. To determine the ideal illumination time, we use the results from the Monte Carlo calculation (from which we deduce the heat generated at each point in the model of skin) and solve the heat transfer equation.

The numerical solution of the heat transfer equation shows us how the temperature distribution within the skin evolves over time. Thus, we can determine the maximum temperature of the endothelial cells which make up the walls of the ectatic capillaries. From this, we see if cell death will occur. These results show that the illumination time should be in the range of 2-8 ms if 578 nm wavelength light is used [13, 14, 18].

3. An automated treatment device

In this section, we describe an automated treatment device which scans the laser beam generated by a copper vapour laser over a port-wine stain.

When a continuous wave laser is used, the light is scanned across the port-wine stain in a rasterscan-like pattern. An illumination time of 5 ms corresponds to a 0.5 mm diameter spot moving at a speed of $10\,\mathrm{cm\,s^{-1}}$ across the surface of the lesion. If the laser light is delivered by a fibre optic, it is not possible to control the position of the fibre by hand to the necessary level of accuracy. For illumination times in the 2-8 ms range, a mechanical device is required to control the position of the spot.

In 1991, Smithies *et al.* [19] described the SCANALL system, an automated treatment device for the treatment of port-wine stains. This device provides a more uniform result than is obtained by manual positioning of the fibre. Illumination times and fluences in the optimal range are obtained when the laser power is around 5 W. To move the spot on the skin a personal computer alters the orientation of a mirror.

The path of light through the SCANALL system is shown in figure 1. The lens, scanning mirror drive unit, camera, fluorescent light, and camera mirror are fixed to an optical bench. The optical bench can slide along the track. Mounted to the track is the angled mirror.

The laser beam is generated within the laser cavity and is directed to the one end of the laser cabinet, where the pillar is mounted. At the base of the pillar, there is a mirror which reflects the laser beam up. At the top of the pillar, the scanner arm is mounted in such a way that it can rotate around the pillar. The angled mirror, which rotates when the scanner arm is rotated, is used to deflect the beam along the scanner arm. The beam travels along the arm of the scanner, where it passes through the lens. The lens has a focal length of 800 mm, and it is used to focus the laser beam.

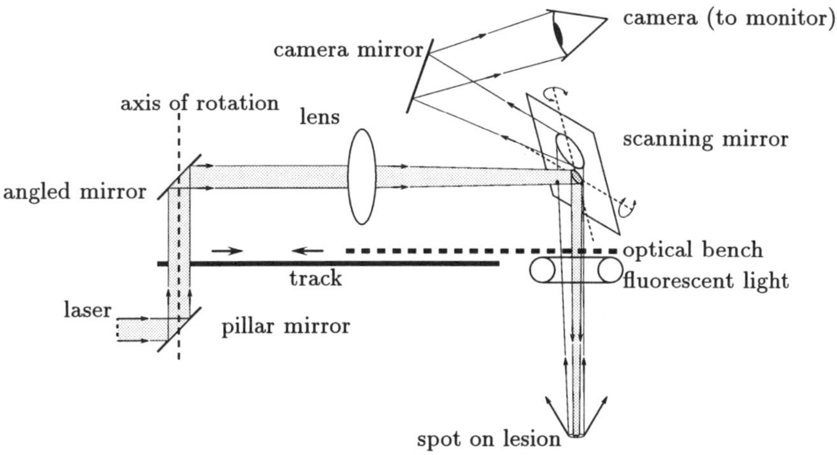

Figure 1. Schematic diagram that shows the path of light through the computer controlled scanning unit. The light is focussed onto the patient by the 800 mm focal length lens and scanned across the lesion by changing the orientation of the stepped mirror. An image of the lesion is obtained by the video camera and displayed on the computer monitor.

At the end of the scanner arm, the scanning mirror is used to deflect the beam downwards onto the lesion that is to be treated.

The patient lies on an adjustable bed with the port-wine stain to be treated directly below the scanning mirror. The distance between the port-wine stain and the scanning mirror is such that the laser beam is focussed on the stain. A laser spot with a diameter of (approximately) 0.3 mm is formed.

The video camera which is mounted above the scanning mirror is used to obtain an image of the lesion. The image is placed on the computer monitor. The computer text and graphics are superimposed over the image of the lesion on the monitor.

The position and shape of the lesion needs to be stored in the computer (Amiga B2000) by the operator. This is achieved by stepping around the image of the lesion on the monitor, using a mouse. The position of the outline is recorded by the computer, which then evaluates the shape of the lesion and determines the path that the laser beam should take.

Once the computer has determined the path the spot should follow, treatment can commence. The shutters open, and the laser beam is incident on the port-wine skin. The laser beam is moved across the port-wine stain when the orientation of the scanning mirror is adjusted by stepper motors, which are controlled by the computer. The area is filled in with a rasterscan like pattern, with no part treated twice. The spot does not travel outside of the shape entered by the operator. At the edge of the lesion, where there is no need to treat, the speed of the spot is increased.

When treating with the automated scanner, the direction of the scan lines are set so that they are parallel to the Lines of Langer in the skin. These lines are the lines along which wrinkles are observed. With this orientation of the scan lines, any gap between scan lines (which may be caused by movement of the patient arising from breathing) is less noticeable.

When flat mirrors were installed in the laser cavity, the beam divergence was in excess of 1 *m*rad and the spot had a diameter of 1 mm. The resultant power density was too small, and consequently the illumination times obtained were in excess of 30 ms. Unstable optics were installed in the laser, which reduced the beam divergence to 0.1 *m*rad. This significantly reduced the laser power available, but increased the power density available as the spot diameter was reduced to about 0.3 mm. With this spot diameter, illumination times of around 4 ms are attained when treating port-wine stains with 5 W of yellow (578 nm) light.

4. The clinical observations

In this section, we describe the observations of the treatment process, and the changes that occur within the skin.

With the automated treatment system described above and a nominal 20 W copper vapour laser, 5 W of yellow light can be deposited onto the port-wine stain. The spot diameter is 0.3 mm, so illumination times around 4 ms are used. With the rare-earth doped safety glasses described by van Halewyn *et al.* [21] the clinician has (nearly) full colour vision. The changes in the skin have been explained by Marini *et al.* [9], who has described the mechanism for the blanching obtained.

When the haemoglobin absorbs the yellow laser light, heat will be conducted from the erythrocytes to the endothelial cells of the capillary walls on a time scale of approximately 3 ms. The next step is that the selective permeability of the endothelial cell membrane is irreparably lost. The sudden influx of water carried by the Na^+ ions redistributing according to their concentration gradient causes the cells to rapidly enlarge. Since the surrounding dermal structure prevents centrifugal expansion, expansion occurs centripetally, blocking the lumen. When the lumen of the capillary is almost entirely occluded, the blanching effect is obtained. This blanching or paling is transient as erythema is observed after treatment. Erythema occurs because the previously empty capillaries have dilated and allowed blood to flow into the skin in response to the thermal injury. Whitening is then observed, as intra-cellular fluid has leaked from the damaged capillaries. This is the "Triple Response", which is described in more detail by

Ganong [3].

If the illumination time (and thus fluence) is increased by 10%, the blanching obtained is permanent. The optical properties of the skin change so more of the incident light is scattered in the shallow layers of the skin [5, 11, 16]. Consequently, the underlying blood filled vessels are no longer visible.

With the computer controlled scanning system that we use, the transient blanching obtained is turned to our advantage. Since the erythema present soon after treatment is comprised of newly dilated capillaries that were empty when the initial treatment was conducted, we have created many new potential targets for the laser light. These capillaries are treated by scanning the laser beam over the area again.

We limit the damage to the non-vascular tissue by cooling the skin after the laser beam has been scanned over the skin. This is the standard 'first-aid' treatment for any superficial burn to the skin. This will help to minimise the unwanted side-effects of the initial laser treatment such as superficial blistering, and basal and dermal cell necrosis. The cooling is insufficient to cause the erythemetous capillaries to constrict, and there is some evidence that cooling may indeed increase the redness of the lesion [4, 8].

5. Clinical Results

In this section we present clinical results which demonstrate the efficacy of the copper vapour laser and automated scanner in the treatment of port-wine stains.

Figure 2a shows a 20 year old female (skin type II) prior to her first treatment with the copper vapour laser. She has a purple, deeply located, port-wine stain on her left cheek. Figure 2b shows the same patient prior to her sixth treatment. The port-wine stain is largely resolved and what remains are the large deep vessels. These were treated and what is left of the lesion can easily be concealed beneath make-up.

Figure 2c shows a 23 year old female (skin type III) with a red port-wine stain on her left forehead. The photograph was taken prior to her first treatment. After four treatments (figure 2d) the superficial vessels have been removed and what remains are the large feeder vessels (arterioles). These were injected with a sclerosing agent (which seals up the large vessels) and treated with the copper vapour laser. We note that there is no hypo-pigmentation of the treated area. This can be a problem when treating patients of skin type III or higher.

A successful treatment does not require that the lesion is 100% clear. Approximately 80% clearance of a typical port-wine stain will result in a

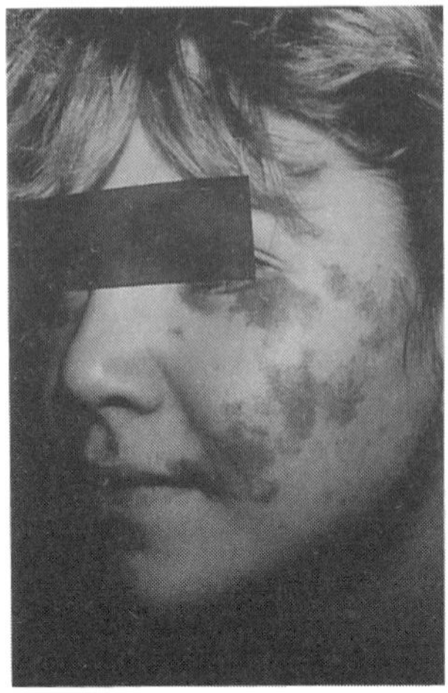

(a) A 20 year old female (skin type II) prior to treatment.

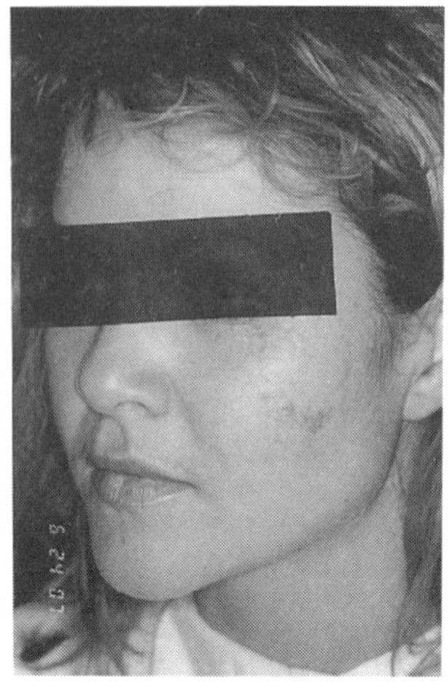

(b) The same woman after 5 treatments.

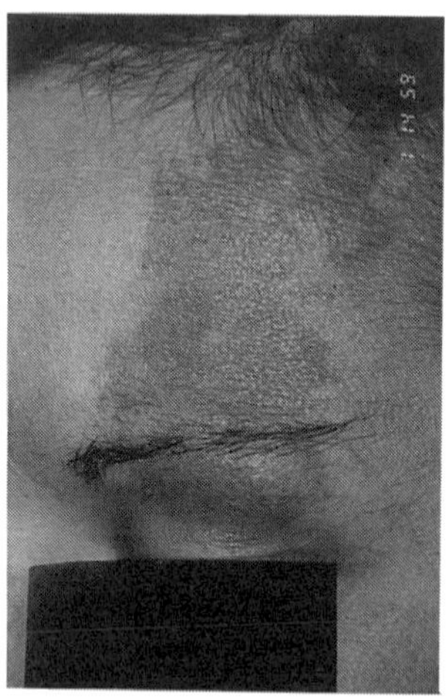

(c) A 23 year old female (skin type III) prior to treatment.

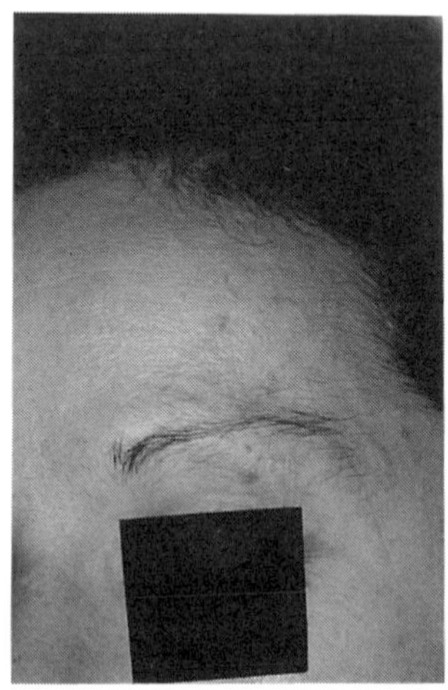

(d) The same woman after 4 treatments.

Figure 2. Two patients who are near completion of copper vapour laser treatment

lesion which is almost invisible from a distance of 2 metres. Additional treatments beyond this point yield little improvement compared to the risk of hypo-pigmentation. Large arterioles are often what remain of the lesion

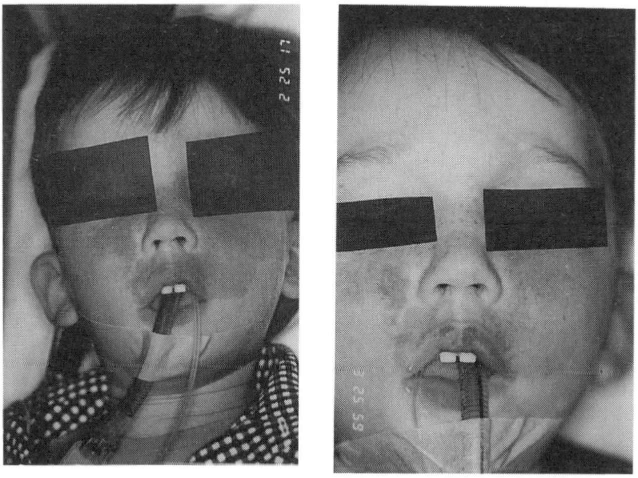

(a) A 4 year old male (skin type II) prior to treatment.

(b) The same boy after 2 treatments.

Figure 3. A patient who recently started copper vapour laser treatment.

at this point. These can be treated by injection sclerotherapy with little risk of causing abnormal pigmentation.

In figure 3a we show a 4 year old boy (skin type II) who has a red port-wine stain covering much of his face. After two treatments (figure 3b) the lesion is significantly paler and the once even colour is becoming irregular. We would anticipate that another four or five treatments will be required to achieve a satisfactory outcome.

6. Discussion

6.1. TREATMENT TIME

In this section, we examine the time required to treat an area of skin, either by scanning the laser beam over the skin or with individual pulses of light.

With a pulsed dye laser, 5 mm spot diameter, approximately 5.5 pulses are required to cover 1 cm^2 [2]. In a private communication, Dr. Petra Koster [8] stated that with a cooperative patient with no anaesthetic, 500 pulses can be delivered in one hour. This is the maximum period of time that a treatment session can be tolerated by such a patient. Thus, in a treatment session 90 cm^2 can be treated. If the port-wine stain is larger than this, the patient must come for repeat treatments so that the remainder of the port-wine stain can be treated.

Some have said that each pulse of light feels like a rubber band snapping against the skin, but Masciarelli describes it as a hot rod being pressed against her skin [10]. For children, this sensation is distressing, particularly when they do not understand why the treatment is being carried out. Consequently, the number of pulses that can be delivered in one hour will be considerably reduced.

With the copper vapour laser and the automated treatment device that we have described an area of 90 cm^2 can be treated in 10 minutes. This includes the time taken to move the patient to the correct position. A further 15 minutes is required to administer the general anaesthetic.

Very large areas can be treated with the system that we use. A port-wine stain that covers an arm can be treated in one session, which is not possible when a pulsed dye laser is used. For small areas ($<30\,cm^2$), the pulsed dye laser may be quicker as no time is spent administering an anaesthetic.

6.2. ALTERNATIVE TREATMENT DEVICES

Instead of scanning the light across the skin and adjusting the illumination time by altering the speed of the spot, we could have shuttered the laser beam so that pulses of light were generated. The shuttered beam would be passed down a fibre which the surgeon moves between pulses so that each pulse of light is incident on a different point of skin. The surgeon would change the exposure time so the desired response is obtained.

There are difficulties with this approach. One assumes that the fibre can be positioned by the surgeon so that the position of the individual laser pulses do not overlap. This problem is overcome with a device such as the Hexascan™ [15], where the position of the fibre is controlled by a microprocessor.

Alternatively, the fibre that the laser beam is passed down is connected to a handheld scanning unit. In the scanning unit, the fibre is moved in a rasterscan like pattern over a region of port-wine stain by the action of two stepper motors. Like the Hexascan™unit, this device has to be held against the port-wine stain, which makes it difficult for the user to observe the changes in the port-wine stain during treatment.

6.3. OTHER CUTANEOUS LESIONS

The copper vapour laser and SCANALL system is also useful for the treatment of other cutaneous lesions. Vascular lesions, such as telangiectasia, capillary haemangioma, and spider naevus, can all be treated successfully. In these cases, injection sclerotherapy followed by laser treatment is used since the vessels involved have large diameters. Telangiectasia and capillary haemangioma may take 2-3 treatments to clear adequately, but spider naevus usually only requires one treatment.

Pigmented lesions can also be treated using the copper vapour laser and automated scanner. In this case, the 511 nm light is used as well as the 578 nm light. This increases the absorption in melanin. Since using both wavelengths increases the available power, the spot is scanned faster than normal to keep the fluence at around 10 J/cm^2. Thus it is still essential to use the SCANALL system.

The laser can also be used for ablation. This is useful in the treatment of viral warts, for example. Here the scanning mirror is kept still, allowing a continuous beam of 511 nm and 578 nm light to be incident on the wart. This burns out the wart, leaving the surrounding tissue sterile and less prone to re-infection.

7. Conclusion

The copper vapour laser, when used with a computerised scanning system, is ideal for the treatment of port-wine stains and other cutaneous lesions. The 578 nm wavelength light generated is strongly absorbed by the haemoglobin, relative to the other skin components. This wavelength is ideal for the treatment of port-wine stains, provided the illumination time is longer than 2 ms.

The copper vapour laser generates sufficient power at the 578 nm wavelength that illumination times in the optimal range of 2-8 ms are obtained when the spot has a diameter of 0.3 mm. The output beam of the copper vapour laser has a minimal divergence, so the beam can be focussed to a spot of this size.

When used in conjunction with a computer controlled scanning device, illumination times in the range of 2-8 ms can be achieved. With these optimal illumination times, it is possible to cause damage to just the enlarged capillaries. The endpoint is transient blanching. The skin whitens as the damaged capillaries contract, removing the blood from the treated region. The undamaged and empty capillaries open up so that blood flows into the treated skin.

Our treatment protocol enables us to take advantage of this response as we then retreat the treated skin. The capillaries which opened up to deliver blood are then treated. In this way, we thermally damage more of the capillaries in the skin in any one treatment session.

References

1. I. R. Chambers, D. Clark, C. Bainbridge, C. Balakrishnan, and T. A. Piggot. Automation of laser treatment of port wine stains (technical note). *Physics in Medicine*

and Biology, 35(7):1025–8, 1990.

2. S. M. Dinehart, S. Flock, and M. Waner. Beam profile of the flashlamp pumped pulsed dye laser: Support of overlap of exposure spots. *Lasers in Surgery and Medicine*, 15:277–80, 1994.
3. W. F. Ganong. *Review of Medical Physiology*. Lange Medical Publications, California, 2 edition, 1965.
4. B. A. Gilchrist, S. Rosen, and J. M. Noe. Chilling port wine stains improves the response to argon laser therapy. *Plastic and Reconstructive Surgery*, 69(2):278–83, 1982.
5. L. Goldman, R. Dreffer, R. J. Rockwell, and E. Perry. Treatment of portwine marks by an argon laser. *Journal of Dermatological Surgery*, 2(5):385–8, 1976.
6. A. H. Jacobs and R. G. Walton. The incidence of birthmarks in the neonate. *Pediatrics*, 58:218–22, 1976.
7. S.-L. Karvonen, P. Vaajalahti, M. Marenk, M. Janas, and K. Kuokkanen. Birthmarks in 4346 Finnish newborns. *Acta Dermato-Venereologica*, 72(1):55–57, 1992.
8. P. Koster. Private Communication, 1994.
9. L. Marini, P. H. Butler, D. J. Smithies, and E. P. Walker. A theoretical model of the blanching response after copper vapour laser treatment of telangiectasia. *British Journal of Dermatology*, 127(2):189–90, 1992.
10. P. C. Masciarelli. Living With a Port-Wine Birthmark. In O. T. Tan, editor, *Management and treatment of benign cutaneous vascular lesions*, chapter 14, pages 180–5. Lea & Febiger, Philadelphia, 1992.
11. R. A. Neumann, R. M. Knobler, H. Leonhartsberger, and W. Gebhart. Comparative histochemistry of port-wine stains after copper vapor laser (578 nm) and argon laser treatment. *Journal of Investigative Dermatology*, 99:160–7, 1992.
12. I. A. Niechajev and L. Clodius. Histology of port-wine stain. *European Journal of Plastic Surgery*, 13:79–85, 1990.
13. J. W. Pickering, P. H. Butler, B. J. Ring, and E. P. Walker. Computed temperature distributions around ectatic capillaries exposed to yellow (578 nm) laser light. *Physics in Medicine and Biology*, 34(9):1247–58, 1989.
14. J. W. Pickering, P. H. Butler, B. J. Ring, and E. P. Walker. Thermal profiles of blood vessels heated by a laser. *Australasian Physical & Engineering Sciences in Medicine*, 12(1):11–5, 1989.
15. G. Rotteleur, S. Mordon, B. Buys, J. P. Sozanski, and J. M. Brunetaud. Robotized scanning laser handpiece for the treatment of port wine stains and other angiodysplasias. *Lasers in Surgery and Medicine*, 8:283–7, 1988.
16. R. A. Sheehan-Dare and J. A. Cotterill. Copper vapour laser treatment of port wine stains: clinical evaluation and comparison with conventional argon laser therapy. *British Journal of Dermatology*, 128:546–9, 1993.
17. D. J. Smithies and P. H. Butler. Modelling the distribution of laser light in port-wine stains with the Monte Carlo method. *Physics in Medicine and Biology*, 40:701–33, May 1995.
18. D. J. Smithies, P. H. Butler, W. T. Day, and E. P. Walker. The effect of the illumination time when treating port-wine stains. *accepted for publication in Lasers in Medical Science*, 1995.
19. D. J. Smithies, P. H. Butler, J. W. Pickering, and E. P. Walker. A computer controlled scanner for the laser treatment of vascular lesions and hyperpigmentation. *Clinical Physics and Physiological Measurement*, 12(3):261–7, 1991.
20. O. T. Tan. Pulsed Dye Laser Treatment of Adult Port-Wine Stains. In O. T. Tan, editor, *Management and treatment of benign cutaneous vascular lesions*, chapter 6, pages 83–99. Lea & Febiger, Philadelphia, 1992.
21. C. N. van Halewyn, C. Miles, P. H. Butler, J. Holdsworth, and E. P. Walker. Rare-earth doped eyewear provide for safer laser surgery (letter). *British Journal of Dermatology*, 120(3):470–1, 1989.

METAL VAPOUR LASER SYSTEMS IN BIOLOGY AND MEDICINE.

V.V.Chvykov, T.P.Kraposhina, S.N.Mazurov,
T.P.Zaharova, K.I.Zemskov.
Cancer Research Centre Academy of Medical Sciences,
Moscow Medical Stomatological Institute,
Lebedev Physical Institute Academy of Sciences
Russia, Moscow

Abstract

The paper presents our 10-years experience of applications metal vapour laser systems in medicine and biology. This experience could be divided on two main branches. First branch is application of metal vapour lasers as sources of irradiation for treatment some kinds of diseases. After long-time biological experiments we had obtained optimum dosage. It allowed us to use such lasers in medicine. On basis of treatment of about thousand patients the results of considerable promise were obtained with copper and gold vapour lasers for gynaecology, dermatology, oncology, gastroentherology and so on. Besides, this experience made it possible to produce high-performance compact laser systems.

The second branch is connected with such important problem of laser medicine as concentrating the laser radiation only onto the pathological tissues. Different approaches to this problem are discussed. In this paper we describe some optical systems providing increased contrast of pathological tissues irradiation due to formation the intensity distribution in laser beam coinciding of shapes and dimensions with pathological zones. Copper and gold vapour amplifiers (wavelengths 510nm, 578nm, and 628nm, respectively) were used for these purposes.

1.Introduction

At the present time advantages and disadvantages of the lasers use in medicine are discussed deeply [1-5]. In common with a plethora of other types of lasers the gas discharge metal vapour lasers [6[occupied the important position in experimental and practical medicine. The known specific properties which are of interest for advanced medical applications of these lasers are: very high gain of their active media, high pulse repetition rates, and good match of

C. E. Little and N. V. Sabotinov (eds.), Pulsed Metal Vapour Lasers, 403–408.

the wavelengths of their laser actions with absorption spectra of haemoglobin and photosensibilizers used usually in photodynamic therapy. These properties of metal vapour lasers pave the way for success of therapy by these type of lasers.

These properties allowed also to propose decision one of problems of large scale application of lasers in biology and medicine. This problem is to prevent the normal tissues surrounding the pathological ones during the laser irradiation [7]. In other words, the problem is to contrast the pathologically altered areas on the background of unaltered healthy tissues. We suggest optical methods to obtain the selective effect to pathological tissues. We used for this purposes systems with metal vapour lasers brightness amplifiers - active optical systems (AOS) [8].

2. Equipment

Copper vapour laser (CVL) is now the most frequently used in dermatology, gynaecology, for treatment of erosion-ulcerative diseases including endoscopic technique. Gold vapour laser (GVL) can be used for photodynamic therapy due to coincidence of its wavelength with maximum absorption of haematoporphyrin's derivatives which are used usually as photosensibilisers. We used three types of CVL with sealed-off discharge tube with average output power of 3W, 8W, 20W, and GVL - up to 2W. Standard equipment of the devices included laser block, power supply, shutter, and fibre optics beam-delivery system. At first the CVL used was a modified commercial model ILGI-101. It operated at the pulse repetition rate about 10 kHz, pulse length was 20-30 ns, average power at the distal end of the optical fibre was about 3 W in both green (510nm) and yellow (578nm) lines. Electrical power consumption was 2.2-2.4 kW. Power supply and laser tube's block were water-cooled (about 2 l/min). The experience of using this type of laser in the medical practice allowed us to design more convenient and compact devices. Now we use, with the same therapeutic efficiency, much more compact (laser part-25x25x85cm/15kg, power supply-35x35x90cm/35kg), air-cooled apparatus with power consumption about 1 kW and output average power in two lines 2.5-3 W. A well-known thyratron system was used as pulse power supply. It is also possible to use either green or yellow line, or both of them altogether for this CVL.

Laser radiation was usually launched from CVL or GVL into a quartz optical fibre with a core diameter of 0.6 mm. Coupling efficiency was about 0.7-0.8. The ends of the fibre were made in such a way that a light spot on target was sufficiently uniform.

Specific properties of metal vapour lasers made them the finest tool for different photodynamic acts in medicine and biology:

- high pulse repetition rates allowed the variations of irradiation dosage from millijoules to thousands joules at the irradiation time as long as some minutes;

- low each pulse energy enabled to reach high accuracy of the dose;

- low beam divergence, achieved with the special unstable resonators, allowed to produce the highest light power densities even under not very high average laser action power;

- high amplification made it possible to create new type of high performance optical systems, using the image brightness amplification, these systems enabled to form on the irradiated object high power density light spots of any shapes [9]. The schematic diagram of one version of such apparatus is shown in Fig. 1.

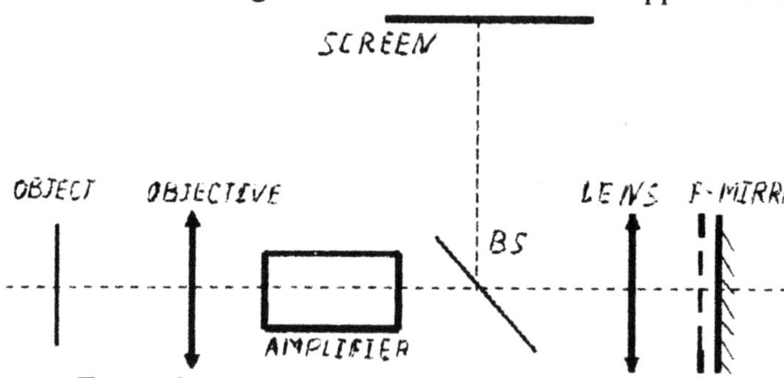

Figure 1. Schematic diagram of experimental set-up.

Superradiance from amplifying element illuminated the object, after single-pass amplification was reflected from beamsplitter, and then produced bright image on the screen. The light passed through the beamsplitter was reflected by optical system consisted of the lens and plane feedback mirror with mask on it. Then the light came into the amplifier again. Therefore this light was amplified by several passes through the active medium after reflections by feedback mirror and object. A geometry of system provided an optical conjugation of object surface and feedback mirror. In this case the object was irradiated by image of mask amplified by active medium.

This AOS was modified during our experiments. We used liquid crystal light modulator as aperture, which could be connected with computer. In order to reduce the illumination power density on the liquid crystal modulator we applied double-pass amplification with polarizer and phase plate [10].

In the experiments copper vapour and gold vapour active media were used as the amplifiers. Operating in the laser mode they provided output average power 20 W and 2 W, respectively. The experimental arrangement described above could provide power density variations from 0.2 to 16 W/cm^2, dosage taken was varied from 60 J to 3 kJ.

3. Treatment

The experiments with some hundred white mice and rats for choosing optimum dosage were carried out. Animals were inflicted experimental standard wounds, and optimum dosage was determined as dosage of more short time wounds healing up.

In experiments with AOS white mace A-Snell were inoculated with embryocarcinoma cells (tumours up to 15-5 mm of sizes and up to 1.5 g of mass were irradiated). Thermal destruction (TD), photodynamic therapy (PDT)and combine TD-PDT modes in dependence on dosage taken and laser power were investigated. The effect was evaluated by ratio of the scar area produced to area of pathology.

A complete thermal destruction of tumours was achieved at the dosage taken about 1 kJ under 10-15 W/cm^2 power density (copper vapour laser was used). It was established that area destroyed was considerably larger than illuminating laser spot under these conditions. It seemed, it was conditioned by a long time of irradiation in order to provide necessary dosage. To optimise the dosage and power density in order to localise the treatment, we irradiated a normal tissues by different dosages. It was obtained, the dosage at which destruction area was equal to area irradiated was about 80 J at the power density of 15 W/cm^2 , and treatment exposure of about 5 sec. The best results were obtained in combined TD-PDT irradiation mode, when AOS was used in addition to HPD-technique. Necessary duration of treatment was 5-10 sec. and full destruction of pathological area was observed in every occasion. The damage of surrounding normal tissues was negligible in comparison with TD-method. After rejecting of necrosis wounds healed up without inflammation. Subsequent epithelization was formed without kelloid scar.

Consequent procedures were carried out for clinical applications of this type of lasers.

All patients were examined ordinary before-treatment tests, such , for example, as clinical and biochemical analysis of blood, endoscopy, colposcopy, X-ray oncocytological and hystological studies, in dependence on pathology localisation. The procedures were carried out with usual endoscopic and gynaecological instruments if it was necessary.

Areas of lesions to be treated varied from 1-3 mm^2 to 1-4 cm^2. Field to be irradiated had, as a rule, very complicated shape, and because of this manual control of fibre was used. When target area was much larger than laser spot with parameters required, slow scanning or "spot by spot" methods were used.

An average power density on the target has been chosen as a parameter which was maintained approximately constant during the treatment. Its values were from 0.8 W/cm^2 to 1.5 W/cm^2. These values were well bellow threshold of tissues destruction under the treatment conditions described. Alternatively, these values enabled to operate under tolerable exposures which provided an energy density on the target tissue about 50-100 J/cm^2 for a time 20-40 sec The values mentioned above were chosen according to the results of our previous experiments with the test animals, as well as several literature references. When slow scanning and "spot by spot" methods were used, the exposures were increased in proper proportions. Rapid hemostasis could also produced, if needed, by the short-time (1-3sec) increase of the irradiating power density up to 50-80 W/cm^2. This offered "welding" of the point or line blooded lesions.

The procedures were carried out every 3-5 days. Quantity of the procedures was defined by treated diseases and it was not more then 12. Detailed patient's examinations were accomplished after 2-nd, 4-th, 8-th, and 12-th laser treatments. There were three cases when aftertreatment was necessary to achieve a complete cure. Aftertreatment was performed within 3 months of the main course. It included in all cases 4 procedures to secure a lasting cure. Total uniform epithelization and normal analyses indicated a complete recovery in cases of cervical lesions. Disappearance of warts was a natural indication of recovery in this case. Post-treatment examinations (more then six months) did not show any evidence of adverse effects.

4. Results

Gastroentherology - about 1000 patients were treated by CVL and GVL with ulcerative diseases of the stomach and the duodenal ulcer. Efficiency was about 90-98 %. Required number of procedures were not more than 8 for CVL, and 5 - for GVL.

Oncology - about 200 patients were treated with inflammatory and erosion-ulcerative after-operation and after-X-ray irradiation diseases, located in the different body parts. Sure positive results were obtained for 87% patients. Post-treatment observation period was from 6 months up to 3 years. Comparison was carried out of treatment efficiency for CVL and He-Ne lasers. As it was shown, efficiency increased by a factor 1.5 for CVL usage. Some evidences of enhancement of immunity were marked by the biochemical studies in approximately 6-months observation period after copper or gold vapour lasers irradiation.

Gynaecology -165 patients were treated with cervical erosion, leukoplakia, vulval warts, herpes infection. Positive therapeutic effect was observed, as a rule, after 3rd procedure. Maximal amount of procedures was 12. Treatment efficiency - more than 95%.

Some other applications - high therapeutic efficiency of CVL use was obtained in treatment of wine-spots, trophical diseases, pus-inflammatory wounds. Recently GVL clinical use in photodynamic therapy was began.

5. Conclusion

The results obtained provide support for the view that MVL offers high therapeutic efficiency. In agreement with classification accepted now as standard for effects of laser radiation on biological tissues, the CVLs described above could be fitted into the medium-intensity sources. It is believed that this enables to combine the effects of low-intensity laser biostimuli and high intensity laser apparatus for strong influence on the tissue's structure. Such combination could provide the new effects and support the new mechanisms of the interaction between the laser light and the tissues irradiated.

6. References

1. Parrish, J.A., Wilson, B.C. (1991) Current and future trends in laser medicine, *Photochemistry and Photobiology* **53**, N 6 , 731-738.

2. Dougherty, T.J. (1989) Photodynamic therapy: status and potential, *Oncology*,.**3**, 67-73.

3. Special issue "Laser in otolaryngology", *Otolaryngology Clin. North. Am.*, **23,** 1-169, (1990).

4. J.Seeger, G.S.Abela, (1990), "Current status of laser angioplasty", *Surg. Annu.*,.**22,** 299-315.

5. D.H.Sliney, M.L.Wolbrasht, (1989), "Future applications of lasers in surgery and medicine: a review", *J. Roy Soc.Med.*,.**82**, 293-296.

6.Petrash, G.G. (1971), "Pulse gas discharge lasers", *Uspekhi Fiz. Nauk*, **105**, N 4, 645-676, (in Russian).

7. Pletnev, S.D. (1981),"Lasers in clinical medical science", Moscow, 31-204,.

8. Kazaryan, M.A., Matveev, V.M., Petrash, G.G., (1982), "Projection system with brightness amplifier and autonomous illuminating source", *Izv. Acad. Nauk SSSR, Ser.Fys.*,.**46**,N 10, 1898-1902.

9. Zemskov, K.I., Ivanov, A.V., et al..,(1989),"Application of metal vapour lasers for selective effect to pathological tissues", *Proceedings SPIE*,.**1041**, 86-90.

10. Vasiliev, Ju.P., Zemskov, K.I, et al..,1990,"Active optical systems with multipasses amplification of brightness images", *Izv. Acad. Nauk SSSR*,.**54,** N 10, 1874-1880.

COPPER VAPOUR LASERS IN ONCOLOGY

M.A.KAZARYAN

P.N.Lebedev Physics Institute, Russian Academy of Sciences

Leninsky Prospect, 53, 117924, Moscow, Russia

1. Abstract

Some results of the copper vapour lasers use in medical practice are presented. An universal capability of dye lasers oscillating in 600-670 nm red region for excitation of the preparates used in photodynamic therapy is demonstrated. The copper vapour lasers are shown also to effectively coagulate pre-tumour neoplasms.

A new method of shaping laser beams fitted to tumour configuration based on quantum optical systems including image brightness amplifiers is described. Possibilities of the copper vapour lasers for tumours drugless phototherapy and the two-step and two-stage methods of the photodynamic therapy are discussed. Some Russian medical systems based on the copper vapour lasers and dye lasers pumped by them are specified in parameters.

The paper is prepared in collaboration with A.V.Armichev from Research and Production Association "ISTOK", and A.V.Ivanov from N.N.Blokhin Cancer Research Center Russian AMS.

2. The Main Peculiarities and Advantages of the Copper Vapour Lasers

Not less than ten groups in Russia are involved in the development and construction of the copper vapour lasers. This is the reason why the devices are widely and variously in practical medicine.

The copper vapour lasers are among the most effective laser sources in the yellow-green spectrum region (510-580 nm). One of their advantages is that the active elements may be very reliably made as long-life sealed-off constructions (lifetimes exceed 1000 hrs) [1].

C. E. Little and N. V. Sabotinov (eds.), Pulsed Metal Vapour Lasers, 409–414.

A peculiar pulse-periodic structure of copper vapour lasers output (the pulses 20-30 ns long are repeated at 5-20 kHz rate) in combination with high powers allows using the lasers for realization of various effects affecting biological objects: molecules photoactivation and cell membrane modification, selective coagulation of some tissues structures, photochemical and photoacoustical destruction of tissues and mineralized forms , photodynamic action into sensified tissue regions.

The copper vapour lasers also allow shaping of the radiation to obtain sufficiently good special distribution that provides effective input of the laser light into quartz monofibers and its application for excitation of dye lasers, which transform the radiation into yellow-red and near-infrared one [2].

3. Some Areas of the Metal Vapour Application in Medicine

3.1. Methods Adequate to the Low-intense Laser Therapy with He-Ne Lasers

A laser system based on the copper vapour lasers is used for endoscopic therapeutic irradiation in erosive-ulcer stomach diseases [3]. It is possible to effectively treat the stomach ulcers by external irradiation of the skin projection of the proper parts of the stomach with dye lasers pumped by copper vapour lasers in the range 620-640 nm [4]. One must note that the mechanism of therapeutic action of the metal vapour lasers pulse-periodic light may essentially differ from that of the He-Ne lasers, can be of photoacoustic character or connected with local photodamages of the membrane [5].

3.2. Microcoagulation Methods

The copper vapour lasers with average power 0.5-3 W with characteristic depth of penetration into the tissues of order a millimeter provide a possibility of effective surface coagulation of soft tissues. It is essential that the coagulation can be selected due to differences in the tissue components absorption indices. The fact is quite effectively used in dermatology when vascular neoplasms or pigmentive spots are treated [6].

The microcoagulation methods also found application for the organs visually or endoscopically reachable in otorynolaringology for polipi removal and in gynecology for the treatment of erosion, inflammations and premalignancies in neck of the womb and external genitalia [7].

3.3. Endoscopic Surgery

The application of the copper vapour laser with average powers from 5 to 15 W allows to resect tissues due to their thermal destruction. But it is impossible to directly tissues with characteristic heat of sublimation about 30 J/cm^2 by action onto them with light of a copper vapour laser which has the light-guide with output energy density 1.5 J/cm^2. In spite of this it turned to be practically possible to effectively destroy the tissue due to photochemical, photothermal or photoacoustic mechanism.

4. Practically Realized Applications of Copper Vapour Lasers in Oncology

The copper vapour lasers are now applied in clinical oncology for:

* photodynamic therapy of the tumours with various localizations,
* pretumour neoplasms coagulation,
* treatment of local primary radiative reaction in the complex radiotherapy of tumours,
* treatment of erosive ulcer and flammatory complications of oncology patients during the post-operational period [8].

As the photodynamic therapy is conducted using the most common photosensifiers, based on the heamatoporphyrine derivatives, a therapeutic procedure includes the tumour irradiation with the light of 630 nm wavelength and power density 0.2-0.3 W/cm^2 (depending on the tumour type). The irradiation is performed during the time necessary for the energy density sum accumulation of 100-300 J/cm^2.

These results have been obtained when the photo-dynamic therapy of gullet and stomach tumours was performed using a dye laser system with copper vapour laser pump, the therapeutic line FWHM was +9 nm.

It was shown that when such broad band radiation is used, it is reasonable its accurate spectral coincidence with heamatoporphyrine absorption band. An additional importance is that such radiation is quite equal in effictivity to the action of narrow-band one, e.g. from gold vapour laser.

5. Selective Laser Irradiation of Tumour with Automated Adjustment of the Radiation Light Power

Among the tasks on the photodynamic therapy of tumour neoplasms a problem exists connected with the necessity of possibly less damaging the tissues around the tumour.

One of the most profitable solution of the problem turned to be application of laser optical systems with self-coupled resonator of Hardy-type [9].

Brightness amplified images of pathologic regions of experimental animals were obtained. Experiments were carried out on photodynamical and photothermal action onto tumours with the beam from a copper vapour laser conformed with tumours.

The experiments performed have demonstrated the profiled beam application allows, in definite irradiation regimes, to elevate the action selectivity, e.g. for the photodestruction, and moreover to significantly shorten the irradiation time, as in the case of photodynamic therapy.

6. Parameters and Characteristics of Some National Copper Vapour Lasers and Dye Lasers Excited by Copper Vapour Lasers

The concrete structure of a medical apparatus is dictated by the corresponding medical aim. The generalized scheme of a system for surgery and therapy may be presented as the scheme of Fig. 1a. The active element of a metal vapour laser 1, fed from a pulsed high voltage supply 2, provided with resonator mirrors 3, generates the beam directed consequently to a high-speed shutter 4, iris diaphragm 5, brancher 6, lines selector 7, focusing system 8 and the end face of a flexible light-guide 9, the operating instrument being its distal part. The shutter is controlled by a programmed unit or time-relay 10. A measurer with thermoelectric transducer 10 controls absolute average power. The output power can be continuously controlled with a photodiode 12 either from the light branches by the element 5, or from the light penetrating through the opaque mirror 13.

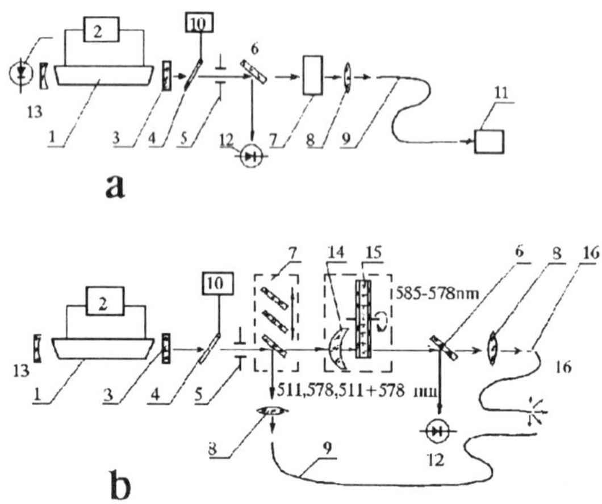

Fig.1.

One of the optimal solutions for universal use dye lasers pumped by copper vapour laser is presented in Fig. 1b. The output of the copper vapour lasers through elements 4 and 5 is directed to a selector-reflector 7 that is a set of accessory interference fitters-mirrors which separate the lines of the copper vapour laser into reciprocally perpendicular directions. This allows, if necessary, to perform spectral coupling with the dye absorption band and to branch the copper vapour laser beam to an additional light-guide 9 where it can be used for another medical aim.

The presented dye-solution laser, opposite to the traditional pumping through schemes, is made on the base of a semiconfocal non-dispensing resonator, which includes a segment of the plane layer of the dye solution and is enveloped into a rotating sealed cuvette 15 [10]. The optical element 14 is used simultaneously for focussing of the copper vapour laser beam into the dye cuvette and for reflection of the light generated by the dye. It is possible due to the difference between the dye laser and copper vapour laser apertures. This arrangement turned to be quite effective for obtaining dye laser output line of +7 nm width with the maximum exactly tuned to a required wavelength in the range 580-67- nm. The light controlled with a photodiode through plate 6, is directed through coupling lens 8 into a multifiber light-guide 16. Such a structure is used in creation of the medical apparatus Yakhroma 2 for tumour photo-dynamic therapy. The dye laser constructed so is suitable also for efficient energetic pulsed excitation of the second harmonic of neodimium glass lasers.

A homeland preparate Photohaem based on haematoporphyrine derivatives has a long wave absorption band with FWHM +15 nm located at the line 630 nm (curve 1) in Fig.2. The technologic deviation of the maximum location can reach 3 nm, that makes reasonable to have more broad-band radiation for its photoexcitation. A metal-complex of phthalocyanine, the base of another homeland sensifier Photosense, has practically one absorption band675 nm with average FWHM about 15 nm (curve 2). A photosensifier on the chlorine base E6 has the absorption band 664+8 nm (curve 3). The curves 4,5,6 in Figure 2 are lasing bands of the copper and fold vapour lasers.

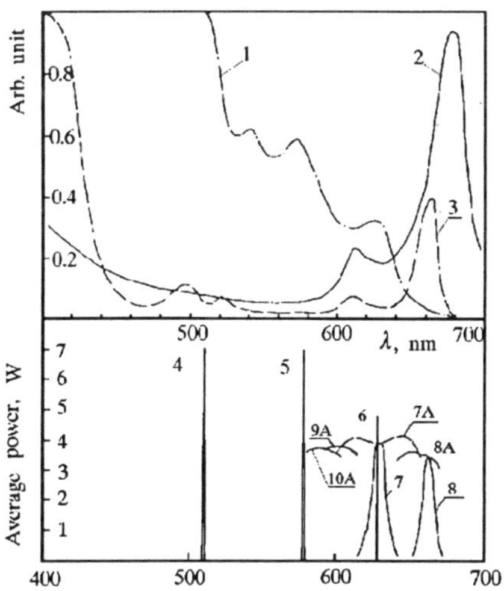

Fig.2.

The generation spectrally coupled with the photo-dynamic preparates absorption can be provided by several types of dye lasers pumped by copper vapour lasers. It is established that 4 types of dyes (Rodamines 6G, B, 101, Phenalemines 430 and 435) produced in Russia allow to fully cover the range 575-670 nm with 22-30 % efficiency with the nondisperse resonator lines, enabling concentration tuning of a line maximum to any point within this spectral range. The maximum efficiency was demonstrated by the Phenalemine 435 solutions (efficiency 29 %). The overloping curves show the possibilities to shift the absorption of the oscillation line maximum when the dyes operational concentrations are changed within the photosensifier absorption bands.

7. Oncological Potentialities of the Metal Vapour Lasers Combined with Dye Lasers

Laser systems consisting of dye lasers pumped from copper vapour lasers can solve the following tasks which are perspective for oncology:

* drugless photodynamic tumour therapy based on the laser light absorption by endogenic photoacceptors with further photodynamic in activation of the tumour cells. For instance, in the papers [11,12] a possibility is demonstrated of direct optical excitation of singlet oxygen at the lines 1265, 762, 637 and 578 nm with FWHM +7 nm. It is essential that the cells sensibilization with haemotoporphyrine diacetate reduced the cytotoxical dose of this radiation only by factor 4-5 [13].

* photodynamic therapy with photosensifier two-step excitation by consequent pulses in singlet and triplet states [14, 15].

two-stage procedure of the photodynamic therapy, when a copper vapour laser forms hollow channels in the tumour with non-graphitized walls, then a light-guide is threaded into them transporting a dye laser radiation that affects tissues through the channel walls. Such walls allow penetration of red range light from a channel into tissue.

Therefore, a medical apparatus built of a dye laser pump by a copper vapour laser following the scheme of Figure 1, with additional light-guide branching of the pump radiation, widens the capabilities of the photodynamic therapy methods.

8. References

1. Zubov, V.V., Chursin, A.D., Lesnoy, M.A., Liabin, N.A., Ugol'nikov, S.A. (1993) Copper vapour lasers with sealed-off active elements, *SPIE* **2110**, 78-89.
2. Sun, W., Tang, C.S., Huge, X.B., Chen, M.S. (1986) Theory and experiments of dye laser longitudinally pumped by copper vapour laser, *Optics Communications* **58**, 196-200.
3. Loginov, A.S., Basov, N.G., Ambatsumyan, R.V. (1988) The efficiency of treatment of chronic gastric ulcers by copper vapour lasers, *Therapeutic Archive* 42-46 (in Russian).
4. Poslavski, M.V., Armichev, A.V., Korochkin, I.M. (1986) *Laser application in surgery*, Alma-Ata, 74-76 (in Russian).
5. Matushichev, V.B., Soldatov A.N., Titov, V.V. (1990) The treatment of gastric and doudenum ulcers with use of copper vapour laser, *Clinical Surgery* 8, 66 (in Russian).
6. Tan, O.T., Murray, S., Kurban, A.K. (1989) Action spectrum of vascular specific injury in pulsed irradiation, *Journal invest. dermatology* **92**, 868-871.
7. Chvykov, V.V., Zazulya, O.I., Zemskov, K.I. (1993) Copper vapour laser in medical practice: gynecology, *SPIE* **2110**, 276-280.
8. Kartashova, E.O., Kuvshinov, Yu.P., Chistiakova, O.V., Chvykov, V.V., Mikhalevich, O.D. (1994) Applications of copper vapour lasers in the treatment of postoperatic distress and cancer of gullet and stomach, *Herald of the Cancer research center of AMS of Russia* 3 36-39 (in Russian}.
9. Vasiliev, Yu.P., Zemskov, K.I., Ivanov, A.V., Kazaryan M.A., Petrash, G.G.,Chvykov, V.V., (1991) Active optical systems: medical and biological applications, *Proc. Lebedev Physical Institute* **206**, 136-148. (in Russian)
10. Armichev, A.V., Alejnikov U.S. (1989) PDT-DYE module with rotation active medium and pumping CVL with connection by fiber, *International conference of photodynamic therapy*. Abstracts, Sofia, 64.
11. Nemtzov, I.Z., Zakharov, S.D., Ivanov, A.V, Armichev, A.V., etc. (1992) Validation of the use of He-Ne laser in medical rehabilitation of patients with atrophy. *Aviakosmicheskaya i Ecologicheskaya Medicina* 2, 26 (in Russian).
12. Danilov, V.P., Eremeev, B.V., Zakharov, S.D., Ivanov, A.V., etc. (1990) Spectral-selective photodynamic effect without exogenic sensibilitators and its possible application in cancer phototherapy and biostimulation, Izv. *Academii Nauk SSSR, Fizika* **54** 1611-1620 (in Russian).
13. Zakharov, A.V.,Ivanov, A.V., etc. (1989) Destroying cancer cells by He-Ne laser radiation. *Reports, Tartu* 29-31 May, 52-54 (in Russian).
14. Armichev, A.V., Alejnikov, V.S., Ivanov, A.V. (1991) Laser systems for biomedical research of step-wise absorption possibilities. *International conference on photodynamic therapy and laser medicine, Beijing, China*, Abstract, 147.
15. Adrioni, A., Gubeddu, R., Svelto, O. (1986) In vivo photochemotherapy with haematoporphyrin derivative via two-step excitation, *SPIE* **701** 308-309.

9. Applications in Science & Technology

APPLICATIONS OF COPPER VAPOUR LASERS IN SPECTROSCOPY

J.P. PIQUE, P. COUTANCE, H. RING and G. SITJA
Laboratoire de Spectrométrie Physique (CNRS UA08), Université Joseph Fourier de Grenoble, B.P. 87, 38402, St Martin d'Héres (France)

Abstract

We discuss in this paper the applications of copper vapour laser (CVL) and Cu/HBr (HyBrId) laser in the field of spectroscopy. After a short review we describe in detail three applications: time-resolved high resolution Fourier transform spectroscopy, spectroscopy of ultra vibrationaly excited states of molecules through very high sensitivity laser induced dispersed fluorescence spectroscopy (LIDFS) and intracavity stimulated emission spectroscopy in the presence of mode-lock. Finally we compare the performance of CVL and Cu/HBr lasers from a spectroscopist point of view.

1. Introduction

The real revolution in the field of spectroscopy came after the development and commercial availability of reliable high power pump laser and the first tunable lasers, that is, the dye lasers and the solid state lasers. We can distinguish two kinds of lasers: cw and pulsed lasers. To simplify, for a long time, cw lasers have been used for very high resolution spectroscopy whereas pulsed lasers have been used to generate high power UV or to produce multiphotonic spectroscopy. Since the seventies, many new spectroscopic techniques have been developed: sub-Doppler spectroscopy, optical-optical double resonance spectroscopy, stimulated emission pumping spectroscopy, intracavity spectroscopy, photoacoustic and photothermal spectroscopy, level-anticrossing spectroscopy in conjunction with high magnetic or electric field, cooled neutral atoms trapped by laser beams... Most of the laser spectroscopy techniques are used for atoms, molecules and solids.

INSPEC database covering the period between 1969 and 1995 retrieved several ten thousands articles (in the title and the abstract) from the keywords « laser » and « spectroscopy », several hundred from the keywords « copper vapor laser » and only a few ten from the keywords « copper vapor laser » and « spectroscopy ».This is a paradox, because for the author the CVL is precisely the laser which can pump tunable lasers which are able to generate high power from the UV to the IR with very high

C. E. Little and N. V. Sabotinov (eds.), Pulsed Metal Vapour Lasers, 417–429.

resolution. Current thinking about the future of a laser is its success in the spectroscopy field. The Ar^+, YAG and Eximer lasers are good examples of such a success. Does it mean that the CVL has no future? Certainly not. In this paper we want to analyze the specificity of the CVL for spectroscopy techniques through some examples. We will then discuss, from a spectroscopist point of view, the improvement which may be done to popularize the copper lasers in the spectroscopist world and we will insist on the huge advantages of the Cu/HBr (or HyBrId) lasers versus the traditional CVL. But before that, let us summarize some beautiful experiments which have been performed with CVL lasers.

Beyond the spectroscopy of uranium which separation of the isotope 235 is the actual main application of the CVL, ultra sensitive laser photoionization spectroscopy of short-lived isotopes and very rare atoms was demonstrated by Letokhov et at [1] The optical part of this experiment consists of three high resolution dye lasers pumped by a CVL and locked on three successive atomic transitions. This three step excitation raises the atom to its autoionization state. This CVL arrangement provides both high power (more than 10 watts of total average power) and high repetition rate (10 kHz). Using this set-up Letokhov et al managed to measure hyperfine structures and isotope shifts. They provided important new interactions about nuclei (their charge radii, spins, and electric quadruple and magnetic dipole moments) for many very rare elements which are of great interest to nuclear physicists. Using the same principle, an instrument for trace analysis based on resonance ionization time-of-flight spectroscopy has been developed [2]. Gadolinium, plutonium as well as technetium has been detected with a very high sensitivity.

The characteristics of the CVL can be of great benefit to atmospheric and environment diagnostics. A fluorescence LIDAR based on a CVL system was used for the study of natural media [3]. Average UV powers at 308 nm approaching 1 W have been realized with UV bandwidths less than 2 GH_z at a repetition rate of 10 kHz. This source has been successfully used to investigate the molecular spectra of OH radicals for tropospheric detection [4]. A method was developed for direct determination of lead in Great Lakes water samples [5]. For this purpose a CVL pumped dye laser was used in the development of a laser excited atomic fluorescence spectrometer. Because of the high sensitivity of the spectrometer, the method does not require the samples to be preconcentrated.

The CVL was shown to have a series of advantages compared with other lasers used for earlier studies of Raman spectral excitation. The form of the first and second-order Raman spectral of sodium nitrate, ammonium chloride and stilbene crystals was discussed in details [6].It has been shown that in multimode Silica fibers up to nine Stokes'orders can be generated when excited by the 511 nm greenline of the CVL [7] with a conversion efficiency greater than 90%. The Raman spectrum of polycrystalline solids with excitation by a CVL have also been discussed. For example, the vibrational spectrum of La_2O_2S has been assigned for the first time [8].

The scope of this paper has been limited mostly to molecular spectroscopy which the author is most familiar with. First we just want to name some experiments in this area before we describe in detail some others. The technique of resonant two and three photoionization has been used to determine highly excited Rydberg states of K_2 molecule, beyond and below the ionization limit [9]. The high power and high repetition rate of copper vapor-laser-pumped dye laser systems were really needed for such an experiment. Information concerning the electronic structure of small metallic clusters, like Na_3 [10], has been carried out by means of CVL molecular spectroscopy. The mechanism for the dissociation of molecular $I_2(BO^+_u)$ energy transfer from $O_2(^1\Delta_g)$ has been studied in detail [11]. $O_2(^1\Delta_g)$ was generated through a microwave discharge and I_2 was excited in the state BO^+_u by a high resolution CVL-dye system.

We are now going to describe in more details some specific experiments in the field of molecular spectroscopy.

2. Time-resolved high resolution Fourier transform spectroscopy

Most of Fourier transform spectrometers use cw sources. Stable sources from the ultraviolet to the infrared allow absorption Fourier transform spectroscopy. Cw laser induced fluorescence spectra on Fourier transform interferometers has been developed and has produced beautiful information in molecular physics [12]. But cw lasers are really limited in terms of spectral range, especially in the UV where most of molecules absorb. Moreover, time resolved Fourier transform spectroscopy has been demonstrated using mainly the timed sampling method and low repetition rate pulsed sources. But, with this method, the time response is usually not better than a few microseconds. We believe that tunable lasers pumped by the CVL associated with a Fourier transform spectrometer is an ideal arrangement to solve the two problems cited above, because of the large spectral range and high repetition rate of such laser systems.

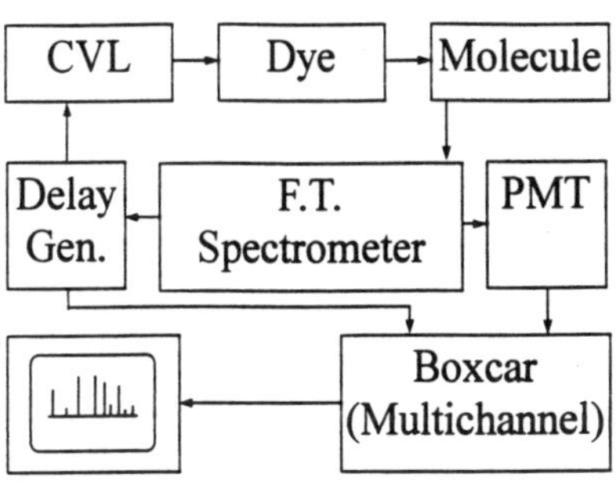

Figure 1: Experimental set-up

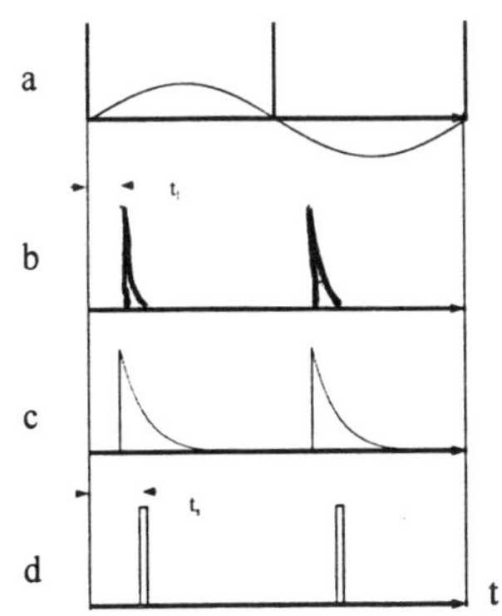

Figure 2: Chronogram

Most of commercial Fourier transform spectrometers are able to slave a laser which runs around 10 kHz.. This technique was demonstrated by Stringat et al [13, 14] with the help of the high resolution CVL pumped dye laser system we developed in Grenoble [15]. The figure 1 and 2 give the experimental set-up and the chronogram of Stringat etal's experiment. The Fourier transform sampling signal (fig 2a) comes from the fringes of the reference HeNe laser. The laser pulse (fig 2b) is synchronized on the Fourier transform sampling signal (fig 2a) with and adjustable delay t_l. The jitter of the CVL system is typically ± 2 ns. Figure 2c represents the laser pulse induced fluorescence of the molecule under study. The Fourier transform spectrometer signal from detector is gated (fig 2d) at an adjustable time t_s by a boxcar (adjustable with a delay from 1 ns to 4 μs). This experimental arrangement allows 3D spectroscopy: frequency, time and intensity. Stringat et al have made a test of this technique using the very well known I_2 molecule. One example sprectum is reported on figure 3.

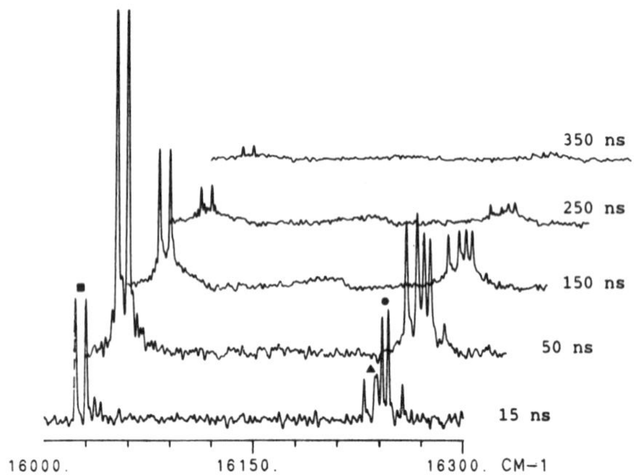

Figure 3: Laser induced Fourier transform spectra of I_2 as a function of time

Because of the large dispersion and resolution of the Fourier transform spectroscopy, this new technique does not require a single level excitation. So, multiplex excited states study can be perform. This powerful technique should find wide applications: multiplex measurements of frequencies and lifetimes of spectral lines, predissociation, dynamical rotational, vibrational and electronic transfers, photofragmentation reactions.

3. Spectroscopy of ultra vibrationaly excited states of molecules through very high sensitivity LIDFS.

It is well known now that molecular chaos appears at very high vibrational excitation. For example in CS_2 molecule a transition to chaos has been observed in the ground electronic state at an energy where the number of vibrational quanta is close to 100 [16]. The problem for the spectroscopist is that the experiments, at such energy, are very difficult due to the weakness of the corresponding transitions. Nevertheless such a study is very exciting because it corresponds to still unexplored spectra. The study of vibrational chaos requires a purely vibrational spectrum of a single symmetry. Moreover, only the ground electronic state of small polyatomic molecules can be modeled with enough precision. Two techniques are able to access at very high energy in the ground electronic state of a molecule: stimulated emission pumping (SEP) spectroscopy and Laser Induced Dispersed Fluorescence spectroscopy (LIDFS) [17].

For a long time only SEP was used for such a study [18]. But this technique is not completely satisfactory because of intrinsic fundamental problems [17]. The technique of LIDFS is in principle the simplest and most reliable, since it is linear. For a long time it was not used for the spectroscopy of highly excited vibrational states because the corresponding spontaneous emission is extremely weak. Table I summarize the advantages of LIDFS versus SEP techniques.

TABLE 1. Comparison between the performances of SEP and LIDFS techniques.

	SEP	LIDFS
- multiplex	no	yes
- laser intensity fluctuations	sensitive	no sensitive
- non linearity	yes	no
- background	laser intensity	zero background
- spectral range	several dyes	from IR to UV
- resolution	laser resolution	Doppler or supersonic jet
- experimental speed	very slow	fast
- multiphotonic effects	yes	no if high repetition, rate lasers
- optimization of the signal	difficult	easy

We have shown that, due to considerable progresses in high repetition rate CVL pumped tunable lasers and near infrared detectors we have succeeded in developing a LIDFS spectrometer with very high sensitivity and resolution [15, 17, 16]. The experimental arrangement is shown in figure 4. At the origin it was based on a Cu60 from Oxford Laser (250 kW peak power for a mean power of 50 W, a repetition rate of 6.5 kHz). For about a year the experiment has used a home made Cu/HBr (HyBrId) laser of 150 W mean power and a repetition rate of 20 kHz [19]. In the last section we will discuss the advantages of the Cu/HBr laser from a spectroscopist point of view. We developed a home made dye lasers and Ti:Sapphire laser pumped by high repetition rate lasers [20, 21]. The dye system consists of an oscillator which is composed of a special holographic grazing incidence grating (Jobin Yvon, 2400 lines/mm), a high refractive index expander consisting of two prisms and a high flow dye cell. The resolution is greater than 2 GHz without etalon. Continuous tuning over the whole gain of the dye is simply achieved by a high precision (10^{-4} degrees) rotation of the mirror. The oscillator beam is then amplified by two or three amplifiers.

For example, the efficiency of our dye laser (using DCM) is 15%. BBO doubling and tripling crystals have allowed us to obtain UV output power of 1 W around 300 nm and 10 mW [22, 23] around 200 nm with a pumping power of 45 W (511+578 nm). The high repetition, as a consequence the relativety low peak power, of such system prevent multiphotonic excitation and dissociation which are really undesirable in our experiments. This is a significant point compared to YAG or Eximer lasers systems. Conversely it is usually difficult to obtain high mean UV power and larger continuous tuning spectral range with cw lasers, whereas it is rather simple with our laser system. We also have calculated and developed [20, 21] a Z cavity Ti: Sapphire laser pumped by CVL. The pump threshold is a few hundred mW and the efficiency is 25%. Several watts of mean output power are possible and a single mode operation by injection of a diode laser can be perform. Again, doubling and tripling crystals allow large UV spectral range. Figure 5 summarizes the capabilities of our actual laser system. The only problem with a Z cavity Ti: Sapphire laser is the destruction of the mirrors coating when the pump power exceeds 10 W. We are developing a new Ti: Sapphire system which solves this limitation and which takes advantage of the excellent beam quality of our home made Cu/HBr laser. OPO pumped Cu/HBr laser is also under consideration.

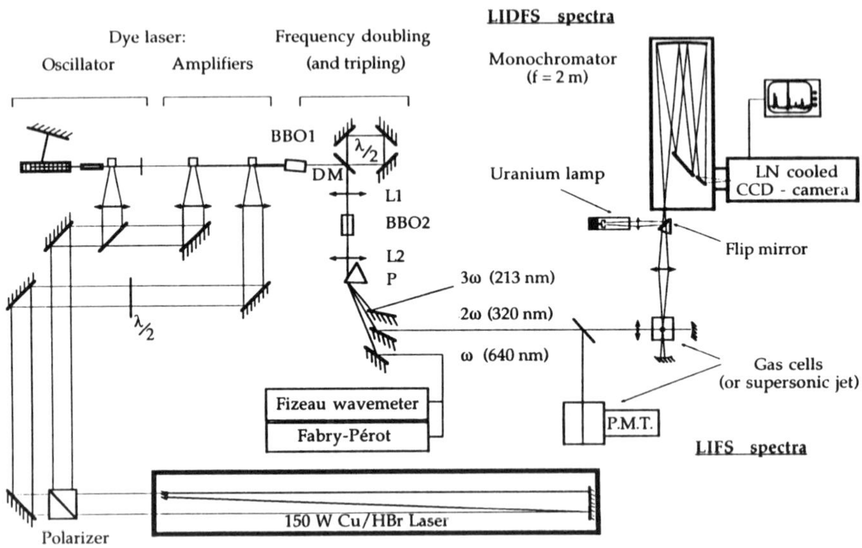

Figure 4: CVL pumped Dye laser induced dispersed fluorescence set-up

Our UV laser system allows efficient preparation of single rovibrational states of many molecules. The fluorescence induced by the laser is then dispersed with a high resolution monochromator (SOPRA) which gives a resolution similar to the linewidth of usual pulse laser systems. The spectrum is observed with the aid of a liquid nitrogen cooled CCD camera (Princeton) which has a good sensitivity between 200 nm and 1100 nm. The quantum efficiency of the detector is larger than 50% and moreover, the low intrinsic noise at liquid nitrogen temperature allows long accumulation times. The intrinsic noise increases to the equivalent of 10 photons after 1 hour. In addition the multichannel aspect (1152 spectral channels) overcomes the problem of intensity fluctuations of the laser (which is in fact only about 1%). The calibration of the spectrum is perform with an uranium hallow cathode which displays enough lines on each spectral window. The high resolution of the uranium atlas [24] and an automatic program we have developed, gives a line position precision of about 0.02 cm^{-1}.

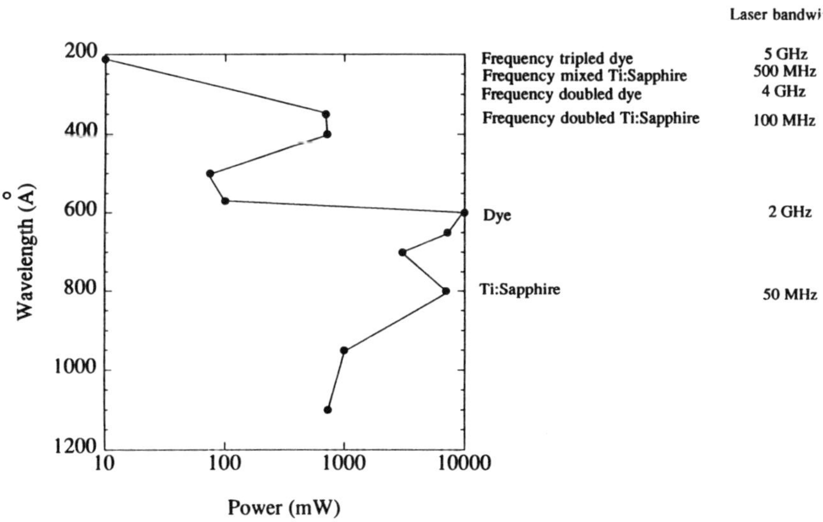

Figure 5: wavelength range and power performance

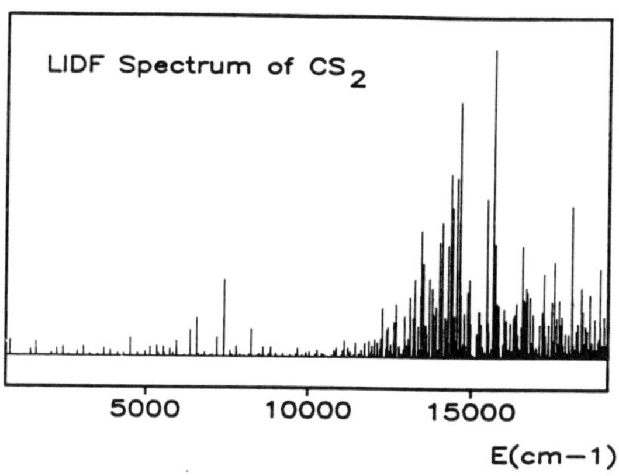

Figure 6: A typical LIDFS CS_2 spectrum

This experimental arrangement allowed us to observe close to 1000 purely vibrational levels (J" = 0) of the ground electronic state $X\Sigma^+_g$ the triatomic CS_2 molecule. [17, 16]. Prior to this work, only 40 levels had been observed with the same technique.

The figure 6 gives an overview of our spectrum. The experiment allow us to observe for the first time a clear transition at 13000 cm^{-1} associated with a transition towards a classically chaotic regime [16]. The accumulated levels function (or staircase function) of figure 7 shows this transition. Using our spectrum, different statistical tests and semiclassical analysis have demonstrated that CS_2 becomes chaotic above 13000 cm^{-1} [17, 16]. We believe that our experimental set-up open the way to the study of unexplored energy range of many molecules.

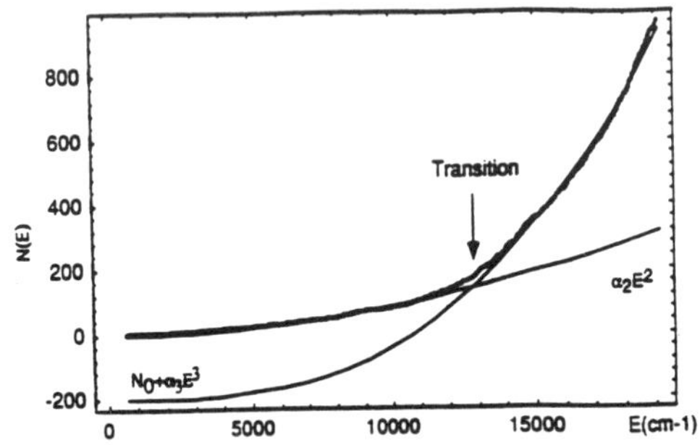

Figure 7: The number of levels below the energy E

4. Intracacity stimulated emission spectroscopy in the presence of mode-lock.

Here we describe an application of a Ti: Sapphire laser pumped by a CVL [21]. The figure 8 shows the experimental set-up. The Z cavity optical length is about 66 cm, giving a round trip time of 4.4 ns. The output mirror has a reflectivity of 97 % and the high reflector mirrors give an operating spectral range from 700 nm to 840 nm. Typically an output power of 1 W is obtained with a pump power of 4 W and with a three plate intracavity Lyot filter, a bandwidth of 1 Å is obtained. The pump threshold is 0.5 W. An efficiency of 25% was measured. An Iodine cell is placed inside the Z Ti: Sapphire cavity. I_2 molecule was pumped either by the 511 nm or 578 nm lines of the CVL. When the CVL power (which pumps the Ti: Sapphire) is just above the threshold (just enough for good stability) the delay between the CVL pulse and the start of Ti: Sapphire oscillation is about 200 ns (figure 9). The Ti: sapphire pulse is very asymmetric with a long tail. Its full width at half maximum is about 300 ns and the whole oscillating time can right several μs. The equivalent length in I_2 cell is then several hundred meters. A high intracavity sensitivity is expected.

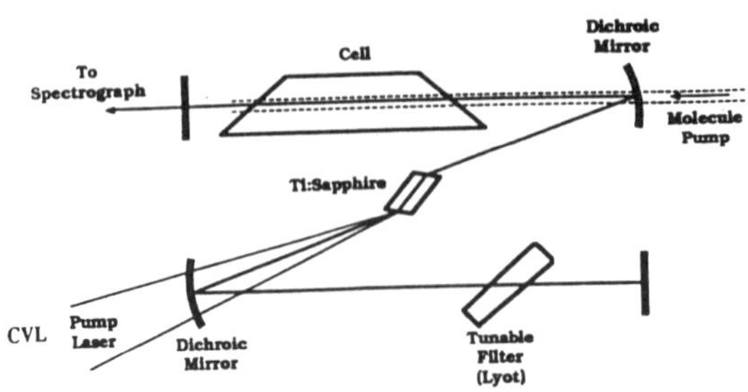

Figure 8: Intracavity set-up

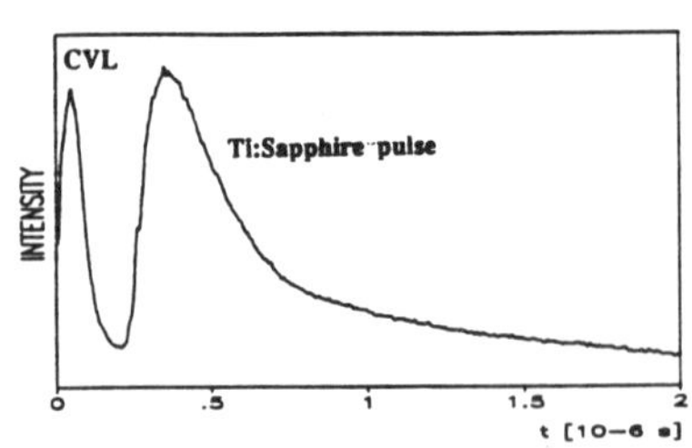

Figure 9: CVL and Ti:Sa pulses

4.1. OPTICAL HEATING OF I_2 MOLECULE

Only when the iodine molecules are pumped at a very low power (< 1 mW) with the 511 nm CVL line we observed an intracavity absorption spectrum. A part of the spectrum is represented on figure 10 together with the corresponding part of the iodine atlas. This absorption lines correspond to transition from high vibrational levels (v" ≈ 30) of the ground electronic state which are not thermally populated at room temperature. The I_2 cell pressure was about 300 mTorr. At this pressure a I_2-I_2 collision appears every 30 ns. During the time delay of 200 ns between the CVL and Ti: Sapphire pulses there is enough time for both emission from the excited electronic state to the ground state and for rotational redistribution by collision. On figure 10 a gain line is also observed. It was identified as belonging to the following successive transitions.

$$X\Sigma_g^+(\upsilon''=0, J''=80) \xrightarrow{abs} BO_u^+\ (\upsilon'=51, J'=81) \xrightarrow{em} X\Sigma_g^+(\upsilon''=33, J''=80)$$

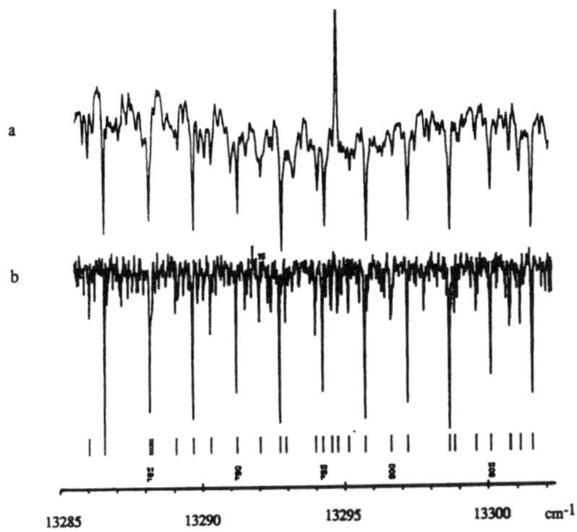

Figure 10: (a) absorption spectrum, (b) from I_2 atlas

This experiment shows that it is possible to study high vibrational states through an optical laser heating process.

4.2. EXTREMELY HIGH SENSITIVE INTRACAVITY EMISSION MODE LOCKED SPECTROSCOPY.

The iodine molecule is now pumped by the CVL yellow line (578 nm). The main iodine transition in coincidence with the CVL yellow line is the P(31)16-1 at 17289.77 cm^{-1}. Largely below the saturation power Ps of this transition (Ps ≈ 10 mW/mm^2 in cw mode) one observed P-R doublet emission lines at 12565 cm^{-1}

The iodine molecule is now pumped by the CVL yellow line (578 nm). The main iodine transition in coincidence with the CVL yellow line is the P(31)16-1 at 17289.77 cm^{-1}. Largely below the saturation power Ps of this transition (Ps ≈ 10 mW/mm^2 in cw mode) one observed P-R doublet emission lines at 12565 cm^{-1} above the laser background (figure 11a). When the iodine pump power increases above Ps, the CVL pumped Ti:Sapphire oscillates only on this P-R doublet (figure 11b). At the same time the delay between the CVL and the Ti:Sapphire pulses is shortened to 40 ns and spontaneously the laser is in a mode locked operation. Iodine acts as a saturated gain. The figure 12 shows the train of impulsions which appears when I_2 is pumped. It shows a recurrence time of 4.4 ns and figure 13 shows a pulsewidth of 300 ps with a 50 GHz sampling oscilloscope. (We measure 180 ps ± 30 ps with an autocorellator). This width corresponds roughly to the Fourier transform of the spectral band of the Doppler iodine linewidth.

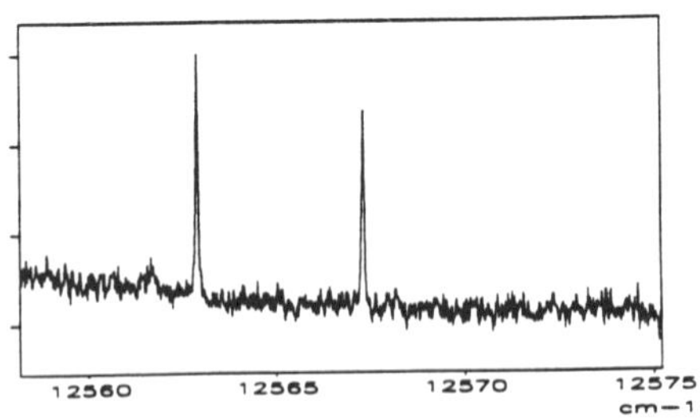

Figure 11a: I_2 P-R doublet emission

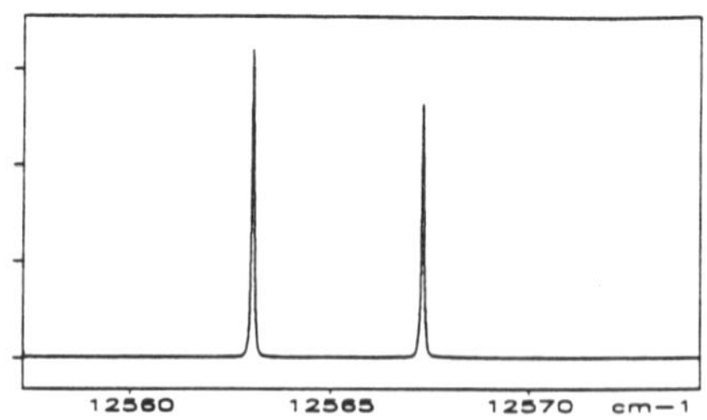

Figure 11b: Idem with saturated pumping

We can see through this rapid example that the dynamic of a Ti: sapphire pumped by a CVL can be of great interest for interesting spectroscopy.

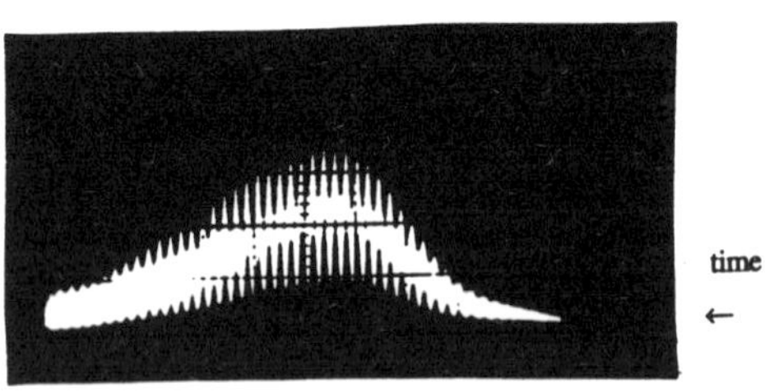

Figure 12: Train of impulsions

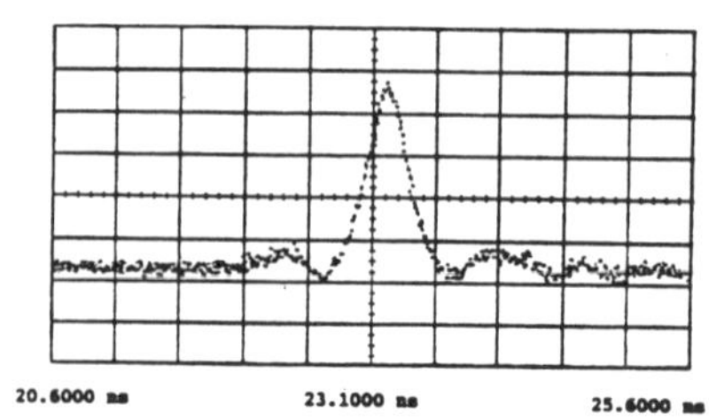

Figure 13: Fourier transform pulse

5. The Cu/HBr laser (HyBrId): an ideal pump for spectrotopists.

The copper "HyBrId" (Cu/HBr) laser developed for the first time at Saint Andrews University is the most efficient laser operating in the visible. To improve the reliability of our experimental set-up we have decided to build our own 150 W Cu/HBr laser [19].

We have several thousand hours of experience with CVL and less than one thousand with our Cu/HBr laser, but it is enough time to appreciate the difference. The first feeling is that the Cu/HBr laser would preserve the nerve system of the experimentalist who must start his laser everyday ! Because spectroscopy experiments need many delicate optical adjustments, we need a fast starting and reliable pump laser. A temperature of 600°C changes fundamentally the technical problems and the excellent beam profile [19] is very useful for pumping a solid state laser or OPO. The table II gives the comparison between our Oxford Lasers Cu60 and our home made Cu/HBr. The table II speak for itself.

TABLE II: comparison between the performance of CVL and Cu/HBr (HyBrId) laser of similar volume.

Laser type	CVL	Cu/HBr
Optical power	40 W	150 W
Plug efficiency	0.7%	1.5%
Repetition rate	6.5 kHz	20 kHz
Temperature	1500°C	600°C
Beam profile	Hat shape	Gaussien
Beam thermal instability	Important	Small
Starting time	2 hours	10 mn
Temperature Sensitivity	yes	no
Running time without maintenance	50 hours	200 hours
Ratio Green/Yellow lines	Variable	Always the same
Power adjustment	not easy	easy
Repetition rate adjustment	not easy	from 5 to 25 kHz

References

1. Letokhov V.S and Mishin V.I. (1987) Ultrasensitive laser photoionization spectroscopy of short-lived isotopes and very rare atoms, *laser spectroscopy VIII*, Springer-Verlag.
2. Ruster, W, Ames, F., Kluge H.J., Otten E.W, Rehklav, D., Scheerer, F., Herrmann, G., Muhleck, C., Riegel, J., Rimke, H., Sattelberger, P., Trautmann, N. (1989) A resonnance ionisation mass spectrometer as an analytical instrument for trace analysis, *Nucl. Inst. and Meth in Phys. Reas.*, section A, **Vol A281**, 547-558.
3. Dudel'zak, A.E., Ippolitov, I.I., Klimkin, V.M., Kudryartsev, S.K., Sokovikov, V.G., Fedorivshchev, V.N., Chikurov, V.A. and Shcheglov, V.B. (1988). Fluorescence LIDAR for the study of natural media, *Pribony i Systemy Upraxleniya*, no **7**, 30-31.
4. Evans, I.J., Webb, C.E. (1994), A 10 kH_z pulsed tunable laser source at 308 nm for tropospheric OH detection, *Chem. Phys. Lett.* **230**, 127-130.
5. Cheam, V., Lechner, J., Sekerka, I. and Desrosiers, R., Nriagu, J. and Lawson, G.(1992), Developpement of a laser-excitated atomic fluorescence spectrometer and a method for the direct determination of lead in Great Lakes water, *An.Chem.Act,*.**269,** 129-136.
6. Mitin, G.G, Maksimov, O.P, Gorelik, V.S, Isaev, A.A., Kaharyan, M.A., Sushchinskii, M.M (1974), The use of copper vapour laser for the study of Raman spectral light in crystals, *Zhurnal Prikladnoi Spektroskopii,* **21**, 332-334, *Translation in J. of Appl.Spect.*
7. Guasti, A., Pini, R., Salimbeni, R. (1991) Stimulated Raman scattering of copper vapour laser radiation in optical fibres, *Opt. and Quant. Elect*, **23**, 5555-5562.
8. Golovin Yu.M, Tkachenko A.A. (1987), The interpretation of the vibrational spectrum and the forcefield of lanthanum oxide sulphide, *Russ. J. of Inaug. Chem.*, **32**, 1682-1683
9. Broyer, M., Chevaleyre, J., Delacretaz, G., Martin,, S. and Wöste, L. (1983), K_2 Rydberg State analysis By two-and three photons ionization, *Chem.Phys.Lett,*.**99**, 206-212.
10. Broyer, M., Delacretaz, G., Labastie, P., Whetten, R.L., Wolf, J.P. and Wöste.,L. (1986) Spectroscopy of Na_3, *Z.Phys. D., Atoms, molecules and clusters*, Springer-Verlag,**3**, 131-136.
11. Crozet, P., Bacis, R., Bouvier, A., Bouvier, A.J., Churassy, S., Pique, J.P. (1987) The singlet molecular oxygen-iodine reaction: population of rovibrational levels of the iodine ground state through inversion of laser excited spectra, *J.de Phys,* **48**,no.c.7,385-387.

12 Vergès, J., and Amiot, C. (1987), Laser induced fluorescence Fourier transform spectroscopy, *J. Mol., Spectrosc,* **126**, 393-397.

13. Stringat, R., Fabre, G., Civio, L.and Ross, A. (1990) High resolution Fourier transform spectroscopy with pulsed laser induced fluorescence, *Opt.Comm,* **81**,179-185.

14. Stringat, R., Fabre, G. and Ross, A. (1991) Time-resolved high resolution Fourier transform spectroscopy with pulsed laser induced fluorescences, *Opt. Comm.*, **83**, 43-48.
15. Pique, J.P., Manners, J., Sitja, G. and Joyeux, M., (1992) Intra-inter polyad mixing and breaking of symmetric-antisymmetric selection rule in the vibrational spectra of CS_2 molecule, *J.Chem.Phys.* **96**, 6495-6508.
16. Sitja, G. and Pique, J.P. (1994) Transition to soft chaos in the vibrational Spectra of the CS_2 molecule, *Phys.Rev.Lett.*, **73**, 232-235
17. Jost, R., Delon, A., Pique, J.P. and Sitja, G.(1993) Laser induced dispersed fluorescence spectroscopy LIDFS of CS2 and NO2, *Adv. Phys. Chem.*, Molecular Dynamics and Spectroscopy by Stimulated Emission Pumping.
18. Pique, J.P. (1990) Molecular dynamics and quantum chaos in small polyatomic molecules (CS_2, C_2H_2) through stimulated-emission pumping experiments and statiscal Fourier-transform analysis, *J.Opt.Soc.Am.B*, **7**, 1816-1828
19. Coutance, P., Naylor, G.and Pique, J.P. (1995) A copper " HyBrId " laser of high beam quality using an axial unstable cavity, *IEEE J. Quantum Electron.*, **31**,1747-1752
20. Ziade, M., Naylor, G.A., and Pique, J.P. (1993) Algebraic solution for Z fold resonators of arbitrary proportions suitable for pulsed lasers, *Opt.Comm,***99**, 274-283
21. Olawsy F.J, Ziadé M., Sitja G. and Pique J.P (1993) Titanium: sapphire Laser pumped by coppor vapour laser, *J.de Phys.IV,* **C7**, 347-351
22. Koprinkov, I., Naylor, G.A., and Pique, J.P. (1994) Generation of 10 mW tunable narrowband radiation around 210 nm using a 6.5 kHz repetition rate copper vapour laser pumped dye laser, *Opt. Comm.*, **104**, 363-368.
23. Naylor, G.A., Koprinkov, I.G., Sitja, G., Ring, H. and Piqué, J.P. (1994) Use of copper lasers for high resolution spectroscopy: generation of 10 mW narrowband tunable laser light at around 210 nm, *J. de Phys.IV*, **C4**, 759-762.
24. Jones, D.R., Maitland, A. and Little, C.E. (1994) Radial and temporal laser pulse profiles in a 0.2 kW average power HyBrId laser, Gas.metal vapour and FEL lasers and applications. *Proc. SPTE*, **vol 2118**, eds V.N. Smiley and F.K. Tittel (SPIE, Bellingham 1994), 42-50.

THE USE OF COPPER-VAPOR LASERS IN GUIDE-STAR APPLICATIONS

R. Q. FUGATE
Starfire Optical Range
USAF Phillips Laboratory
Kirtland AFB, NM 87117-5776, USA

1. Introduction

This paper discusses the use of copper-vapor lasers as sources of artificial guide stars for ground-based telescope adaptive-optics systems. In particular, the system in use at the 1.5-m telescope at the USAF Phillips Laboratory Starfire Optical Range near Albuquerque, NM, USA is described, and experiences with its use are discussed.

As optical astronomers well know, the randomly-inhomogeneous variations in the refractive index of turbulent air create phase distortions in propagating electromagnetic waves, robbing large telescopes of the diffraction-limited resolution they deserve. Adaptive optics is a real-time technique for correcting wave-front aberrations by reflection from a mirror deformed to exactly match the distortions present. The process requires an optical sensor and analyzer that samples the reflected wave front for residual errors so that further correction can be made by re-shaping the deformable mirror.

An essential component of the wave-front sensing process is the beacon—a source of light which ideally is emitted from or reflected by the object being imaged. In many cases of high interest to astronomers, however, the object is too faint to provide adequate signal-to-noise ratio in the wave-front sensor in an integration time short (~1 millisecond) compared to the rate at which the atmospheric distortion is changing (~10 milliseconds).

In these cases, an artificial beacon or guide star can be created along the line of sight to the object of interest by focusing a laser in the atmosphere and sensing the distortions on wave fronts created by Rayleigh backscatter.[1, 2, 3] The 'Rayleigh beacon' is not ideal since it is not a point source at infinity. Clearly, Rayleigh backscatter requires the beacon be located within the atmosphere at some finite range from the telescope whereas astronomical objects are infinitely distant. The

C. E. Little and N. V. Sabotinov (eds.), Pulsed Metal Vapour Lasers, 431–440.

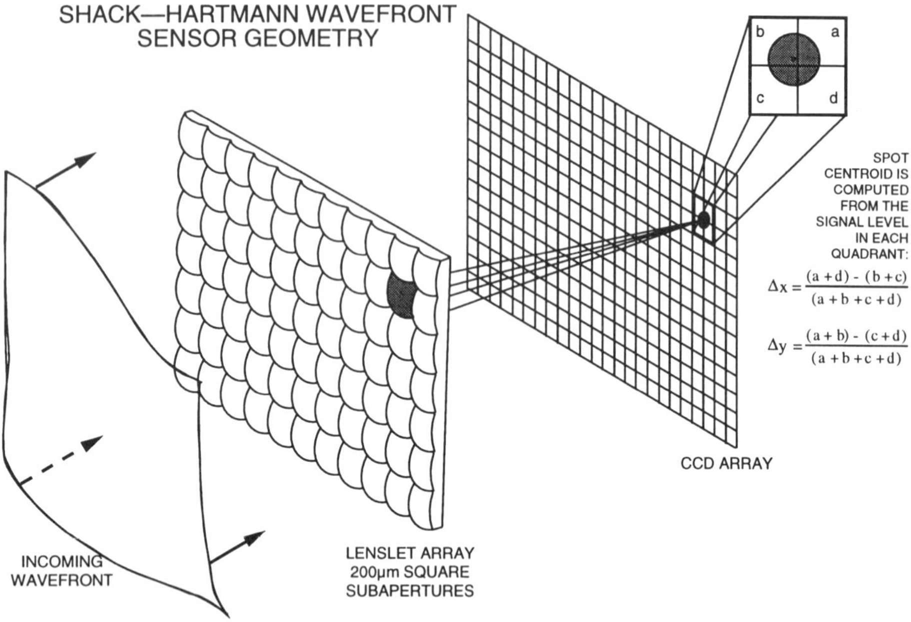

Figure 1. Shack-Hartmann wave-front sensor concept.

beacon is defined as the scattering volume along a length as large as several kilometers centered about the range to best focus. The exact limits of the scattering length are set by range-gating the wave-front sensor so that light outside the gate is not detected.

2. Laser beacon requirements for wave-front sensing

Copper-vapor lasers have high average power, short pulse widths, and high pulse repetition rates. These properties make them well suited as sources for artificial laser beacons. General disadvantages of copper-vapor lasers are poor beam quality and the fact that their power is split nearly evenly between the green and yellow lines, making aperture sharing with other instruments less efficient.

What drives the requirements of the copper-vapor laser in terms of the wave-front sensing requirements? A common type of wave-front sensor in use in adaptive-optics systems is the Shack-Hartmann sensor, shown schematically in Fig.1. Typically, the primary mirror of the telescope is re-imaged on an array of small lenslets (200 μm square in our system) which focus the sampled wave front on a CCD array. The optics and CCD scale are usually arranged so that the spot from each lenslet falls on the intersection of a 2x2 (or 4x4) array of pixels to facilitate the computation of the centroid. As the wave-front distortions change in time, the spots move around on the CCD array. The positions of the spots are therefore a measure of the wave-front tilts over each subaperture.

The performance of the wave-front sensor depends on on the angular size and brightness of the Shack-Hartmann spots. The rms error, σ_θ, in the subaperture wave-front tilt measurement is given by [4]

$$\sigma_\theta \cong 0.6 \frac{(\lambda/d)}{\mathrm{SNR}_V}, \tag{1}$$

where d is size of the subaperture as measured in the output space of the telescope. SNR_V is the voltage signal-to-noise ratio of a quad-cell sensor formed by 4 pixels in the CCD array, and is given approximately by [5]

$$\mathrm{SNR}_V = \frac{N_s}{\sqrt{N_s + 4(n_B + n_D + n_e^2)}}, \tag{2}$$

where n_B is number of background electrons/pixel, n_D is the number of dark current electrons/pixel, n_e is the number of read noise electrons/pixel, and N_s is the total number of signal electrons.

Typical state-of-the-art CCD arrays have read noise figures, n_e, of 5 to 10 electrons , requiring N_s to be of the order of 150 detected electrons per subaperture (4 pixels) to obtain adequate noise performance.

Equation 1 also calls attention to the fact that if the angular size of the beacon exceeds λ/d, the angular measurement error will increase and additional signal will be needed to reduce the measurement error to acceptable levels. Consequently, a very important criterion for the laser is that it have adequate beam quality to allow the projecting telescope to focus the beacon to a size that cannot be resolved by an individual subaperture. In other words, we require that the size of the focused beam be limited by propagation through the turbulence, not by the beam quality of the laser. For instance, in our system we are required to expand the laser to the full aperture of the 1.5-m telescope in order to geometrically reduce the natural beam divergence to a value of approximately λ/r_0, where r_0 is the coherence diameter of the atmosphere (roughly 7 cm at our site). In general, propagation through the corrected aperture is of no benefit since the beam has so many modes.

3. Required pulse energy for wave-front sensing

With these criteria in mind, it is instructive to calculate the required pulse energy to generate 150 photo-detected electrons in each subaperture of the wave-front sensor.

The number of photo-detected electrons per subaperture for Rayleigh scattering is given by the lidar equation:

$$N_{\mathrm{pde}} = \eta_{QE} T_t T_r T_{\mathrm{atm}}^2 \frac{A_{\mathrm{sub}} \beta_{BS} \Delta l}{R^2} \frac{E_p \lambda}{hc}, \tag{3}$$

where

E_p is the energy per pulse,
T_t is the laser transmitter optical transmission,
T_r is the optical transmission of the wave-front sensor,
T_{atm} is the one way transmission of the atmosphere,
A_{sub} is the area of a subaperture in the wave-front sensor,
β_{BS} is the fraction of incident laser photons scattered in the backward propagation direction per meter of scattering volume ($\mathrm{sr}^{-1}\mathrm{m}^{-1}$),
Δl is the length of the scattering volume (the range gate),
η_{QE} is the quantum efficiency of the wavefront sensor detection process,
λ is the wavelength,
R is the range to the midpoint of the scattering volume,
h is Planck's constant, and
c is the speed of light.

The volume back scattering function β_{BS} is given approximately by [6]

$$\beta_{BS}(h) = 2.8 \cdot 10^{-32} \frac{P(h)}{T(h)} sr^{-1} m^{-1}, \tag{4}$$

where $P(h)$ and $T(h)$ are the atmospheric pressure and temperature at altitude h respectively and where we have used the sea level parameters from the US Standard Atmosphere.[7] At an altitude of 10 km β_{BS} is approximately $5.1 \cdot 10^{-7}\ \mathrm{sr}^{-1}\mathrm{m}^{-1}$. This means a one km long beacon returns only 0.05% of the incident photons per steradian toward the telescope.

We can get more signal by increasing the scattering length (opening up the range gate). However, we must limit the length of the beacon based on the criteria that subapertures at the edge of the telescope aperture are not able to resolve the projected length of the beacon. A simple geometrical analysis leads to the result

$$\Delta l_{\mathrm{max}} = 2 \frac{\lambda H^2}{D r_0}, \tag{5}$$

where D is the aperture diameter of the wave-front sensing (and imaging) telescope. Note that the maximum range-gated length of the laser beacon is proportional to the square of the average beacon height.

Equation 3 has been evaluated assuming the subaperture size is r_0^2, N_{pde} is 150 electrons, η_{QE} is 0.50, and the transmit and receive optics efficiencies are each 0.30. Results are shown in Fig. 2 for atmospheric seeing of 7 cm and imaging at 0.5 μm and for atmospheric seeing of 15 cm with imaging at 2.2 μm (allowing much larger subapertures for wave-front sensing). For compensation at an imaging wavelength of 0.5 μm, we must operate the adaptive optics system at a closed loop bandwidth of approximately 100 Hz, meaning we must sample the wave-front error at 1 kHz or faster. Consequently, the average power of the beacon laser is approximately 50 watts for a 10 km high beacon and over 200 watts for a beacon at 20 km range.

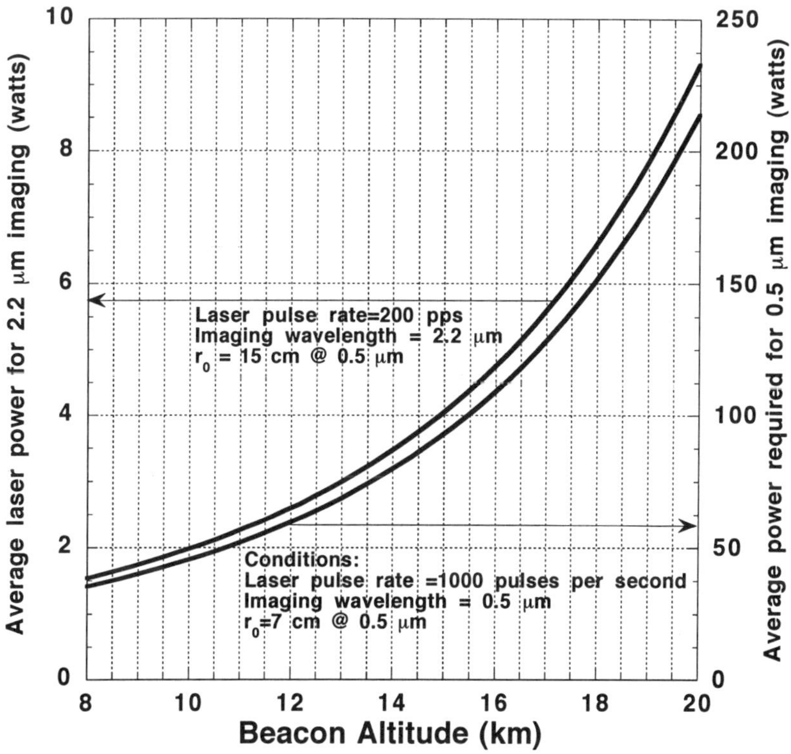

Figure 2. Copper vapor laser average power requirements for two seeing conditions and two imaging wavelengths.

4. Starfire Optical Range copper-vapor laser adaptive optics system

The system in operation at the USAF Phillips Laboratory Starfire Optical Range has been described previously.[8] Emphasis here will be on the integration and use of the copper-vapor laser. This system was first demonstrated in closed loop operation in February, 1989. A closed loop high bandwidth laser beacon adaptive optics system could not have been demonstrated at that time without copper-vapor laser technology. Figure 3 is a schematic cross section of the 1.5-m telescope facility showing the general arrangement of the telescope, optics laboratory, copper-vapor laser laboratory, and electronics processing and control room.

Figure 4 is a block diagram of the copper-vapor laser system designed and built by Oxford Lasers, Ltd, Oxford, England and sold in the United States by Oxford Lasers, Inc. The system consists of an injection seeded oscillator, and three single pass amplifiers, one 40 and two 100 watt units. The system is operated at 5,000 pulses per second, has a 60 mm output beam and is normally operated at an average power of 150-170 watts. Only 60-70 watts reach the atmosphere out of the telescope as a result of the optical losses in the coudé feed and aperture sharing

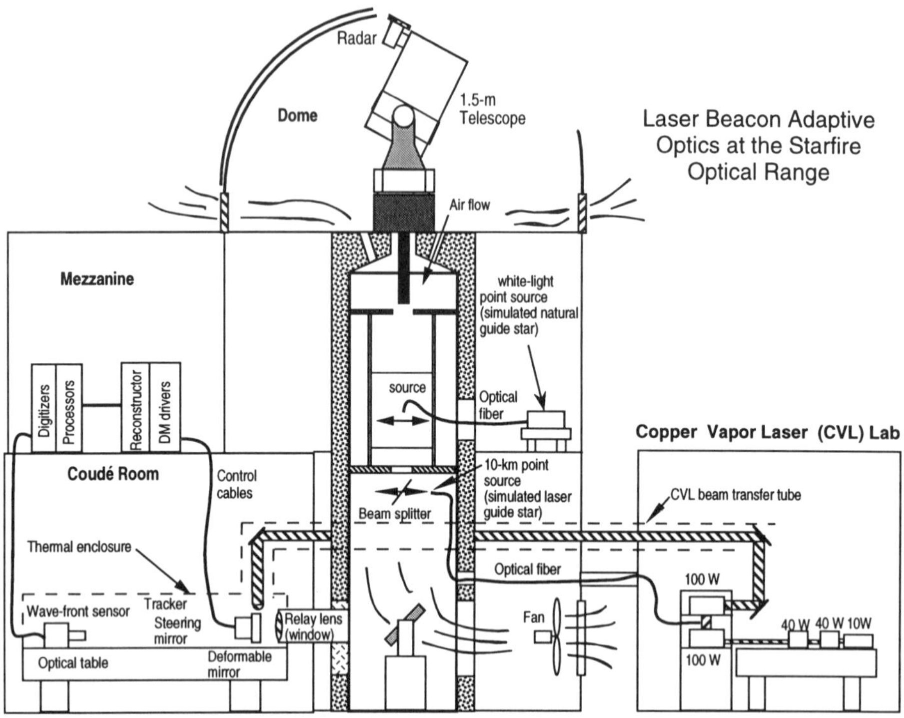

Figure 3. Starfire Optical Range 1.5-m telescope facility.

optics.

Figure 5 is a schematic layout of the coudé room showing how the copper-vapor laser is injected into the optical feed to the telescope. The beam is brought into this room through a pipe near the ceiling and down onto the optical bench by means of an optical periscope mounted to the bench. The laser passes under the main beam path, through a $\lambda/2$ wave plate polarizer, a negative lens and up into a custom made polarizing beam splitter cube. The negative lens is designed to just fill the 1.5-m aperture and focus the beam at 14 km range. The polarizing beam splitter and a $\lambda/4$ plate form an efficient optical duplexer. Back-scattered light passes through the beam splitter to the wave-front sensor. Light from natural stars in a slightly longer wavelength band is sensed by another wave-front sensor for performance comparisons.

However, the optical duplexer is not capable of isolating specular reflections in the optics (the secondary of the telescope is a big contributor) from the outgoing pulse. We have therefore installed precision mechanical chopper wheels synchronized with the laser to block the wave-front sensors and avalanche photo-diode track sensor while a laser pulse is within a few kilometers of the telescope.[9] These are shown on the layout in Fig. 5.

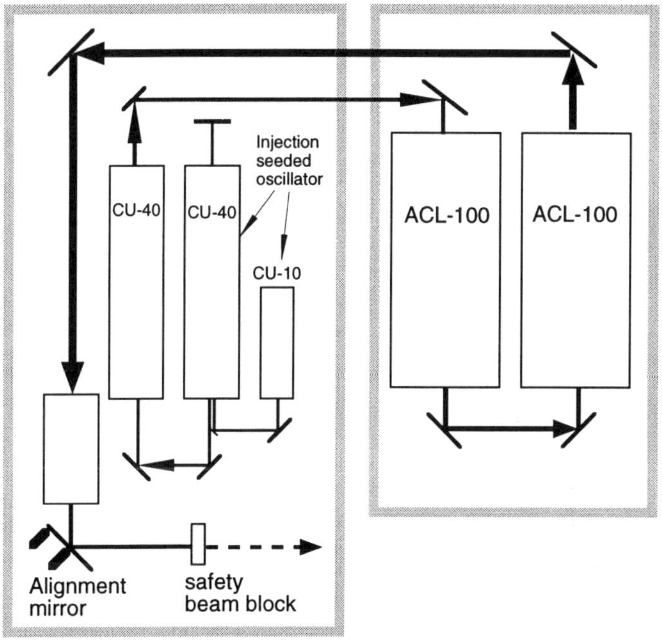

Figure 4. Block diagram of copper-vapor laser provided by Oxford Lasers.

Even though we are able to eliminate direct scattering with the mechanical choppers, there is another effect which has imposed some limitations on the performance of the system. The laser induces a prompt fluorescence in the materials in the optical train and a longer lived phosphorescence that has a decay time constant of the order of hundreds of microseconds. The prompt fluorescent emission is blocked by the chopper wheels. The phosphorescence is equivalent to an 8th magnitude star as sensed by the avalanche photo-diode track sensor. The theoretical limit for the tracker is 13th magnitude. The phosphorescence has reduced the potential usefulness of the system by several stellar magnitudes.

After many tests on optical materials, we believe it is difficult to avoid this problem. The only certain approach is to project the beam out of a separate telescope behind the secondary mirror to avoid beacon image elongation in the wave-front sensor. In order for this to work, the beam quality of the laser must be such that it can be focused to a λ/d-sized spot (where d is the subaperture size). In addition there is the complexity of getting the beam around the gimbal axes of the mount through a separate coudé path.

Another requirement for good performance is minimum jitter in the wave-front sensor range gate. Jitter in edges of the gate creates a corresponding jitter in the focus term of the wave-front measurement. A mechanical shutter does not have adequate precision for this purpose. The system at the SOR uses a Pockels cell

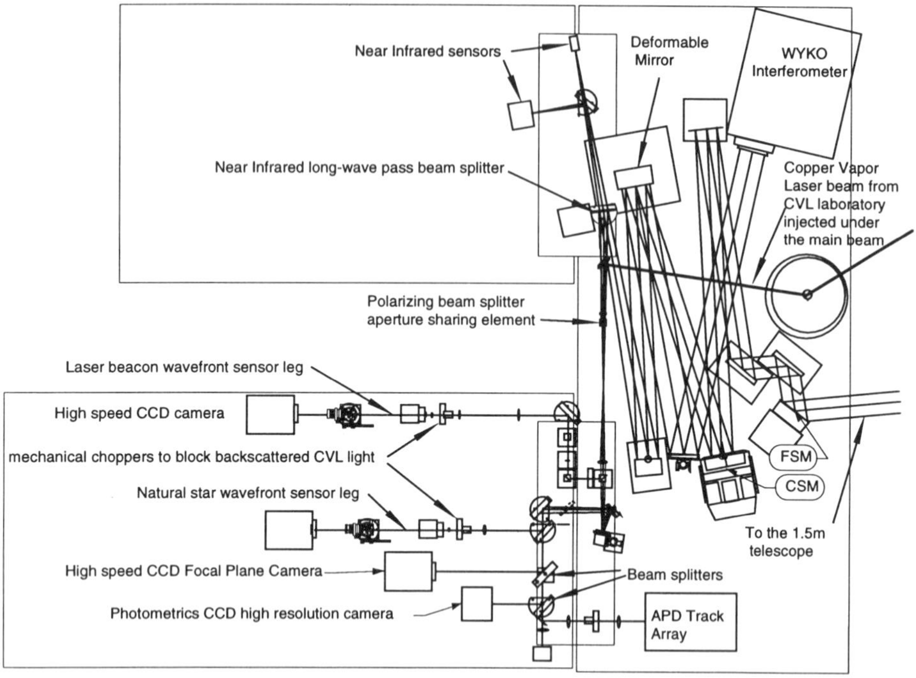

Figure 5. Method of injecting copper-vapor laser into telescope optics.

in the laser beacon wave-front sensor leg that is gated on and then off while the mechanical shutter is open. In normal operation, the Pockels cell is gated on three successive pulses of the laser return and then the CCD is read out. The time between wave front samples is therefore 600 μsec or a rate of 1667 Hz. This is required since the CCD focal plane has a maximum read rate of 2000 frames per second.

5. Example results obtained with the system

Figures 6 and 7 are example results obtained with the laser beacon adaptive-optics system, where in each example comparison is made with uncorrected images. Figure 6 is the the binary star β-Del, 0.2 arc sec separation in a 60 sec exposure (camera attenuated) to demonstrate the diffraction limited capability of the corrected telescope. Each star image in the binary star pair is 0.13 arc sec full-width-half-maximum. The value of λ/D at 0.85 μm is 0.12 arc sec. The peak intensity of the uncompensated image is 384 counts whereas it is 3260 counts in the compensated image.

Figure 7 shows two images of the planet Saturn. The left image was made without adaptive optics and the right with the copper-vapor laser beacon adaptive optics. This is a 3 second exposure covering the spectral region 0.65 to 0.95 μm.

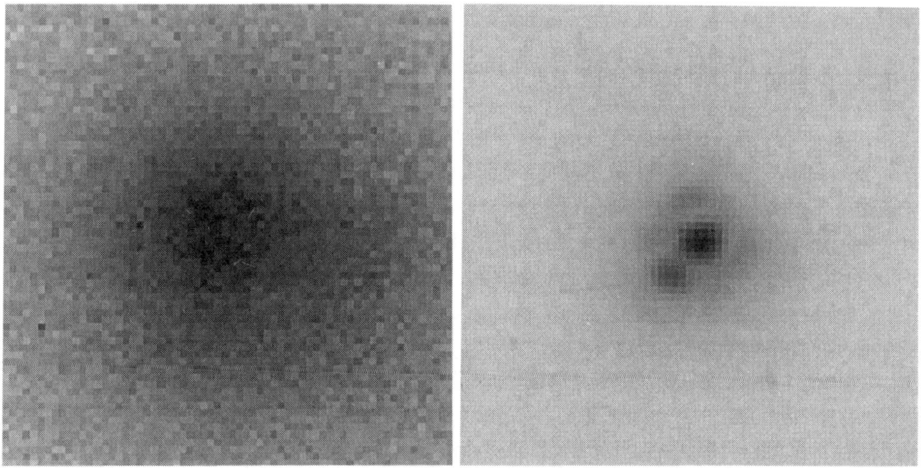

Figure 6. Images of the binary star β–Delphinius. Left panel without adaptive optics. Right panel with laser beacon adaptive optics. The separation of the stars is 0.199 arc sec and the full-width-half-maximum image size of each star is 0.13 arc sec. 60 sec exposure at 0.85 μm center wavelength.

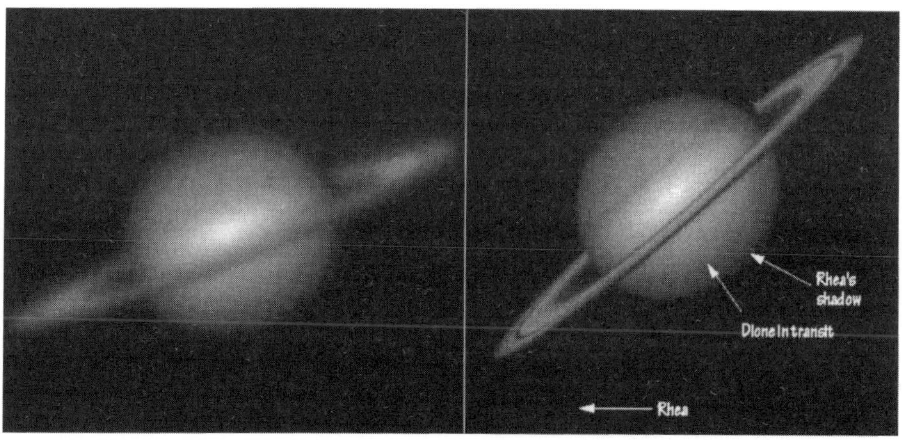

Figure 7. Images of the planet Saturn. Left panel without adaptive optics. Right panel with the laser beacon adaptive optics. In the original, fine detail can be seen in the rings, planet's atmosphere, one moon can be seen in transit, and another's shadow is visible.

The original image shows significantly more detail than can be reproduced here. The shadow of the moon Rhea is clearly visible on the surface of the planet and the moon Dione is observed in transit against the planet background. We know of no other ground based telescope capable of this resolution. The beacon was focused at 14 km range for these observations. A range-gate of 16 μsec (2.4 km) produced approximately 150 photo-detected electrons per sample of 3 laser back-scattered pulses.

6. Summary

This paper has described the use of a 200 watt class copper-vapor laser for creating artificial beacons by Rayleigh backscatter. The laser has been successfully employed at the Starfire Optical Range 1.5-m telescope to make diffraction limited images of stars and extended objects. The laser is focused at 10 to 15 km range and shares the full aperture of the telescope. The principal limitation encountered in the system is long-lived phosphorescence which contaminates the photon-counting track sensor, limiting tracking to objects $\sim 8^{th}$ magnitude or brighter. This could be prevented in future systems by improving the beam quality of the copper-vapor laser so that it may be projected without significant divergence from behind the secondary mirror of the telescope. The copper-vapor laser was probably the only laser technology available enabling an early demonstration of closed-loop laser beacon adaptive optics performance.

References

1. Fugate, R.Q., Fried, D.L., Ameer, G.A., Boeke, B.R., Browne S.L., Roberts, P.H., Ruane, R.E, Tyler, G.A., and Wopat, L.M. (1991) Measurement of atmospheric wavefront distortion using scattered light from a laser guide-star, *Nature*, Vol. 276, pp. 314-316
2. Fugate, R.Q. (1993) Laser beacon adaptive optics, *Optics and Photonics News*, Vol. 4, pp.14-19
3. Fugate, R.Q. (1994) Laser Beacon Adaptive Optics—Boom or Bust?, in J.C. Dainty (ed.) *Current Trends in Optics*, Academic Press Ltd., London, pp. 289-304
4. Tyler, G.A. and Fried, D.L. (1982) Image position error associated a quadrant detector, *J. Opt. Soc. Am.* , Vol. 72, pp. 804-808
5. Howell, S.B. (1992) Introduction to differential time-series astronomical photometry using charge-coupled devices, in S.B. Howell (ed.), *Astronomical CCD Observing and Reduction Techniques*, ASP Conference Series, Vol 23, Astronomical Society of the Pacific, San Francisco, pp. 105-129
6. Penndorf, R. (1957) Tables of the Refractive Index for Standard Air and the Rayleigh Scattering Coefficient for the Spectral Region between 0.2 and 20.0 μ and Their Application to Atmospheric Optics, *J. Opt. Soc. Am.* Vol. 47, pp. 176-182
7. *U. S. Standard Atmosphere, 1976*, National Oceanic and Atmospheric Administration, National Aeronautics and Space Administration, and the U. S. Air Force, Washington (1976).
8. Fugate, R.Q., *et al.* (1994) Two generations of laser-guide star adaptive-optics experiments at the Starfire Optical Range, *J. Opt. Soc. Am. A*, Vol. 11, pp 310-324
9. Fugate, R.Q., *et al.* (1993) Laser beacon adaptive optics with an unintensified CCD wavefront sensor and fiber optic synthesized array of silicon avalanche photodiodes, in F. Merkle (ed.), *Active and Adaptive Optics*, European Southern Observatory, Garching, pp. 487-492

INTERACTION OF CVL RADIATION WITH SOLIDS: APPLICATION TO MASS SPECTROMETRY

I.I.OPACHKO
Department of Physics, Uzhgorod State University,
Voloshin St., 54, Uzhgorod, 294000, Ukraine

The purpose of this work was to conduct the continuation early experimental studies [1-4] on the effect of CVL radiation upon surface of solids. Examination of main mechanisms of surface ionisation study of ion emission peculiarities and some attendant effects occuring when CVL radiation is focused upon the modified experimental setup is shown in Fig.1. The time-of-flight mass-spectrometer with energy focusing similar to mass reflectrons MCX-6 mass-spectrometer.

The laser part of the setup consists of three copper vapour self heating gas discharge tubes, a high voltage pulsed power source a resonator a laser radiation focusing device upon the target a projecting aspherical lens a recording system of optical radiation parameters with FEK-16 coaxial vacuum photodiode and 6LOR multichannel oscillograph IMO calorimetric detector laser tubes used in the experiments were of 0,5-1,0 m long, and had a discharge channel diameter ∅ = 2 - 3 cm copper powder and buffer neon gas being filled at the pressure of 50 - 200 Torr.

The pulsed power source is connected as a thyristor-choke circuit of accumulative capacitance charging with their subsequent switching to active elements by 4 thyratrons. The consumed power from the rectifier was 5-7 kW. The charging efficiency of accumulative capacitances $\eta = U_c/2U_R = CU^2_c/2/(U\cdot Q)$ may change from 0,5 (diode-resistor charging) to 1,0 (diode-choke charging), and it is expressed in a general way

$$\eta = \frac{1}{2}[1 + \sqrt{1 + \frac{\varphi^2}{4}} \exp(-\frac{\pi\varphi}{\sqrt{4-\varphi^2}} + \frac{\varphi}{\sqrt{4-\varphi^2}} \text{arctg} \frac{\varphi}{\sqrt{4-\varphi^2}})] \quad (1)$$

where $\varphi = R/\sqrt{L/C}$ -relative active resistance; U_c -charge voltage of accumulative capacitance C; U_R -rectifier voltage; η changes from 0,5 at $\varphi > 2$ to 1,0 at $\varphi = 0$;
$Q = \int_0^t I(t)dt$ - charge on capacitance.

In our case the charging efficiency was η = 0,7-0,8 and it was mainly determined by resistance of p-n transition for the charging thyristor (diode)

C. E. Little and N. V. Sabotinov (eds.), Pulsed Metal Vapour Lasers, 441–444.

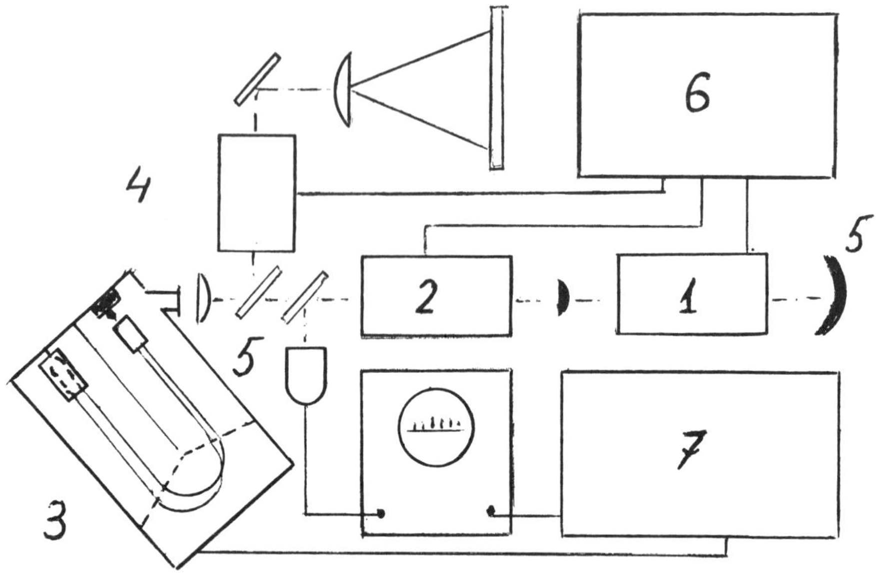

Fig.1 Scheme of experimental setup.
1,2.-CVL with amplifier; 3.-mass-reflectron MCX-6; 4.-laser projecting microscope (quantum luminance amplifier); 5.-optical elements; 6.-power supply; 7.-ion signal recording system.

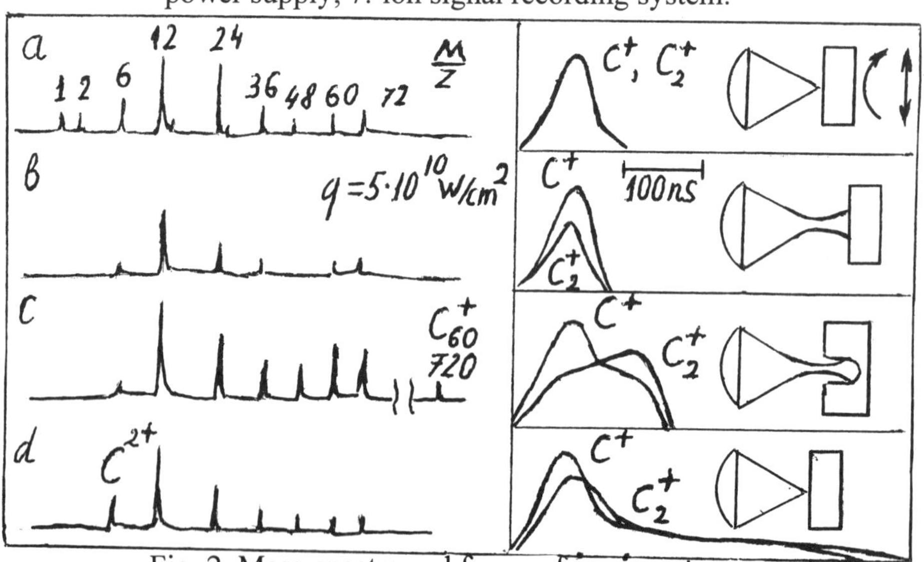

Fig. 2. Mass-spectra and forms of ion impulses.
a) "explosive" regime; b) "multiphoton" regime; c) "cummulative" regime;
d) "explosive" and thermoemissione regime.

and active resistance of choke windings. An active power of 1,5-2 kW to be released on these elements was removed by a forced cooling.

A laser projecting microscope based on active media of the CVL (qantum luminance amplifier) was developed for the experiments with used a graphite target.

Two independent active elements available allowed us to obtain light beams with divergence of ~ 10^{-4} rad to measure the active medium gain temporal relaxation of active medium parameters to head the surface by radiation of one element, and using the second one as the projecting microscope. A mean radiation power of the oscillator amplifier system was 15-20 W λλ 510,6; 578,2 nm at pulse repetition frequencies from 5-10 kHz with a pulse duration of 15-20 ns.

In most cases the measurements have been made at 510,6 nm to be selected by a light filter placed into the resonator. A mean radiation at this wavelength during optimal operation mode achieved 70-75% of the total power. A maximum laser pulse at λ 510,6 nm was 1,8-2,0 mJ. The laser radiation divergence was $\alpha = 10^{-4}$ rad that when employing an aspherical focusing lens with a 5-10 cm focal distance allowed us to focus it easily into the spot with dimensions of 10-30 μm. At this case place the pulse radiant flux density may achieve $q = 10^{10}$-10^{11} W/cm^2. At such flux densities an explosive evaporation from the target surface is achieved, and some nonlinear effects reveal themselves.

Various regiments were discussed in the early work [3,4] of ionization of this surface impulse- periodical radiation CVL. The use of mass-reflrctron of time of flight (TOF) mass-spectrometer with electrostatic analiser allowed to study the emission of ion components in dynamics with time resolution ~50 ns.

The following regimes have been realized.

1. An "explosive" regime.
2. The regime of multifoton ionization.
3. A "cummulative" regime with focusing radiation into the crater.
4. The regime connected with appearance of selffocusing and thermoemission in the interimpulse period.

Fig.2 presents mass-spectra and form of ion emission impulses under various regimes of processing.

"Explosive" and "multifoton" regimes (Fig.2a,b) are characterised with uninertion, absence of emission in interimpulse period. The pure "explosive" regime is realised when the target is rotated.

The multifoton regime is realised when the target is heated with high frequency repetition CVL radiation to high temperature > 3600K (that are controlled with the qantum luminance amplifier) and continual emission of neutral components in the interimpulsive period. The multiphoton ions emission sharply depended on the power density with photons necessary for the multiphoton ionization and depended while being placed "contraction" of the focus system's caustic at 1-2 mm distance from the surface.

The "cummulative" regime (Fig.2c) is realized when the CVL radiation in the crater is formed with multipulse (10^5-10^6 impulses) one area target radiation. Thus, cluster ion emission C_2^+, C_3^+, C_4^+, C_5^+, C_6^+, C_{60}^+ lasts 100-150 ns, and is explained by inertia of cummulative processes development.This regime can be realised too when we have monoimpulse.

Regime 4 (Fig.2d) realizes two mechanisms of ionisation: "explosive" syncronized with laser impulse, and continuation of emission during 500-700 ns caused by thermoionization in the interimpulse period.

The duration of thermoemission impulse correlated with the duration of the surface heated state in interimpulse period that is controlled with quantum luminance amplifier.

The result of the research proves the presence of various mechanisms of surface ionization with the help of CVL radiation. These various mechanisms lead to expanding ion impulses, and to reducting of time of flight mass-spectrometers resolving ability.

Acknowledgements

The author wish to thank prof. G.G.Petrash for his help in the discussion of the experimental results.

References

1. Opachko I.I. (1989) The effects of interaction nanosecond impulse-periodic radiation with surface observed with laser projecting microscope. *Teplofizika Visokikh Temperatur,* v.27, 1020-1022.

2. Opachko I.I., Shimon L.L. (1991) Non- linear processes of ionization and selffocusing at pulse periodic copper vapour laser effect on the surface. *Book of proceedings 21st European Conference on Laser Interaction with Matter,* Warsaw, October 21-25, 1991, 184-186.

3. Opachko I.I., Shimon L.L. (1993) Interaction of CVL radiation with surface of solids. *Proc. SPIE, CIS Selected Papers,* v.2110, 250-267.

4. Opachko I.I., Shimon L.L. (1994) Interaction of copper vapour laser radiation with solid surface. *Ukrainski Fizicheski Zhurnal.* v.34, N2, 221-226.

APPLICATIONS OF PULSED METAL VAPOUR LASERS IN ATMOSPHERIC OPTICS

G.S. EVTUSHENKO

Institute of Atmospheric Optics

Siberian Branch of the Russian Academy of Science

634055, Tomsk, Russia

Abstract

Parameters of metal vapour lasers for remote sensing of atmospheric aerosol and ozone are examined. The lasers under review are also useful for ship beacon systems operating in poor visibility conditions and for laboratory experiments aimed at simulation of characteristics of laser radiation propagation through different media.

1. Introduction

The metal vapour lasers take the special place in a wide class of laser sources due to the unique set of output data. Already by the middle of eighties researches of a number of scientific groups [1] established the fundamental physical principles of operation of these lasers and possible areas of its practical application.

One of such areas is the atmospheric optics: spectroscopy, gas analysis, remote sensing of atmospheric parameters. So, as distinct from conventional laser sources (solid-state lasers and high-power gas lasers) the metal vapour lasers can sound the atmosphere by pulses of small power (units of mJ) with repetition rate of units of kHz. Moreover, remote sensing can be carried out at several wavelengths simultaneously, and in that spectral region where the atmosphere possesses a good transparency, optical elements of the lidar system are simple and reliable, and radiation receivers have high sensitivity and time resolution.

Typical pulse duration of the metal vapour lasers (10^{-8} s) is capable to provide the required spatial resolution when sounding the atmosphere up to

C. E. Little and N. V. Sabotinov (eds.), Pulsed Metal Vapour Lasers, 445–452.

tens of kilometers. In this case sounding range determines the desired repetition rate of lasing pulse (Table 1).

In simulation experiments on laser radiation propagation through the atmosphere (along the short optical paths) the required spatial resolution can be provided when sounding pulse has a duration of 10^{-9} s only. Principal possibility exists to obtain such a duration for the metal vapour lasers, but development of resonator and pump schemes is to be necessary.

The metal vapour lasers, the copper and gold vapour lasers in the first place, are prospective for creation visual navigation systems using a principle of scanning by the narrow laser beam in the space too, if the pulse repetition rate will be more than 20 kHz for ship beacon systems and more than 100 kHz for landing devices.

All requirements for the lasers are formulated in Table 1.

TABLE 1. Requirements for the metal vapour lasers operating in the atmosphere

1. Spectral region	in the near UV and near IR regions 5-7 lines are used for the multifrequency sounding; in the visual regions 2-3 lines different sharply by color are used for the navigation and show-systems, for example, 510.6, 578.2 nm of CuI and 678.2 nm of AuI
2. Spectral line width	< 1 nm for aerosol sounding $<10^{-3}$ nm for differential absorption sounding of gases, spectroscopy, and gas analysis
3. Pulse repetition rate (sounding range)	< 5 kHz (up to 30 km) < 2.5 kHz (up to 60 km) > 20 kHz (for navigation) > 100 kHz (for landing)
4. Pulse power	$>10^{-4}$ J (photon counting regime)
5. Pulse duration (spatial resolution)	$3 \cdot 10^{-9}$ s (1 m) $3 \cdot 10^{-7}$ s (100 m)
6. Divergence	$2 \cdot 10^{-4}$ rad (sounding) $2 \cdot 10^{-3}$ rad (navigation)
7. Stability of the directional pattern axes	$3 \cdot 10^{-4}$ rad
8. Stability of the power parameters	10 %
9. Resource	$> 10^2 - 10^3$ hours
10. Continuous operation time	> 8 hours

It should be noted analyzing the presented parameters that far from all of them are realized in the metal vapour laser produced by different firms and some of these parameters do not realized in laboratory experiments even.

Therefore, some of results of our efforts directed to achieve the desired parameters are presented shortly in the first part of the paper and the examples of application of the created metal vapour lasers in the specific atmospheric-optical devices are given in the second part.

2. Obtaining of the Required Parameters

Peculiarities of lasing at the increased rate of pump pulse repetition (more than 30 kHz) in the gold vapour lasers and metal vapour multicomponents medium lasers are investigated in the paper. Optimal and limit rates of excitation pulse repetition are determined for gas discharge pumping method for such media. So, for the gold vapour laser the pulse repetition rate of 150 kHz (λ = 627.8 nm) was first attained by decrease of power contribution into discharge in the gas discharge tube of small diameter (8 mm) [2], as it was done for the copper vapour laser [1]. Metal vapour excitation in the gas discharge tube with spacing of active zones of several elements was shown to be taken place in the experiment preliminary in a zone corresponding to position of this element. Simultaneous effective lasing was carried out in the following combination of vapours: Cu + Au (λ = 627.8, 578.2, and 510.6 nm), Ba + Mn (λ = 0.54, 0.61, 1.13, 1.3, and 1.5 μm), and Ba + Pb (λ = 0.61, 0.72, 1.13, and 1.3 μm). By addition of the temperature zones up to three zones simultaneous lasing at the five atomic and one ionic transitions (Cu + Ba + Pb) were realized. Mixing of the different metal vapours in the laser operation process was shown to be ignored that determined a service life of unsoldered active elements with vapour of two or more metal at the same time [3].

Possibility is demonstrated in the paper to create the laser with controllable radiation pulse duration. The copper vapour laser is an example of such a laser. Shortening of the lasing pulse was obtained by the passive methods (elongation of resonator and change in the temperature regime of operation of gas discharge tube) as well as active methods (covering the additional controlling pulses with the duration of 3-5 μs) which did not cause lasing. The minimum pulse radiation duration (λ = 510.6 nm) was of 1.5 ns (Fig. 1a). The described method of control makes it possible to change a ratio of lasing powers at the wavelengths λ = 510.6 nm and λ = 578.2 nm too, getting a primary lasing at the required transition for the specific application [4]. Increase in the lasing pulse duration (λ = 510.6 nm) up to 230 ns (Fig. 1b) was attained by forming of the special pump pulse with limiting of current during the discharge development process.

To extend the functional feasibility the metal vapour active media and obtain the discrete-tunable radiation in UV spectral region the effective conversion of the copper vapour radiation was carried out in the nonlinear BBO-crystal. For the pump power 2 W in every line of the copper vapour laser constructed by the scheme "generator-amplifier" 0.5 W of the mean power was obtained in the β-BaB_2O_4 crystal in turn at three wavelengths (255, 271, and 289 nm). Maximum conversion efficiency was of 25 % for the second harmonic lasing λ = 510.6 nm and λ = 578.2 nm of the copper vapour laser and 13.6 % for lasing at the sum frequency. Smaller efficiency of conversion into lasing at the sum frequency was caused by the time mismatch between the radiation pulses of green and yellow lines of the copper vapour laser [5].

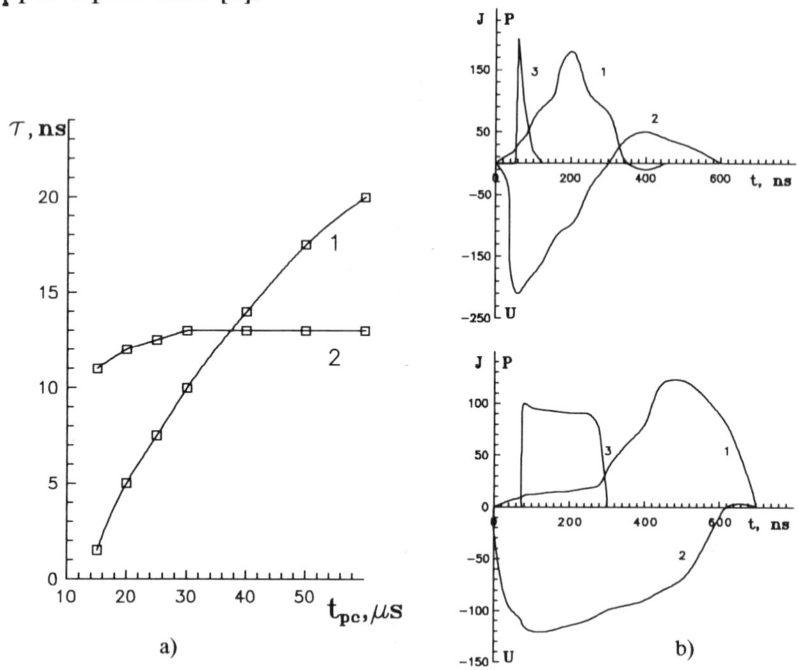

Figure 1. Control of radiation pulse duration τ of the copper vapour laser:
a) regime of short pulses,
t_{pc} is the delay between the main pulse (pumping) and controlling one,
the length of active volume is of 30 cm, diameter is of 1 cm,
pulse repetition rate is of 5 kHz, λ = 510.6 nm (curve 1), λ = 578.2 nm (curve 2);
b) regime of long pulses,
J(t) is the discharge current (curve 1), *U(t)* is the voltage (curve 2),
P(t) is lasing at the wavelength λ = 510.6 nm (curve 3).

3. Examples of Application of the Metal Vapour Lasers to Solve Problem of the Atmospheric Optics

The laser sounding method can provide monitoring of the atmospheric conditions. The absorptive coefficient profile can be reconstructed and aerosol concentration can be estimated (with the use of a priory information) by singe-frequency sounding. The multifrequency sounding method allows the aerosol particle concentration, particles dimensions, and nature of this aerosol to be determined and also aerosol correction of sounding results to be carried out, in particular, for ozone in the troposphere and stratosphere [6].

To study the weather and climate and control of the general ecological situation above industrial installations the complex lidar station was constructed at the Institute of Atmospheric Optics of Siberian Branch of the Russian Academy of Sciences. Basis of this station is the lidar with the main mirror of 2.2 m in diameter and set of lasers including the cooper, gold, and lead vapour lasers with radiation conversion in the nonlinear crystals among others.

Figure 2 presents the results of the simultaneous elevation sounding of atmosphere by three laser sources including the gold vapour laser (a) and sounding of the industrial aerosol with the use of the copper vapour laser (λ = 510.6 nm) (b) along one of horizontal path above the industrial zone of Tomsk [7].

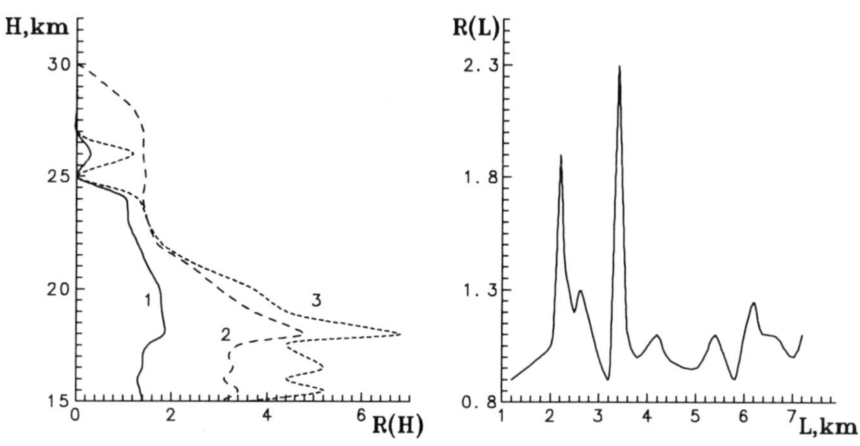

Figure 2. Scattering ratio:
a) vertical profile, λ = 353 nm (curve 1), λ = 532 nm (curve 2), λ = 628 nm (curve 3);
b) horizontal profile.

Elevation profiles of the scattering ratio R(H) show the aerosol situation in the stratosphere which was characteristic for the West-Siberian region in spring 1992. This situation was caused by aerosol disturbance by products of Pinatubo volcano eruption in summer 1991 on Philippine.

The scattering ratio $R(H)$ is defined by the formula

$$R(H) = [\beta_a(H) + \beta_m(H)]/\beta_m(H),$$

where $\beta_a(H)$ and $\beta_m(H)$ are the coefficients of the back aerosol and molecular scattering, H is the altitude.

The obtained data allowed the aerosol correction of ozone evaluation profiles obtained by the differential absorption sounding technique in UV region of spectrum to be carried out.

For horizontal sounding along the path of L length the scattering ratio $R(L)$ characterizes the relative content of the aerosol (industrial, in the main) and molecular components. Spikes of the aerosol signal at the distances 2.5, 3.5, and 6.5 km well correlate with the industrial installations and smoke plumes located along the sounding path.

The lidar system based of the copper and gold vapour lasers was used in the investigations of Bulgarian researchers [8] for measurement of the aerosol distribution profiles in troposphere and stratosphere.

The Raman lidar created at the Institute of Atmospheric Optics by Yu.F. Arshinov and colleagues [9] on the base of the copper vapour laser allows the information on the content of the gas components to be obtained, the temperature and humidity profiles to be reconstructed, and the wind velocity and wind direction in the ground atmospheric layer to be measured up to the units of kilometers.

The results of numerical simulation of tropospheric ozone sounding on the differential absorption technique [10] at the copper vapour laser wavelengths converted in the BBO-crystal are presented in Fig. 3a. The sounding altitude is plotted on the abscissa and sum sounding error depending on apparatus errors, background radiation, absorption by water and other gases in the given spectral region is plotted on the ordinate. Figure 3b presents the echo-signals recorded experimentally under the mean power of sounding pulses of 0.5 W and lasing pulse repetition rate of 7 kHz [11].

The obtained results show that with the use of the wavelength pair 271 and 289 nm ozone can be sounded in the altitude interval 2-5 km by day and 2-6 km by night when the signal accumulation time is of 10 min (photon counting regime). Figure 4 presents the results of the time deformation of the short pulse of green line of the copper vapour laser when passing through the dense optical medium simulated in the Big aerosol camera of the Institute. The results show that the quantitative characteristics of the time distortions of the initial pulse (duration of 4 ns) can be surely recorded when the optical thickness (k = æ l) of the medium is about 20 for the polarized component (I_{II}) and about 10 for depolarized one ($I_{\perp}$). For the optical density of the medium ~ 70 the radiation is depolarized completely [12].

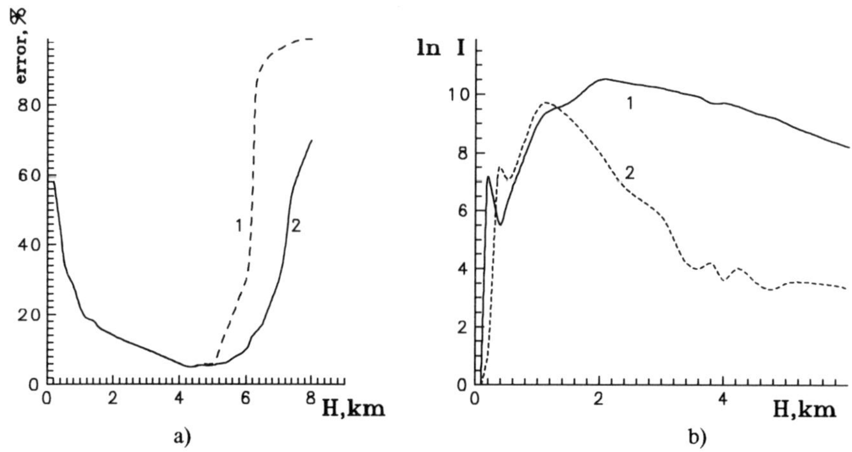

Figure 3. Feasibility of UV laser system (copper vapour laser + BBO-crystal) for sounding of tropospheric ozone:
a) simulation, day (curve 1), night (curve 2);
b) experiment, λ = 289 nm (curve 1), λ = 271 nm (curve 2).

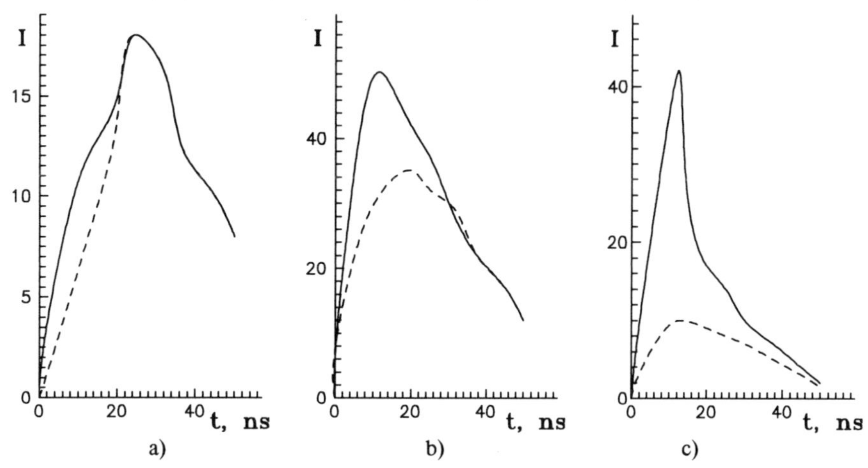

Figure 4. Oscillograms of the copper vapour laser pulses passed through the scattering media with the different optical thickness; solid curves - I_{II}, dashed curves - $I_{\perp}$:
a) k = 71, b) k = 55, c) k = 38.

4. Conclusion

The investigations and developments show an expediency to use of the pulse periodic metal vapour lasers in the atmospheric optical devices. These studies are a part of the complex program of the Institute of Atmospheric Optics. This program is to research the atmospheric optical characteristics and includes the numerical calculations, simulation and natural experiments.

5. References

1. Soldatov A.N., Solomonov V.I. (1985) *Gas Discharge Lasers at the Self-limited Transitions in Metal Vapours*, Nauka, Novosibirsk.
2. Evtushenko G.S., Polunin Yu.P., and Fedorov B.F. (1987) Study of pulse-periodic generation in the gold vapour under the high frequencies, *J. Appl. Spectroscopy* **46**, 1009-1011.
3. Evtushenko G.S., Kirilov A.E., Polunin Yu.P., Soldatov A.N., and Fedorov B.F. (1983) Study of the laser at the copper and gold vapour with the spaced apart active media, *J. Appl. Spectroscopy* **39**, 939-944.
4. Evtushenko G.S., Kirilov A.E., Kruglyakov V.L., Polunin Yu.P., Soldatov A.N., and Filonova N.A. (1988) Control of the pulse duration of the copper vapour laser, *J. Appl. Spectroscopy* **49**, 745-751.
5. Evtushenko G.S. and Troitskii V.O. (1994) Effective conversion of copper vapour laser emission in β-BaB_2O_4 crystals, *J. Russian Laser Research* **15**, 28-33.
6. Zuev V.E. and Zuev V.V. (1992) *Optical Remote Sensing of the Atmosphere*, Gidtometeoizdat, Cent Petersburg.
7. Burlakov V.D., Zuev V.V., Evtushenko G.S. et al. (1993) Metal vapour lasers for remote sensing of atmospheric aerosol, *Atmospheric and Oceanic Optics* **6**, 326-331.
8. Stoyanov D.V., Donchev A.K., Kolarov J.V., and Mitsev A. (1988) Lidar system with copper and gold vapours lasers for study of troposphere and stratosphere, *Atmospheric Optics* **1**, 109-116.
9. Arshinov Yu. F., Bobrovnikov S.M., Zuev V.E., and Mitev V.M. (1983) Atmospheric temperature measurements using a pure rotational Raman lidar, *Appl. Optics* **22**, 2984-2990.
10. Burlakov V.D., Zuev V.V., Evtushenko G.S. et al. (1994) Lidar for tropospheric ozone sounding using nonlinear conversion of copper vapour laser emission, *Atmospheric and Oceanic Optics* **7**, 1614-1618.
11. Burlakov V.D., Zuev V.V., Evtushenko G.S. et al. (1995, in press) Copper vapour lidar for tropospheric ozone sounding, *Proc. SPIE.*
12. Vergun V.V., Kirilov A.E., Kokhanenko G.P. et al. (1986) Application of copper vapour laser for observation the time distortions of scattered radiation, *Dep. VINITI*, № 2568-B86 Dep.

OPTICAL SYSTEMS WITH METAL VAPOUR BRIGHTNESS AMPLIFIERS.

K.I.ZEMSKOV, G.G.PETRASH, V.V.CHVYKOV
P.N.Lebedev Physical Institute, Academy of Sciences,
Russia, Moscow

1. Introduction

The problem of production of bright optical image of relatively low-illuminated object is connected closely with the known optical form of the energy conservation law: the output intensity of optical system may only be less (or equal in ideal system) than initial one. Great progress that has been made last few decades in electronics enabled to solve partially the problem. Images produced by modern TV-screens, PC-displays, information tables, night-vision devices achieved high quality. But this technique is naturally limited by the losses of significant part of initial optical information in light-electron-light transformations. However, as it was clear from the beginning of quantum electronics development, stimulated emission by its nature could serve to image amplification without, in principle, losses of information about amplitude and phase distributions to be amplified. Thus, the problem was to provide such characteristics of some quantum electronic device which could satisfy simultaneously requirements to both electronic and optical imaging systems as: high amplification, large enough output power, high efficiency, steady-state or high repetition rate operation, high image quality, including optical contrast, sharpness, large enough number of resolved elements over a field of view at the size of each resolved element close to the diffraction limit, visual spectral range operation, and so on. There were also some design requirements, for example, to the mass and size of amplifying element, which could allow to place it inside the optical system of practical interest. As it is known now, metal vapour lasers active media and the amplifying elements based on them are the most suitable to satisfy all requirements mentioned above.

Table 1 presents some amplifying parameters of metal vapour amplifying elements. They were measured in a laser projection microscope [1] - the first optical system using metal vapour brightness amplifier, or as it also called "active optical systems" (AOS). Schematic diagram of the simplest version of it is shown in Fig.1.

The principles of its operation are the common ones for the most AOS [2]: amplified spontaneous emission, passing through the amplifying element 3 and objective 2, illuminates the object 1. Light reflected and scattered by the object produces an intermediate image 4 behind the amplifying element, and then the image is projected

C. E. Little and N. V. Sabotinov (eds.), Pulsed Metal Vapour Lasers, 453–458.

to the screen 6 through the use of the projecting system 5. The functions of elements 7-11 will be discussed below.

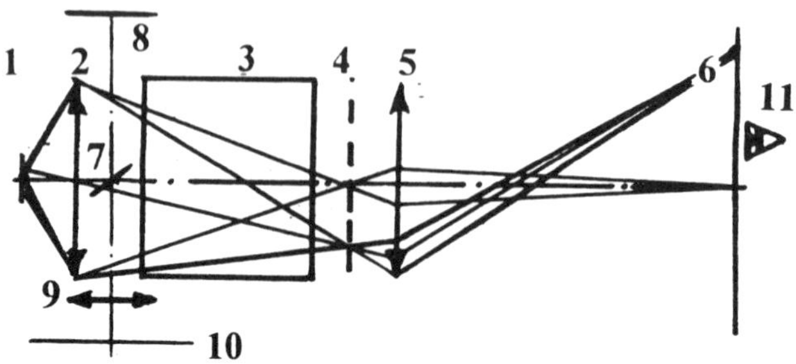

Figure 1. Schematic diagram of experimental set-up: 1-object, 2-objective, 3-amplifying element, 4-plane of intermediate image, 5-projecting lens, 6,8,10-screens, 7-beamsplitter, 9-lens, 11-photodiode.

TABLE 1. Parameters of metal vapour amplifiers

Active media	Wave-length, λ, nm	Repetition rate kHz	$(D^2/\lambda L)^2$, 10^5	Small signal gain	Ratio amplifier/laser average power (*	Maximal measured effective amplification (**
Cu	510.5	10	12	0.2	0.44	$1.6 \cdot 10^4$
CuCl	510.5	10	32		0.56	
CuBr	510.5	14	32		0.59	$6.5 \cdot 10^3$
Au	627.8	10	5.4	0.11	0.34	$1 \cdot 10^4$
Mn	green	7	5.7	0.11	0.18	$1 \cdot 10^3$
Mn	IR	7	1.1	0.13	0.2	$3.5 \cdot 10^3$
Pb	722.9	12	3.1	0.12	0.52	$3 \cdot 10^3$
Ba	1500	8	2.1	0.17	0.82	$2.7 \cdot 10^3$

D, and L are inner diameter and length of active element

*) Laser average output power was measured with flat-flat resonator, output mirror was uncovered quartz plate, another mirror - Al-covered plate.

**) Effective amplification was defined as the ratio of output and input average power of the amplifying element.

This system became the subject of experimental studies which allowed to conclude, that metal vapour amplifying elements have not produced considerable distortions of amplified images. But energetic characteristics differed essentially from those of the usual optical systems. Effective amplification of more than 15000 was achieved for a small input signal, and of about 1000, when output power was high enough to produce bright images on a screen of more than 10 m^2 area. As it was measured, the output power of an AOS practically did not depend on the parameters of optical system at the amplifying element input, for example, a change of objective magnifications from 1x up to 40x led to variations in output power less than 10%. Linear magnification on the screen increased in this case by a factor more than twenty.

This would be lead to dramatic decrease of image brightness on the screen of usual optical systems.

Even the simplest AOS described above have found some useful applications: in biology - to observe on a large screen the images of leaving biological objects and record them with a good temporal resolution using high repetition rate of amplifying system[3, 4], in experiments with optical fibre systems [5], in IR-microscopy [6, 7], especially with anti-Stocks multi-photon visualisators. The most comprehensive descriptions of the specific features , applications, and full bibliography on laser projection systems were presented in [1, 2, 8].

All the ensuing development of AOS with metal vapour amplifiers could be described as two main lines: research of characteristic properties of such systems and investigations closely connected with applications of these systems to the solution of some practical problems in science, medicine, and technology.

2. Influence of amplifier non-linearity (saturation of active medium)

2.1. SATURATION OF AMPLIFICATION

In order to achieve good efficiency amplifying element must operate close to saturation of active medium. This means that under typical conditions of operation the amplifier is non-linear and that there can be some interaction between light beams from different points of an object [9]. This is quite unusual situation for instrumental optics. It results in some unusual properties of AOS.

2.2. CONTRAST OF AMPLIFIED IMAGES

It is clear, and it was demonstrated in experiments that in the case when light beams from different points of the object independently propagate through a saturated active medium the contrast in an amplified image may be lost completely. However in the case when light beams from the different points of the object intersect in a saturated amplifier the influence of the saturation can be different. It is difficult to predict it theoretically, so special experiments were carried out [2] for typical laser projection microscope. Schematic diagram of experimental arrangement is shown in Fig.1. In this case strong overlapping of light beams from different points of an object exists in an amplifier. Contrast of the amplified image was measured in the image plane 6 with and without amplification in a saturated copper vapour amplifier.

It was shown that in this case the contrast of an amplified image is close to that of the image obtained without amplification and to initial contrast of the object. Our long-time experience with AOS using CVL and other metal vapour amplifiers showed that under usual operating conditions amplified images have good contrast and high optical quality. Resolution of laser projection microscope is normally determined by the optical elements used.

2.3. NEGATIVE IMAGES

Saturation of active medium does not substantially decrease the quality of amplified images, but nevertheless it leads to some unusual phenomena non existent in conventional optical systems. One of these phenomena is formation of negative images. It was observed that when intermediate image of an object is inside the amplifying medium, in the region of strong saturation, then a more or less distinct negative image of the object is produced in light beam illuminating an object in a laser projection microscope [10]. It can be observed with the help of beamsplitter (7) and additional lens (9) (see Fig.1).

Explanation of the phenomenon may be as follows. When the intermediate image of an object is located inside amplifying medium spatial modulation of intensity in the image results, due to saturation, in spatial modulation of gain and laser level populations. This means that information about the object is recorded in the active medium. This information is "read out" by seed spontaneous emission propagating from the far end of the active element through it toward the objective and object. As stronger intensity in image leads to stronger saturation, this results in formation of the negative image of the object.

Negative images can be also formed in the beams propagating to a screen if active medium is saturated by the polarised beams but observation is carried out in crossed polarisation [11].

If an active medium has two laser transitions with common upper or lower laser level then saturation of one transition leads to modulation of another one and formation of negative image in other wavelength. Using this it was possible to visualise IR-images [12]. Manganese vapour laser active medium was used in these experiments.

3. Systems with resonators

If in an optical system with brightness amplifier, for example in laser projection microscope shown in Fig.1, a reflector is placed reflecting light beam back to the object through amplifying element (feedback mirror (6) in Fig 2), a resonator can be formed. In this resonator another reflector can be an object itself. Usually gain of pulsed metal vapour lasers is so high that even in a resonator with very high losses a beam of considerable power can be formed. Spatial intensity distributions in this beam depend on what feedback reflector and in what position is introduced in the system. This means that designing an optical system with high gain brightness amplifier one must be careful to prevent all undesirable reflections toward the amplifying element because they can produce intense beams and spoil an object. On the other hand, introduction of a feedback reflector can be used for additional controllable illumination of some points on the object. If the power of the beam formed in a resonator is high enough it can be used for processing of an object.

As it was shown in many experiments [2,9,13,14] formation of intense beam in such system does not spoil considerably optical quality of images obtained. So it is possible simultaneously to observe an object on a screen with high linear magnification and to process the object by a powerful beam formed in the system using one amplifying element.

3.1. SYSTEM WITH CONJUGATE RESONATOR

Due to limitation of space we'll consider here briefly only one system with so-called conjugate resonator. Schematic diagram of this system is shown in Fig.2.

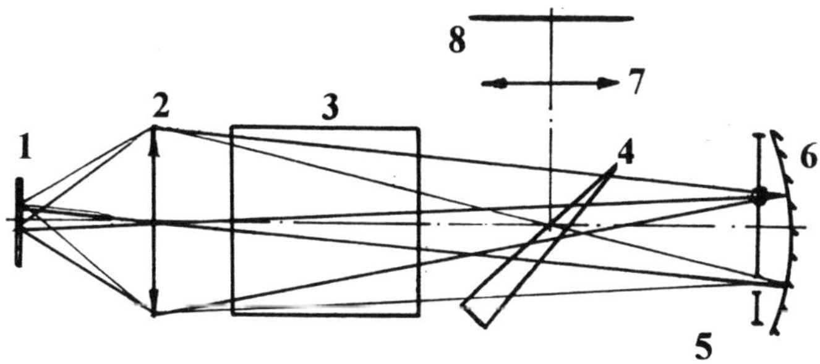

Figure 2. Schematic diagram of AOS with conjugate resonator: 1-object, 2-objective, 3-amplifying element, 4-beamsplitter, 5-mask, 6-feedback mirror, 7-projection lens, 8,-screen.

Here the resonator consists of the object 1 and feedback mirror 6. They are located in optically conjugated planes, i.e. objective 2 forms an image of the object on the feedback mirror and *vice versa.* The feedback mirror is spherical and its radius of curvature is equal to the distance from the feedback mirror to the objective 2. The object is supposed to be scattering. In this resonator any ray from any point of the viewed field of the object is returned back to the same point after reflection from the feedback mirror (in geometrical optics approximation). If there is a mask in front of the feedback mirror then an intense laser beam is formed in the resonator. And the spot of this beam on the object is determined by the shape of openings in the mask. So the object in such system can be processed in every pulse by a laser beam of rather complicated form. The form and position of the spot on the object, and the process of treatment itself can be observed visually on the screen 8 with large enough magnification.

The advantages of these systems in comparison with conventional projection systems used now usually for laser material processing are:

1. Only beams used for treatment are generated;
2. There are no losses on the mask, and mask can not be overheated by light;
3. Electronically controlled (i.g. program-driven) spatial light modulator can be used as mask. This open possibility to design automated processing laser systems with high productivity.
4. Treatment process, position, and shape of processing beam can be controlled on a screen with large magnification and high image quality without additional imaging optical system.

4. Conclusion

Pulsed metal vapour laser amplifying elements can be successfully used as brightness amplifiers in different optical systems. High gain and high output power of these amplifiers allow to obtain bright images of small objects on large screens keeping illumination of the object at acceptable level. Such systems can be used in laser projection microscopy; for projection of dynamic pictures; projection TV for large screens; computer displaying; entertainment advertisements and show-business.

Many pulsed metal vapour laser amplifiers can be used in different optical systems providing high enough gain in visible, near IR and UV spectral ranges. But still there is no laser with good enough characteristics for the blue part of spectrum. This is the main problem in creating full-colour active optical system.

Introduction of additional reflectors in AOS gives some new possibilities non-existing in conventional optical systems. This enables to produce intense laser beams of arbitrary configurations including very complicated ones. These systems are successfully used for material processing by a light spot of complex configuration. Using spatial-temporal light modulators it is possible to display on large screen dynamic information, and design automated systems for laser material processing.

Up to date, only pulsed metal vapour laser amplifiers can be used in practical optical systems. There is no competition in this field of application from the side of other type lasers.

5. References.

1. Zemskov, K.I., Isaev, A.A., Kazaryan, M.A., *et al.* (1976) Investigations of principal characteristics of laser projection microscope, *Sov.J.Quant.Electron.*. **6**(1), 17-22.
2. Zemskov, K.I., Kazaryan, M.A., Petrash, G.G., (1991) Image brightness amplifiers in projection optical systems, *Trudy FIAN* **206**, 3-62.
3. Bunkin, F.V., Zemskov, K.I., Kazaryan, M.A., *et al.* (1978) Projection system with brightness amplifier for biology and medicine, *Doklady AN SSSR*, **243**, N 6, 1568-1570, (In Russian).
4. Ambartscumyan, R.V., Zakharov, S.D., Zemskov, K.I., *et al.* (1988) Joint motion of particles in suspension, which is induced by laser pulses in laser projection microscope, *Kratkie soob. po fizike FIAN,* N 8, 35-37.(In Russian).
5. Alejnikov, V.S., Bunkin, F.V., Dianov, E.M., *et al.* (1984) Optical fibres in the system with brightness amplifier, *Doklady AN SSSR,* **274**, N 6, 1373-1376.
6. Arapova, E.Ja., Isaev, A.A., Kazaryan, M.A., *et al.* (1975) Infrared laser projection microscope, *Sov.J.Quant.Electron.*. **5**(7), 849-851.
7. Zemskov, K.I., Kazaryan, M.A., Petrash, G.G., *et al.* (1980) Laser projection microscope with a barium vapour amplifier and luminescent screens for IR-images visualisation, *Sov.J.Quant. Electron.* 10(11), 1428-1431.
8. Zemskov, K.I., Kazaryan, M.A. (1988) "Optical systems with brightness amplifiers" (eds. by V.I.Bespalov, and G.A.Pasmanik), IPF, Gor'kii.
9. Zemskov, K.I., Kazaryan, M.A., Matveev, V.M., *et al.* (1991) Image transformation in brightness amplifiers, Trudy FIAN 206, 63-100.
10. Zemskov, K.I., Kazaryan, M.A., Matveev, V.M., *et al.* (1983) Investigations of characteristics of negative images in optical copper vapour systems with image amplifiers, *Sov.J.Quant. Electron.*. **13**(11), 1480-1483.
11. Zemskov, K.I., Kazaryan, M.A., Petrash, G.G. (1986), Negative images formation by polarised light beams in optical systems with brightness amplifiers, *Kratkie soob. po fizike FIAN,* N 4, 17-18.
12. Zemskov, K.I., Kazaryan, M.A., Petrash, G.G. (1985), Visualisation of intensified IR-images in a non-linear interaction of beams in saturable amplifiers, *JETP Lett.*, **42**(6), 320-323.
13. Zemskov, K.I., Kazaryan, M.A., Petrash, G.G., Skripnichenko, A.S. (1986) Laser beam formation for intra-cavity processing of objects, *Sov.J.Quant.Electron.*, **16**(10), 1380-1384.
14. Gorbarenko, V.A., Onoukhov, N.A., Skripnichenko, A.S. (1978) Optical systems with laser image brightness amplifier, *Electronic technics, ser.3 Microelectronics,* N 2, 61-65, (In Russian).

10. Collated Characteristics of Some Commercial Pulsed Metal Vapour Lasers

Information on commercial copper vapour lasers was made available to the editors by representatives of eight laser manufacturers in attendance at the ARW. In alphabetical order, they are

1. BML (Bulgarian Metal Lasers) Ltd:
3020 Street, No. 34, Sofia 1360, Bulgaria,
Ph. (+359) 2 22 95 53, Fax (+359) 2 52 20 06

2. CILAS (Compagnie Industrielle des Lasers):
Route de Nozay, PO Box 27, 91460 Marcoussis, France,
Ph. (+33) 1 64 54 48 00, Fax (+33) 1 69 01 37 39

3. Joint Venture "Copris & M":
Tamanskaja 2a, Moscow 123103, Russia,
Ph. (+7) 095 943 0095, 978 5022, Fax (+7) 095 943 0096

4. Dynamic Light Ltd:
9/42 Leighton Place, Hornsby NSW 2077, Australia,
Ph. (+61) (0)2 482 1580, Fax (+61) (0)2 482 1581

5. EEV Ltd:
Power Tubes Division, Waterhouse Lane, Chelmsford, Essex CM1 2QU, UK,
Ph. (+44) (0)1245 493493, Fax (+44) (0)1245 353472

6. Tomsk State University, Prof. A N Soldatov, Laser and Crystal Physics Lab.,
26 Lenin Prospect, Tomsk 634050, Russia,
Ph. (+7) 3822 25 85 89, Fax (+7) 3822 22 61 62

Small Business Company "Lasery", 8-39, 30 ulitsa Let Pobedy, Tomsk 634055, Russia,
Ph: (+7) 3822 25 85 89

7. Norseld Pty Ltd:
9 Claxton St, Adelaide, South Australia 5000, Australia,
Ph. (+61) (0)8 231 9000, Fax (+61) (0)8 231 9009

8. Oxford Lasers Ltd:
Abingdon Science Park, Barton Lane, Abingdon, Oxon. OX14 3YR, UK,
Ph. (+44) (0)1235 554211, Fax (+44) (0)1235 554311

C. E. Little and N. V. Sabotinov (eds.), Pulsed Metal Vapour Lasers, 459–462.

The specifications of selected models of low-power (5-10 W) copper lasers offered for sale by the manufacturers listed on the previous page. (Data taken from available data sheets.) P_{av}, P_{pk} = average & peak laser powers. * Divergence for plane-plane cavity. ** Metal refill time for gas-flow systems.

	Model	Laser Medium	P_{av} (W)	P_{pk} (kW)	Pulse Energy (mJ)	PRF (extended) (kHz)	Pulse Dur. FWHM (ns)	Div.* FWHM (mrad)	Green: Yellow Ratio	Beam Diameter (mm)	Tube Life Time** (hours)
BML	CBL-5	CuBr	5	-	-	20	50	2.5	1.5	14	>500
Copris & M	CVL 00.12	Cu	6-8	13.5	0.3-0.4	16-20	15-20	3	-	10-12	>1000
Dynamic Light	VC-10	Cu	10	-	0.8	15 (4-20)	24	5.1	-	20	>600 Cu refill
EEV (tube only)	XL7000	Cu	8	-	-	12-18	10-30	-	-	11	1000 (>3 W)
Lasery	MALAKHIT	Cu	5	16-66	0.5-1.0	5-10	15-30	3	2	20	1000
		or CuBr	8-10	6-16	0.3-0.5	20-30	30-50	3	1.5	20	>500
Norseld	CuB-D10	CuBr	>10	~20	~0.65	16	~30	<6	2	18	1000
Oxford Lasers	CU10-A	Cu	10	50	up to 2	8-14 (2-20)	10-40	4	2	25	>300 Cu refill

	Weight incl. Power Supply (kg)	Dimensions Laser Head only L x W x B m^3	Cooling	Gas Flow/ Sealed	Gas Consumption (litre.atm.hr^{-1})	Input Power (kW)	Warm-Up Time lasing/full power (min.)	Other Models Avail. (W)
BML	80	1.12 x 0.36 x 0.23	air	sealed	0	2, 1φ	-	2
Copris & M	-	0.93 x 0.24 x 0.17	air	sealed	0	2.2, 1φ	30/90	1.5, 20, 50
Dynamic Light	-	-	air	gas flow	<0.5	2.5, 1φ	35/50	3, 20, 30, 40
EEV (tube only)	3.6	0.935 L x 0.064 dia.	air	sealed	0	<2, -	25/50	-
Lasery	70 67	1.0 x 0.35 x 0.25 1.1 x 0.16 x 1.6	air air	sealed sealed	0 0	1.6, 1φ 1.6, 1φ	35/60 15/25	0.3, 1, 3 30, 50
Norseld	45	1.2 x 0.3 x 0.3	air	sealed	0	1.4, 1φ	-/10	-
Oxford Lasers	180	1.86 x 0.31 x 0.19	air	gas flow	0.5	2.5, 1φ	-	15, 25, 35, 45, 100, 120

The specifications for high-power (≥100 W) copper lasers offered for sale by two of the manufacturers who attended the NATO Workshop. (Data taken from available data sheets.) P_{av}, P_{pk} = average & peak laser powers. * Divergence for plane-plane cavity.

	Model	Laser Medium	P_{av} (W)	P_{pk} (kW)	Pulse Energy (mJ)	PRF (extended) (kHz)	Pulse Dur. FWHM (ns)	Div.* FWHM (mrad)	Green: Yellow Ratio	Beam Diameter (mm)	Cu Refill Time (hours)
CILAS	ICL 100	Cu	120	-	-	5	40	-	>1.4	50	-
Oxford Lasers	ACL120	Cu	120	400 max.	20 max.	4-6 (4-10)	30-60	8	1.3	65	>300

	Weight incl. Power Supply (kg)	Dimensions Laser Head only L x W x B m^3	Cooling	Gas Flow/ Sealed	Gas Consumption (litre.atm.hr^{-1})	Input Power (kW)	Warm-Up Time lasing/full power (min.)	Other Models Avail. (W)
CILAS	670	3.9 x 0.7 x 0.5 (incl. power supply)	water	gas flow	1.2	15, 3ϕ	-	30
Oxford Lasers	526	3.26 x 0.72 x 0.30	water	gas flow	0.5	15, 3ϕ	-	10, 15, 25, 35, 45, 100

11. Contact List of Participants

Dr Yaakov Achiam
Laser Dept., Nuclear Research Centre - Negev
PO Box 9001, Beer-Sheva 84190, Israel
Ph: +972 57 968268. Fax: +972 57 31578

Dr Michael Amit
R & D Dept., Nuclear Research Centre - Negev
PO Box 9001, Beer-Sheva 84190, Israel
Ph: +972 7 567672. Fax: +972 7 567665.

Dr Dimo N Astadjov
Department of Metal Vapor Lasers, Institute of Solid State Physics,
Bulgarian Academy of Sciences
Tzarigradsko Chaussee 72, Sofia 1784, Bulgaria
Ph: +359 2 756009. Fax: +359 2 757032.

Prof. Dr-Ing. Hans W Bergmann
Institut für Werkstoffwissenschaften, Universität Erlangen-Nürnberg
Martinsstrasse 5, 91058 Erlangen, Germany
Ph: +49 9131 8577520. Fax: +49 9131 857515.

Dr R Bhatnagar
Centre for Advanced Technology
PO CAT, Rajendra Nagar, Indore 452 012, India
Ph: +91 731 481078. Fax: +91 731 488401/481426/481525.
E-mail: rbcvl@cat.ernet.in

Dr Pinhas Blau
Laser Department, Nuclear Research Centre - Negev
POB 9001, Beer Sheva, Israel
Ph: +972 7 567701. Fax: +972 7 568770/554848.
E-mail: pblau@bgumail.bgu.ac.il

Prof. Peter A Bokhan
Institute of Semiconductor Physics
Lavrentyev Ave 13, Novosibirsk, Russia
Ph: +7 3832 354266. Fax: +7 3832 356811.
E-mail: bokhan@isph.nsk.su

Jean Marc Borgard
CEA/DTE/SLC
BP 111, 26702 Pierrelatte cedex, France
Ph: +33 75504324. Fax: +33 75504302.

Dr Boris L Borovich
Special Design Bureau of High Energy Lasers "Granat"
Volokolamskoye Chaussee 95, Moscow 123424, Russia
Ph: +7 095 4916100. Fax: +7 095 4915021.

Dr Danny J W Brown
Centre for Lasers and Applications, Macquarie University
NSW 2109, Australia
Ph: +61 2 850 8911. Fax: +61 2 850 8983.
E-mail: danny@macadam.mpce.mq.edu.au

Ass. Prof. Philip H Butler
Department of Physics & Astronomy, University of Canterbury
Private Bag 4800, Christchurch, New Zealand
Ph: +64 3 3642541. Fax: +64 3 3642469.
E-mail: p.butler@phys.canterbury.ac.nz

Dr Robert J Carman
Centre for Lasers and Applications, Macquarie University
NSW 2109, Australia
Ph: +61 2 850 7058. Fax: +61 2 850 8983.
E-mail: PH_CARMAN@HOPE.OCS.MQ.EDU.AU

Dr Jim J Chang
Copper Laser Program, Lawrence Livermore National Laboratory
P.O. Box 5508, Livermore, California 94551-9900, USA
Ph: +1 510 422 4064. Fax: +1 510 422 1413.
E-mail: Jim.Chang@quickmail.llnl.gov

Dr Volodya Chvykov
Cancer Research Centre
Kashirskoe Chaussee, Moscow, Russia
Fax: +7 095 2302450.
E-mail: zems@sci.fian.msk.su

Dr David W Coutts
Clarendon Laboratory, University of Oxford
Parks Road, Oxford OX1 3PU, UK
Ph: +44 1865 272205. Fax: +44 1865 272400.
E-mail: d.coutts@physics.oxford.ac.uk

Dr Peter Davis
Norseld Pty Ltd
9 Claxton Street, Adelaide, SA 5000, Australia
Ph: +61 8 2319000. Fax: +61 8 2319009.

Dr Geoff J Dickinson
Urenco (Capenhurst) Ltd
Capenhurst, Chester CH1 6ER, UK
Ph: +44 151 3473285. Fax: +44 151 3473790.
E-mail: gjd@cap.urenco.co.uk

Dr Denis Doizi
CEA CE Saclay, DPE/SPL
Bât 467, 91191 Gif Sur Yvette Cedex, France
Ph: +33 1 69086852. Fax: +33 1 69088691.
E-mail: DOIZI@BORDEAUX.SACLAY.CEA.FR

Dr Gennadii Evtushenko
Institute of Atmospheric Optics, Siberian Division, RAS
Akademichevskii Ave 1, Tomsk 634055, Russia
Ph: +7 3822 259989. Fax: +7 3822 259086.
E-mail: qel%asd.tomsk.su@asd.tomsk.su

Dr Steve A Fairlie
School of Physics and Astronomy, University of St Andrews
North Haugh, St Andrews, Fife KY16 9SS, UK
Ph: +44 1334 463181. Fax: +44 1334 463104.
E-mail: saf@st-and.ac.uk

Dr Peter Foster
Norseld Pty Ltd
9 Claxton Street, Adelaide, SA 5000, Australia
Ph: +61 8 2319000. Fax: +61 8 2319009.
E-mail: pfoster@physics.adelaide.edu.au

Dr Robert Q Fugate
Starfire Optical Range, Phillips Laboratory
3550 Aberdeen Ave SE, Kirtland AFB, NM 87117-5776, USA
Ph: +1 505 846 4712 X314/216. Fax: +1 505 846 2213/846 0439.
E-mail: fugate@plk.af.mil

Ass. Prof. Ji-hong Geng
Shanghai Institute of Optics and Fine Mechanics
PO Box 800-211, Shanghai 201800, China
Ph: +86 21 9534890/475. Fax: +86 21 9528812/9528885

Mr Ludovic Gigant
CEA/DTE/SLC
BP 111, 26702 Pierrelatte cedex, France
Ph: +33 75504324. Fax: +33 75504302.

Dr Geoff P Hogan
Clarendon Laboratory, University of Oxford
Parks Road, Oxford OX1 3PU, UK
Ph: +44 1865 272259. Fax: +44 1865 272400.
E-mail: HOGANGP@VAX.OX.AC.UK

Dr Anatolii A Isaev
P N Lebedev Physical Institute
Leninsky Prospect 53, 117924 Moscow, Russia.
Ph: +7 095 132 6730. Fax: +7 095 938 2251.
E-mail via: zems@sci.fian.msk.su

Dr Borislav L Ivanov
Dept of Semiconductors, Higher Institute of Chemical Technology
8 Kliment Ohridski Street, Sofia 1756, Bulgaria
Ph: +359 2 6254649. Fax: +359 2 522006.
E-mail: blivanov@cserv.mgu.bg

Dr David R Jones
School of Physics and Astronomy, University of St Andrews
North Haugh, St Andrews, Fife KY16 9SS, UK
Ph: +44 1334 463181. Fax: +44 1334 463104.
E-mail: drj@st-and.ac.uk

Dr Thomas Karras
Martin Marietta, Valley Forge
PO Box 8555, Philadelphia, PA 19101, USA
Ph: +1 610 3543928. Fax: +1 610 3542045.

Dr Mishik A Kazaryan
P N Lebedev Physical Institute
Leninsky Prospect 53, 117924 Moscow, Russia
Ph: +7 095 1357890. Fax: +7 095 1352408.
E-mail: optdep@sci.fian.msk.su

Dr Andrew J Kearsley
Oxford Lasers Ltd
Abingdon Sci. Park, Barton L., Abingdon, Oxon. OX14 3YR, UK
Ph.: +44 1235 554211. Fax: +44 1235 554311.

Dr Chikara Konagai
Nuclear Engineering Laboratory, Toshiba Corporation
8 Shinsugita, Isogo-ku, Yokohama 235, Japan
Ph: +81 45 7702379. Fax: +81 45 7702308.
E-mail: 0000137DE504@tg-mail.toshiba.co.jp

Mr Ivan Kostadinov
Pulssvet Company
Bl. 463-A, Ap. 50, Nadezhda-4, Sofia 1324, Bulgaria
Ph: +359 2 688022. Fax: +359 2 522006.
E-mail via: MARKOVA@BGEARN.BITNET

Dr Evgeny L Latush
Physics Department, Rostov State University
Zorge St 5, Rostov-on-Don 344104, Russia
Ph: +7 8632 220824. Fax: +7 8632 285044.
E-mail: latush@cosmos.physfac.rnd.su

Dr Pascal Lemaire
CEA/DTE/SLC
BP 111, 26702 Pierrelatte cedex, France
Ph: +3375504705. Fax: +33 75504302.
E-mail: LEMAIRE@PIEDTE.CEA.FR

Dr Michail Lesnoy
Scientific and Production Association "Istoc"
Fryazino 141120, Moscow Region, Russia.
Ph: +7 095 1326730.
E-mail via: zems@sci.fian.msk.su

Dr Chris E Little
School of Physics and Astronomy, University of St Andrews
North Haugh, St Andrews, Fife KY16 9SS, UK
Ph: +44 1334 463115. Fax: +44 1334 462176.
E-mail: cel@st-and.ac.uk

Mrs Laura Little
School of Physics and Astronomy, University of St Andrews
North Haugh, St Andrews, Fife KY16 9SS, UK
Ph: +44 1334 463227. Fax: +44 1334 462176.
E-mail: lb2@st-and.ac.uk

Dr Ewan S Livingstone
EEV Ltd, Power Tubes Division
Waterhouse Lane, Chelmsford, Essex CM1 2QU, UK
Ph: +44 1245 493493, X3547. Fax: +44 1245 353472

Dr A Ian McIntosh
Dynamic Light Ltd
9/42 Leighton Place, Hornsby NSW 2077, Australia
Ph: +61 2 4821580. Fax: +61 2 4821581.
E-mail: AIMc@eworld.com

Dr Jacques Maury
CEA/DTE/SLC
BP 111, 26702 Pierrelatte cedex, France.
Ph: +33 75504324. Fax: +33 75504302.

Dr Roland Mayerhofer
Institut für Werkstoffwissenschaften, Universität Erlangen-Nürnberg
Martinsstrasse 5, 91058 Erlangen, Germany.

Dr Jean B Metzen
CILAS (Compagnie Industrielle des Lasers)
Route de Nozay BP 27, 91460 Marcoussis, France.
Ph: +33 1 64544800. Fax: +33 1 69013739.

Mr Richard Mildren
Centre for Lasers and Applications, Macquarie University
NSW 2109, Australia
Ph: +61 2 850 89. Fax: +61 2 850 8983.
E-mail: PH_MILDREN@hope.ocs.mq.edu.au

Prof. Jerzy Mizeraczyk
Institute of Fluid Flow Machinery, Polish Academy of Sciences
Fiszera 14, 80-315 Gdansk, Poland.
Ph: +48 58 411271. Fax: +48 58 416144.
E-mail: jmiz@imppan.imp.pg.gda.pl

Dr Takashige Omatsu
Department of Image Science, Chiba University
1-33 Yayoi-cho, Inage-ku, Chiba 263, Japan
Ph: +81 43 2903477. Fax: +81 43 2903477/90.
E-mail: omatsu@tcom.tech.chiba-u.ac.jp

Prof. I Opachko
Department of Physics, Uzhgorod State University
Voloshin St. 54, Uzhgorod 294000, Ukraine
Ph: +380 31 36187. Fax: +380 31 30656

Prof. Gueorgii G Petrash
P N Lebedev Physical Institute, Russian Academy of Sciences
Leninsky Prospect 53, 117924 Moscow, Russia
Ph: +7 095 132 5088. Fax: +7 095 938 2251.
E-mail: gpetrash@petg.fian.msk.su

Prof. Jim A Piper
Centre for Lasers and Applications, Macquarie University
NSW 2109, Australia
Ph: +61 2 850 8911. Fax: +61 2 850 8983/850 8115.
E-mail: jpiper@macadam.mpce.mq.edu.au

Prof. Jean Paul Pique
Laboratoire de Spectrométrie Physique (UA 08),
Université Joseph-Fourier de Grenoble
B.P. 87 - 38402 Saint-Martin-d'Heres cedex, France
Ph: +33 76 51 47 45. Fax: +33 76 51 45 44.
E-mail: pique@grenet.fr

Prof. Nikola V Sabotinov
Department of Metal Vapor Lasers,
Institute of Solid State Physics, Bulgarian Academy of Sciences
Tzarigradsko Chaussee 72, Sofia 1784, Bulgaria
Ph: +359 2 756009. Fax: +359 2 757032.
E-mail: sabotino@bgearn.bitnet

Prof. Renzo Salimbeni
Instituto di Electronica Quantistica del CNR
Via Panciatichi 56/30, 50127 Firenze, Italy.
Ph: +39 55 416128. Fax: +39 55 414612.
E-mail: salimbeni@mailbox.ieq.fi.cnr.it

Prof. Miroslav F Sem
Physics Department, Rostov State University
Zorge St 5, Rostov-on-Don 344104, Russia
Ph: +7 8632 220824. Fax: +7 8632 640833.
E-mail: sem@cosmos.physfac.rnd.su

Dr Yurij O Shpenik
Dept of Quantum Electronics, Institute of Electron Physics
Universitetskaya St. 21, Uzhgorod 294016, Ukraine
Ph: +380 3122 43518/43881. Fax: +380 3122 43650.
E-mail: IEP@IEP.UZHGOROD.UA

Prof. William T Silfvast
CREOL-University of Central Florida
12424 Research Parkway, Suite 400, Orlando FL 32826, USA
Ph: +1 407 658 6855. Fax: +1 407 658 6880.
E-mail: silfvast@mail.creol.ucf.edu

Mrs Sunita Singh
Laser and Plasma Technology Division, BARC
Trombay, Bombay, India
Ph: +91 22 5563060X2425 or 5519947. Fax: +91 22 5560750.
E-mail: ukchat@magnum.barct1.ernet.in

Dr Israel Smilanski
Nuclear Research Center Negev
P O Box 9001, Beer Sheva, Israel
Ph: +972 7 567116. Fax: +972 7 568770.

Prof. Anatoly Soldatov
Tomsk State University
26 Lenin Prospect, Tomsk 634010, Russia
Ph: +7 3822 258589. Fax: +7 3822 226162.
E-mail: office@tsu.tomsk.su

Dr Kiwamu Takehisa
Ion Beam and Laser Application Group,
Electric Power Delivery Equipment
Department, Hitachi Ltd
1-1-1 Kokubu-cho, Hitachi-shi, 316 Japan.
Ph: +81 294 365159. Fax: +81 294 368399.

Dr Rod C Tobin
Dept of Physics, Monash University
Wellington Rd, Clayton, Victoria 3168, Australia.
Ph: +61 3 905 3647. Fax: +61 3 905 3637.
E-mail: Rod.Tobin@sci.monash.edu.au

Dr Nikolay K Vuchkov
Department of Metal Vapor Lasers,
Institute of Solid State Physics, Bulgarian Academy of Sciences
Tzarigradsko Chaussee 72, Sofia 1784, Bulgaria
Ph: +359 2 756009. Fax: +359 2 757032.
E-mail via: sabotino@bgearn.bitnet

Mr William Wadsworth
Clarendon Laboratory, University of Oxford
Parks Road, Oxford OX1 3PU, UK
Ph: +44 1865 272259. Fax: +44 1865 272400.
E-mail: w.wadsworth1@physics.oxford.ac.uk

Prof. William T Walter
Laser Consultants Inc.
344 West Hills Rd, Huntington, NY 11743, USA
Ph: +1 516 595 5465. Fax: +1 516 595 3328.
E-mail: WWALTER@AIL.COM

Dr Bruce E Warner
Copper Laser Program, Lawrence Livermore National Laboratory
P.O. Box 5508, L-467, Livermore, California 94551-9900, USA
Ph: +1 510 422 9237. Fax: +1 510 423 7651.
E-mail: warner2@llnl.gov

Prof. Colin E Webb
Clarendon Laboratory, University of Oxford
Parks Road, Oxford OX1 3PU, UK
Ph: +44 1865 272210. Fax: +44 1865 272400.
E-mail: webb@vax.ox.ac.uk

Mr Colin G Whyte
School of Physics and Astronomy, University of St Andrews
North Haugh, St Andrews, Fife KY16 9SS, UK
Ph: +44 1334 463181. Fax: +44 1334 463104.
E-mail: cgw@st-and.ac.uk

Dr Michael Withford
Centre for Lasers and Applications, Macquarie University
NSW 2109, Australia
Ph: +61 2 850 89. Fax: +61 2 850 8983.
E-mail: withford@macadam.mpce.mq.edu.au

Dr Konstantin I Zemskov
P N Lebedev Physical Institute, Russian Academy of Sciences
Leninsky Prospect 53, 117924 Moscow, Russia.
Ph: +7 095 132 6730. Fax: +7 095 938 2251.
E-mail: zems@sci.fian.msk.su

INDEX